Ulrich Krause

Positive Dynamical Systems in Discrete Time

De Gruyter Studies in Mathematics

Volume 62

Ulrich Krause

Positive Dynamical Systems in Discrete Time

Theory, Models, and Applications

DE GRUYTER

Mathematics Subject Classification 2010
15B48, 37B55, 39B12, 47B65, 47H07, 47N10, 60J10

Author
Prof. Dr. Dr. Ulrich Krause
Universität Bremen
Fachbereich 03 – Mathematik/Informatik
Bibliothekstr. 1
28359 Bremen
Germany
krause@math.uni-bremen.de

ISBN 978-3-11-036975-5
e-ISBN (PDF) 978-3-11-036569-6
e-ISBN (EPUB) 978-3-11-039134-3
Set-ISBN 978-3-11-036571-9
ISSN 0179-0986

Library of Congress Cataloging-in-Publication Data
A CIP catalog record for this book has been applied for at the Library of Congress.

Bibliographic information published by the Deutsche Nationalbibliothek
The Deutsche Nationalbibliothek lists this publication in the Deutsche Nationalbibliografie; detailed bibliographic data are available on the Internet at http://dnb.dnb.de.

Typesetting: PTP-Berlin, Protago TEX-Production GmbH
Printing and binding: CPI books GmbH, Leck
♾Printed on acid-free paper
Printed in Germany

www.degruyter.com

To Carola and Daniel

Preface

Positive dynamical systems come into play when relevant variables of a system take on values which are nonnegative in a natural way. This is the case, for example, in fields as biology, demography and economics, where the levels of populations or prices of goods are positive. Positivity comes in also if the formation of averages by weighted means is relevant since weights, for example probabilities, are not negative. This is the case in quite diverse fields ranging from electrical engineering over physics and computer science to sociology. Thereby averaging takes place with respect to signals in a sensor network or in a swarm (of birds or robots) or with respect to velocities of particles or the opinions of people. In the fields mentioned the dynamics is often modeled by difference equations which means that time is treated as discrete. Thus, in reality one meets a huge variety of positive dynamical systems in discrete time.

In many cases these systems can be captured by a linear mapping given by a nonnegative matrix. The dynamics (in discrete time) then is given by the powers of the matrix or, equivalently, by the iterates of the linear mapping which maps the positive orthant into itself. A powerful tool then is the Perron–Frobenius Theory of nonnegative matrices (including the asymptotic behavior of powers of those matrices) which has been successful since its inception by O. Perron and G. Frobenius over about hundred years ago. Concerning theory as well as applications there are two insufficient aspects of Perron–Frobenius Theory which later on drove this theory into new directions. The first aspect is that this theory is not just about nonnegative matrices but applies happily also to certain matrices with negative entries. This means that the theory should be understood as dealing with linear selfmappings of convex cones in finite dimensions not just of the standard cone, the positive orthant. The second aspect is that even simple positive dynamical systems are not linear. Thus, what is needed is an extension of classical Perron–Frobenius Theory to nonlinear selfmappings of convex cones in finite dimensions. Moreover, with respect to theory as well as applications, such an extension is needed also in infinite dimensions. Since classical Perron–Frobenius Theory has already so many applications one can imagine the great variety of applications such an extension to nonlinear selfmappings in infinite dimensions will have.

It is the aim of the present book to provide a systematic, rigorous and self–contained treatment of positive dynamical systems based on the analysis of the iterations of nonlinear selfmappings of a convex cone in some real vector space. To pursue this task, help comes from a beautiful approach developed for the linear case by G. Birkhoff considering Jentzsch's Theorem in infinite dimensions and, independently, by H. Samelson considering Perron–Frobenius Theory in finite dimensions. The crucial point of this approach is the translation of a strong positivity property of the linear mapping into a contractivity property with respect to some metric internal to the convex cone. This metric has been used long before by D. Hilbert within the completely different area of the foundations of geometry and is called Hilbert's projective metric (a quasi–metric, actually). The extension of this approach, also called the Birkhoff program, to nonlinear selfmappings of convex cones is a cornerstone of the present book. As it turns out the investigation of the nonlinearity is made easier by having it based on a convex cone and its analysis. Since the convex cone reflects the positivity of the system one might say that positivity helps to tame nonlinearity. Many beautiful results are available which are impossible without positivity.

The following paragraphs sketch briefly the content of each of the eight chapters of the book.

Chapter 1 motivates the study of positive dynamical systems (in discrete time) by means of examples from biology and economics. As for biology a nonlinear extension of the classical Leslie model used in population dynamics and demography is presented by taking population pressure into account. Considering economics, for the likewise classical Leontief model of commodity production a nonlinear extension is treated which captures the choice of techniques. There are much more examples of nonlinear positive dynamical systems. The example of opinion dynamics under bounded confidence has recently attracted much attention and will be investigated in the last chapter of the book.

Chapter 2 on "Concave Perron–Frobenius Theory" presents an extension of classical Perron–Frobenius Theory from linear to concave mappings (including linear ones). In the proofs Hilbert's projective metric makes its first appearance. The point thereby is that for this metric concave mappings are contractions and the interior of the standard cone is complete. By this Perron–Frobenius theorems can be proved using Banach's contraction mapping principle. Though only a particular form of nonlinearity concavity covers the nonlinearities in the models of Leslie and Leontief. Whereas in later chapters more general nonlinearities will be tackled on, this chapter concentrates just on concave mappings since for these a variety of results is possible comparable to those of classical Perron–Frobenius Theory. It should be noted, however, that even concave selfmappings of the standard cone in finite dimensions exhibit already spectral properties in sharp contrast to the linear case in that there may be infinitely many eigenvalues.

Whereas in the first two chapters positivity is restricted to the standard cone in finite dimensions, theory as well as applications in later chapters require more general convex cones in infinite dimensions.

Chapter 3 on "Internal metrics on convex cones" treats general convex cones in topological vector spaces with a focus on internal metrics. The latter are (quasi–) metrics solely determined by the cone's convex structure. Hilbert's projective metric and the Thompson metric or part metric are the most relevant internal metrics but there are much more. Besides certain geometrical properties of internal metrics the chapter concentrates on criteria for a convex cone to be complete for an internal metric. For later use the topology of the vector space is related to the one induced by an internal metric and criteria for internal completeness are obtained in terms of the vector space topology. A particular case is the result obtained first by G. Birkhoff that the positive cone of a Banach lattice – as well as its interior – are complete for Hilbert's projective metric. Extending the method applied in Chapter 2 for finite dimensions, by Chapter 3 selfmappings of a convex cone can be looked at as selfmappings of a complete metric space. Since later on contractivity with respect to internal metrics will play a role, *Chapter 4* on "Contractive dynamics on metric spaces" investigates various types of contractivity in general metric spaces. Conditions are specified which guarantee pointwise convergence of the iterates of a selfmapping to a fixed point. An important principle states that this global property applies already if it holds locally in case of power–lipschitzian mappings (including nonexpansive mappings). For later applications to nonautonomous positive systems the composition of infinitely many selfmappings and its asymptotic behaviour is analyzed.

Both, Chapter 3 and Chapter 4 supply in a general setting tools needed in subsequent chapters. Beside this, both chapters present known and new results which are interesting in itself.

Chapter 5 on "Ascending dynamics in convex cones of infinite dimension" presents a far–reaching extension of Chapter 2 to convex cones in infinite dimensions and corresponding selfmappings including concave ones. An ascending operator is, roughly speaking, a selfmapping of a convex cone, the values of which, on a subset of the cone, increase with respect to the cone's ordering on vectors as well as with respect to the common order on positive scalars. It is an important feature of ascending operators to be positive without being necessarily monotone.

In the linear case the uniformly positive linear operators introduced by G. Birkhoff are examples. Nonlinear examples are the u_0–concave operators studied by M. Krasnoselskii and his collaborators. Whereas these mappings need to be monotone, this is, however, not the case for ascending operators in general. Using Hilbert's projective metric for ascending operators relative stability is proven, meaning the iterates of the normalized operators do converge to an eigenvector. Using the part metric for weakly ascending operators, absolute stability is shown that is the iterates converge to a fixed point. Applications concern nonlinear difference equations and a nonlinear version of Jentzsch's Theorem on integral operators, including an approximation algorithm to compute the unique solution.

Chapter 6 on "Limit set trichotomy" investigates a fundamental phenomenon of positive dynamical systems which means that either all orbits tend to infinity or all orbits tend to zero or all orbits tend to a fixed point in the interior of the cone. Various conditions for this phenomenon to happen are specified. Limit set trichotomy can be used in many ways, it guarantees, for example, the existence of a globally stable fixed point in the interior if there exists an orbit positively bounded from below and above. For the case of differentiable selfmappings in finite dimensions easy to check conditions for limit set trichotomy are given. An application is to nonlinear difference equations including a generalized nonlinear Fibonacci equation. Another application considers cooperative systems of differential equations with a biochemical control circuit as a particular example.

Chapter 7 on "Nonautonomous positive systems" deals with the asymptotic behavior of compositions of infinitely many selfmappings of a convex cone. Various kinds of behaviour as path stability, asymptotic proportionality, weak and strong ergodicity are analyzed. The result on concave weak ergodicity is an extension of the famous (linear) Coale–Lopez Theorem in demography. Another nonlinear extension concerns the classical strong ergodicity theorem for nonnegative matrices. Furthermore, a beautiful theorem of H. Poincaré on nonautonomous linear difference equations is extended to include nonlinear difference equations. Also, the nonlinear versions of the models of Leslie and Leontief introduced in the first chapter are investigated for survival reates dependent on time and for time–dependent technical change, respectively. Finally, for populations being under enforcement from the environment conditions on population pressure are given which still yield path stability.

The last and longest chapter, *Chapter 8*, is on the "Dynamics of interaction: Opinions, mean maps, multi–agent coordination, and swarms". It is the aim of this chapter to develop a systematic and rigorous analysis for the dynamics of several fascinating kinds of interaction. Such interactions have been addressed recently in a widespread and fastly growing literature by researchers from quite different fields which range from electrical engineering over physics and computer science to sociology and economics. The leading question thereby asks under what conditions a group of agents, being it roboters or humans or other kinds of animals, is able to coordinate themselve to reach a consensus. Mathematically, the latter means for a dynamical system with several components whether these converge all to the same state. In its most simple case one considers a nonnegative matrix with all rows summing up to one and asks for conditions under which the powers of the matrix converge to a matrix having all its rows equal. The answer in this special case is that this happens precisely if the matrix has a power which is scrambling. This is (a sharpened version of) the well–known Basic Limit Theorem for Markov Chains. Already simple cases of interaction, however, are nonlinear (or time–variant) as for the model of opinion dynamics under bounded confidence (also known as Hegselmann–Krause model in the literature) which has attracted many researchers in recent years. A nonlinear analogue of a (row–) stochastic matrix is a mean map. Concerning time–variance one considers a sequence of stochastic matrices. Both cases lead to positive dynamical systems as considered in previous chapters. Facing the particular type of convergence to consensus tools adapted to that are developed in Chapter 8. In case of time–variance these are tools to handle infinite products of stochastic matrices. What is needed are conditions on the structure and intensity of interaction which make the infinite product convergent to a matrix with equal rows. An often used tool, the theorem of Wolfowitz, is generalized. The chapter concludes with an application to the dynamics of swarms of birds. The recently much discussed Cucker–Smale model is treated and a new model of swarming is developed where birds are forming swarms under some weaker conditions on interaction.

Each chapter is subdivided into sections the material of which is illustrated by examples and contains exercises ranging from simple verifications over additional topics to open problems. Remarks comment on results obtained and provide links to the literature. In each chapter results, examples, remarks are consecutively numbered by a.b.c where a refers to the chapter, b to the section and c to the particular item. Each chapter appended is a bibliography specific to it. A list of notations and an index conclude the book.

The book is directed to researchers from various disciplines and graduate students, too, who are interested in positive dynamical systems.

The book is self–contained and organized in a manner such that its material can also be used in courses and seminars. Chapters 1 and 2 require only a basic knowledge in linear algebra and analysis and could be used for an introductory course in nonlinear Perron–Frobenius Theory including applications. Chapters 3, 4, and 5 could serve as material in a course or seminar for graduate students and require some familiarity with fundamental concepts in topology and functional analysis. The same applies to Chapters 6 and 7 which could be used as material in an advanced course. The last Chapter 8 can be read independently of the previous ones and could serve as an introduction into recent applications of positive dynamical systems. The fascinating topics are suitable for graduate students to work on, analytically as well as by doing computer simulations.

This book grew out of several courses and seminars I held over the years at the University of Bremen. It was a great experience to share with the students the enthusiasm for a field which is just in the beginning. I like to thank all the students for their contributions and I want to mention in particular Tim Nesemann and Jan Lorenz. The reader will consult the references given in the book to their work and that of other students as well as to the work of researchers I enjoyed to write joint papers with. Here I like to thank Christian Bidard, Rainer Hegselmann, Diederich Hinrichsen, Takao Fujimoto, Tim Nesemann, Roger Nussbaum, Mihály Pituk, Peter Ranft, Dietrich Weller.

Furthermore, I want to thank Birgit Feddersen from the Department of Mathematics for her experienced and nice translation of the manuscript into LaTex, including the figures.

For many careful and helpful comments I have to thank the three anonymous reviewers of the manuscript.

Finally, I want to thank the publisher De Gruyter and in particular Friederike Dittberner and Silke Hutt, who have been most helpful in the process of publication.

The book I dedicate to my wife Carola and to our son Daniel who stayed so friendly to someone who lived with a desk for days, months and years.

Bremen, November 2014 Ulrich Krause

Contents

Notation

$\mathbb{N} = \{0, 1, 2, 3 \ldots\}$
$\mathbb{R}$ field of real numbers
$\mathbb{R}_+ = \{x \in \mathbb{R} | x \geq 0\}$
$\mathbb{R}^n = \{x = (x_1, \ldots, x_n) | x_i \in \mathbb{R}, 1 \leq i \leq n\}$
$\mathbb{R}^n_+ = \{x = (x_1, \ldots, x_n) | x_i \in \mathbb{R}_+, 1 \leq i \leq n\}$ positive orthant
$\text{int}\, \mathbb{R}^n_+ = \{x \in \mathbb{R}^n_+ \mid x_i > 0, 1 \leq i \leq n\}$
$\mathbb{R}_+ x = \{rx \mid r > 0\}$ ray through $x \in \mathbb{R}^n$
$|x| = (|x_1|, |x_2|, \ldots, |x_n|)$ absolute value of $x = (x_1, \ldots, x_n) \in \mathbb{R}^n$
$\tilde{T}x$ normalized/rescaled operator, 25
$G(T)$, graph associated to T, 42, 151
$\lambda(x, y)$ order function/extraction grade for cones, 77
$\mu(x, y)$, 77
$xCy, x \sim y$ component, part relation, 84
$d(x, y)$ projective Hilbert metric, 85
$p(x, y)$ Thompson metric or part metric, 85
$h(x, y)$ Harnack metric, 85
$g(x, y)$ Gleason metric, 85
$b(x, y)$ Bear metric, 85
$k(x, y)$ Kobayashi metric, 86
$\beta(x, y)$ order function for convex sets, 91
$\alpha(x, y)$, 91
$[x, y]$ interval, 95
$B_m(x, r)$ closed ball for internal metric m, 95
$C(T)$ space of continuous functions on T, 99
$\||\cdot\||$ sup–inf norm, 99
$\omega(x)$ (omega) limit set, 118
$\omega_s(x)$ nonautonomous (omega) limit set, 123
$\text{int}\, K = \overset{\circ}{K}$ interior of cone K, 142
$x < y$ strict order relation for $y - x \in \overset{\circ}{K}$, 142
$D(T)$ ascending domain of T, 147
$W(T)$ weak ascending domain of T, 156
$N(P)$ set of nonexpansive selfmappings on part P, 179
$c(T)$ contraction constant of T for part metric, 188
$J_T(x)$ Jacobian of T at x, 200
$\delta(T)$ 200
$F(t, x)$ cooperative system of differential equations, 204
ΔM diameter of M, 259
$c(A)$ scrambling factor of matrix A, 259
$s(M)$ 264

$I(i,x)$ confidence set, 271
diagS^n diagonal of S^n, 275
$\bar{c}(x)$ consensus, 275
$\dot{S}^n$ complement of diagonal, 275
$N(i,t)$ neighbors of i at t, 279
$U(i,t)$ neighborhood of i at t, 279
$M(t,s)$ matrix product, 291
$\rho_A(x)$ 293
$B(I)$ matrix product, 294
$S_k(M)$ 297
$\mu(A)$ strength of matrix A, 301
$\lambda(A), \delta(A), \tau(A)$ coefficients of ergodicity, 303
$M(i)$ smallest saturated set, 314
v valuation, 319
$\mathfrak{T}$ set of all mean maps, 335

List of Figures

1 How positive discrete dynamical systems do arise

1.1 Non-linear population dynamics in one dimension

Consider a population of individuals, which could be plants, animals or human beings. living in a fixed environment. In the course of time the number of individuals may increase or decrease or stay constant. Let $p(t)$ denote the number of individuals living at time t with t being measured in discrete steps like days, months or years that is $t \in \mathbb{N}: = \{0, 1, 2, 3, \ldots, \}$. The growth rate of the population at time t is by definition

$$g(t) = \frac{p(t+1) - p(t)}{p(t)}. \tag{1.1.1}$$

Of course, $0 \leq p(t)$ and $-1 \leq g(t)$.

Let us first see what happens if we assume the growth rate to be constant over time, i.e., $g(t) = g$ for all t. The dynamics of the population is then given by

$$p(t+1) = (1+g)p(t), t \in \mathbb{N}, \tag{1.1.2}$$

where $p(t) \in \mathbb{N}$ and $0 \leq 1 + g$. The solution of the difference equation 1.1.2 is easily obtained by iteration as

$$p(t) = (1+g)^t p(0) \text{ for all } t \in \mathbb{N}. \tag{1.1.3}$$

From this one concludes that the dynamic behavior of the population must be of one of the following three types:

- If $g > 0$ (and $p(0) > 0$) then there holds exponential growth without limits, i.e. $\lim\limits_{t\to\infty} p(t) = \infty$.
- If $g < 0$ then the population decreases to zero, i.e. $\lim\limits_{t\to\infty} p(t) = 0$.
- If $g = 0$ then the population stays constant, i.e. $p(t) = p(0)$ for all t.

In particular, it is impossible for the population to approach in the long run a finite number which is (strictly) positive and different from the initial population level. Although a population can show a behavior of the above types for a while it is very unrealistic that one single type will last forever. The unrealistic dynamic behavior in this model stems, of course, from the assumption that the growth rate is the same all of the time. (By the way, the same thing happens if we choose to model in continuous time instead in discrete time, obtaining a differential equation instead of difference equation (1.1.2). There are many reasons, discussed in detail in the biological literature for the growth rate not being constant (Edelstein–Keshet [7], Hoppensteadt [14], Murray [24], Pielou [27], Pollard [28]). Among others, the growth rate will depend on the actual level of population $p(t)$ due to **population pressure.** The latter means that by limitations in food and living space the growth rate will decrease if the population level is increasing. Therefore our model (1.1.2) has to be replaced by a model of the type

$$p(t+1) = f(p(t)), \quad t \in \mathbb{N}, \tag{1.1.4}$$

where $f\colon \mathbb{R}_+ \longrightarrow \mathbb{R}_+$ is the so called **reproduction function** (or curve).

Since the growth rate is, according to (1.1.1), given by $\frac{f(p(t))}{p(t)} - 1$, population pressure means that

$$\frac{f(x)}{x} \text{ is (strictly) decreasing in } x \text{ (for } x > 0\text{).} \tag{1.1.5}$$

Equation (1.1.4) together with condition (1.1.5) constitutes a **positive discrete dynamical system in one dimension**. An equation as (1.1.4) is called a *difference equation* of first order. (For difference equations see Elaydi [8], Kocic and Ladas [15].) The relevant magnitude $p(t)$ is not only a (real) number but a positive number (including 0). Furthermore, the 'law of motion' f maps positive numbers into positive numbers and satisfies a condition (1.1.5) which also employs the ordering relation $\leq$ of real numbers. The dynamics of (1.1.4) is given by the iterates $f^t = f \circ \ldots \circ f$ (t-times) of mapping f, namely

$$p(t) = f^t(p(0)) \quad \text{for all} \quad t \in \mathbb{N}. \tag{1.1.6}$$

In contrast to (1.1.3), however, it is not easy to find out what types of dynamic behavior are concealed in equation (1.1.6). Actually, depending on the particular function f, it might be *very* difficult to determine the dynamic behavior for (1.1.6) which even for simple reproduction functions can be very complicated.

Let us discuss two examples on the extreme, whereas the general equation (1.1.4) will be taken up again in a later chapter. Of course, our ealier equation (1.1.2) is a special case of (1.1.4), namely the linear case $f(x) = (1 + g)x$. In taking care of population pressure, however, we have to turn to non-linear selfmappings f of $\mathbb{R}_+$. An example of such a mapping is given by the reproduction function

$$f(x) = \frac{\rho x}{x + K} \tag{1.1.7}$$

discussed by biologists and going back to the early mathematical biologist P.F. Verhulst (Edelstein–Keshet [7], Pielou [27]). The function f, which obviously satisfies condition (1.1.5), depends on two parameters, where $\rho > 0$ is the maximal possible size of the population and where $K > 0$ is a measure for the strength of the populaton pressure. Of special interest is a possible *equilibrium population*, measuring a population level p^* which does not change through time. In spite of (1.1.4) this amounts to

$$p^* = f(p^*) = \frac{\rho p^*}{p^* + K},$$

which admits two solutions $p^* = 0$ and $p^* = \rho - K$, the latter being meaningful only if $\rho \geq K$. The main question considering the dynamic behavior is whether the system will approach an equilibrium, provided there exists one. In one dimension often a graphic procedure may help which is called ***graphic iteration*** or ***cobwebbing*** and which, in our particular example, goes as follows:

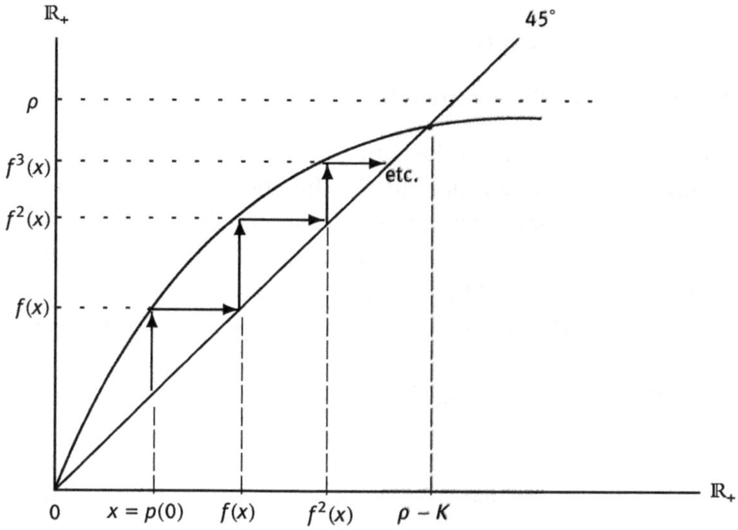

Fig. 1.1. Graphic iteration for $f(x) = \frac{\rho x}{x+K}$.

This graphic iteration shows that in case of $\rho > K$ the iterates $f^t(x)$, for $x > 0$, converge to the equilibrium $\rho - K$ for $t \to \infty$. In case of $\rho \le K$, the reproduction curve is below the 45° line and graphic iteration then shows that $f^t(x)$ converges to 0 for $t \to \infty$. (Of course, for $x = p(0) = 0$ the population stays at 0 all the time.) In any case, the population level $p(t)$ approaches an equilibrium if t tends to infinity. Contrary to the constant growth case we now observe the possiblity that the population approaches a (strictly) positive equilibrium. Since such a graphical method is only heuristic and not fully convincing (not to think of higher dimensions), one should be able to demonstrate the above observations in an analytic manner. For the case $\rho > K$ this can be done as follows. Since $f(x) = \rho x(K + x)^{-1} = \rho(\frac{K}{x} + 1)^{-1}$ the function f is increasing. Hence for $p^* = \rho - K$:

$$0 \le x \le p^* \Longrightarrow \rho(K + x)^{-1} \ge 1 \Longrightarrow x \le f(x) \le f(p^*) = p^*.$$

By iterating we get $f^n(x) \le f^{n+1}(x) \le p^*$. As an increasing sequence which is bounded from above the sequence $(f^n(x))_{n\ge 0}$ must converge to some $q \in \mathbb{R}_+$. By continuity of f, $f(q) = q$ and $q = p^*$ if $x > 0$. Thus,

$$\lim_{n\to\infty} f^n(x) = p^* \quad \text{for all} \quad 0 < x \le p^*.$$

Similarly,

$$p^* \le x \Longrightarrow \rho(K + x)^{-1} \le 1 \Longrightarrow p^* = f(p^*) \le f(x) \le x.$$

By iteration $p^* \le f^{n+1}(x) \le f^n(x)$ and the sequence $(f^n(x))_{n\ge 0}$ is decreasing and bounded from below. As above, this implies $\lim_{n\to\infty} f^n(x) = p^*$ for all $p^* \le x$. For the population level, we, therefore, obtain by using (1.1.6)

$$\lim_{t\to\infty} p(t) = \rho - K$$

if there is any initial population at all, i.e. $p(0) > 0$.

The reproduction function (1.1.7) discussed is just one possibility to model population pressure, there are many others. Another reproduction function proposed for population pressure models the decreasing growth rate by " subtraction", that is $\frac{f(x)}{x} = \rho(P-x)_+$, where $\rho > 0$, $P > 0$ are parameters, P being a maximal possible population level and where $r_+ = \max\{r, 0\}$ for any real number r. Note that $\frac{f(x)}{x} = \rho(K+x)^{-1}$ as in (1.1.7) models the decreasing growth rate by "division". In other words, consider the model

$$p(t+1) = f(p(t)) = \rho p(t)(P - p(t))_+. \tag{1.1.8}$$

Introducing $x(t) = \frac{p(t)}{P}$ we obtain

$$x(t+1) = ax(t)(1 - x(t)), \tag{1.1.9}$$

where $a = \rho P$ and $x(t)$ is in the unit interval $[0, 1]$ for all t provided $0 \le a \le 4$. Equation (1.1.9) is the famous *logistic difference equation* which is known to generate for certain values of parameter a very complicated dynamics dubbed **chaotic dynamics** (May [22], Peitgen, Jürgens, and Saupe [26], Zaslavskiĭ [35]). An impression of that dynamics can be obtained by doing graphic iteration:

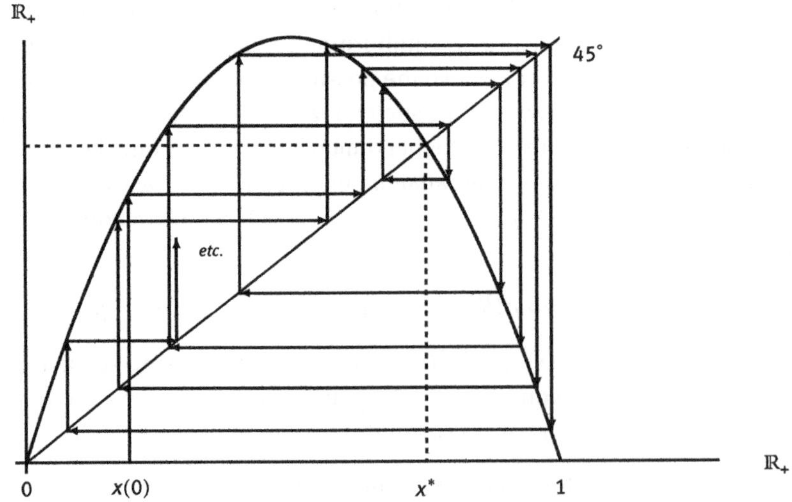

Fig. 1.2. Graphic iteration for the logistic difference equation.

For this model there exists a unique (strictly) positive equilibrium $x^* = \frac{a-1}{a}$ (for $a > 1$) but $x(t)$, and hence $p(t) = Px(t)$, does for certain values of the parameter a not approach the equilibrium x^* or $p^* = Px^*$, respectively. The population model underlying the logistic difference equation has been sometimes also attributed to Verhulst. It seems, however, that Verhulst used a logistic differential equation, namely $\frac{dp(t)}{dt} = \rho p(t)(P - p(t))$, the solution of which approaches P for $t \to \infty$ for all possible values $\rho > 0$, $P > 0$, in contrast to what happens for the logistic difference equation. (See Exercises to 1.1, Problem 4.)

Although the models (1.1.7) and (1.1.8) both picture population pressure the resulting dynamics are completely different – the kind of modeling does matter. (For various dynamic models see Beltrami [1], Farina and Rinaldi [9], Krause and Nesemann [18], Luenberger [21], Sandefur [30].) Equation (1.1.8) is a rather extreme case of modeling negative growth rates. It has been argued by biologists that a reproduction function

$$f(x) = \lambda x(1 + ax)^{-b} \quad \text{with parameters } \lambda, a, b > 0 \tag{1.1.10}$$

gives an empirical description of density limited population growth. (Edelstein–Keshet [7]; Hassel [12]; see Cull [4] for a mathematical investigation of one dimensional models admitting negative growth rates.) Consider as a special case of (1.1.10) the function

$$f(x) = 5x(1 + x)^{-2}. \tag{1.1.11}$$

This function has a unique positive equilibrium $x^* = \sqrt{5} - 1$. In contrast to the logistic model, this reproduction function tends smoothly to 0 for population levels above the equilibrium level.

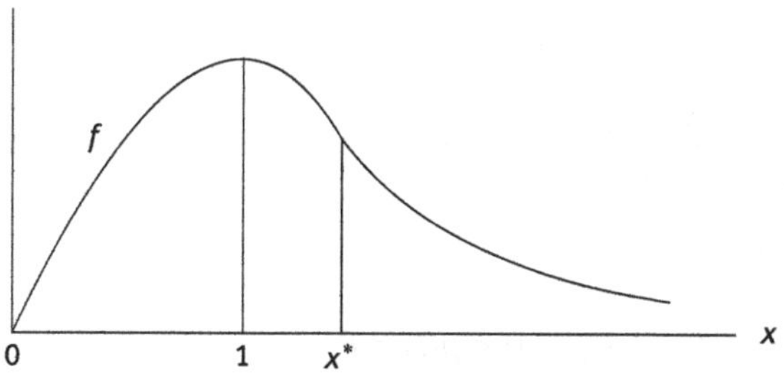

Fig. 1.3. Density limited population growth.

It turns out that for this model $\lim_{t\to\infty} p(t) = \sqrt{5} - 1$ for all $p(0) > 0$. In case the reproduction function is differentiable the equation (1.1.5) expressing population pressure amounts to

$$xf'(x) < f(x) \text{ for all } x > 0. \tag{1.1.12}$$

In a later section (Section 5.3, Exercise 7 (d)) we will see that if there holds the stronger condition

$$x \mid f'(x) \mid < f(x) \quad \text{for all} \quad x > 0 \tag{1.1.13}$$

and if there exists a (strictly) positive equilibrium p^* then we must have $\lim_{t\to\infty} p(t) = p^*$ for all $p(0) > 0$. In example (1.1.11) we have that

$$x \mid f'(x) \mid = 5x \mid 1 - x \mid (1 + x)^{-3} < 5x(1 + x)^{-2} = f(x) \quad \text{for all} \quad x > 0,$$

and, hence, condition (1.1.13) is satisfied. This condition is also satisfied for the function given by (1.1.7) but it is not satisfied for the logistic model (1.1.8). For the latter, (1.1.12) does hold but not (1.1.13) (where, of course, x is restricted to $0 < x \leq 1$).

Although much more can be said about population pressure in one dimension we stop its discussion for turning to multidimensional situations which are more realistic by taking the age structure of populations into account.

Exercises

1. Prove for the reproduction function given by equation (1.1.7) that

$$\lim_{t\to\infty} p(t) = 0 \quad \text{for all} \quad x \geq 0,$$

provided $0 \leq \rho \leq K$.

2. Show for the selfmapping of $\mathbb{R}_+$ given by $f(x) = \sqrt{x}$

$$\lim_{t\to\infty} f^t(x) = 1 \text{ for all } x > 0.$$

3. Find for the logistic difference equation (1.1.9) with $a = 4$ a 2-cycle that is some $x \in [0, 1]$ such that $f^2(x) = x$ but $f(x) \neq x$.

4. Solve the logistic differential equation

$$\frac{dp(t)}{dt} = \rho p(t)(P - p(t)), \quad 0 \leq p(t) \leq P$$

and show that $\lim_{t\to\infty} p(t) = P$ for all $p(0) > 0$.

5. Find a direct argument showing for the reproduction function $f(x) = 5x(1 + x)^{-2}$ that $\lim_{t\to\infty} p(t) = \sqrt{5} - 1$ for all $p(0) > 0$.

6. By using condition (1.1.13) prove that for the logistic equation (1.1.9) $\lim_{t\to\infty} x(t)$ exists for all $x(0) \in [0, \frac{2}{3}]$ if the value of the parameter a lies between 0 and $\frac{8}{3}$.

1.2 The density dependent Leslie model

The earliest population model at all is the one formulated by Leonardo di Pisa, better known as Fibonacci, in the early 13th century about the reproduction of rabbits. Suppose rabbits produce in pairs in such a way that any pair being at least two months old produces each month a new pair without any rabbit dying. Measuring time $t \in \mathbb{N}$ in number of months and denoting by $p(t)$ the number of pairs at time t we then have that

$$p(t+2) = p(t+1) + p(t) \quad \text{for all} \quad t \in \mathbb{N}. \tag{1.2.1}$$

Assuming $p(0) = p(1) = 1$, that is starting with one (newborn) pair of rabbits, this linear difference equation of second order generates the famous **Fibonacci numbers** $1, 1, 2, 3, 5, 8, 13$ etc. Setting $x_1(t) = p(t)$, $x_2(t) = p(t+1)$ equation (1.2.1) may be rewritten as $x_1(t+1) = x_2(t), x_2(t+1) = p(t+2) = p(t+1) + p(t) = x_1(t) + x_2(t)$ or, in matrix notation,

$$x(t+1) = \begin{bmatrix} 0 & 1 \\ 1 & 1 \end{bmatrix} x(t) \quad \text{for all} \quad t \geq 0, \tag{1.2.2}$$

where $x(t) = (x_1(t), x_2(t)) \in \mathbb{R}^2_+$, $x(0) = (1,1)$ and $F = \left[\begin{smallmatrix} 0 & 1 \\ 1 & 1 \end{smallmatrix}\right]$ is the *Fibonacci matrix*. (Vectors $x \in \mathbb{R}^n$ will always be understood to be column vectors; a row vector will be denoted by the transposed vector x'. $\mathbb{R}^n_+$ is the positive orthant in $\mathbb{R}^n$, $\mathbb{R}^n_+ = \{x \in \mathbb{R}^n \mid x = (x_1, \ldots, x_n),\ 0 \leq x_i \text{ for } 1 \leq i \leq n\}$.)

This two-dimensional representation reflects the underlying age structure, that is, $x_1(t)$ is the number of young rabbits (in pairs) which are less than 2 months old and $x_2(t)$ is the number of old rabbits (in pairs) which are at least 2 months old. The equation $x_1(t+1) = x_2(t)$ reflects the assumption that every old pair produces a new pair one month later and $x_2(t+1) = x_1(t) + x_2(t)$ means that old pairs next month stem from old or young pairs this month. By iteration, equation (1.2.2) has as solution

$$x(t) = F^t x(0), \quad t \in \mathbb{N},$$

which is uniquely determined by $x(0)$. To know the solution for arbitrary $x(0)$ means to know all the matrix powers F^t. Though the model is linear a constant growth rate g as in (1.1.2), i.e., $p(t+1) = (1+g)\,p(t)$ for all t, is not always possible. For, this would mean $x(t+1) = (1+g)x(t)$ and, hence, $Fx(0) = \lambda x(0)$ with $\lambda = 1+g$. But $x(0) = (1,1)$ is not an eigenvector of the matrix F. By choosing, however, initial vectors $x(0)$ which are eigenvectors of F one obtains two possibilities for constant growth. Matrix F has as its eigenvalues $\lambda_1 = \frac{1+\sqrt{5}}{2}$ and $\lambda_2 = \frac{1-\sqrt{5}}{2}$ with corresponding eigenvectors $x^1 = (1, \lambda_1)$ and $x^2 = (1, \lambda_2)$, respectively. An arbitrary given initial vector $x(0) = (a, b)$ we can obtain as a linear combination of the two linearly independent vectors x^1 and x^2, namely

$$x(0) = \frac{b - a\lambda_2}{\sqrt{5}} x^1 + \frac{a\lambda_1 - b}{\sqrt{5}} x^2.$$

For the uniquely determined solution starting in $x(0)$ this implies

$$\begin{aligned} x(t) &= F^t x(0) = \frac{b - a\lambda_2}{\sqrt{5}} F^t x^1 + \frac{a\lambda_1 - b}{\sqrt{5}} F^t x^2 \\ &= \frac{b - a\lambda_2}{\sqrt{5}} \lambda_1^t x^1 + \frac{a\lambda_1 - b}{\sqrt{5}} \lambda_2^t x^2. \end{aligned}$$

For the components of $x(t)$ this means

$$\begin{aligned} x_1(t) &= (\sqrt{5})^{-1}(a\lambda_1\lambda_2\,(\lambda_2^{t-1} - \lambda_1^{t-1}) + b(\lambda_1^t - \lambda_2^t)) \\ x_2(t) &= (\sqrt{5})^{-1}(a\lambda_1\lambda_2\,(\lambda_2^t - \lambda_1^t) + b(\lambda_1^{t+1} - \lambda_2^{t+1})). \end{aligned}$$

On the other hand, $x(t) = F^t x(0)$, and by using $\lambda_1\lambda_2 = -1$ we conclude that for all $t \geq 1$

$$F^t = (\sqrt{5})^{-1} \begin{bmatrix} \lambda_1^{t-1} - \lambda_2^{t-1} & \lambda_1^t - \lambda_2^t \\ \lambda_1^t - \lambda_2^t & \lambda_1^{t+1} - \lambda_2^{t+1} \end{bmatrix}.$$

Thus, we have determined the powers of the Fibonacci matrix F in terms of the eigenvalues λ_1, λ_2 of F. Hence, we can compute the solution $x(t)$ for any given $x(0)$, which in case of the Fibonacci numbers yields

$$\begin{aligned} p(t) &= (\sqrt{5})^{-1}(\lambda_1^{t-1} + \lambda_1^t - (\lambda_2^{t-1} + \lambda_2^t)) \\ &= (\sqrt{5})^{-1}(\lambda_1^{t+1} - \lambda_2^{t+1}) \end{aligned}$$

that is the *Binet formula*

$$p(t) = (\sqrt{5})^{-1}\left(\left(\frac{1+\sqrt{5}}{2}\right)^{t+1} - \left(\frac{1-\sqrt{5}}{2}\right)^{t+1}\right) \quad \text{for} \quad t \in \mathbb{N}. \tag{1.2.3}$$

This formula shows, as one would expect, that the number of rabbit pairs tends to infinity for $t \to \infty$ but it also shows that the ratio of old rabbits to young rabbits, i.e., $\frac{x_2(t)}{x_1(t)} = \frac{p(t+1)}{p(t)}$ stabilizes to the *golden mean* $\frac{1+\sqrt{5}}{2}$.

As has been argued in the previous section, population pressure should be taken into account leading thereby to a non-linear model.

Hence instead of (1.2.1) we should better consider a **non-linear Fibonacci model**, as, e.g.,

$$p(t+2) = \sqrt{p(t+1)} + \sqrt{p(t)} \quad \text{for} \quad t \in \mathbb{N}. \tag{1.2.4}$$

In this model the survival rates are no longer constantly equal to 1 but decrease by population pressure with an increase in the population size. The non-linear difference equation of second order (1.2.4) cannot easily be handled by graphic iteration. For the corresponding two-dimensional system equation (1.2.2) has to be replaced by

$$x(t+1) = \begin{bmatrix} 0 & 1 \\ (x_1(t))^{-\frac{1}{2}} & (x_2(t))^{-\frac{1}{2}} \end{bmatrix} x(t), \tag{1.2.5}$$

where the matrix now depends on the state $x(t)$. The matrix analysis done for the Fibonacci matrix is no longer applicable. In a later section we shall show that for the above model the size of the rabbit population will not tend to infinity but will approach a positive equilibrium. What has been said can be extended from two age classes to an arbitrary number of age classes. This more realistic model is known as the *Leslie model* and will be described in the following. (For the history and theory of classical Leslie models see Leslie [20], Caswell [3], Hansen [11], Hoppensteadt [14], Pollard [28].)

Consider a population for which $n \geq 1$ age classes are to be distinguished and denote by $x_i(t)$ the number of individuals in age class i at period $t \in \mathbb{N}$. (Being concerned with individuals capable of reproduction, the individuals usually will be females or will be taken in pairs.) All classes are assumed to contribute with certain birth rates to class 1 representing the youngest group in the population. The members of class i will survive with a certain rate to become members of class $i+1$ in the next period. Due to population pressure birth rates b_i and survival rates s_i ($= 1 - m_i$, m_i mortality rate) depend on the population levels of the various classes. Furthermore, those rates may depend in addition explicitly on time t. Denoting by $x(t) = (x_1(t), \ldots, x_n(t)) \in \mathbb{R}^n_+$ the population vector at period $t \in \mathbb{N}$, the assumptions made amount to the equations

$$\begin{aligned} x_1(t+1) &= \sum_{i=1}^{n} b_i(t, x(t))x_i(t) \\ x_{i+1}(t+1) &= s_i(t, x(t))x_i(t) \text{ for all } 1 \leq i \leq n-2 \\ x_n(t+1) &= s_{n-1}(t, x(t))x_{n-1}(t) + s_n(t, x(t))x_n(t). \end{aligned} \tag{1.2.6}$$

Here, of course, $0 \leq b_i, s_i \leq 1$. The last equation in (1.2.6) means that with a certain rate s_n members of the oldest group remain when becoming older in this group. If $s_n = 1$, as in the Fibonacci model, then members of class n will never die. If $s_n = 0$, as we shall often assume, members of class n will die out in the next period. In matrix notation (1.2.6) becomes

$$x(t+1) = L(t, x(t))x(t) \text{ for } t \in \mathbb{N}, \tag{1.2.7}$$

where, for $x \in \mathbb{R}^n_+$

$$L(t,x) = \begin{bmatrix} b_1(t,x) & b_2(t,x) & \ldots & b_{n-1}(t,x) & b_n(t,x) \\ s_1(t,x) & 0 & \ldots & 0 & 0 \\ 0 & s_2(t,x) & \ldots & 0 & 0 \\ \vdots & \vdots & & \vdots & \vdots \\ 0 & 0 & \ldots 0 & s_{n-1}(t,x) & s_n(t,x) \end{bmatrix} \tag{1.2.8}$$

denotes the **(generalized) Leslie matrix.**

In case the Leslie matrix $L(t,x)$ does not explicitly depend on time t it is called the **density-dependent Leslie matrix** $L(x)$; in case $L(t,x)$ does neither depend on t nor on x, it is called the *constant Leslie matrix*

$$L = \begin{bmatrix} b_1 & b_2 & \dots & b_{n-1} & b_n \\ s_1 & 0 & \dots & 0 & 0 \\ 0 & s_2 & \dots & 0 & 0 \\ \vdots & \vdots & & \vdots & \vdots \\ 0 & 0 & \dots 0 & s_{n-1} & s_n \end{bmatrix}.$$

The Fibonacci matrix is a special case of a constant Leslie matrix, and the non-linear Fibonacci matrix in (1.2.5) is a special case of a density-dependent Leslie matrix. The model (1.2.7) is an example of a **non-autonomous positive discrete dynamical system**

$$x(t+1) = T(t)x(t) \text{ for all } t \in \mathbb{N},\ x(0) \in \mathbb{R}^n_+, \tag{1.2.9}$$

where the selfmapping $T(t)$ of $\mathbb{R}^n_+$ is given by the Leslie matrix, $T(t)x = L(t,x)x$. If $T(t) = T$ for all t, i.e. T is given by the Leslie matrix $L(x)$, the system (1.2.9) is called *autonomous*. The dynamical system (1.2.9) is non-linear for a density-dependent Leslie matrix $L(x)$ and it is linear for a constant Leslie matrix L. The most interesting questions considering the dynamical system (1.2.9) are if there exists an equilibrium and whether this is stable or not; more generally, one wants to know the behavior of the system (1.2.9) for $t \to \infty$. As for equilibria in the autonomous case one has to find the non-trivial *fixed points* x^* of T in $\mathbb{R}^n_+$, i.e. $0 \neq x^* \in \mathbb{R}^n_+$ such that $Tx^* = x^*$. As already shown by the simple (linear) example of Fibonacci's rabbits such a non-trivial equilibrium need not exist. In this example, however, there exists for the matrix $F = \left[\begin{smallmatrix} 0 & 1 \\ 1 & 1 \end{smallmatrix}\right]$ an eigenvalue $\lambda_1 = \frac{1+\sqrt{5}}{2} > 0$ with eigenvector $x^1 = (1, \lambda_1) \in \mathbb{R}^2_+$. By $Fx^1 = \lambda_1 x^1$ it holds $x(t) = \lambda_1^t x^1$ for all t, that is a constant growth solution which may be considered a generalization of a (stationary) equilibrium; the latter corresponds to an eigenvalue equal to 1. Hence, it will be important also in the general case to find solutions to the **non-linear eigenvalue problem**

$$Tx^* = \lambda^* x^* \text{ with } 0 \neq x^* \in \mathbb{R}^n_+ \text{ and } \lambda^* > 0. \tag{1.2.10}$$

Introducing the **normalized mapping**

$$\tilde{T}x = \frac{Tx}{\|Tx\|}, \tag{1.2.11}$$

where $\|z\| = |z_1| + \dots + |z_n|$ denotes the sum-norm on $\mathbb{R}^n$, the eigenvalue problem (1.2.10) with a normalized eigenvector, i.e. $\|x^*\| = 1$, can be formulated as $\tilde{T}x^* = x^*$, that is as a fixed point problem for the mapping $\tilde{T}$. This normalization is related to the age structure in our population model. The *age structure* at time t

may be described by comparing the number $x_i(t)$ of individuals in age class i to the total population number $\|x(t)\| = x_1(t) + \cdots + x_n(t)$, i.e. by $y_i(t) = \frac{x_i(t)}{\|x(t)\|}$. To know the numbers $y_i(t)$ for $1 \le i \le n$ is, of course, equivalent to know the ratios $\frac{x_i(t)}{x_j(t)}$ for all $1 \le i,j \le n$. As already remarked, in the Fibonacci case $\lim_{t\to\infty} \frac{x_2(t)}{x_1(t)} = \lambda_1$, and, hence, for $y(t) = (y_1(t), y_2(t))$ one has that $\lim_{t\to\infty} y(t) = y^*$ with $y^* = ((1+\lambda_1)^{-1}, \lambda_1(1+\lambda_1)^{-1})$. Since $y^* = \frac{x^1}{\|x^1\|}$ it follows from $Tx^1 = \lambda_1 x^1$ that $\tilde{T}y^* = \frac{Tx^1}{\|Tx^1\|} = \frac{x^1}{\|x^1\|} = y^*$. Thus, in the Fibonacci case there exists a stable equilibrium age structure which, after normalization, is the unique (non-trivial) fixed point of $\tilde{T}$. The above convergence to y^* can be expressed also by the normalized operator as

$$\lim_{t\to\infty} \tilde{T}^t y(0) = y^*, \quad y(0) \text{ being any initial age structure.}$$

What has been said for the (linear) Fibonacci model can be extended to the Leslie model with a constant Leslie matrix L. An elegant way of doing this is to employ the so called **Perron–Frobenius theory** for non-negative matrices (see Gantmacher [10], Seneta [32]). But how to handle the density-dependent Leslie model? In the next chapter we will develop a **concave Perron–Frobenius theory** which generalizes many results of the (linear) Perron–Frobenius theory and which will prove to be useful in handling non-linear Leslie models. Since in that chapter we will obtain the most important results of the standard Perron–Frobenius theory as a by-product it is not required that the reader has some prior knowledge of that theory.

Exercises

1. Determine all eigenvalues and eigenvectors of the Fibonacci matrix $\left[\begin{smallmatrix} 0 & 1 \\ 1 & 1 \end{smallmatrix}\right]$.
2. Find by graphic iteration the dynamic behavior of the ratio $\frac{p(t+1)}{p(t)}$ for the Fibonacci numbers $p(t)$, $t \in \mathbb{N}$.
3. Examine the behavior of the powers L^t for $t \to \infty$ for the following Leslie matrix

$$L = \begin{bmatrix} 0 & 1 & 1 \\ 1 & 0 & 0 \\ 0 & 1 & 1 \end{bmatrix}.$$

4. How could the method of graphic iteration be extended to illustrate the dynamics of the following non-linear Fibonacci equation

$$p(t+2) = \sqrt[2]{p(t+1)} + \sqrt[4]{p(t)}\,?$$

5. Find a direct proof showing for the non-linear Fibonacci equation of Exercise 4 that $\lim_{t\to\infty} p(t)$ does exist for all $p(0) \ge 0$.

1.3 Non-linear price dynamics in one dimension

In economic theory sometimes pretty stylized models are analyzed which involve one single good only. This method was already used by the classical economists in their corn model (cf. Ricardo [29]) and it reappeared later on in the one sector growth models involving one single capital good. Consider the production of one single good, say corn, by using corn as seed together with labor. Assume the production of one bushel of corn needs as inputs a bushels of corn and l hours of labor with $0 < a < 1$ and $0 < l$. If p is the price per bushel corn and w is the wage for one hour of labor then the cost of production for one bushel of corn is given by $c(p) = ap + lw$. Assume also that time is measured in discrete steps $t \in \mathbb{N}$ and that the price $p(t+1)$ for the next period $t+1$ is given by the cost of the present period t, that is

$$p(t+1) = c(p(t)) = ap(t) + lw. \tag{1.3.1}$$

To simplify, we assume w to be constant over time and a possible profit to be included in w. Starting with an arbitrary initial price $p(0)$, the price $p(t)$ then is given by

$$p(t) = \left(\sum_{i=0}^{t-1} a^i\right) lw + a^t p(0) \quad \text{for} \quad t \geq 1. \tag{1.3.2}$$

This immediately yields $\lim_{t\to\infty} p(t) = p^*$ with $p^* = (1-a)^{-1}lw$. Thus, in this simple (affine) linear model there exists a unique price equilibrium p^*, defined by $p^* = c(p^*)$, and the prices set according to the **positive discrete dynamical system** (1.3.1) approach the equilibrium for $t \to \infty$, irrespective of the initial price $p(0) \geq 0$. In reality, however, there is not just one single technique of production available but there are often several possibilities. In our example, to grow corn one method may require less corn for seed than another method by doing seeding more carefully, that is by employing more labor. A technique being described by a pair (a, l) of inputs, suppose a set of techniques $\{(a_i, l_i) \mid 1 \leq i \leq m\}$ is available among which the producer can choose. Of course, for a given price the producer will choose a technique which minimizes the cost of production Our little model, enriched by a choice of techniques then becomes

$$p(t+1) = \min\{a_i p(t) + l_i w_i \mid 1 \leq i \leq m\} \tag{1.3.3}$$

This model is no longer (affine) linear, but we may try graphic iteration as in Section 1.1.

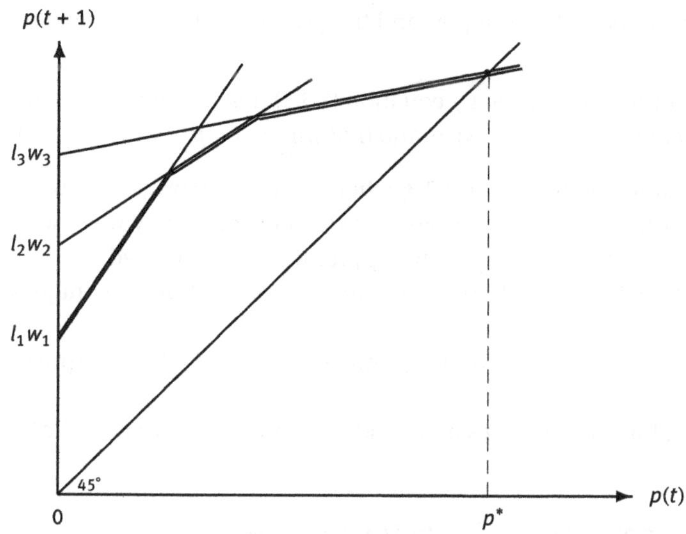

Fig. 1.4. Choice of techniques.

In the above figure there are three techniques possible and the line starting in $l_i w_i$ represents the cost of technique i. As is obvious from the figure, depending on the price indeed different techniques will be choosen. It may happen that a technique with $a_i > 1$ will be choosen provided the wage cost $l_i w_i$ is low enough. Analytically, since the minimum in (1.3.3) is concave in $p(t)$ (the boldfaced curve in the above figure), there will be a unique equilibrium price p^* as long as $a_i < 1$ for at least one technique i. Also by concavity, a similar argument as in section 1.1 shows that $\lim_{t\to\infty} p(t) = p^*$ for all $p(0) > 0$. (An explicit formula for $p(t)$ like (1.3.2) is possible but not very transparent.) Concave cost curves are quite common in economics and, different from the reproduction curve in population dynamics, the cost $c(p)$ must increase with p. Since the cost function in (1.3.3) is not differentiable in all points the earlier criterion (1.1.13) cannot be applied to check the behavior of prices in the long run. Next we turn to the more realistic situation where more than one good is involved.

Exercises

1. Suppose that a producer has the three techniques $(2, 1)$, $(0.8, 3.4)$ and $(0.5, 4)$ at his disposal for producing a single good.
   (a) Determine for $w = 1$ the price intervals for each of the above techniques to be choosen.
   (b) Determine the equilibrium price p^* and show by cobwebbing that $\lim_{t\to\infty} p(t) = p^*$ for all $p(0) \geq 0$.

(c) Examine how the choice of techniques and the price equilibrium depend on the wage rate w.

2. Show analytically for the model given by equation (1.3.3) that under the assumption $\min_{1\le i\le n} a_i < 1$ a price equilibrium p^* exists and that $\lim_{t\to\infty} p(t) = p^*$ for all $p(0) \ge 0$.
3. Explore the generalization of the model (1.3.3) where a minimum wage w_0 is guaranteed and where for $w \ge w_0$ the real wage $\frac{w}{p}$, instead of the money wage w, is held constant. That is, let $w = w(p) = \max\{w_0, cp\}$ for $w_0 \ge 0, c \ge 0$ fixed.
   (a) Find the asymptotic behavior of prices $p(t)$ for $w_0 = 1, c = 1$ and techniques given as in Exercise 1.
   (b) Determine all values of minimum wage w_0 and real wage c for which the prices $p(t)$ are converging for $t \to \infty$.
   (c) Is it possible that for certain values of w_0 and c the prices behave periodic?

1.4 The Leontief model with choice of techniques

Consider finitely many producers who are interdependent in that each of them produces a specific good by means of the goods produced by all the other producers and by employing (homogeneous) labor. Periodically each producer may set a new price for his product according to his cost of production which depends possibly on all prices set by the other producers one period before.

We want to know if there exist equilibrium prices on all markets and, even more important, whether the process of price setting will lead the producers towards equilibrium prices. More specifically, denote by $\{1, \dots, n\}$ the set of producers where any producer i can be identified with the specific good he is producing. Every producer may choose a technique within a certain technology set. A technique is described by a pair (a, l) where a is a vector in $\mathbb{R}^n_+$ with components $a_j, 1 \le j \le n$, specifying the input of good j used for producing one unit of the producers good and where $l \ge 0$ is the labor input required thereby, measured, say, in hours. Let $A_i(t)$ denote the (nonempty) set of techniques which producer i has at his disposal and which may depend on time due to technological development. By $p_i(t) \ge 0$ we denote the price of one unit of good i in period t, by $p(t)$ the price vector with components $p_i(t)$ for $1 \le i \le n$ and by $w_i > 0$ the constant money wage per hour paid by producer i. Forced by competition each producer will produce for a given vector of prices $p \in \mathbb{R}^n_+$ at *minimum cost* which is for producer i in period $t \in \mathbb{N}$ given by

$$c_i(p, t) = \inf\{pa + w_i l \mid (a, l) \in A_i(t)\}, \tag{1.4.1}$$

where $pa := \sum_{j=1}^n p_j a_j$ is the inner product.

From the various rules of price setting one could think of we shall adopt the classical economists *rule of prices driven by cost,* that is

$$\frac{p_i(t+1)}{p_j(t+1)} = \frac{c_i(p(t),t)}{c_j(p(t),t)} \quad \text{for} \quad 1 \le i,\ j \le n,\ t \in \mathbb{N}.$$

This rule does not require prices of the next period to be equal to cost of this period but requires only proportionality, that is $p(t+1) = k(t)c(p(t),t)$ with a factor $k(t) > 0$ which may depend on time and where $c(p(t),t)$ is the vector having components $c_i(p(t),t)$. Introducing the **cost operator** $T(t)\colon \mathbb{R}^n_+ \longrightarrow \mathbb{R}^n_+$ by $T(t)p = c(p,t)$ with $c_i(p,t)$ given by (1.4.1) we arrive at the **positive discrete dynamical system**

$$p(t+1) = k(t)T(t)p(t), \quad t \in \mathbb{N},\ p(0) \in \mathbb{R}^n_+. \tag{1.4.2}$$

This system is non-autonomous and non-linear. If the technology sets $A_i(t)$ do not depend on t, that is if technological development is not taken into account, then there is just one cost operator $T\colon = T(t)$ for all t; if in addition $k(t)$ is constant, in particular if $k(t) = 1$ for all t, then the system (1.4.2) is autonomous. Disregarding technological development, the non-linear operator T defined by $Tp = c(p)$ (omitting now variable t) is *concave*, namely for $0 \le \alpha \le 1$ and $p, q \in \mathbb{R}^n_+$ we have

$$\begin{aligned} c_i(\alpha p + (1-\alpha)q) &= \inf\{\alpha(pa) + (1-\alpha)(qa) + \alpha w_i l + (1-\alpha)w_i l \mid (a,l) \in A_i\} \\ &\ge \alpha \inf\{pa + w_i l \mid (a,l) \in A_i\} + (1-\alpha)\inf\{qa + w_i l \mid (a,l) \in A_i\} \\ &\ge \alpha c_i(p) + (1-\alpha)c_i(q), \end{aligned}$$

and, hence, $T(\alpha p + (1-\alpha)q) \ge \alpha Tp + (1-\alpha)Tq$ componentwise, that is T is concave.

Concavity comes in very naturally by taking a choice of techniques into account.

If no choice of techniques is admitted then (in the autonomous case) A_i must consist of one technique (a^i, l_i) only and, hence, the i-th component T_i of T is given by $T_i p = c_i(p) = pa^i + w_i l_i$. Thus, T is an affine-linear mapping given by $Tp = Ap + b$ where A is the matrix having $a^1, \dots, a^n$ as its rows and b is a column vector with components $w_1 l_1, \dots, w_n l_n$. In this linear case, matrix A is called the *Leontief matrix* and the model of production the **Leontief model** or the input-output model of production. (For linear models of Leontief type or Sraffa type see Cassels [2], Nikaido [25], Schwartz [31] and, taking choice of techniques into account, Kurz and Salvadori [19], Woods [34].) A model as the above which combines a Leontief model of production with a choice of techniques may be called *Morishima model* because the dynamics of such a model has been first analyzed in (Morishima [23]).

In economics it is often more meaningful to consider relative prices instead of absolute prices. If $\|p\| = p_1 + \cdots + p_n$ for $p \in \mathbb{R}^n_+$ then the *relative price* in period t is given by $q(t) = p(t)\|p(t)\|^{-1}$. To know the vector $q(t)$ is equivalent to know all the ratios $p_i(t)(p_j(t))^{-1}$ for $1 \le i,j \le n$. Casting system (1.4.2) in relative prices amounts to $q(t+1) = k(t)T(t)q(t)$. Because of $\|q(t+1)\| = 1$ this implies

$q(t+1) = T(t)q(t)(\|T(t)q(t)\|^{-1})$. Introducing the *normalized cost operator* $\tilde{T}(t)$ defined by $\tilde{T}(t)p := T(t)p(\|T(t)p\|^{-1})$ (provided $T(t)p \neq 0$), we obtain the new system

$$q(t+1) = \tilde{T}(t)q(t),\ t \in \mathbb{N},\ \|q(0)\| = 1. \tag{1.4.3}$$

This equation describes a non-autonomous and non-linear positive discrete dynamical system on the positive unit sphere $S = \{p \in \mathbb{R}^n_+ \mid \|p\| = 1\}$. (For systems of this type see Krause [16].) In general, the systems described by (1.4.2) and (1.4.3) respectively, need not be equivalent. The equations, however, model the same system if we assume a constant real wage instead of a constant money wage w_i. Here, a constant real wage is a basked of goods $b^i \in \mathbb{R}^n_+$ such that $w_i = pb^i$ (inner product) for the ruling price vector p. By this change the minimum cost becomes

$$c_i(p,t) = \inf\{p(a + lb^i) \mid (a,l) \in A_i(t)\}$$

and, hence, $c_i(\lambda p, t) = \lambda c_i(p,t)$ for arbitrary scalars $\lambda > 0$. This implies $T(t)(\lambda p) = \lambda T(t)p$ for $\lambda > 0$ and equation (1.4.2) becomes

$$\begin{aligned} q(t+1) = p(t+1)(\|p(t+1)\|^{-1}) &= k(t)T(t)p(t)(k(t)\|T(t)p(t)\|)^{-1} \\ &= T(t)(p(t)\|p(t)\|^{-1})\Big(\|T(t)(p(t)\|p(t)\|^{-1})\|\Big)^{-1} \\ &= T(t)q(t)(\|T(t)q(t)\|^{-1}) = \tilde{T}(t)q(t), \end{aligned}$$

which is equation (1.4.3).

From the *concave Perron–Frobenius theory* which we shall develop in the next chapter it will follow for the autonomous case that under certain assumptions, roughly by assuming producers to be interdependent strongly enough, system (1.4.3) has a unique equilibrium price vector q^*, i.e. $q^* = \tilde{T}q^*$, such that $\lim_{t\to\infty} q(t) = q^*$ for all $q(0) \in S$. This is an important finding because it means that the producers who selfishly minimize own costs without being guided by some external central agency are nevertheless able to find a joint price equilibrium by setting prices according to costs.

Exercises

1. Consider two producers each equipped with two technologies. Suppose producer 1 (producing good 1) can use a technique with $a = (0,1)$ and $l = 1$ or a technique with $a = (0.5, 0.5)$ and $l = 1$. Similarly, producer 2 (producing good 2) can use a technique $a = (1,0)$ and $l = 2$ or $a = (0.1, 0)$ and $l = 1$.
   Assume further for both producers a given real wage $b = (1,1)$, that is $w = p_1 + p_2$ for prices (p_1, p_2) given.
   (a) Calculate the cost operator $T\colon \mathbb{R}^2_+ \longrightarrow \mathbb{R}^2_+$ and show that $\lim_{t\to\infty} T^t p = \infty$ for all $p \in \mathbb{R}^2_+, p \neq 0$.

(b) Show there exists a unique positive equilibrium q^* in relative prices, i.e. $\tilde{T}q^* = q^*$, $\|q^*\| = 1$ (take $\|x\| = |x_1| + |x_2|$).

(c) Develop an argument demonstrating that relative prices must converge to q^*, i.e. $\lim_{t\to\infty} \tilde{T}^t q = q^*$ for all $q \in \mathbb{R}^2_+$, $\|q\| = 1$.

2. Consider two producers equipped with techniques as in Exercise 1. Different, however, from 1. assume a variable real wage of the type $b = (s, 0)$ with $s \geq 0$, which means that the real wage is measured in terms of good 1. That is to say, assume for both producers a wage-price relationship of $w = sp_1$.
   (a) Determine the scalar $\lambda > 0$ for the equilibrium $Tp = \lambda p$ in dependence on the real wage s.
   (b) Interprete the relationship between s and λ as a so called *wage-profit curve*, that is as a relationship between s and r whereby r is the uniform rate of profit given by $r = \frac{p_i - c_i(p)}{c_i(p)}$ for $i = 1, 2$.
   Determine the maximal possible values for the rate of profit and the real wage, respectively.
   (c) Discuss (a) and (b) above for the case that the real wage is measured in terms of good 2, that is $w = sp_2$.
   (d) What can be said for a varying real wage s with respect to the questions asked in Exercise 1?

3. Let A be a non-negative $n \times n$-matrix which is *productive*, i.e. there exists some $x \in \mathbb{R}^n_+$ such that $Ax < x$ (where $<$ is with respect to all components). Prove that
   (a) $\lim_{n\to\infty} A^n = 0$
   and
   (b) $(I-A)^{-1}$ exists and is given by the Neumann series, $(I-A)^{-1} = I + A + A^2 + \cdots$ (I the $n \times n$-identity matrix).

4. Let $T\colon \mathbb{R}^n_+ \longrightarrow \mathbb{R}^n_+$ be concave and suppose that for some $x \in \mathbb{R}^n_+$ $Tx < x$ and that for all $0 < \varepsilon \leq 1$ $T(\varepsilon x) \leq \varepsilon Tx$. Prove that
   (a) $\lim_{t\to\infty} T^n = 0$ (T^n the n-th iterate of T, 0 the zero mapping)
   and
   (b) $S\colon = \sum_{n=0}^{\infty} T^n$, $S\colon \mathbb{R}^n_+ \longrightarrow \mathbb{R}^n_+$ exists and $S - S \circ T = I$. (I the identity map on $\mathbb{R}^n_+$.)

5. Find examples of non-(affine-)linear mappings $T\colon \mathbb{R}^2_+ \longrightarrow \mathbb{R}^2_+$ such that
   (a) T satisfies the assumptions of Exercise 4
   and
   (b) T satisfies in addition $S \circ (I - T) = (I - T) \circ S = I$.

1.5 Opinion dynamics under bounded confidence

Consider a group of n experts $i = 1, \dots, n$ who have to assess a certain magnitude like the world's wheat production in the year 2030. Each of the experts has his own expertise but is open to revise it by taking into account expertises by colleagues he trusts in. This process of revision iterates and the question arises whether the experts will oscillate in their assessments or run into disagreement or reach a consesus.

Let $t \in \mathbb{N}$ denote a round of the discussion and let $x_i(t)$ the assessment of expert i in round t. Considering trust among the experts let

$$\{1 \le j \le n \mid |x_i(t) - x_j(t)| \le \epsilon\}$$

denote those experts the expert i has trust in where $\epsilon > 0$ is a certain *confidence level*. Denote this *confidence set* by $I(i, x(t))$ where $x(t) = (x_1(t)), \dots, x_n(t))$ is the collection of the assessments in round t. Among the many ways to model the iterative formation of assessments a rather simple one is given by

$$x_i(t+1) = |I(i, x(t))|^{-1} \sum_{j \in I(i,x(t))} x_j(t) \text{ for } t \in \mathbb{N}, x(0) \in \mathbb{R}^n_+, \tag{1.5.1}$$

that is, the revised assessment is made by taking the arithmetical mean of those one trusts in. Instead of an assessment of some (positive) magnitude, $x_i(t)$ could be any **opinion** as long as it can be measured by a real number.

System (1.5.1) describes opinion dynamics under bounded confidence as it has been developed in [13] and [17]. This system is a **positive dynamical system** in discrete time. The positivity lies not so much in the state variables $x_i(t)$, which could be negative, but in the positivity of the weights given to other experts.

The system (1.5.1) is non-linear in the state variables and it is not easy to analyze. Alternatively, one could put system (1.5.1) as a linear but non-autonomous system. To see this define a $n \times n$-matrix $A(t)$ with entries $a_{ij}(t) = |I(i, x(t))|^{-1}$ for $j \in I(i, x(t))$ and $a_{ij}(t) = 0$, otherwise. Then system (1.5.1) is equivalent to

$$x(t+1) = A(t)x(t), t \in \mathbb{N}, x(t) \in \mathbb{R}^n_+. \tag{1.5.2}$$

In later chapters we shall develop methods to handle non-linear as well as non-autonomous positive dynamical systems. System (1.5.2), however, has the advantage that $A(t)$ is a (row-) stochastic matrix, that is each row sums up to one. By this methods especially adapted to stochastic matrices can be used which will be done in full detail in Chapter 8. Developing methods to treat infinite products of stochastic matrices we will be able in Chapter 8 to answer in particular the questions raised above considering consensus or disagreement.

Exercises

1. Examine the model of opinion dynamics under bounded confidence for $n = 2$ and different confidence levels ϵ_1 and ϵ_2 of the two experts.
   (a) Show that convergence to consensus holds if and only if $|x_1(0) - x_2(0)| \leq \max\{\epsilon_1, \epsilon_2\}$.
   (b) Show that in case of convergence to consensus the latter is reached for a number of rounds which is given by the smallest natural number above $\log_2 \frac{|x_1(0)-x_2(0)|}{\min\{\epsilon_1,\epsilon_2\}}$.
2. Investigate the system (1.5.1) for $n = 3$ by computer simulations.
   (a) What can be said about the dependence of the dynamics on the initial conditions $x(0)$?
   (b) What can be said about the dependence of the dynamics on the confidence level ϵ?
3. Find an example of system (1.5.1) which converges to a consensus which is not the arithmetic mean of $x_1(0), \ldots, x_n(0)$.
4. Investigate system (1.5.1) for initial conditions given by $x_i(0) = (i-1)\epsilon$, $1 \leq i \leq n$.
   (a) Check small values of n.
   (b) Try big values by computer simulations.
   (See also Exercise 16 to Chapter 8.)

Bibliography

[1] E. Beltrami. *Mathematics for Dynamic Modeling.* Academic Press, New York, 1987.
[2] J. W. S. Cassels. *Economics for Mathematicians.* Cambridge University Press, Cambridge, 1981.
[3] H. Caswell. *Matrix Population Models.* Sinauer Associates, Sunderland, 1989.
[4] P. Cull. Global stability of population models. *Bull. Math. Biol.*, 43:47–58, 1981.
[5] J. M. Cushing. *An Introduction to Structured Population Dynamics.* CBMS-NSF Series, vol. 71, SIAM, Philadelphia, 1998.
[6] K. Deimling. *Nonlinear Functional Analysis.* Springer, Berlin, Heidelberg, New York, Tokyo, 1985.
[7] L. Edelstein-Keshet. *Mathematical Models in Biology.* Random House/ Birkhäuser, New York, 1988.
[8] S. N. Elaydi. *An Introduction to Difference Equations.* Springer, New York, 1996.
[9] L. Farina, S. Rinaldi. *Positive Linear Systems: Theory and Applications.* Wiley-Interscience, New York, 2000.
[10] F. R. Gantmacher. *The Theory of Matrices*, volume I, II. Chelsea, New York, 1959.
[11] P. E. Hansen. Raising Leslie matrices to power: a method and applications to demography. *J. Math. Biol.*, 18:149–161, 1983.
[12] M. P. Hassell. Density dependence in single-species populations. *J. Anim. Ecol.*, 44:283–295, 1975.

[13] R. Hegselmann and U. Krause. Opinion dynamics and bounded confidence: Models, analysis, and simulation. *J. Artificial Societies and Social Simulation* 5 (3), 2002. online http://jasss.soc.surrey.ac.uk/5/3/2.html.
[14] F. . Hoppensteadt. *Mathematical Methods of Population Biology*. Cambridge University Press, Cambridge, 1982.
[15] V. L. Kocic and G. Ladas. *Global Behavior of Nonlinear Difference Equations of Higher Order with Applications* Kluwer, Dordrecht, 1993.
[16] U. Krause. Positive non-linear systems in economics. In T. Maruyama and W. Takahashi, editors, *Nonlinear and Convex Analysis in Economic Theory* pp, 181–195. Springer, Berlin, 1995.
[17] U. Krause. A discrete non-linear and non-autonomous model of consensus formation. In S. Elaydi et al., editor, *Communications in Difference Equations*, Gordon and Breach Publ., Amsterdam 2000, pp. 227–236.
[18] U. Krause and T. Nesemann. *Differenzengleichungen und Diskrete Dynamische Systeme*. 2. Auflage. De Gruyter, Berlin etc., 2012.
[19] H. D. Kurz and N. Salvadori. *Theory of Production*. Cambridge University Press, Cambridge, 1995.
[20] P. H. Leslie. On the use of matrices in certain population mathematics. *Biometrika*, 35:213–245, 1948.
[21] D. G. Luenberger. *Introduction to Dynamic Systems*. Wiley & Sons, New York, 1979.
[22] R. M. May. Simple mathematical models with very complicated dynamics. *Nature*, 261:459–467, 1976.
[23] M. Morishima. *Equilibrium, Stability, and Growth*. Clarendon Press, Oxford, 1964.
[24] J. D. Murray. *Mathematical Biology*. Springer, Berlin, 1993.
[25] H. Nikaido. *Convex Structures and Economic Theory*. Academic Press, New York, 1968.
[26] H. O. Peitgen, H. Jürgens, and D. Saupe. *Chaos and Fractals*. Springer, New York, 1992.
[27] E. C. Pielou. *An Introduction to Mathematical Ecology*. Wiley Interscience, New York, 1969.
[28] J. H. Pollard. *Mathematical Models of the Growth of Human Populations*. Cambridge University Press, Cambridge, 1973.
[29] D. Ricardo. *Principles of Political Economy and Taxation*. Cambridge University Press, Cambridge, 1951.
[30] J. T. Sandefur. *Discrete Dynamical Systems*. Clarendon, Oxford, 1990.
[31] J. T. Schwartz. *Lectures on the Mathematical Method in Analytical Economics*. Gordon and Breach, New York, 1961.
[32] E. Seneta. *Non-negative Matrices and Markov Chains*. Springer, Berlin, Revised Printing, 2006.
[33] H. R. Thieme. *Mathematics in Population Biology*. Princeton University Press, Princeton, 2003.
[34] J. E. Woods. *Mathematical Economics*. Longman, London, 1978.
[35] B. G. Zaslavskiĭ. Chaos in a population. *Soviet Math. Dokl.*, 23:549–552, 1981.

2 Concave Perron–Frobenius theory

In this chapter we develop an extension of the (linear) Perron–Frobenius theory which is applicable to the concave operators encountered in Chapter 1. No use will be made of classical Perron–Frobenius theory, on the contrary, many results of the latter will turn out to be special cases of our approach which at the same time provides new proofs for some classical results.

Concave mappings are attractive in that on the one hand they represent a first step in capturing non-linear phenomena and on the other hand they still admit some systematic theory. There is, however, not such a strong tool as it is linear algebra for linear mappings. Actually, whereas a linear mapping on a finite dimensional space can be described by finitely many parameters in form of a matrix one needs in general infinitely many parameters even in one dimension, to describe a concave mapping. (See Exercise 8 of Section 2.1.)

2.1 Iteration of normalized concave operators

Let $\mathbb{R}^n: = \{x = (x_1, \dots, x_n) \mid x_i \in \mathbb{R} \text{ for } 1 \le i \le n\}$ denote the n-dimensional Euclidean space. For $x, y \in \mathbb{R}^n$ we employ the following notations:

$$\begin{aligned} x \le y \quad &\text{iff} \quad x_i \le y_i \quad \text{for all} \quad i, \\ x \lneq y \quad &\text{iff} \quad x \le y \quad \text{but} \quad x \ne y, \\ x < y \quad &\text{iff} \quad x_i < y_i \quad \text{for all} \quad i. \end{aligned}$$

Let $K = \mathbb{R}^n_+: = \{x \in \mathbb{R}^n \mid 0 \le x\}$ denote the cone of non-negative vectors in $\mathbb{R}^n$. For simplicity we denote the elements of $\mathbb{R}^n$ by row vectors. A different use, as with respect to matrices, will be mentioned explicitly.

Definition 2.1.1. A subset $D \subset \mathbb{R}^n$ is **convex** if for any $x, y \in D$ and $\alpha \in [0, 1]$ it holds that $\alpha x + (1-\alpha)y \in D$. For a convex subset $D \subset \mathbb{R}^n$ an operator (mapping) $T: D \longrightarrow \mathbb{R}^m$ is *concave* if for any $x, y \in D$ and $\alpha \in [0, 1]$

$$\alpha Tx + (1 - \alpha)Ty \le T(\alpha x + (1 - \alpha)y).$$

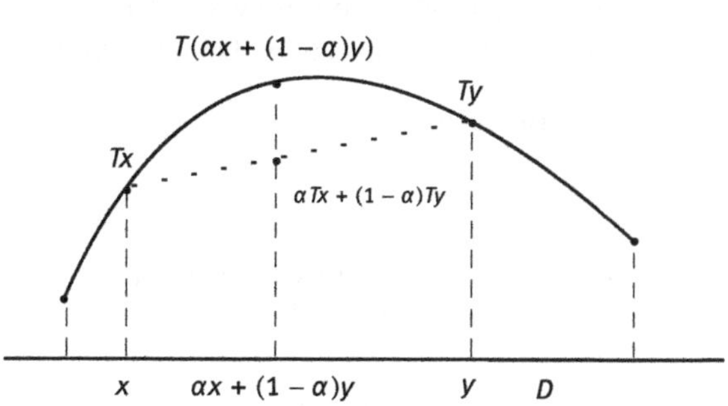

Fig. 2.1. Concave mapping.

An operator $T\colon D \longrightarrow \mathbb{R}^m$ is concave iff all its components T_i are concave, where $T_i\colon D \longrightarrow \mathbb{R}_+$ is defined by $T_i x = (Tx)_i$. Furthermore, Definition 2.1.1 yields immediately that for a concave operator

$$\sum_{i=1}^{k} \alpha_i T x^i \le T\Big(\sum_{i=1}^{k} \alpha_i x^i\Big)$$

for arbitrary $k \in \mathbb{N}$, $x^i \in D$, $\alpha_i \ge 0$ with $\sum_{i=1}^{k} \alpha_i = 1$.

Definition 2.1.2. An operator $T\colon D \longrightarrow \mathbb{R}^m, D \subset \mathbb{R}^n$, is **monotone** if

$$x \le y \quad \text{implies} \quad Tx \le Ty.$$

Lemma 2.1.3. *A concave operator $T\colon K \longrightarrow K$ is monotone.*

Proof. For $x, y \in K$ with $x \le y$ and $k \in \{1, 2, \ldots\}$ one has that $x + k(y - x) \in K$ and $y = (1 - k^{-1})x + k^{-1}(x + k(y - x))$. Concavity of T implies $Ty \ge (1 - k^{-1})Tx + k^{-1}T(x + k(y - x))$ for all k and, hence, $Ty \ge Tx$. □

Remark 2.1.4. A concave operator $T\colon D \longrightarrow K$ for D convex in K, $D \ne K$ need **not** be monotone.

Examples 2.1.5. (i) T **affine-linear**, i.e. $T\colon K \longrightarrow K$, $Tx = Ax + a$ where $a \in K$ and A is a non-negative $n \times n$-matrix, i.e. $a_{ij} \ge 0$ for all entries a_{ij} of A. (Here the elements of K are to be understood as column vectors.)

(ii) $T\colon K \longrightarrow K$ is of **Verhulst type**, i.e.

$$T_i x = \sum_{j=1}^{n} r_{ij} x_j (x_j + s_{ij})^{-1} \quad \text{for} \quad x \in K, i \in \{1, \ldots, n\}$$

with constants $r_{ij} \ge 0, s_{ij} > 0$.

(iii) $T\colon K \longrightarrow K$ is the **infimum of affine linear mappings**, i.e.

$$T_i x = \inf\Big\{(A(j)x)_i + a_i(j) \Big| j \in J\Big\} \quad \text{for} \quad x \in K;\ i \in \{1, \ldots, n\}$$

where for an arbitrary (non-empty) index set J and $j \in J$ $A(j)$ is a non-negative $n \times n$-matrix and $a(j) \in K$. (Elements of K are to be understood as column vectors.)
Since in all three cases $T\colon K \longrightarrow K$ is concave it is also monotone by Lemma (2.1.3).

Definition 2.1.6. A (vector) norm $\|\cdot\|$ on $\mathbb{R}^n$ is **monotone** if

$$0 \le x \le y \quad \text{implies} \quad \|x\| \le \|y\|.$$

Examples of monotone norms on $\mathbb{R}^n$ are the *max-norm* $\|x\| = \max\{|\, x_i \,|\, |1 \le i \le n\}$, the *sum-norm* $\|x\| = \sum_{i=1}^{n} |\, x_i \,|$ and the *Euclidian norm* $\|x\| = (\sum_{i=1}^{n} x_i^2)^{\frac{1}{2}}$.

For a monotone norm we rescale the operator T as we did already in Section 1.4.

Definition 2.1.7. For an operator $T: D \longrightarrow K, D \subset \mathbb{R}^n$, and any monotone norm $\|\cdot\|$ on $\mathbb{R}^n$ the **normalized or rescaled operator** $\tilde{T}$ is given by

$$\tilde{T}x = (Tx)(\|Tx\|^{-1}) \quad \text{for} \quad x \in D \quad \text{with} \quad Tx \neq 0.$$

Geometrically, normalizing an operator means to project its images on the unit sphere of the norm.

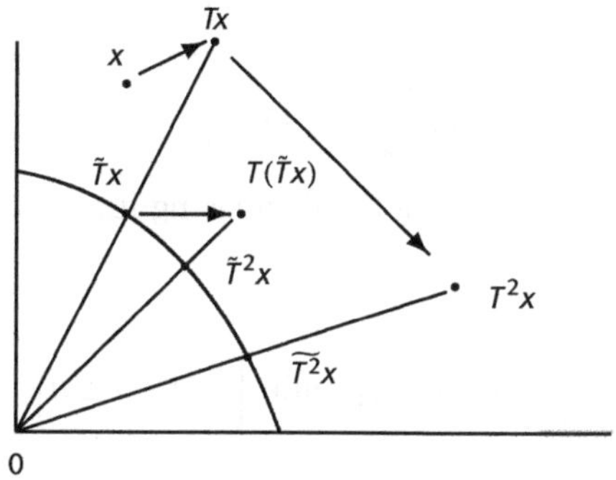

Fig. 2.2. Normalized operator (Euclidean norm).

It is important, as indicated in Fig. 2.2, to distinguish between $(\tilde{T})^k$ and $\widetilde{(T^k)}$. A simple example where these two iterates are different is given by $T: \mathbb{R}^2_+ \longrightarrow \mathbb{R}^2_+, Tx = (1 + x_1, 1)$ (see Exercises to 2.1, Problem 3).

To prove our main result in this section we need to define a special metric for which we shall apply Banach's contraction mapping principle.

Definition 2.1.8. On $\mathring{K} = \{x \in \mathbb{R}^n \mid 0 < x\}$ Hilbert's projective quasi-metric or **Hilbert's metric** for short, is defined for $x, y \in \mathring{K}$ by

$$d(x,y) = -\log\left(\min\left\{\frac{x_i}{y_i}\Big|1 \le i \le n\right\} \cdot \min\left\{\frac{y_i}{x_i}\Big|1 \le i \le n\right\}\right) \tag{2.1.1}$$

The following lemma confirms that by this definition d is a quasi-metric, where $d(x,y) = 0$ for $x \neq y$ may happen.

Lemma 2.1.9. *d as defined by (2.1.1) has the following properties for $x, y, z \in \mathring{K}$:*
(i) $d(x,y) \in \mathbb{R}_+$;
(ii) $d(x,y) = 0$ *iff* $x = ry$ *for some* $r > 0$;
(iii) $d(x,y) = d(y,x)$;
(iv) $d(x,z) \le d(x,y) + d(y,z)$;
(v) $d(rx, sy) = d(x,y)$ *for arbitrary scalars* $r, s > 0$
and
$d(z \cdot x, z \cdot y) = d(x,y)$ *where* $z \cdot x = (z_1x_1, \ldots, z_nx_n)$.

Proof. (i) Since the product of the two minima in (2.1.1) is strictly positive and must be less than or equal to $\frac{x_i}{y_i} \cdot \frac{y_i}{x_i} = 1$ we have that $d(x,y) \in \mathbb{R}_+$.

(ii) If $x = ry$ then $d(x,y) = -\log(r \cdot r^{-1}) = 0$. Conversely, if $d(x,y) = 0$ and $r = \min\{\frac{x_i}{y_i} | 1 \le i \le n\}$ then $r \cdot \min\{\frac{y_i}{x_i} | 1 \le i \le n\} = 1$. Hence, $r > 0$ and $\min\{\frac{y_i}{x_i} | 1 \le i \le n\} = r^{-1}$. This shows $r \le \frac{x_i}{y_i}$ as well as $\frac{x_i}{y_i} \le r$ for all i, that is $x_i = ry_i$ for all i.

(iii) Immediate from (2.1.1).

(iv) From

$$\min\left\{\frac{x_i}{y_i}\middle|1 \le i \le n\right\} \cdot \min\left\{\frac{y_i}{z_i}\middle|1 \le i \le n\right\} \le \min\left\{\frac{x_i}{z_i}\middle|1 \le i \le n\right\}$$

and

$$\min\left\{\frac{y_i}{x_i}\middle|1 \le i \le n\right\} \cdot \min\left\{\frac{z_i}{y_i}\middle|1 \le i \le n\right\} \le \min\left\{\frac{z_i}{x_i}\middle|1 \le i \le n\right\}$$

it follows by multiplying the two inequalities and applying the decreasing function $-\log$ to the result that $d(x,y) + d(y,z) \ge d(x,z)$.

(v) The first equation follows from

$$\min\left\{\frac{rx_i}{sy_i}\middle|1 \le i \le n\right\} \cdot \min\left\{\frac{sy_i}{rx_i}\middle|1 \le i \le n\right\} = \min\left\{\frac{x_i}{y_i}\middle|1 \le i \le n\right\} \cdot \min\left\{\frac{y_i}{x_i}\middle|1 \le i \le n\right\};$$

the second equation follows from

$$\min\left\{\frac{z_i x_i}{z_i y_i}\middle|1 \le i \le n\right\} = \min\left\{\frac{x_i}{y_i}\middle|1 \le i \le n\right\}. \qquad \square$$

To prove our first version of a concave Perron Theorem we need the following Lemma.

Lemma 2.1.10. *The set* $X = \{x \in \mathring{K} | \ \|x\| = 1\}$ *equipped with Hilbert's metric is a complete metric space.*

Proof. By Lemma 2.1.9 d is a metric on X. First we show for $x, y \in X$ the inequality

$$\|x - y\| \le 3\Big(1 - \exp(-d(x,y))\Big). \tag{2.1.2}$$

Let $a = \min\{\frac{x_i}{y_i} | 1 \le i \le n\}$, $b = \min\{\frac{y_i}{x_i} | 1 \le i \le n\}$ and $c = d(x,y) = -\log(ab)$. Since $ay \le x$, $bx \le y$ and $\|x\| = \|y\| = 1$ we must have $0 < a \le 1$, $0 < b \le 1$. This gives

$$\exp(-c) = ab \le a \quad \text{and} \quad \exp(c) = (ab)^{-1} \ge b^{-1}.$$

Therefore

$$\exp(-c)y \le x \le \exp(c)y \tag{$*$}$$

and

$$0 \le (x - y) + y(1 - \exp(-c)) \le x(1 - \exp(-c)) + y(1 - \exp(-c)).$$

By the monotonicity of the norm

$$\|(x - y) + y(1 - \exp(-c))\| \le 2(1 - \exp(-c)),$$

and applying the triangle inequality to the left hand side we arrive at (2.1.2). Now, if $(x^k)_k$ is a Cauchy sequence for d in X then by inequality (2.1.2) $(x^k)_k$ must be a Cauchy sequence for $\|\cdot\|$. Therefore, $(x^k)_k$ converges for $\|\cdot\|$ to some $x^* \in \mathbb{R}^n_+$ with $\|x^*\| = 1$. Since $(x^k)_k$ is a Cauchy sequence for d, to $\epsilon > 0$ given there exists $N \in \mathbb{N}$ such that with the help of $(*)$

$$\exp(-\epsilon)x^k \le x^l \le \exp(\epsilon)x^k \quad \text{for all} \quad k, l \ge N.$$

Taking $l \to \infty$ it follows that

$$\exp(-\epsilon)x^k \le x^* \le x^k \exp(\epsilon) \quad \text{for all} \quad k \ge N.$$

Therefore, $x^* > 0$ that is $x^* \in X$, and

$$d(x^k, x^*) \le \log\left(\exp(\epsilon) \cdot \exp(\epsilon)\right) = 2\epsilon \quad \text{for all} \quad k \ge N.$$

This shows that $(x^k)_k$ converges to x^* also with respect to d. Hence, (X, d) is a complete metric space. □

Theorem 2.1.11 (First Concave Perron Theorem). *For any concave operator $T\colon K \longrightarrow K$ such that $Tx > 0$ for all $x \gneqq 0$ the following properties hold ($\|\cdot\|$ being any monotone norm on $\mathbb{R}^n$).*

(i) *The conditional eigenvalue problem*

$$Tx = \lambda x \text{ with } \lambda \in \mathbb{R} \text{ and } x \text{ restricted to } x \in K, \|x\| = 1 \tag{2.1.3}$$

has a unique solution $x = x^*, \lambda = \lambda^*$; moreover, $x^* > 0$ and $\lambda^* > 0$.

(ii) *For the iterates of the normalized operator one has (with respect to $\|\cdot\|$)*

$$\lim_{k\to\infty} \tilde{T}^k x = x^* \quad \text{for all} \quad 0 \ne x \in K. \tag{2.1.4}$$

Proof. By Lemma 2.1.10 we have that (X, d) is a complete metric space where $X = \{x \in \mathring{K} \mid \|x\| = 1\}$.

(i) In a first step we show that $\tilde{T}\colon X \longrightarrow X$ is a contraction for d. Choose for each $1 \le i \le n$ a vector $e^i \in K, \|e^i\| = 1$ and such that all the components other than the i-th component are 0. Every $x \in \bar{X} = \{x \in K \mid \|x\| = 1\}$ has a representation $x = \sum_{i=1}^n x_i e^i$ with $x_i \ge 0$, and by monotonicity of $\|\cdot\|$

$$0 \le x_i = \|x_i e^i\| \le \|x\| = 1.$$

This implies $x \le e$ for $e := \sum_{i=1}^n e^i$.

Also for $x \in \bar{X}$,

$$1 = \|x\| \le \sum_{i=1}^n x_i \|e^i\| = \sum_{i=1}^n x_i$$

and, hence,

$$\Big(\sum_{j=1}^n x_j\Big)^{-1} \sum_{i=1}^n x_i e^i \le x.$$

Since T is concave and, by Lemma 2.1.3, monotone it follows for all $x \in \bar{X}$ that

$$\min\{Te^i \mid 1 \le i \le n\} \le \left(\sum_{j=1}^{n} x_j\right)^{-1} \sum_{i=1}^{n} x_i Te^i \le Tx \le Te.$$

By assumption $Te^i > 0$ for all $1 \le i \le n$ and there exists some $0 < r < 1$ such that $rTe \le Te^i$ for all $1 \le i \le n$. For $u, v \in \bar{X}$ arbitrary we must have, therefore,

$$rTu \le Tv. \tag{$*$}$$

Let $x, y \in X$ and suppose $\lambda = \min\{\frac{y_i}{x_i} | 1 \le i \le n\} < 1$. It follows that $z = y - \lambda x \in K$ and $y = \lambda x + (1-\lambda)\frac{z}{1-\lambda}$ which, by concavity of T, implies

$$Ty \ge \lambda Tx + (1-\lambda)T\left(\frac{z}{1-\lambda}\right).$$

Inequality ($*$) together with $1 = \|y\| \le \lambda + \|z\|$ implies that

$$rTx \le T\frac{z}{\|z\|} \le T\frac{z}{1-\lambda}$$

and we arrive at

$$Ty \ge \lambda Tx + (1-\lambda)rTx = ((1-r)\lambda + r)Tx. \tag{$**$}$$

In the case where $\lambda = \min\{\frac{y_i}{x_i} | 1 \le i \le n\} \ge 1$ we must have by $\|x\| = \|y\| = 1$ that $\lambda = 1$ and $x \le y$. Thus, for this case ($**$) holds by monotonicity of T. The function log being concave from ($**$) it follows that

$$\log\left(\min\left\{\frac{T_i y}{T_i x}\middle| 1 \le i \le n\right\}\right) \ge \log((1-r)\lambda + r \cdot 1) \ge (1-r)\log\lambda.$$

Since $x, y \in X$ are arbitrary we may exchange the roles of x and y to obtain altogether

$$\log\left(\min\left\{\frac{T_i y}{T_i x}\middle| 1 \le i \le n\right\} \cdot \min\left\{\frac{T_i x}{T_i y}\middle| 1 \le i \le n\right\}\right)$$
$$\ge (1-r)\log\left(\min\left\{\frac{y_i}{x_i}\middle| 1 \le i \le n\right\} \cdot \min\left\{\frac{x_i}{y_i}\middle| 1 \le i \le n\right\}\right)$$

and, according to Definition 2.1.8, $d(Tx, Ty) \le (1-r)d(x, y)$ for all $x, y \in X$.

Lemma 2.1.9 (v) finally yields

$$d(\tilde{T}x, \tilde{T}y) \le (1-r)d(x, y) \quad \text{for all} \quad x, y \in X \tag{2.1.5}$$

that is, $\tilde{T}$ is a contraction for d, $0 < 1 - r < 1$ being the contraction constant.

(ii) By step (1) and Lemma 2.1.10 we can apply Banach's contraction mapping principle (for this see, e.g., Deimling [9]) to the space (X, d) and the mapping $\tilde{T}: X \longrightarrow X$ to obtain

$$\lim_{k\to\infty} \tilde{T}^k x = x^* \quad \text{for all} \quad x \in X,$$

x^* being the unique fixed point of $\tilde{T}$ in X. $\tilde{T}x^* = x^*$ implies $Tx^* = \lambda^* x^*$ with $\lambda^* = \|Tx^*\| > 0$. Conversely, $Tx = \lambda x$ with $x \in K$, $\|x\| = 1$, $\lambda \in \mathbb{R}$ implies that $\tilde{T}x = \lambda x(\|\lambda x\|^{-1}) = x$ and, hence, $x = x^*$ and $\lambda = \|Tx\| = \|Tx^*\| = \lambda^*$.

This shows part (i) of Theorem 2.1.11. Furthermore, by (2.1.2) the convergence of $\tilde{T}^k x$ to x^* for d implies convergence for $\|\cdot\|$. Finally, if $0 \neq x \in K$ then by assumption $\tilde{T}x \in X$ and, hence,

$$\lim_{k\to\infty} \tilde{T}^k x = \lim_{k\to\infty} \tilde{T}^k(\tilde{T}x) = x^* \quad (\text{for } \|\cdot\|).$$ □

Remark 2.1.12. The first step in the proof of Theorem 2.1.11 yields by Banach's contraction principle the estimate

$$d(\tilde{T}^k x, x^*) \le \frac{c^k}{1-c} d(\tilde{T}x, x) \quad \text{for all} \quad x \in K \setminus \{0\}, \quad \text{all} \quad k \in \mathbb{N},$$

where $c = 1 - \min\{\frac{T_j e^i}{T_j e} \mid 1 \le i, j \le n\}$. Inequality (2.1.2) yields the estimate $\|\tilde{T}^k x - x^*\| \le 3\frac{c^k}{1-c} d(\tilde{T}x, x)$.

Also by the first step in the proof of Theorem 2.1.11, any concave operator $T\colon K \longrightarrow K$ is monotone and non-expansive for d, i.e. $d(Tx, Ty) \le d(x, y)$ for all $x, y \in K \setminus \{0\}$.

Beside concavity, the assumption $Tx > 0$ for $x \gneqq 0$ is crucial in Theorem 2.1.11. As it is obvious from the identity map $Tx = x$, this assumption cannot be simply relaxed. From Theorem 2.1.11, however, it follows that the existence statement of part (i) remains true for any concave operator on T. To show this, we need the following well-known fact about concave operators.

Lemma 2.1.13. *Any concave operator $T\colon D \longrightarrow \mathbb{R}^m$ is continuous on the interior $\mathring{D}$ of $D \subset \mathbb{R}^n$.*

Proof. It suffices to show that a concave function $f\colon D \longrightarrow \mathbb{R}$ is continuous on $\mathring{D} \neq \emptyset$. To $x \in \mathring{D}$ fixed there exists $d > 0$ such that

$$B = \{z \in \mathbb{R}^n \mid \|z - x\| \le d\} \subset D$$

for the max-norm $\|\cdot\|$. The cube B has finitely many vertices and every point of B is a convex combination of these vertices. If m denotes the minimum of f on the vertices we have $f(z) \ge m$ for all $z \in B$. For $y \in B, y \neq x$ let $\alpha = \frac{1}{d}\|x - y\|$. Then $0 < \alpha \le 1$ and $u = \frac{1}{\alpha}(y - x) + x \in B$, $v = \frac{1}{\alpha}(x - y) + x \in B$. Therefore, $y = (1-\alpha)x + \alpha u$ and $x = \frac{1}{1+\alpha} y + \frac{\alpha}{1+\alpha} v$ and, by concavity of f,

$$f(y) \ge (1-\alpha) f(x) + \alpha f(u) \ge (1-\alpha) f(x) + \alpha m$$

and

$$f(x) \ge \frac{1}{1+\alpha} f(y) + \frac{\alpha}{1+\alpha} f(v) \ge \frac{1}{1+\alpha} f(y) + \frac{\alpha}{1+\alpha} m,$$

yielding

$$f(x) - f(y) \le \alpha\big(f(x) - m\big)$$

and

$$f(y) - f(x) \leq \alpha(f(x) - m).$$

Putting together,

$$| f(x) - f(y) | \leq \frac{1}{d}\|x - y\|(f(x) - m),$$

which proves the continuity of f in $x \in \mathring{D}$. □

Now, we are ready to prove

Theorem 2.1.14 (Concave Perron–Frobenius Theorem). *For any concave operator $T: K \longrightarrow K$ the eigenvalue problem $Tx = \lambda x$ has at least one solution $x \in K \setminus \{0\}$, $\lambda \in \mathbb{R}_+$.*

Proof. (1) We first prove that the inequality $Tx \leq \lambda x$ has a solution $x \in K \setminus \{0\}$, $\lambda \in \mathbb{R}_+$ of a kind such that $Tx = \lambda x$ in case of $x > 0$. Let $e = (1, \ldots, 1) \in K$ and

$$T(k)x = Tx + \frac{1}{k}e \quad \text{for} \quad k \in \mathbb{N} \setminus \{0\},\ x \in K$$

The operator $T(k): K \longrightarrow K$ is concave with $T(k)x \geq \frac{1}{k}e > 0$ for all $x \in K$. Hence, by Theorem 2.1.11 for every k the equation $T(k)x = \lambda x$ has a solution $x = x(k)$, $\|x(k)\| = 1$, $\lambda = \lambda(k) \in \mathbb{R}_+$($\|\cdot\|$ the max-norm). By monotonicity of T and $x(k) \leq e$ we have $Tx(k) \leq Te$ and, hence,

$$\lambda(k) = \|T(k)x(k)\| \leq \|Tx(k)\| + 1 \leq \|Te\| + 1,$$

i.e., the sequence $(\lambda(k))_k$ is bounded.

This together with $\|x(k)\| = 1$ for all k allows us to assume without loss of generality that

$$\lim_{k\to\infty} \lambda(k) = \lambda \quad \text{and} \quad \lim_{k\to\infty} x(k) = x,\ \|x\| = 1.$$

To $\epsilon > 0$ given there exists $N \in \mathbb{N}$ such that $x \leq (1 + \epsilon)x(k)$ for all $k \geq N$. Concavity of T yields for any $\lambda \geq 1$, $x \in K$ $Tx = T(\frac{1}{\lambda}\lambda x + (1 - \frac{1}{\lambda})0) \geq \frac{1}{\lambda}T(\lambda x)$, that is $T(\lambda x) \leq \lambda Tx$. Taking $\lambda = 1 + \epsilon$ this gives for all $k \geq N$
$Tx \leq T((1 + \epsilon)x(k)) \leq (1 + \epsilon)Tx(k) \leq (1 + \epsilon)T(k)x(k) = (1 + \epsilon)\lambda(k)x(k)$. Letting $k \longrightarrow \infty$ we arrive at $Tx \leq (1 + \epsilon)\lambda x$ and, hence, $Tx \leq \lambda x$ since $\epsilon > 0$ was arbitrary. Furthermore, suppose that $x > 0$. Since T is continuous on $\mathring{K}$ by Lemma 2.1.13, we obtain

$$\lim_{k\to\infty} T(k)x(k) = \lim_{k\to\infty}\left(Tx(k) + \frac{1}{k}e\right) = Tx$$

and, from $T(k)x(k) = \lambda(k)x(k)$,

$$Tx = \lim_{k\to\infty} \lambda(k)x(k) = \lambda x.$$

(2) The assertion of the theorem we now prove by induction over the dimension n of $K = \mathbb{R}^n_+$. For $n = 1$ the assertion holds trivially with, e.g., $x = 1$, $\lambda = T1$. Suppose

the assertion holds for $\dim K \leq n$, n fixed, and let $T\colon \mathbb{R}_+^{n+1} \longrightarrow \mathbb{R}_+^{n+1}$ be any concave operator. By step (1) there exist $x^* \in \mathbb{R}_+^{n+1} \setminus \{0\}$, $\lambda^* \in \mathbb{R}_+$ such that $Tx^* \leq \lambda^* x^*$. Consider $I = \{1 \leq i \leq n+1 \mid x_i^* > 0\}$. Obviously, $I \neq \emptyset$ and, by step (1), the assertion holds true for $n+1$ if $|I| = n+1$. Assume, therefore, $1 \leq |I| \leq n$. If $x \in \mathbb{R}_+^{n+1}$ with $x_j = 0$ for $j \notin I$ then there exists $\lambda \geq 1$ such that $x \leq \lambda x^*$. Concavity of T implies $Tx \leq \lambda Tx^* \leq \lambda\lambda^* x^*$ and, hence, $(Tx)_j = 0$ for $j \notin I$. This allows us to define an operator $S\colon \mathbb{R}_+^I \longrightarrow \mathbb{R}_+^I$ by $(Sy)_i = (T\bar{y})_i$ for $y \in \mathbb{R}_+^I$, $i \in I$, where $\bar{y} \in \mathbb{R}_+^{n+1}$ is given by $\bar{y}_i = y_i$ for $i \in I$ and $\bar{y}_j = 0$ for $j \notin I$. The operator S is concave since $y \longmapsto \bar{y}$ is linear. Because of $|I| \leq n$ there exists by assumption some $y \in \mathbb{R}_+^I \setminus \{0\}$ and $\lambda \in \mathbb{R}_+$ such that $Sy = \lambda y$. From this we obtain

$$(T\bar{y})_i = (Sy)_i = \lambda y_i = \lambda \bar{y}_i \quad \text{for} \quad i \in I$$

and

$$(T\bar{y})_j = 0 = \lambda \bar{y}_j \quad \text{for} \quad j \notin I,$$

that is $T\bar{y} = \lambda \bar{y}$ where $\bar{y} \in \mathbb{R}_+^{n+1} \setminus \{0\}$, $\lambda \geq 0$. □

As remarked already, one has to distinguish $(\tilde{T})^k$, the iterates of the normalized operator, from $(\widetilde{T^k})$, the normalization of the iterates. In Theorem 2.1.11 only the former ones are relevant. It may happen that also $(\widetilde{T^k})$ converges for $k \to \infty$ but then not necessarily to an eigenvector, as shown by the following example.

Example 2.1.15. The operator $T\colon \mathbb{R}_+^2 \longrightarrow \mathbb{R}_+^2$, $Tx = (1 + x_1, 1)$ is concave with $Tx > 0$ for $x \geq 0$. By Theorem 2.1.11 therefore, $\lim_{k\to\infty} \tilde{T}^k x = x^*$ for all $x \geq 0$ and $Tx^* = \lambda^* x^*$ with $x^* \geq 0$. Taking the sum-norm the unique solution is given by

$$x^* = \left(\frac{2}{1+\sqrt{5}}, \frac{2}{3+\sqrt{5}}\right), \quad \lambda^* = \frac{3+\sqrt{5}}{2}.$$

On the other hand,

$$T^k x = (k + x_1, 1) \text{ and } (\widetilde{T^k})x = \left(\frac{k + x_1}{1 + k + x_1}, \frac{1}{1 + k + x_1}\right).$$

The latter converges for $k \to \infty$ to $(1, 0)$, independently of x, which is different from x^* and which is not an eigenvector of T.

The next example illustrates the case where Theorem 2.1.14 is applicable but not Theorem 2.1.11.

Example 2.1.16. Let $T\colon \mathbb{R}_+^3 \longrightarrow \mathbb{R}_+^3$ be given by

$$Tx = \begin{cases} (\sqrt{x_1 x_2},\ 1 + x_2, x_3) & \text{for} \quad x_1 > 0, x_2 > 0, x_3 > 0 \\ (\min\{x_1, x_2\}, 1, \min\{x_2, x_3\}) & \text{for} \quad x_1 = 0 \quad \text{or} \quad x_2 = 0 \quad \text{or} \quad x_3 = 0. \end{cases}$$

The operator T is concave but we do not have $Tx > 0$ for $x \gneqq 0$. Thus, Theorem 2.1.11 is not applicable. Indeed, $Tx = \lambda x$ has no solution with $x > 0$ and $\lambda \geq 0$. According to Theorem 2.1.14, however, there must be at least one solution $x \gneqq 0$, $\lambda \geq 0$. Indeed, there are three solutions of this kind namely (for the sum-norm)

$$\lambda = 1 \quad \text{with} \quad x = (0, 1, 0)$$

and

$$\lambda = \frac{1 + \sqrt{5}}{2} \quad \text{with} \quad x = \left(\frac{1}{\lambda^2}, \frac{1}{\lambda}, 0\right) \text{ or } \left(0, \frac{1}{\lambda}, \frac{1}{\lambda^2}\right)$$

By Lemma 2.1.13 the operator T has to be continuous in the interior of $\mathbb{R}^3_+$. T is, however, not continuous on the whole of $\mathbb{R}^3_+$, e.g. in none of the above eigensolutions T is continuous.

Remark 2.1.17. The concave Perron–Frobenius theorem applies in particular to linear self-mappings given by non-negative matrices, yielding the existence of a (maximal) non-negative eigenvalue (see also Section 2.4 below). This existence can be obtained very easily also as a consequence of Brouwer's fixed point theorem. (Another approach is developed in [19, 66].) The latter theorem for arbitrary but continuous mappings was used also by the economists Solow and Samuelson and, later on, for a more general model, by the economists Morishima, Nikaido to study what seems to be the first nonlinear version ever of the Perron–Frobenius Theorem ([43, 47, 62]; for an approach employing differentiability and the Kuhn–Tucker theorem see [44]). Observe, however, that in Theorem 2.1.14 the self-mapping need not be continuous on the whole cone. Indeed, Example 2.1.16 exhibits a case where the concave Perron–Frobenius theorem guarantees the existence of a (maximal) non-negative eigenvalue which cannot be obtained from Brouwer's fixed point theorem. (In addition, Theorem 2.1.14 is constructive in that it rests on induction and iteration.) In the recent monograph [37] selfmappings are examined which are monotone with $\alpha Tx \leq T(\alpha x)$ for $\alpha \in [0, 1]$ and, hence, are more general than concave ones.

Exercises

1. Find a (vector) norm on $\mathbb{R}^n$ which is not monotone and depict its unit ball.
2. Let d be Hilbert's metric on $\mathring{K}$ for $K = \mathbb{R}^n_+$.
   (a) Show that
   $$d(x, y) = -\log(\lambda(x, y) \cdot \lambda(y, x)),$$
   where $\lambda(x, y) = \sup\{\lambda \in \mathbb{R}_+ \mid \lambda x \leq y\}$ for $x, y \in \mathring{K}$.
   (b) Describe geometrically the balls for d.
   (c) For the case $n = 2$ show that $\frac{1}{2}d(x, y)$ equals the area determined by the rays $\mathbb{R}_+x$, $\mathbb{R}_+y$ and the standard hyperbola $u \cdot v = 1$.
3. Consider the operator $T\colon \mathbb{R}^2_+ \longrightarrow \mathbb{R}^2_+$ given by $Tx = (1 + x_1, 1)$.
   (a) Compute all eigenvalues $\lambda \in \mathbb{R}$ with $Tx = \lambda x, x \gneqq 0$.

(b) Show that for all $k \in \mathbb{N}$

$$\tilde{T}^k \neq \widetilde{T^k}.$$

(c) Compute the smallest contraction constant $c \geq 0$ such that $d(Tx, Ty) \leq cd(x, y)$ for all x, y in the interior of $\mathbb{R}^2_+$.

4. Consider the operator $T\colon \mathbb{R}^2_+ \longrightarrow \mathbb{R}^2_+$ given by

$$Tx = \begin{cases} (x_1 + x_2, x_2) & \text{for} \quad x_1 > 0, x_2 > 0 \\ 0 & \text{for} \quad x_1 = 0 \quad \text{or} \quad x_2 = 0. \end{cases}$$

(a) Show that T is concave.
(b) Determine all points in which T is not continuous.
(c) Find all solutions $x^* \in \mathbb{R}^2_+ \setminus \{0\}$, $\lambda^* \geq 0$ of the eigen equation $Tx = \lambda x$.
(d) Check whether the unique solution $\left(x(k), \lambda(k)\right)$ of the eigen equation $T(k)x = \lambda x$ for $T(k)x = Tx + \frac{1}{k}(1, \ldots, 1)$, where $x(k) \in \mathbb{R}^2_+$, $\|x(k)\| = 1$ ($\|\cdot\|$ max-norm), $\lambda(k) \geq 0$, converges to a solution (x^*, λ^*) of $Tx = \lambda x$.

5. Let $T\colon \mathbb{R}^2_+ \longrightarrow \mathbb{R}^2_+$ be given by $Tx = (x_1 + x_2, x_1)$.
(a) Show that $T^2 x > 0$ for all $x \gneqq 0$ and $\tilde{T}^k = \widetilde{T^k}$ for all $k \in \mathbb{N}$.
(b) Show that $\lim_{n \to \infty} \|T^n x\|^{-1} T^n x = x^*$ for all $x \gneqq 0$, where $\|\cdot\|$ is any monotone norm and x^* uniquely determined by the conditional eigenvalue problem $Tx = \lambda x$, $x \gneqq 0$, $\|x\| = 1$, $\lambda \geq 0$.
(c) Obtain from (b) that for the *Fibonacci sequence* $(f_n)_n$ given by $f_{n+2} = f_{n+1} + f_n$ for $n \in \mathbb{N}$ with $f_0 = f_1 = 1$ it holds that

$$\lim_{n \to \infty} \frac{f_{n+1}}{f_n} = \frac{1 + \sqrt{5}}{2}.$$

6. Let $T\colon \mathbb{R}^2_+ \longrightarrow \mathbb{R}^2_+$ be given by $Tx = \left(\sqrt{x_1} + \sqrt{x_2}, \sqrt{x_1}\right)$.
(a) Show that $T^2 x > 0$ for all $x \gneqq 0$ and $\tilde{T}^k = \widetilde{T^k}$ for all $k \in \mathbb{N}$.
(b) Show that

$$\lim_{n \to \infty} \|T^n x\|^{-1} T^n x = x^* \text{ for all } x \gneqq 0,$$

where $\|\cdot\|$ is any monotone norm and x^* uniquely determined by the conditional eigenvalue problem $Tx = \lambda x, x \gneqq 0, \|x\| = 1, \lambda \geq 0$.
(c) Derive from (b) that

$$\lim_{n \to \infty} T^n x = \bar{x} \text{ for all } x \gneqq 0,$$

where $\bar{x}$ is the unique solution of $Tx = x$, $\|x\| = 1$ for some suitable monotone norm $\|\cdot\|$.
(d) Obtain from (c) that the *root Fibonacci sequence* $(r_n)_n$ given by $r_{n+2} = \sqrt{r_{n+1}} + \sqrt{\sqrt{r_n}}$ converges for arbitrary $(r_0, r_1) \gneqq 0$ to the same limit $\bar{r}$ and compute $\bar{r}$.

7. Let T be a selfmapping of $\mathbb{R}^n_+$ given by

$$T_i x = \sum_{j=1}^{n} a_{ij}\phi(x_j), \quad 1 \le i \le n$$

where $A = (a_{ij}) > 0$ and ϕ is a continuous selfmapping of $\mathbb{R}_+$ with $\phi(r) > 0$ for $r > 0$.
   (a) Show that for each $a > 0$ the conditional eigenvalue problem

   $$Tx = \lambda x, \lambda > 0, x \in \mathbb{R}^n_+, \|x\| = a \quad (\|\cdot\| \text{ any norm})$$

   has at least one solution.
   (b) Show that for ϕ concave the solution of the conditional eigenvalue problem in (a) is unique (for each $a > 0$).
   (c) Find a function ϕ and $a > 0$ such that the conditional eigenvalue problem in (a) has two solutions (λ, x), (μ, y) with $\lambda \ne \mu, x \ne y$.
8. Let C be the convex cone of all concave selfmappings of $\mathbb{R}_+$ and let D be the set of differences from C restricted on the fixed interval $[a, b]$ for $0 < a < b$.
   (a) Show that D is a linear subspace of the space $\mathcal{C}[a, b]$ of all real valued continuous functions on $[a, b]$ which separates the points of $[a, b]$ and contains the constant functions.
   (b) Show that for f in D the function $|f|$ defined by $|f|(x) = |f(x)|$ belongs to D.
   (c) Apply the Stone–Weierstraß theorem to obtain that D is dense in $\mathcal{C}[a, b]$ with respect to the supremum norm.
   (d) Conclude that the convex cone C is not finitely generated whereas the convex cone of all linear selfmappings of $\mathbb{R}_+$ has just one generator.

2.2 Indecomposability and primitivity for ray-preserving concave operators

The two kinds of iterates $\tilde{T}^k$ and $\widetilde{T^k}$ which are different in general turn out to be equal if the operator T maps rays into rays. As before by K we denote the cone $\mathbb{R}^n_+$.

Definition 2.2.1. The operator $T\colon K \longrightarrow K$ is **homogeneous of degree d** for $d \in \mathbb{R}$ if

$$T(\lambda x) = \lambda^d Tx \quad \text{for all} \quad x \in K, \quad \text{all} \quad \lambda > 0 \quad \text{and} \quad T0 = 0.$$

T is **positively homogeneous** if it is of degree $d = 1$.

T is **ray-preserving** if for every $x \in K$ and $\lambda > 0$ there exists some $\lambda' = \lambda'(x, \lambda) > 0$ such that

$$T(\lambda x) = \lambda' Tx \quad \text{and} \quad T0 = 0.$$

Geometrically, T ray-preserving means that T maps a *ray* $\mathbb{R}_+ x := \{\lambda x \mid \lambda \in \mathbb{R}_+\}$ into a ray again, namely into the ray $\mathbb{R}_+ Tx$.

Examples 2.2.2. (i) $T\colon K \longrightarrow K$, $K = \mathbb{R}^n_+$, $Tx = Ax$, A a non-negative $n \times n$-matrix. T is positively homogeneous.

(ii) $T\colon \mathbb{R}^2_+ \longrightarrow \mathbb{R}^2_+$, $Tx = (\sqrt{x_1 x_2}, x_2)$, is not linear, (i.e., not of the type in (i)) but it is still positively homogeneous.

(iii) $T\colon \mathbb{R}^2_+ \longrightarrow \mathbb{R}^2_+$, $Tx = (\sqrt{x_1}, \sqrt{x_2})$ is ray-preserving with $\lambda'(x,\lambda) = \sqrt{\lambda}$ but it is not positively homogeneous.

(iv) $T\colon \mathbb{R}^2_+ \longrightarrow \mathbb{R}^2_+$, $Tx = \begin{cases} (x_1, x_2), & x_1 \le x_2 \\ (\sqrt{x_1}, \sqrt{x_2},) & x_1 > x_2. \end{cases}$

$T(\lambda x) = \lambda'(x,\lambda) Tx$ where $\lambda'(x,\lambda) = \begin{cases} \lambda, & x_1 \le x_2 \\ \sqrt{\lambda}, & x_1 > x_2 \end{cases}$

does depend on both, λ and x.

(v) $T\colon \mathbb{R}^2_+ \longrightarrow \mathbb{R}^2_+$, $Tx = (\sqrt{x_1}, x_2)$ is not ray-preserving; it maps the rays $\mathbb{R}_+(1,0)$ and $\mathbb{R}_+(0,1)$ into itself but destroys all other rays. The mapping $T\colon \mathbb{R}^2_+ \longrightarrow \mathbb{R}^2_+$ $Tx = (1 + x_1, 1)$ does not map any ray at all ($\ne \{0\}$) into a ray again.

Definition 2.2.3. For $T\colon K \longrightarrow K$ and $\|\cdot\|$ a norm on $\mathbb{R}^n$ the operator $T_{\|\cdot\|}$ defined by

$$T_{\|\cdot\|}x = \|x\| T \frac{x}{\|x\|} \quad \text{for} \quad x \ne 0 \quad \text{and} \quad T_{\|\cdot\|}0 = 0$$

is the **homogenized operator for *T*.**

It follows directly from the definition that $T_{\|\cdot\|}(\lambda x) = \lambda T_{\|\cdot\|}x$ for all $x \in K$, $\lambda \in \mathbb{R}_+$ and, hence, the homogenized operator is always positively homogeneous. E.g., for the self-mapping of $\mathbb{R}^2_+$ given by $Tx = (1 + x_1, 1)$ the homogenized operator for the sum norm is

$$T_{\|\cdot\|}x = (x_1 + x_2) T \frac{x}{x_1 + x_2} = (2x_1 + x_2, x_1 + x_2) = \begin{bmatrix} 2 & 1 \\ 1 & 1 \end{bmatrix} x \quad (x \text{ a column vector}),$$

which is even linear. (See also Exercise 1.) It may happen, however, that a concave operator when homogenized remains no longer concave whatever norm is choosen. (See Exercise 2.) Because of

$$T_{\|\cdot\|}x = \lambda x \quad \text{if and only if} \quad T\frac{x}{\|x\|} = \lambda \frac{x}{\|x\|}$$

the eigenvalue problems for T and $T_{\|\cdot\|}$ correspond to each other. In the next lemma also the dynamics of T and $T_{\|\cdot\|}$ will be related.

Lemma 2.2.4. *Let $T\colon K \longrightarrow K$ and $\|\cdot\|$ be a norm on $\mathbb{R}^n$.*

(i) *If T is ray-preserving and $x \in K$ such that $T^k x \ne 0$ for all $k \in \mathbb{N}$ then*

$$(\tilde{T})^k x = (\widetilde{T^k})x \quad \textit{for all} \quad k \in \mathbb{N}.$$

(ii) *If $Tx \ne 0$ for all $x \ne 0$ then*

$$\widetilde{(T_{\|\cdot\|})^k}x = \tilde{T}^k x \quad \textit{for all} \quad x \in K \quad \textit{with} \quad \|x\| = 1.$$

Proof. (i) For x fixed we prove the equality by induction over k.

For $k = 1$, the assertion is trivial. Suppose $(\tilde{T})^k x = (\widetilde{T^k})x$ for some $k \geq 1$.

Since $(\widetilde{T^k})x = \frac{T^k x}{\|T^k x\|} \neq 0$ we have that $(\tilde{T})^k x \neq 0$. Therefore $(\tilde{T})^{k+1}x$ is defined and by induction hypothesis

$$(\tilde{T})^{k+1}x = \tilde{T}((\tilde{T})^k x) = \frac{T((\tilde{T})^k x)}{\|T((\tilde{T})^k x)\|} = \frac{T(\frac{T^k x}{\|T^k x\|})}{\|T(\frac{T^k x}{\|T^k x\|})\|}.$$

For T ray-preserving there exists some $\lambda' > 0$ (possibly dependent on k and x) such that $T(\frac{T^k x}{\|T^k x\|}) = \lambda' T(T^k x) = \lambda' T^{k+1}x$. Thus

$$(\tilde{T})^{k+1}x = \frac{\lambda' T^{k+1}x}{\|\lambda' T^{k+1}x\|} = \frac{T^{k+1}x}{\|T^{k+1}x\|} = (\widetilde{T^{k+1}})x.$$

(ii) We have that $T_{\|\cdot\|}x \neq 0$ for all $x \neq 0$ because of $Tx \neq 0$ for all $x \neq 0$. Therefore $(T_{\|\cdot\|})^k x \neq 0$ for all $x \neq 0$ and by positive homogeneity of $T_{\|\cdot\|}$ if follows from (i) that

$$(\widetilde{T_{\|\cdot\|}})^k x = \widetilde{(T_{\|\cdot\|})^k} x \quad \text{for} \quad x \in K \setminus \{0\}.$$

By definition of $T_{\|\cdot\|}$, $T_{\|\cdot\|}x = Tx$ for $\|x\| = 1$. Therefore, $\widetilde{T_{\|\cdot\|}}$ and $\tilde{T}$ coincide on $\{x \in K \mid \|x\| = 1\}$ which implies the equality of $(\widetilde{T_{\|\cdot\|}})^k$ and $\tilde{T}^k$ on $\{x \in K \mid \|x\| = 1\}$.

Putting together we obtain the equality of $\widetilde{(T_{\|\cdot\|})^k}$ and $\tilde{T}^k$ on $\{x \in K \mid \|x\| = 1\}$. □

Remark 2.2.5. Using Lemma 2.1.9 (ii) we may reformulate the convergence statement (2.1.4) of Theorem 2.1.11 also as

$$\lim_{k\to\infty} \frac{(T_{\|\cdot\|})^k x}{\|(T_{\|\cdot\|})^k x\|} = x^* \quad \text{for all} \quad x \in K \setminus \{0\}.$$

This statement does not hold in general given the assumptions of Theorem 2.1.11, if $T_{\|\cdot\|}$ is replaced by T as we have seen from the example $Tx = (1 + x_1, 1)$. If T is assumed to be ray-preserving, however, by Lemma 2.2.4 (i) $T_{\|\cdot\|}$ may be replaced by T in the above convergence statement.

Definition 2.2.6. $T\colon K \longrightarrow K$ is **indecomposable** if for any index set $\emptyset \subsetneq I \subsetneq \{1,\ldots,n\}$ ($n \geq 2$ given) there exist indices $i \in I$, $j \notin I$ such that $T_i e_j > 0$. (T_i the i-th component map of T, $e_j \in K$ the j-th unit vector.)

$T\colon K \longrightarrow K$ is **weakly indecomposable** if for any indices $1 \leq i,j \leq n$ there exists some $p = p(i,j) \in \mathbb{N}$ such that $T_i^p e_j := (T^p)_i e_j > 0$.

Lemma 2.2.7. *Let $T\colon K \longrightarrow K$ be ray-preserving and monotone. If T is indecomposable then T is weakly indecomposable.*

Proof. (1) For $1 \leq j \leq n$ let $I_j = \{1 \leq i \leq n \mid T_i^p e_j > 0 \text{ for some } p \in \mathbb{N}\}$. We show that for T ray-preserving and monotone we must have that for arbitrary indices i, j, k

$$i \in I_j \quad \text{and} \quad j \in I_k \quad \text{imply} \quad i \in I_k. \tag{$*$}$$

For this, suppose $T_i^p e_j > 0$ and $T_j^q e_k > 0$ for some $p, q \in \mathbb{N}$. $T_j^q e_k > 0$ implies $T^q e_k \geq c e_j$ for some $c > 0$ and, hence, $T^{p+q} e_k \geq T^p(c e_j)$. Since T is ray-preserving it follows by iteration that $T^p(c e_j) = c(p) T^p e_j$ with some $c(p) > 0$. Thus, $T^{p+q} e_k \geq c(p) T^p e_j$ and by assumption $T_i^{p+q} e_k \geq c(p) T_i^p e_j > 0$. This proves (∗).

(2) Suppose T is indecomposable and i and j are given indices. Because of $n \geq 2$ we have that $I_j \neq \emptyset$. Suppose that $I_j \neq \{1, \dots, n\}$. Applying the definition of indecomposability to $I = \{1, \dots, n\} \setminus I_j$ there exist $k \in I, l \notin I$ such that $T_k e_l > 0$ and, hence, $k \in I_l$. Since $k \in I_l$ and $l \in I_j$ it follows from (∗) of step (1) that $k \in I_j$ – which contradicts $k \in I$. Therefore, we must have $I_j = \{1, \dots, n\}$ and $i \in I_j$. □

Remarks 2.2.8. (1) Lemma 2.2.7 fails if T is not monotone or not ray-preserving (see Exercise 3).

(2) In contrast to the linear case (see below) it may occur that an operator T, even if it is positively homogeneous and concave, is weakly indecomposable but not indecomposable. Consider $T\colon \mathbb{R}_+^3 \longrightarrow \mathbb{R}_+^3$ $Tx = (\sqrt{x_2 x_3}, x_1 + x_2 + x_3, x_1 + x_2 + x_3)$. T is positively homogeneous and concave (and, a fortiori, monotone). T is weakly indecomposable because $T_i e_j > 0$ for $2 \leq i \leq 3$ and $1 \leq j \leq 3$ and, since $T_1^2 x = x_1 + x_2 + x_3$, $T_1^2 e_j > 0$ for $1 \leq j \leq 3$. T is, however, not indecomposable because $T_1 x = \sqrt{x_2 x_3}$ implies $T_1 e_j = 0$ for $2 \leq j \leq 3$.

Definition 2.2.9. $T\colon K \longrightarrow K$ is **primitive** if there exists $p \in \mathbb{N}$ such that $T^m x > 0$ for all $m \geq p$ and all $x \in K \setminus \{0\}$.

Lemma 2.2.10. *Let $T\colon K \longrightarrow K$ be ray-preserving and monotone. If T is weakly indecomposable and there exists some h such that $T_h e_h > 0$ then T is primitive.*

Proof. Since T is weakly indecomposable there exists $p_{ij} \in \mathbb{N}$ such that $T_i^{p_{ij}} e_j > 0$ for all $1 \leq i, j \leq n$. Let $p = 2q$ with $q = \max\{p_{ij} \mid 1 \leq i, j \leq n\}$. For $c = T_h e_h$ obviously $T e_h \geq c e_h$. Since T is ray-preserving for any given $r, s \in \mathbb{N}$ there exist constants $c', c'', c''' > 0$ such that

$$T^{r+s}(c e_h) = T^r(T^s(c e_h)) = T^r(c' T^s e_h) \geq T^r(c'' e_h) = c''' T^r e_h,$$

and, hence,

$$T^{r+s}(e_h) \geq \tilde{c} T^r e_h \quad \text{with} \quad \tilde{c} > 0.$$

If $m \geq q$ it follows with $r = p_{ih}$ and $s = m - p_{ih}$ that $T^m(e_h) \geq \tilde{c} T^{p_{ih}} e_h$ and, hence, $T_i^m(e_h) \geq \tilde{c} T_i^{p_{ih}} e_h > 0$.

Thus we obtain

$$T^m e_h > 0 \quad \text{for all} \quad m \geq q. \tag{∗}$$

On the other hand, $T_h^{p_{hj}} e_j > 0$ implies $T^{p_{hj}} e_j \geq c(j) e_h$ with $c(j) > 0$. For $m \geq p = 2q$, $m' = m - p_{hj}$ it follows that

$$T^m e_j = T^{m'}(T^{p_{hj}} e_j) \geq T^{m'}(c(j) e_h) \geq c(j)' T^{m'} e_h \quad \text{with} \quad c(j)' > 0.$$

Since $m' = m - p_{hj} \geq m - q \geq q$ by $(*)$ we have

$$T^m e_j > 0 \quad \text{for all} \quad 1 \leq j \leq n.$$

If $x \in K \setminus \{0\}$ then there exists an index j such that $x \geq x_j e_j$ with $x_j > 0$. The above then implies $T^m x \geq x_j' T^m e_j > 0$ for all $m \geq p$, i.e., T is primitive. □

The above Lemmas and the concepts involved are extensions from the linear case of non-negative matrices to certain non-linear situations. A *non-negative matrix* $A = (a_{ij})_{1\leq i,j\leq n}$, $a_{ij} \geq 0$ for all $1 \leq i, j \leq n$ is called **indecomposable** (or irreducible) if for any index set $\emptyset \subsetneqq I \subsetneqq \{1, \dots, n\}$ there exist $i \in I$, $j \notin I$ such that $a_{ij} > 0$. It is well-known (cf. [61]) that a non-negative matrix is indecomposable iff for any two indices i and j there exists a $p = p(i,j)$ such that the (i,j)-entry of the matrix power A^p is strictly positive. Therefore, for linear operators $Tx = A \cdot x$, x being a column vector, it is not necessary to introduce the notion of weak indecomposability. We have seen, however, that for non-linear operators the two notions do not coincide. A non-negative matrix A is called **primitive** if some power of A is a strictly positive matrix. Thus, in the matrix case Lemma 2.2.10 states that a non-negative and indecomposable matrix is primitive provided that there is at least one strictly positive element on the diagonal of A. The latter condition cannot be omitted as can be seen from the simple example $A = \left[\begin{smallmatrix} 0 & 1 \\ 1 & 0 \end{smallmatrix}\right]$.

In the literature concerning non-linear extensions of the Perron–Frobenius theory there occur several different notions extending those of indecomposability and primitivity of matrices (cf. [28, 32, 42–44, 46, 47, 50, 51]). Dealing with continuous selfmappings $T\colon K \to K$ which are monotone and positively homogeneous, M. Morishima and H. Nikaido introduced the following notions [42, 43, 46, 47]. Call T *MN-indecomposable* (indecomposable in the sense of Morishima and Nikaido) if for any index set $\emptyset \subsetneqq I \subsetneqq \{1, \dots, n\}$ and any $x, y \in K$ with $x_i = y_i$ for all $i \in I$ and $x_j < y_j$ for all $j \notin I$ it follows for some $k \in I$ that $T_k x \neq T_k y$. Furthermore, T is *globally primitive* if for every $x \in K$ there exists $s = s(x) \geq 1$ such that $x \lneqq y$ for $y \in K$ implies that $T^s x < T^s y$.

If T is linear then T is MN-indecomposable iff T is indecomposable, and T is globally primitive iff T is primitive and, hence, the notions coincide with the ones above for the linear case. In general, however, the notions are quite different. In particular, it is easy to give examples, even for T continuous, monotone and positively homogeneous, where T is indecomposable and primitive but neither MN-indecomposable nor globally primitive (see Exercise 8).

Another extension of the concept of an indecomposable matrix to monotone and homogeneous (of degree 1) mappings is made by S. Gaubert and J. Gunawardena in [18]; see also [37]. (A monotone homogeneous mapping need not be concave and the latter need not be homogeneous.) In [18] and [37] a graph $G(T)$ associated to T is required to be strongly connected which, in case of a matrix, is equivalent to indecomposability. If T is indecomposable in the sense of Definition 2.2.6 then $G(T)$ is strongly connected but not the other way round (see Exercise 9). In [18] and [37] a generalized

Perron–Frobenius Theorem is proven for monotone and homogeneous mappings under the assumption that $G(T)$ is strongly connected. This theorem provides even an eigenvector in the interior of $\mathbb{R}^n_+$. Later on, when dealing with ascending operators, we will obtain as a particular case an extension of the above theorem to monotone and subhomogeneous mappings (see Corollary 5.2.5 part (ii)(b)).

Now we are ready to prove a refined version of the Concave Perron Theorem in which for a ray-preserving operator the positivity assumption is weakened and the conclusions are strengthened at the same time.

Theorem 2.2.11 (Second Concave Perron Theorem). *For any concave operator $T: K \longrightarrow K$ which is ray-preserving, weakly indecomposable with $T_h e_h > 0$ for some $1 \le h \le n$ the following properties hold ($\|\cdot\|$ being an arbitrary monotone norm on $\mathbb{R}^n$):*

(i) *The eigenvalue problem $Tx = \lambda x$ with $\lambda \in \mathbb{R}$ and $x \in K\setminus\{0\}$ has a solution $x = x^* > 0$ with $\|x^*\| = 1$ and $\lambda = \lambda^* > 0$. Moreover, for any solution $(x, \lambda) \in K\setminus\{0\} \times \mathbb{R}$ $x = rx^*$ for some $r > 0$ and $\lambda > 0$. If, in addition, T is positively homogeneous then $\lambda = \lambda^*$.*

(ii)
$$\lim_{k\to\infty} \frac{T^k x}{\|T^k x\|} = x^* \quad \textit{for all} \quad x \in K \setminus \{0\}$$

(Convergence with respect to $\|\cdot\|$). This statement is equivalent to

$$\lim_{k\to\infty} \frac{(T^k)_i x}{(T^k)_j x} = \frac{x_i^*}{x_j^*} \quad \textit{for all} \quad x \in K \setminus \{0\}, \quad \textit{all} \quad 1 \le i,\ j \le n.$$

Proof. We first prove the second part of the Theorem.

(ii) By Lemma 2.2.10 T is primitive (as already remarked, a concave operator has to be monotone). Hence, there exists $p \in \mathbb{N}$ such that for $S = T^p$ we have $Sx > 0$ for all $x \in K \setminus \{0\}$. Since any iterate of a concave operator is concave again we can apply Theorem 2.1.11 to S to obtain

$$\lim_{k\to\infty} \tilde{S}^k x = x^* \quad \text{for all} \quad x \in K \setminus \{0\},$$

where $x^* > 0$, $\|x^*\| = 1$. Since any iterate of a ray-preserving operator is ray-preserving too, by Lemma 2.2.4 we have that

$$\tilde{S}^k = \widetilde{S^k} = \widetilde{T^{kp}} \quad \text{for all} \quad k \in \mathbb{N}.$$

Therefore,

$$\lim_{k\to\infty} \frac{T^{kp} x}{\|T^{kp} x\|} = x^* \text{ for all } x \in K \setminus \{0\}. \tag{$*$}$$

Consider $x = T^i y$ for $y \in K \setminus \{0\}$ and $1 \le i < p$. If $x = 0$ then $T^p y = T^{p-i} x = T^{p-i} 0 = 0$ since T is ray-preserving; this, however, contradicts the primitivity of T and we must have $x = T^i y \neq 0$. Applying $(*)$ to starting points $x = T^i y$ yields

$$\lim_{m\to\infty} \frac{T^m x}{\|T^m x\|} = x^* \quad \text{for all} \quad x \in K \setminus \{0\}.$$

Obviously, this implies

$$\lim_{m\to\infty}\frac{(T^m)_i x}{(T^m)_j x}=\frac{x_i^*}{x_j^*}.$$

Conversely, the latter, for some $x^*>0$ with $\|x^*\|=1$, implies

$$\lim_{m\to\infty}\frac{(T^m)_i x}{\sum\limits_{j=1}^{n}(T^m)_j x}=\frac{x_i^*}{\sum\limits_{j=1}^{n}x_j^*}$$

or, $|\cdot|$ denoting the sum-norm,

$$\lim_{m\to\infty}\frac{T^m x}{|T^m x|}=\frac{x^*}{|x^*|}.$$

Since $\|\cdot\|$ is continuous it follows

$$\lim_{m\to\infty}\frac{\|T^m x\|}{|T^m x|}=\frac{1}{|x^*|},$$

and, hence,

$$\lim_{m\to\infty}\frac{T^m x}{\|T^m x\|}=x^*.$$

This proves part (ii) of the theorem.

(i) From Theorem 2.1.11, part (i), we also have that $Sx=\mu x$ with $x\in K$, $\|x\|=1$, $\mu\in\mathbb{R}$ iff $x=x^*$. Since T is ray-preserving and $Tx^*\neq 0$ we have that

$$S\left(\frac{Tx^*}{\|Tx^*\|}\right)=\rho T^{p+1}x^*=\rho T(Sx^*)=\rho T(\mu x^*)=\kappa\frac{Tx^*}{\|Tx^*\|}$$

where $\rho>0$, $\kappa>0$.

It follows that $\frac{Tx^*}{\|Tx^*\|}=x^*$ and, therefore, $Tx=\lambda x$ has the solution $x=x^*>0$, $\|x^*\|=1$, $\lambda=\lambda^*=\|Tx^*\|>0$.

Suppose $Tx=\lambda x$ with $x\in K\setminus\{0\}$, $\lambda\in\mathbb{R}$. By the primitivity of T we have $\lambda=\|x\|^{-1}\|Tx\|>0$ and, therefore, $S\frac{x}{\|x\|}=T^p\frac{x}{\|x\|}=\rho T^p x=\kappa\frac{x}{\|x\|}$ with $\rho>0$, $\kappa>0$, which implies $\frac{x}{\|x\|}=x^*$.

If, in addition, T is positively homogeneous then $Tx=\lambda x$ implies

$$\lambda(\|x\|x^*)=T(\|x\|x^*)=\|x\|Tx^*$$

and $\lambda=\lambda^*=\|Tx^*\|$. □

The following remarks will illustrate that none of the assumptions in Theorem 2.2.11 can simply be omitted.

Remarks 2.2.12. (1) It is clear from part (ii) of the proof of the Theorem that weak indecomposability and $T_h e_h>0$ for some h was used to get the primitivity of T by Lemma 2.2.10. Hence the Theorem remains true if these conditions are replaced by that of primitivity.

(2) The identity map $T\colon K \longrightarrow K$ satisfies all assumptions of Theorem 2.2.11 with the exception of weak indecomposability. Since none of the conclusions of Theorem 2.2.11 does hold for the identity map the assumption of weak indecomposability cannot simply be omitted.

(3) The map $T\colon \mathbb{R}^2_+ \longrightarrow \mathbb{R}^2_+$, $T(x_1,x_2) = (x_2,x_1)$ satisfies all assumptions of the Theorem with the exception that $T_h e_h = 0$ for all h. Whereas conclusion (i) of the Theorem holds for this map, conclusion (ii) does not hold. Thus, the assumption $T_h e_h > 0$ for some h cannot simply be omitted.

(4) For the map $T\colon \mathbb{R}^2_+ \longrightarrow \mathbb{R}^2_+$, $T(x_1,x_2) = (1 + x_1, 1)$ it holds that $T_i e_j > 0$ for all i,j. Therefore, T satisfies all assumptions of the Theorem except that T is not ray-preserving. Considering conclusion (i), $Tx = \lambda x$ for $(x,\lambda) \in (K \setminus \{0\}) \times \mathbb{R}$ is equivalent to $x = x(\lambda) = ((\lambda - 1)^{-1}, \lambda^{-1})$ for arbitrary $\lambda > 1$. Hence, all solutions belong to different rays, in contrast to (i). In Example 2.1.15 we have already seen that for all $x \in K$ the sequence of $\|T^k x\|^{-1} T^k x$ converges to $(1,0)$ for $k \to \infty$ which is not a solution of the eigenvalue problem. Thus, conclusion (ii) of the Theorem does not hold in this case. Thus, none of the conclusions (i) and (ii) holds in this case, showing that the assumption of T being ray-preserving cannot simply be omitted.

(5) The map $T\colon \mathbb{R}^2_+ \longrightarrow \mathbb{R}^2_+$, $T(x_1,x_2) = (\sqrt{x_1} + \sqrt{x_2}, \sqrt{x_1} + \sqrt{x_2})$ does satisfy all assumptions of Theorem 2.2.11. Hence, for any solution $(x,\lambda) \in (K \setminus \{0\}) \times \mathbb{R}$ of $Tx = \lambda x$ it holds that $x = rx^*$ for some $r > 0$ and $\lambda > 0$; for the sum-norm one has $x^* = (\frac{1}{2}, \frac{1}{2})$. The value of λ, however, is not uniquely determined. Indeed, for any $\lambda > 0$ one has $Tx(\lambda) = \lambda x(\lambda)$ with $x(\lambda) = \frac{8}{\lambda^2} x^*$.

Thus, in contrast to the linear case, a non-linear self-mapping of the standard cone in n dimensions may have more than n eigenvalues – it may even be a continuum. In case the self-mapping is monotone and positively homogeneous, however, there are only finitely many non-negative eigenvalues. (Cf. Exercise 7.)

The following example meets all the conditions appearing in Theorem 2.2.11.

Example 2.2.13. Let $T\colon \mathbb{R}^2_+ \longrightarrow \mathbb{R}^2_+$ be defined by $T(x_1,x_2) = (4x_1 + 2x_2 + \sqrt{x_1 x_2}, \min\{x_1 + x_2, 2x_1\})$. Obviously T is concave and even positively homogeneous. Furthermore, $Te_1 = (4,1)$ and $Te_2 = (2,0)$. To check weak indecomposability we calculate $T^2 e_2 = T(2,0) = (8,2)$. Thus, T is weakly indecomposable and $T_h e_h > 0$ for $h = 1$. Solving the eigenvalue problem means to find $x \in \mathbb{R}^2_+ \setminus \{0\}$ and λ such that

$$4x_1 + 2x_2 + \sqrt{x_1 x_2} = \lambda x_1 \quad \text{and} \quad \min\{x_1 + x_2, 2x_1\} = \lambda x_2. \tag{$*$}$$

Obviously, we must have $x_1 > 0$ and $x_2 > 0$.

Suppose first that $2x_1 \le x_1 + x_2$, i.e., $x_1 \le x_2$. The second equation then becomes $x_1 = \frac{\lambda}{2} x_2$ and putting this into the first equation we obtain $2\lambda + 2 + \sqrt{\frac{\lambda}{2}} = \frac{\lambda^2}{2}$ and, hence, in particular $2 < \frac{\lambda^2}{2}$, that is $\lambda > 2$. This, however, contradicts $x_1 \le x_2$. Thus, this case is impossible and we must have that $2x_1 > x_1 + x_2$, i.e., $x_1 > x_2$. From $(*)$ we obtain $x_1 + x_2 = \lambda x_2$, i.e., $x_1 = (\lambda - 1)x_2$, and $4(\lambda - 1) + 2 + \sqrt{\lambda - 1} = \lambda(\lambda - 1)$

which equation we have to solve for λ. Putting $\mu = \sqrt{\lambda - 1}$ the equation becomes $\mu^4 - 3\mu^2 - \mu - 2 = 0$. By inspection we find a root $\mu^* = 2$ and conclude that all other roots must satisfy $\mu^3 + 2\mu^2 + \mu + 1 = 0$. That is, $\mu^* = 2$ is the only positive root and hence we must have that $\lambda^* = 1 + {\mu^*}^2 = 5$. Employing the sum-norm, from $x_1 = (\lambda^* - 1)x_2 = 4x_2$ we obtain $x^* = (\frac{4}{5}, \frac{1}{5})$. The Theorem yields the stability statement (ii) which is not easy to verify directly due to the non-linear character of T.

The above Theorem 2.2.11 holds also for certain operators which are not necessarily concave, provided that the relative magnitudes behave like those of a concave operator. More precisely, we have the following consequence of Theorem 2.2.11.

Corollary 2.2.14. *Suppose $T: K \longrightarrow K$ is an operator as in Theorem 2.2.11 and define $Rx = r(x)Tx$ where $r: K \longrightarrow \mathbb{R}_+$ with $r(x) > 0$ for $x \neq 0$. Then statements* (i) *and* (ii) *of Theorem 2.2.11 hold also for R with a solution (y^*, μ^*) linked to that of T by $y^* = x^*$ and $\mu^* = r(x^*)\lambda^*$.*

Proof. For the rescaled operators we have that $\tilde{R}x = \frac{Rx}{\|Rx\|} = \frac{Tx}{\|Tx\|} = \tilde{T}x$ for $x \neq 0$. Furthermore, for $\lambda \geq 0$ we have

$$R(\lambda x) = r(\lambda x)T(\lambda x) = r(\lambda x)\lambda' Tx = \frac{r(\lambda x)\lambda'}{r(x)} Rx,$$

that is, R is also ray-preserving. By Lemma 2.2.4

$$\frac{R^k x}{\|R^k x\|} = \widetilde{R^k}x = \tilde{R}^k x = \tilde{T}^k x = \widetilde{T^k}x = \frac{T^k x}{\|T^k x\|}$$

and part (ii) of Theorem 2.2.11 gives $\lim_{k\to\infty} \frac{R^k x}{\|R^k x\|} = x^*$ for all $x \in K \setminus \{0\}$. Furthermore, $Rx^* = r(x^*)Tx^* = r(x^*)\lambda^* x^* = \mu^* x^*$ with $\mu^* = r(x^*)\lambda^*$. If $Ry = \mu y$ with $y \in K \setminus \{0\}$ and $\mu \in \mathbb{R}$ then $Ty = r(y)^{-1}\mu y$ and $y = rx^*$ for some $r > 0$ by part (i). □

Simple examples of non-concave operators of the type admitted in Corollary 2.2.14 can be obtained from linear operators by choosing

$$r(x) = \Big(\sum_{i=1}^{n} a_i x_i\Big)^{-1}\Big(\sum_{i=1}^{n} b_i x_i\Big)$$

with certain positive weights a_i, b_i. It is, however, easy to construct non-concave operators which are not of the type considered in Corollary 2.2.14 but satisfy all other assumptions appearing in Theorem 2.2.11 and for which none of the conclusions of Theorem 2.2.11 does apply. (Cf. Exercise 6.)

Exercises

1. Let $T: K \longrightarrow K$, $Tx = Ax + a$ be affine-linear (cf. Example 2.1.5, $K = \mathbb{R}^n_+$).
   (a) Find a norm $\|\cdot\|$ such that the homogenized operator for T is linear.
   (b) Use (a) to describe the unique solution (x^*, λ^*) of $Tx = \lambda x$, $x \in K$, $\|x\| = 1$, $\lambda \in \mathbb{R}$ in terms of matrices.
   (c) Illustrate (a) and (b) for the special case $T: \mathbb{R}^2_+ \longrightarrow \mathbb{R}^2_+$, $Tx = (1 + x_1, 1)$.
2. Find a concave operator $T: K \longrightarrow K$ such that for all norms $\|\cdot\|$ the homogenized operator $T_{\|\cdot\|}$ is *not* concave.
3. Find for each of the following cases an operator T which is indecomposable but not weakly indecomposable.
   (a) T is monotone but not ray-preserving.
   (b) T is positively homogeneous but not monotone.
4. Let $T: \mathbb{R}^n_+ \longrightarrow \mathbb{R}^n_+$ be defined by

$$Tx = \Big(\sum_{i=1}^{n} x_i, x_1, x_2, \ldots, x_{n-1}\Big).$$

   (a) Show that T is primitive.
   (b) Calculate the *primitivity index* of T that is the smallest $p \in \mathbb{N}$ such that $T^p x > 0$ for all $x \in \mathbb{R}^n_+ \setminus \{0\}$.
5. Let $T: \mathbb{R}^2_+ \longrightarrow \mathbb{R}^2_+$ be defined by

$$Tx = \begin{cases} (2x_1 + x_2, \quad 3x_1 + x_2) & \text{for} \quad x_1 \le x_2 \\ (x_1 + 2x_2, \quad x_1 + 3x_2) & \text{for} \quad x_1 > x_2. \end{cases}$$

   (a) Show that T is concave, positively homogeneous and weakly indecomposable with $T_h e_h > 0$ for some h.
   (b) Use Theorem 2.2.11 to calculate a solution (x^*, λ^*) of $Tx = \lambda x, x \in \mathbb{R}^2_+$, $\|x\| = 1$, $\lambda \in \mathbb{R}$ ($\|\cdot\|$ being the sum-norm) and show that it is unique.
   (c) Illustrate the statement $\lim\limits_{k\to\infty} \frac{T^k x}{\|T^k x\|} = x^*$ by means of computer simulations for some particular values of $x \in \mathbb{R}^2_+ \setminus \{0\}$.
6. Construct an example of an operator $S: \mathbb{R}^2_+ \longrightarrow \mathbb{R}^2_+$ which is positively homogeneous with $Sx > 0$ for $x \in \mathbb{R}^2_+ \setminus \{0\}$ and which is not of the form $Sx = s(x)Tx$ for T concave and for which none of the conlusions (i), (ii) of Theorem 2.2.11 holds true.
7. [43] Let T be a self-mapping of $\mathbb{R}^n_+$ that is monotone and positively homogeneous. Show that T has only finitely many non-negative eigenvalues (with non-negative eigenvectors).
8. Let T be a selfmapping of K.
   (a) Show that for T with $T0 = 0$
      – globally primitive implies primitive,

- MN-indecomposable, positively homogeneous and convex implies indecomposable.

(b) Show that for T positively homogeneous and concave
- primitive implies globally primitive,
- indecomposable implies MN-indecomposable and this implication cannot be reversed.

(c) Find a continuous, monotone and positively homogeneous selfmapping of $\mathbb{R}_+^2$ which is indecomposable and primitive but neither MN-indecomposable nor globally primitive.

9. Let $K = \mathbb{R}_+^n$ and $T\colon \overset{\circ}{K} \to \overset{\circ}{K}$. For $u > 0, 1 \le j \le n$, define $u(j) \in \overset{\circ}{K}$ by $u(j)_i = u$ if $i = j$ and $u(j)_i = 1$, otherwise. Let $G(T)$ the graph with vertices $1, 2, \ldots, n$ and an arc from i to j if $\lim\limits_{u\to\infty} T_i u(j) = \infty$. $G(T)$ is strongly connected if there is a directed path between any two distinct vertices. (Cf. [18, p. 4932] and [37, p. 131].)

(a) Let T be a monotone and homogeneous (of degree 1) selfmapping of K which is indecomposable (Definition 2.2.6). Show that T maps $\overset{\circ}{K}$ into itself and that $G(T)$ is strongly connected. Find a monotone and homogeneous selfmapping T of K, mapping $\overset{\circ}{K}$ into itself, for which $G(T)$ is strongly connected but T is not indecomposable.

(b) Let T be a selfmapping of $\overset{\circ}{K}$ for which $G(T)$ is strongly connected. Suppose $\lambda x \le y$ implies $\lambda Tx \le Ty$, where $\lambda \in [0, 1], x, y \in \overset{\circ}{K}$. Prove the following property for T: To $c > 0$ given there exists $\tilde{c} > 0$ such that for $e = (1, \ldots, 1) \in \overset{\circ}{K}$ and each $x \in \overset{\circ}{K}$ with $\|x\| = 1$ for the l_1-norm $\|\cdot\|$

$$Tx \le cx \text{ implies } x \le \tilde{c}\lambda\,(e, x)e. \qquad (*)$$

(Cf. [18] for arguments in a topical framework and [37, pp. 131/132].)

(c) Let T be a continuous selfmapping of K mapping $\overset{\circ}{K}$ into itself and such that $\lambda x \le y$ implies $\lambda Tx \le Ty (\lambda \in [0, 1], x, y \in \overset{\circ}{K})$. Show that property $(*)$ in (b) implies that for each $c > 0$ the set $\{x \in \overset{\circ}{K} \mid \|x\| = 1, Tx \le cx\}$ is closed for the l_1-norm $\|\cdot\|$.

2.3 Concave operators which are positively homogeneous

Dealing with concave operators $T\colon K \longrightarrow K$ for $K = \mathbb{R}_+^n$ in this chapter, until now we obtained two concave versions of Perrons theorem. In the first version, Theorem 2.1.11, we obtained a convergence result for the iterates of the normalized operator T by assuming a strong positivity assumption, viz. $Tx > 0$ for $x \gneq 0$. This assumption was weakened in the second version, Theorem 2.2.11, which gave us the convergence of $\frac{T^k x}{\|T^k x\|}$ by assuming that T is ray-preserving. Now we will strengthen the latter assumption to positive homogeneity to obtain convergence results which are much more pleasant.

Theorem 2.3.1. *Let $T\colon K \longrightarrow K$ be concave, primitive and positively homogeneous. Then there exist $\lambda^* > 0$ and $x^* > 0$, $\|x^*\| = 1$ such that the solutions $(x, \lambda) \in K \times \mathbb{R}$ of the eigenvalue problem $Tx = \lambda x$ are given precisely by $x = rx^*$ for some $r \geq 0$ and $\lambda = \lambda^*$. The solution (x^*, λ^*) has the following properties:*

(i) *For each $x \in K$ the limit $Sx := \lim_{k\to\infty} \frac{T^k x}{\lambda^{*k}}$ exists (with respect to $\|\cdot\|$) and defines a mapping $S\colon K \longrightarrow \mathbb{R}_+ x^*$ which is concave and positively homogeneous with $Sx > 0$ for $x \gneqq 0$. Furthermore, the mapping S satisfies the equations $ST = TS = \lambda^* S$ and $SS = S$ on K.*

(ii) *For any $x \gneqq 0$ there holds*

$$\lim_{k\to\infty} \frac{\|T^{k+1}x\|}{\|T^k x\|} = \lim_{k\to\infty} \|T^k x\|^{\frac{1}{k}} = \lambda^*.$$

Proof. We will apply Theorem 2.2.11 (cf. Remark 2.2.12 (1)). From this we get x^* and λ^*, where by positive homogeneity λ is uniquely determined because $Tx = \lambda x$ implies that $rTx^* = T(rx^*) = \lambda(rx^*)$ and, hence, $\lambda = \|Tx^*\| = \lambda^*$.

(i) From Theorem 2.2.11 we have for fixed $x \gneqq 0$ that $\lim_{k\to\infty} \frac{T^k x}{\|T^k x\|} = x^*$. Since $x^* > 0$ to $\varepsilon > 0$ given, there exists $N(\varepsilon)$ such that

$$(1-\varepsilon)\|T^k x\| x^* \leq T^k x \leq (1+\varepsilon)\|T^k x\| x^* \quad \text{for all} \quad k \geq N(\varepsilon).$$

Applying T to these inequalities, yields by monotonicity and positive homogeneity of T that

$$(1-\varepsilon)\|T^k x\| \lambda^{*l} x^* \leq T^{k+l} x \leq (1+\varepsilon)\|T^k x\| \lambda^{*l} x^* \quad \text{for all} \quad k \geq N(\varepsilon), \quad \text{all} \quad l \in \mathbb{N}.$$

Combining these two sets of inequalities, yields

$$\left(\frac{1-\varepsilon}{1+\varepsilon}\right)\frac{T^k x}{\lambda^{*k}} \leq \frac{T^{k+l}x}{\lambda^{*(k+l)}} \leq \left(\frac{1+\varepsilon}{1-\varepsilon}\right)\frac{T^k x}{\lambda^{*k}} \quad \text{for all} \quad k \geq N(\varepsilon), \quad \text{all} \quad l \in \mathbb{N}. \qquad (*)$$

Setting $x_k = \frac{T^k x}{\lambda^{*k}}$, $(*)$ means that

$$\frac{1-\varepsilon}{1+\varepsilon} x_k \leq x_{k+l} \leq \frac{1+\varepsilon}{1-\varepsilon} x_k$$

and by the monotonicity of $\|\cdot\|$ it follows that

$$\begin{aligned}
\|x_{k+l} - x_k\| &\leq \left\|x_{k+l} - \frac{1-\varepsilon}{1+\varepsilon} x_k\right\| + \frac{2\varepsilon}{1+\varepsilon}\|x_k\| \\
&\leq \left(\frac{1+\varepsilon}{1-\varepsilon} - \frac{1-\varepsilon}{1+\varepsilon}\right)\|x_k\| + \tfrac{2\varepsilon}{1+\varepsilon}\|x_k\|. \\
&\leq \frac{6\varepsilon}{1-\varepsilon^2}\|x_k\| \quad \text{for all} \quad k \geq N(\varepsilon), \quad \text{all} \quad l \in \mathbb{N}.
\end{aligned}$$

This shows, in particular, that the sequence $(x_k)_k$ is bounded in K for $\|\cdot\|$ and, hence, it must be a Cauchy sequence for all $\|\cdot\|$. Thus, $(x_k)_k$ converges to some element of K, that is $Sx := \lim_{k\to\infty} \frac{T^k x}{\lambda^{*k}} \in K$ for $x \gneqq 0$ and, trivially, $S0 = 0$. If $Sx \gneqq 0$ then $\frac{Sx}{\|Sx\|} = \lim_{k\to\infty} \frac{T^k x}{\|T^k x\|} = x^*$ and, therefore, $S\colon K \longrightarrow \mathbb{R}_+ x^*$. By definition S is concave, positively homogeneous, and by primitivity of T, it follows from $(*)$ that $Sx > 0$ for $x \gneqq 0$. Furthermore, $S(Tx) = \lambda^* \lim_{k\to\infty} \frac{T^{k+1}x}{\lambda^{*(k+1)}} = \lambda^* Sx$ and, because T is continuous on the interior of K by Lemma 2.1.13 we have for $x \gneqq 0$ that $T(Sx) = \lim_{k\to\infty} \frac{T^{k+1}x}{\lambda^{*k}} = S(Tx)$. Thus, $ST = TS = \lambda^* S$. From $Sx^* = x^*$ and $S\colon K \longrightarrow \mathbb{R}_+ x^*$ it follows that $SS = S$ on K.

(ii) From (i) we have for $x \gneqq 0$ that $Sx = \lim_{k\to\infty} \frac{T^k x}{\lambda^{*k}}$ and $\lambda^* Sx = \lim_{k\to\infty} \frac{T^{k+1}x}{\lambda^{*k}}$ which implies $\lim_{k\to\infty} \frac{\|T^{k+1}x\|}{\|T^k x\|} = \frac{\lambda^*\|Sx\|}{\|Sx\|} = \lambda^*$.

Finally, $\|Sx\| = \lim_{k\to\infty} \frac{\|T^k x\|}{\lambda^{*k}}$ for $x \gneqq 0$ implies $\lim_{k\to\infty} (\frac{\|T^k x\|}{\lambda^{*k}})^{\frac{1}{k}} = 1$, that is $\lim_{k\to\infty} \|T^k x\|^{\frac{1}{k}} = \lambda^*$. □

Remarks 2.3.2. (1) Without primitivity it may happen that for all $x \gneqq 0$ $\lim_{k\to\infty} \frac{\|T^{k+1}x\|}{\|T^k x\|}$ and $\lim_{k\to\infty} \|T^k\|^{\frac{1}{k}}$ exist and are equal but different from λ^* and that $Sx = \lim_{k\to\infty} \frac{T^k x}{\lambda^{*k}}$ also exists for all $x \geq 0$ but S does not project on the one dimensional ray $\mathbb{R}_+ x^*$. This is the case, e.g., for T being the identity map.

(2) Without primitivity it may also happen that conclusion (ii) of the Theorem holds true but not conclusion (i). This is the case, e.g., for the mapping $T\colon \mathbb{R}^2_+ \longrightarrow \mathbb{R}^2_+$ defined by $T(x_1, x_2) = (x_2, x_1)$. This mapping is concave, positively homogeneous and indecomposable but not primitive. $Tx = \lambda x$ has the unique solution $\lambda^* = 1$, $x^* = (\frac{1}{2}, \frac{1}{2})$ (up to a positive multiple for x^*) and for any $x \gneqq 0$ $\lim_{k\to\infty} \frac{\|T^{k+1}x\|}{\|T^k x\|} = 1$ and $\lim_{k\to\infty} \|T^k x\|^{\frac{1}{k}} = 1$. The first statement, however, does not hold for all monotone norms, in contrast to the second statement. The sequence $(\frac{T^k x}{\lambda^{*k}})_k$ does not converge for $x = (x_1, x_2)$ with $x_1 \neq x_2$.

(3) Without positive homogeneity it may happen that none of the conclusions (i) and (ii) holds true, although all the limits do exist. For $T(x_1, x_2) = (1 + x_1, 1)$, which is primitive but not even ray-preserving, $\lim_{k\to\infty} \frac{\|T^{k+1}x\|}{\|T^k x\|} = \lim_{k\to\infty} \|T^k x\|^{\frac{1}{k}} = 1 \neq \lambda^*$ (for the sum-norm) for all $x \gneqq 0$ and $\lim_{k\to\infty} \frac{T^k x}{\lambda^{*k}} = 0$ for all $x \geq 0$. Similar for the mapping $T(x_1, x_2) = (\sqrt{x_1} + \sqrt{x_2}, \sqrt{x_1} + \sqrt{x_2})$ which is primitive and ray-preserving but not positively homogeneous. (Cf. Exercise 1 below.)

Example 2.3.3. Let us look again at Example 2.2.13 with $T(x_1, x_2) = (4x_1 + 2x_2 + \sqrt{x_1 x_2}, \min\{x_1 + x_2, 2x_1\})$. As we have seen already T is concave, primitive and positively homogeneous with $\lambda^* = 5$ and $x^* = (\frac{1}{5}, \frac{4}{5})$. In addition to what we obtained by Theorem 2.2.11 we now ave by Theorem 2.3.1 that $Sx = \lim_{k\to\infty} \frac{T^k x}{5^k} = c(x)\,(\frac{1}{5}, \frac{4}{5})$ for $x \gneqq 0$ and some $c(x) > 0$. That is, for every starting point $x \gneqq 0$ the path $k \longmapsto T^k x$ grows with the factor 5. In particular, the only bounded path is the trivial one staying at 0. To determine the mapping S one has to calculate $c(x)$ which, however, is not an easy task. By Theorem 2.3.1 the mapping $c(\cdot)\colon \mathbb{R}^2_+ \longrightarrow \mathbb{R}_+$ is concave and positively homogeneous. To determine $c(\cdot)$ we also have the useful property that $c(Tx) = 5c(x)$, for all $x \geq 0$, which follows immediately from $STx = \lambda^* Sx$. For T linear $c(\cdot)$ also must be linear and can easily be computed from T as we shall see in the next section. (Cf. also Exercise 4 below.) In the above example, however, $c(\cdot)$ cannot be linear as the following considerations show. From $Te_2 = 2e_1$ we get $2c(e_1) = c(2e_1) = c(Te_2) = 5c(e_2)$. Linearity of $c(\cdot)$ would imply that $c(re_1 + se_2) = rc(e_1) + sc(e_2) = (r + \frac{2}{5}s)c(e_1)$ for all $r, s \geq 0$. But we have for $r = 4$ and $s = 1$ that $c(4e_1 + e_2) = c(Te_1) = 5c(e_1) \neq (4 + \frac{2}{5})c(e_1)$.

As remarked already, Theorem 2.3.1 ceases to hold if the assumption of positive homogeneity is omitted. Also, it is not enough to replace the latter assumption by homogeneity of degree d for $d \neq 1$. It follows easily, however, that for $0 \leq d < 1$ the iterates of T itself converge as the following Corollary shows.

Corollary 2.3.4. Let $T\colon K \longrightarrow K$ be concave, primitive and homogeneous of degree d with $0 \leq d < 1$. Then the fixed point equation $Tx = x$ has a unique solution x^* in $K \setminus \{0\}$. Moreover, $x^* > 0$ and $\lim_{k\to\infty} T^k x = x^*$ for all $x \in K \setminus \{0\}$.

Proof. Obviously, T is ray preserving and by Theorem 2.2.11 (cf. Remarks 2.2.12 (1)) there exist $\bar{x} > 0$, $\|\bar{x}\| = 1$, $\bar{\lambda} > 0$ such that $T\bar{x} = \bar{\lambda}\bar{x}$. Furthermore, $Tx = \lambda x$ for $x \gneq 0$ implies $x = r\bar{x}$ with $r > 0$ and $\lambda > 0$. For $x^* = \bar{\lambda}^{\frac{1}{1-d}}\bar{x}$ it holds that $x^* > 0$ and $Tx^* = \bar{\lambda}^{\frac{d}{1-d}} T\bar{x} = \bar{\lambda}^{\frac{d}{1-d}}\bar{\lambda}\bar{x} = x^*$.

Concerning uniqueness, suppose $Tx = x$ for $x \gneq 0$. It follows that $x = r\bar{x}$ with $r > 0$ and, hence, $r^d\bar{\lambda}\bar{x} = r\bar{x}$. Therefore, $r = \bar{\lambda}^{\frac{1}{1-d}}$ and $x = x^*$. Moreover, by Theorem 2.2.11 (ii) it holds for $x \gneq 0$ given that $\lim_{k\to\infty} \frac{T^k x}{\|T^k x\|} = \bar{x}$ and, hence, to $\epsilon > 0$ there exists $N(\epsilon)$ such that

$$(1-\epsilon)\|T^k x\|\bar{x} \leq T^k x \leq (1+\epsilon)\|T^k x\|\bar{x} \quad \text{for all} \quad k \geq N(\epsilon).$$

Applying the monotone mapping T to this inequality l times gives

$$T^l((1-\epsilon)\|T^k x\|\bar{x}) \leq T^{k+l} x \leq T^l((1+\epsilon)\|T^k x\|\bar{x}) \quad \text{for all} \quad k \geq N(\epsilon).$$

For any given $\lambda > 0$ induction over l yields

$$T^l(\lambda\bar{x}) = \lambda^{d^l}\bar{\lambda}^{s(l)}\bar{x} \quad \text{with} \quad s(l) = \sum_{i=0}^{l-1} d^i \quad \text{for} \quad l \geq 1.$$

By assumption $0 \leq d < 1$ and we obtain $\lim_{l\to\infty} T^l(\lambda\bar{x}) = \bar{\lambda}^{\frac{1}{1-d}}\bar{x} = x^*$ for arbitrary $\lambda > 0$. Thus, from the inequalities we obtain $x^* \leq \lim_{l\to\infty} T^{k+l}x \leq x^*$ for all $k \geq N(\epsilon)$ and, hence, $\lim_{k\to\infty} T^k x = x^*$. □

Example 2.3.5. Let $T\colon \mathbb{R}^2_+ \longrightarrow \mathbb{R}^2_+$ be given by $T(x_1, x_2) = (\min\{\frac{1}{5}\sqrt{x_1} + \sqrt{x_2}, \frac{1}{2}\sqrt{x_1} + \frac{1}{4}\sqrt{x_2}\}, 2\sqrt{x_1} + \sqrt{x_2})$. Obviously, T is concave and $Tx > 0$ for $x \gneq 0$, i.e., T is primitive. Furthermore, $T(\lambda(x_1, x_2)) = \lambda^{\frac{1}{2}} T(x_1, x_2)$ and T is homogeneous of degree $d = \frac{1}{2}$. By Corollary 2.3.4, therefore, $\lim_{k\to\infty} T^k x = x^*$ for any $x \gneq 0$. To compute the unique fixed point x^* of T consider for $x \gneqq 0$ the equation $Tx = x$, that is $\min\{\frac{1}{5}\sqrt{x_1} + \sqrt{x_2}, \frac{1}{2}\sqrt{x_1} + \frac{1}{4}\sqrt{x_2}\} = x_1$ and $2\sqrt{x_1} + \sqrt{x_2} = x_2$. Suppose $x_1 \leq x_2$. It follows that $\frac{1}{2}\sqrt{x_1} + \frac{1}{4}\sqrt{x_2} \leq \frac{3}{4}\sqrt{x_2} \leq \frac{1}{5}\sqrt{x_1} + \sqrt{x_2}$ and $\frac{1}{2}\sqrt{x_1} + \frac{1}{4}\sqrt{x_2} = x_1$. This implies $x_2 = 2\sqrt{x_1} + \sqrt{x_2} = 4x_1$ and $4x_1 = 2\sqrt{x_1} + \sqrt{4x_1} = 4\sqrt{x_1}$. Thus, $x_1 = 1$ and $x_2 = 4$, that is $x^* = (1,4)$. (This we infer from the uniqueness of the fixed point but one could argue also that $Tx = x$ has no solution for $x_1 > x_2$.) Thus, we obtain $\lim_{k\to\infty} T^k x = (1,4)$ for all $x \gneqq 0$.

Similarly as in the previous section, from Theorem 2.3.1 we obtain the following result considering operators which are not necessarily concave.

Corollary 2.3.6. Suppose $T\colon K \longrightarrow K$ is an operator as in Theorem 2.3.1 and define $Rx = r(x)Tx$ where $r\colon K \longrightarrow \mathbb{R}_+$ with $r(x) > 0$ for $x \gneqq 0$. Then the solutions $(y,\mu) \in (K \setminus \{0\}) \times \mathbb{R}$ of the eigenvalue problem $Ry = \mu y$ are given by $y = rx^*$ with $r > 0$ and $\mu = r(x^*)\lambda^*$, where (x^*, λ^*) is the solution of the eigenvalue problem for T according to Theorem 2.3.1. Moreover, the following statements hold.

(i) If $r\colon K \longrightarrow \mathbb{R}_+$ is continuous on the interior of K with $r(cx) = r(x)$ for all $c > 0$, all $x \in K$ then for all $x \gneqq 0$

$$\lim_{k\to\infty} \frac{\|R^{k+1}x\|}{\|R^k x\|} = \lim_{k\to\infty} \|R^k x\|^{\frac{1}{k}} = r(x^*)\lambda^*.$$

(ii) If, in addition to the assumptions of (i), $r(Tx) \geq r(x)$ for all $x \in K$ then for all $x \in K$

$$Sx\colon = \lim_{k\to\infty} \frac{R^k x}{(r(x^*)\lambda^*)^k}$$

exists and defines a mapping $S\colon K \longrightarrow \mathbb{R}_+ x^*$ with $SR = RS = r(x^*)\lambda^* id$ and $SS = S$.

(iii) If, in addition to the assumptions of (ii), the non-negative sum $\sum_{k=0}^{\infty} r(x^*) - r(T^k x)$ is finite then $S(x) > 0$.

Proof. The statement about the solutions (y,μ) is immediate from Theorem 2.3.1.

(i) From $Rx = r(x)Tx$ it follows by induction, using positive homogeneity of T, that

$$R^k x = r(R^{k-1}x)\dots r(Rx)r(x)T^k x \text{ for } k \geq 1,\ x \in K. \tag{2.3.1}$$

By the assumptions on $r(\cdot)$ and Corollary 2.2.14 one has

$$\lim_{k\to\infty} r(R^k x) = \lim_{k\to\infty} r\left(\frac{R^k x}{\|R^k x\|}\right) = r\left(\lim_{k\to\infty} \frac{R^k x}{\|R^k x\|}\right) = r(x^*).$$

Hence,

$$\lim_{k\to\infty} \frac{\|R^{k+1}x\|}{\|R^k x\|} = \lim_{k\to\infty} r(R^k x) \cdot \lim_{k\to\infty} \frac{\|T^{k+1}x\|}{\|T^k x\|} = r(x^*)\lambda^*.$$

Furthermore,

$$\|R^k x\|^{\frac{1}{k}} = \left(r(R^{k-1}x)\cdots r(Rx)r(x)\right)^{\frac{1}{k}} \|T^k x\|^{\frac{1}{k}}. \tag{2.3.2}$$

For $a_k = r(R^k x)$ we just proved that $\lim_{k\to\infty} a_k = r(x^*)$ and, hence, to $\varepsilon > 0$ given there exists $N = N(\varepsilon) \in \mathbb{N}$ such that $r(x^*) - \varepsilon \le a_k \le r(x^*) + \varepsilon$ for all $k \ge N$. Thus, for $k \ge N$

$$(r(x^*) - \varepsilon)^{1-\frac{N}{k}} \left(\prod_{i=0}^{N-1} a_i\right)^{\frac{1}{k}} \le \left(\prod_{i=0}^{k-1} a_i\right)^{\frac{1}{k}} \le \left(\prod_{i=0}^{N-1} a_i\right)^{\frac{1}{k}} (r(x^*) + \varepsilon)^{1-\frac{N}{k}},$$

which implies

$$r(x^*) - \varepsilon \le \liminf_{k\to\infty} \left(\prod_{i=0}^{k-1} a_i\right)^{\frac{1}{k}} \le \limsup_{k\to\infty} \left(\prod_{i=0}^{k-1} a_i\right)^{\frac{1}{k}} \le r(x^*) + \varepsilon.$$

Since $\varepsilon > 0$ was arbitrary chosen it follows that $\lim_{k\to\infty} (\prod_{i=0}^{k-1} a_i)^{\frac{1}{k}} = r(x^*)$ and by (2.3.2) therefore

$$\lim_{k\to\infty} \|R^k x\|^{\frac{1}{k}} = \lim_{k\to\infty} \left(\prod_{i=0}^{k-1} a_i\right)^{\frac{1}{k}} \cdot \lim_{k\to\infty} \|T^k x\|^{\frac{1}{k}} = r(x^*)\lambda^*,$$

using the result for T from Theorem 2.3.1.

(ii) For $a_k = r(R^k x)$ we have by (2.3.1) that $a_k = r(T^k x)$. The assumption $r(Tx) \ge r(x)$ for all $x \in K$ implies that $a_k \le a_{k+1}$ for all k and, using $\lim_{k\to\infty} a_k = r(x^*)$ from step (i), we obtain that $a_k \le r(x^*)$ for all k. This shows that

$$u(x) := \lim_{k\to\infty} \frac{\prod_{i=0}^{k-1} a_i}{r(x^*)^k} \tag{2.3.3}$$

exists for all $x \in K$. Putting together we obtain

$$Sx := \lim_{k\to\infty} \frac{R^k x}{r(x^*)^k \lambda^{*k}} = \lim_{k\to\infty} \left(\frac{\prod_{i=0}^{k-1} a_i}{r(x^*)^k} \cdot \frac{T^k x}{\lambda^{*k}} \right) = u(x) \lim_{k\to\infty} \frac{T^k x}{\lambda^{*k}} = u(x)c(x)x^*$$

by using Theorem 2.3.1 for T with $c(x) > 0$ for $x \gneq 0$. The remaining properties for S follow as in Theorem 2.3.1.

(iii) We have to show that $u(x) > 0$ for $x \gneq 0$ where $u(\cdot)$ is defined by (2.3.3). If $\sum_{k=0}^{\infty} r(x^*) - r(T^k x)$ is finite the same is true for $\sum_{k=0}^{\infty} \frac{r(x^*)-r(T^k x)}{r(x)}$ and, because of $r(x) \le r(T^k x)$ for all k, we must have that $\sum_{k=0}^{\infty} \frac{r(x^*)-r(T^k x)}{r(T^k x)}$ is finite. Since $r(T^k x) \le r(x^*)$ for all k and $\log(1 + t) \le t$ for non-negative real numbers t we obtain

$$\log \prod_{k=0}^{\infty} \frac{r(x^*)}{r(T^k x)} = \sum_{k=0}^{\infty} \log \frac{r(x^*)}{r(T^k x)} \le \sum_{k=0}^{\infty} \frac{r(x^*) - r(T^k x)}{r(T^k x)}$$

and, hence, $\prod_{k=0}^{\infty} \frac{r(x^*)}{r(T^k x)}$ must be finite. By the definition of $u(\cdot)$ therefore $u(x)$ cannot be zero, i.e., $u(x) > 0$. □

Example 2.3.7. Let $T: \mathbb{R}_+^2 \longrightarrow \mathbb{R}_+^2$ be the linear operator defined by $T(x_1, x_2) = (\frac{1}{3}x_1 + \frac{2}{3}x_2, \frac{2}{3}x_1 + \frac{1}{3}x_2)$ and $r: \mathbb{R}_+^2 \longrightarrow \mathbb{R}_+$ be given by $r(x_1, x_2) = \frac{x_1+x_2}{\max\{x_1,x_2\}}$ for $x \gneq 0$ and

$r(0) = 0$. The solution of the eigenvalue problem $Tx = \lambda x$ is given by $x^* = (\frac{1}{2}, \frac{1}{2})$ (normed with respect to the sum norm) and $\lambda^* = 1$. Therefore, the solution of the eigenvalue problem for $Rx = r(x)Tx$ is given by $y = rx^*$, $r > 0$, and $\mu = r(x^*)\lambda^* = 2$. Furthermore, $r(\cdot)$ is continuous on the interior of $\mathbb{R}^2_+$, $r(cx) = r(x)$ for $c > 0$ and

$$r(Tx) = \frac{x_1 + x_2}{\max\{\frac{1}{2}x_1 + \frac{2}{3}x_2, \frac{2}{3}x_1 + \frac{1}{3}x_1\}} \geq \frac{x_1 + x_2}{\max\{x_1, x_2\}} = r(x).$$

Thus, the conclusions of Corollary 2.3.6, parts (i) and (ii), hold. Obviously, the operator R is not concave; actually, none of the components of R is concave in any of the variables. Considering part (iii) of the Corollary we have for $x \in \mathbb{R}^2_+$ with $x_1 + x_2 = 1$ that $\max\{T_1x, T_2x\} = \frac{1}{3} + \frac{1}{3}\max\{x_1, x_2\}$. By induction this implies

$$\max\{T_1^k x, T_2^k x\} = \frac{1 - \frac{1}{3^k}}{2} + \frac{1}{3^k}\max\{x_1, x_2\} \tag{2.3.4}$$

for all $x \in \mathbb{R}^2_+$, $x_1 + x_2 = 1$ and $k \in \mathbb{N}$. Since

$$r(x^*) = \frac{\frac{1}{2} + \frac{1}{2}}{\max\{\frac{1}{2}, \frac{1}{2}\}} = 2,$$

this implies

$$r(x^*) - r(T^k x) = 2 - \frac{2}{1 + \frac{1}{3^k}m(x)} = 2\frac{3^{\frac{1}{k}m(x)}}{1 + \frac{1}{3^k}m(x)}$$

where $m(x) = 2\max\{\frac{x_1}{x_1+x_2}, \frac{x_2}{x_1+x_2}\} - 1 \geq 0$ for $x \gneqq 0$. Since $m(x) \leq 1$ it follows that $r(x^*) - r(T^k x) \leq \frac{2}{3^k}$ for all $x \gneqq 0$ and $k \in \mathbb{N}$. Thus $\sum_{k=0}^{\infty} r(x^*) - r(T^k x)$ is finite for all $x \gneqq 0$ and, by part (iii) of the corollary, it holds that $Sx > 0$ for all $x \gneqq 0$. Actually, in this Example the iterates $T^k x$ can be computed explicitly and, hence, the iterates $R^k x = r(T^{k-1}x)\dots r(Tx)r(x)T^k x$ and the operator $Sx = \lim_{k\to\infty} \frac{R^k x}{2^k}$, too. (Cf. Exercise 3 below.)

The results of Theorem 2.3.1 and Corollary 2.3.6 provide quite strong properties for positively homogeneous concave operators in case these operators are primitive. As mentioned already, convergence for the iterates of the operator cannot be expected if the operator is only assumed to be indecomposable instead of being primitive. There are, however, some interesting properties as, in particular, certain dominance properties for the eigenvalue λ^* which apply already to indecomposable operators. For these operators the statement made by the Concave Perron–Frobenius Theorem (Theorem 2.1.14) on the existence of a (non-trivial) solution of the eigenvalue problem can be considerably sharpened as shown in the following theorem.

Theorem 2.3.8. *Let $T\colon K \longrightarrow K$ be concave and indecomposable, where $K = \mathbb{R}^n_+$.*

(i) *The eigenvalue problem $Tx = \lambda x$ has a strictly positive solution (x, λ), i.e., $x > 0$ and $\lambda > 0$, and any solution (x, λ) with $x \gneqq 0$ must be strictly positive.*

(ii) *If T is positively homogeneous then the eigenvalue problem $Tx = \lambda x$ has a solution (x^*, λ^*) with $x^* > 0$ and $\lambda^* > 0$ and such that the following properties hold.*

(a) *If $\lambda x \leq Tx$ holds for some $x \gneq 0$ then $\lambda \leq \lambda^*$.*

(b) *If $\lambda^* x \leq Tx$ holds for some $x \gneq 0$ then $x = rx^*$ with $r > 0$.*

(c) *The solution of the non-negative eigenvalue problem is essentially unique, i.e., the solutions of $Tx = \lambda x$ with $x \gneq 0$ and $\lambda \geq 0$ are given by $\lambda = \lambda^*$ and $x = rx^*$ with $r > 0$.*

(iii) *Suppose T is positively homogeneous and can be extended to $\mathbb{R}^n$, $T\colon \mathbb{R}^n \longrightarrow \mathbb{R}^n$. Consider the eigenvalue problem $Tx = \lambda x$ for arbitrary $x \in \mathbb{R}^n$, $x \neq 0$ and $\lambda \in \mathbb{R}$. The eigenvalue λ^* possesses the following* **dominance property***:*
If $|Tx| \leq T|x|$ for all $x \in \mathbb{R}^n$ then $|\lambda| \leq \lambda^*$ and if, moreover, $|T^p x| < T^p|x|$ for some $p \in \mathbb{N}$ and all $x \in \mathbb{R}^n$ with $x \neq |x|$ then $\lambda = \lambda^*$ or $|\lambda| < \lambda^*$.
(Here $|x|$ is understood to be componentwise, i.e., $|x| = (|x_1|, |x_2|, \ldots, |x_n|)$ for $x = (x_1, x_2 \ldots, x_n)$.)

Proof. (1) Let $Tx = \lambda x$ with $x \gneq 0$. Suppose the set $I = \{1 \leq k \leq n \mid x_k = 0\}$ is non-empty. Then $\emptyset \subsetneq I \subsetneq \{1, \ldots, n\}$ and by indecomposability of T there exist $i \in I$ and $j \notin I$ such that $T_i e_j > 0$. For $\alpha = \min\{x_j, 1\}$ we have that $0 < \alpha \leq 1$ because of $j \notin I$, and $x = \alpha e_j + (1 - \alpha)y$ with $y \in K$. Concavity of T implies that $T_i x \geq \alpha T_i e_j > 0$ which contradicts $T_i x = \lambda x_i = 0$. Therefore, we must have that $I = \emptyset$ and, hence, $x > 0$. Similarly, we must have $\lambda > 0$. Otherwise $Tx = 0$ and indecomposability of T would imply for $I = \{1\}$ that $T_1 x \geq \alpha T_1 e_j > 0$ for some $j \neq 1$ which is a contradiction. Finally, by Theorem 2.1.14 there exist $x \gneq 0$ and $\lambda \geq 0$ such that $Tx = \lambda x$.

(ii) Suppose first that T is even primitive and let (x^*, λ^*) be a solution of the eigenvalue problem according to Theorem 2.3.1. If $\lambda x \leq Tx$ for some $x \gneq 0$ and $\lambda \geq 0$ then $\lambda^k \|x\| \leq \|T^k x\|$ for all $k \in \mathbb{N}$. Theorem 2.3.1 implies that $\lambda^* = \lim_{k\to\infty} \|T^k x\|^{\frac{1}{k}} \geq \lambda \lim_{k\to\infty} \|x\|^{\frac{1}{k}} = \lambda$. This proves property (a). Property (c) follows immediately from Theorem 2.3.1. To show property (b) let $\lambda^* x \leq Tx$ for some $x \gneq 0$ and $u = Tx - \lambda^* x$. By Theorem 2.3.1 it suffices to show that $u = 0$. Assume this is not the case, i.e., $u \gneq 0$. Primitivity of T implies $T^p u > 0$ for some $p \in \mathbb{N}$. Since T^p is concave and positively homogeneous we obtain that

$$T^{p+1}x = T^p(u + \lambda^* x) \geq T^p u + \lambda^* T^p x > \lambda^{*(p+1)} x.$$

Hence, there exists $\mu > \lambda^{*(p+1)}$ with $T^{p+1}x \geq \mu x$. This implies, as in the proof of property (a), that $\mu \leq \lambda^{*(p+1)}$ – a contradiction. Thus, we must have that $u = 0$.

Now we shall relax the primitivity assumption on T. For this let T be concave, indecomposable and positively homogeneous and consider the operator $S\colon K \longrightarrow K$ defined by $Sx = x + Tx$. This operator also is concave, indecomposable and positively homogeneous. Moreover, S is primitive by Lemma 2.2.7 and Lemma 2.2.10 because $S_h e_h = 1 + T_h e_h > 0$ for each $h \in \{1, \ldots, n\}$. Thus, by the above there is a solution (x^*, μ^*) of $Sx = \mu x$ with $x^* > 0$, $\mu^* > 0$ and such that properties (a) - (c) hold with respect to this solution. Obviously, $Tx^* = \lambda^* x^*$ for $\lambda^* = \mu^* - 1$ and $\lambda^* > 0$. This shows property (a) for T. If $\lambda^* x \leq Tx$ for $x \gneq 0$ then $\mu^* x = (\lambda^* + 1)x \leq Sx$ and we obtain $x = rx^*$ for some $r > 0$. This shows property (b) for T. In the same way property (c) does follow for T.

(iii) If $Tx = \lambda x$ with $x \in \mathbb{R}^n \setminus \{0\}$, $\lambda \in \mathbb{R}$ then by assumption

$$|\lambda||x| = |\lambda x| = |Tx| \leq T|x|.$$

Property (a) of part (ii) implies that $|\lambda| \leq \lambda^*$. Suppose now $|T^p x| < T^p|x|$ for some p and all $x \neq |x|$ and let $Tx = \lambda x$ with $x \in \mathbb{R}^n \setminus \{0\}$ and $\lambda \in \mathbb{R}$ such that $|\lambda| = \lambda^*$. Then $\lambda^*|x| = |Tx| \leq T|x|$ and by property (b) of part (ii) we have that $|x| = rx^*$ for some $r > 0$. If $x \neq |x|$ then $r\lambda^{*p}x^* = \lambda^{*p}|x| = |\lambda^p x| = |T^p x| < T^p|x| = rT^p x^* = r\lambda^{*p}x^*$ – a contradiction. Thus, we must have $x = |x|$ which implies $\lambda x = Tx = T|x| = \lambda^*|x| = \lambda^* x$ and, hence, $\lambda = \lambda^*$. □

By the following examples we shall illustrate some of the assumptions and statements made in Theorem 2.3.8.

Examples 2.3.9. (i) Consider the mapping $T\colon \mathbb{R}^2 \longrightarrow \mathbb{R}^2$ defined by $Tx = (\min\{2x_1 + x_2, x_1 + 2x_2\}, x_1 + x_2)$. T maps $\mathbb{R}^2_+$ into itself and restricted to $\mathbb{R}^2_+$ the mapping is concave, positively homogeneous, indecomposable and even primitive. The eigenvalue problem $Tx = \lambda x$ means $\min\{2x_1 + x_2, x_1 + 2x_2\} = \lambda x_1$ and $x_1 + x_2 = \lambda x_2$.

For $x_1 \leq x_2$ this means $2x_1 + x_2 = \lambda x_1$ and $x_1 + x_2 = \lambda x_2$. As eigenvalues and related eigenvectors in $\mathbb{R}^2$ one obtains in this case (up to a scalar for eigenvectors) $\lambda_1 = \frac{1}{2}(3 + \sqrt{5})$ with $x^1 = -(\frac{1}{2}(1 + \sqrt{5}), 1)$ and $\lambda_2 = \frac{1}{2}(3 - \sqrt{5})$ with $x^2 = (\frac{1}{2}(1 - \sqrt{5}), 1)$.

For $x_1 > x_2$ the eigenvalue problem amounts to $x_1 + 2x_2 = \lambda x_1$ and $x_1 + x_2 = \lambda x_2$, that is $\lambda_3 = 1 + \sqrt{2}$ with $x^3 = (\sqrt{2}, 1)$ and $\lambda_4 = 1 - \sqrt{2}$ with $x^4 = (\sqrt{2}, -1)$.

The (essentially) unique positive solution of the eigenvalue problem according to Theorem 2.3.8 is given by $\lambda^* = 1 + \sqrt{2}$ with $x^* = (\sqrt{2}, 1)$. It holds that $|\lambda_2| \leq \lambda^*$ and $|\lambda_4| \leq \lambda^*$ but $|\lambda_1| > \lambda^*$. Indeed, the corresponding assumption in (iii), viz. $|Tx| \leq T|x|$, is not fulfilled since, e.g., for $x = (-1, 0)$ one has $Tx = (-2, -1)$ but $T|x| = (1, 1)$.

(ii) The next example addresses the strict dominance of λ^* according to part (iii) of the Theorem. Let $T\colon \mathbb{R}^2 \longrightarrow \mathbb{R}^2$ be defined by

$$Tx = \begin{cases} (x_1 + x_2)(1, 1) & \text{for} \quad x \in \mathbb{R}^2_+ \\ (x_1 - x_2)(-1, 1) & \text{for} \quad x \notin \mathbb{R}^2_+. \end{cases}$$

Obviously, $T\colon \mathbb{R}^2_+ \longrightarrow \mathbb{R}^2_+$ is concave, positively homogeneous, indecomposable and even primitive. The eigenvalue problem $Tx = \lambda x$ amounts for $x \in \mathbb{R}^2_+$ to $x_1 + x_2 = \lambda x_1$ and $x_1 + x_2 = \lambda x_2$ and, hence, $\lambda_1 = 2$ with $x^1 = (1, 1)$. For $x \notin \mathbb{R}^2_+$ the eigenvalue problem amounts to $x_1 - x_2 = -\lambda x_1$ and $x_1 - x_2 = \lambda x_2$ and, hence, $\lambda_2 = 0$ with $x^2 = (1, 1)$ and $\lambda_3 = -2$ with $x^3 = (1, -1)$. The unique positive solution according to the Theorem is given by $\lambda^* = 2$ and $x^* = (1, 1)$ (up to a positive scalar). It holds that $|\lambda_2| \leq \lambda^*$ and $|\lambda_3| \leq \lambda^*$ and, indeed, the corresponding assumption of part (iii) is fulfilled because for $x \notin \mathbb{R}^2_+$, $|Tx| = |x_1 - x_2|(1, 1) \leq (|x_1| + |x_2|)(1, 1) = T|x|$. Strict dominance, however, does not hold for λ^* since $\lambda_3 \neq \lambda^*$ and $|\lambda_3| = \lambda^*$. The corresponding assumption of part (iii) is not fulfilled because, e.g., for $x = (1, -1)$ one has $x \neq |x|$ but $|Tx| = (2, 2) = T|x|$.

(iii) A non-linear operator $T\colon \mathbb{R}^2 \longrightarrow \mathbb{R}^2$ for which all assumptions of the Theorem are satisfied is given by

$$Tx = \begin{cases} (x_1 + x_2)(1,1) & \text{for} \quad x \in \mathbb{R}^2_+ \\ \frac{1}{2}(x_1 + x_2)(1,1) & \text{for} \quad x \notin \mathbb{R}^2_+ . \end{cases}$$

T is not linear because, for $x \in \mathbb{R}^2_+ \setminus \{0\}$, $T(-x) = \frac{1}{2}(x_1 + x_2)(1,1) = \frac{1}{2}Tx \neq -Tx$. Obviously, $T\colon \mathbb{R}^2_+ \longrightarrow \mathbb{R}^2_+$ is concave, positively homogeneous, indecomposable and even primitive. For $x \notin \mathbb{R}^2_+$ one has $|Tx| = \frac{1}{2}(|x_1 + x_2|)(1,1) < (|x_1| + |x_2|)(1,1) = T|x|$ and, obviously, $|Tx| \le T|x|$ for $x \in \mathbb{R}^2_+$. For eigenvalues and eigenvectors, respectively, one obtains for $x \in \mathbb{R}^2_+$ $\lambda_1 = 2$ with $x^1 = (1,1)$ and for $x \notin \mathbb{R}^2_+$ $\lambda_2 = 0$ with $x^2 = (1,-1)$ and $\lambda_3 = 1$ with $x^3 = (-1,-1)$. Thus, $\lambda^* = 2$ and $x^* = (1,1)$ (up to a positive scalar) and $|\lambda_i| \le \lambda^*$ with $|\lambda_i| < \lambda^*$ for $\lambda_i \neq \lambda^*$, according to the Theorem.

Remarks 2.3.10. Hilbert's metric (cf. Definition 2.1.8) was initially introduced by Hilbert [23] in his investigations on the foundations of geometry for convex bodies in finite dimensional space. Sometimes this metric has been also associated with the names of A. Cayley and F. Klein. Birkhoff [5] applied this metric to cones in infinite dimensions and to linear operators mapping a cone into itself. (See also Ostrowski [52].) About the same time Samelson [57] used the Cayley–Hilbert metric to give a short proof of Perron's theorem on positive matrices. Hilbert's metric then became a useful tool to investigate, in finite as well as in infinite dimensions, linear operators as well as non-linear ones which leave a cone invariant. See, e.g., [6, 14, 28–30, 34, 63, 65] and, in particular, in [48] and [49]. For the extension of various parts of classical Perron–Frobenius theory to certain non-linear mappings see [13, 26, 37, 42, 44, 47, 48, 50, 51, 63]. The concave version of Perron–Frobenius theory as presented in this chapter has its roots in [29, 30, 35]. In Chapter 3 we will come back in a more general setting to Hilbert's metric and other metrics intrinsic to a cone.

In the next section we shall see how many statements of the common linear Perron–Frobenius Theory appear as special cases of the results we obtained for concave operators.

Exercises

1. Demonstrate for the mapping $T\colon \mathbb{R}^2_+ \longrightarrow \mathbb{R}^2_+$, $T(x_1, x_2) = (\sqrt{x_1} + \sqrt{x_2}, \sqrt{x_1} + \sqrt{x_2})$ the following properties.
   (a) T is concave, primitive but not positively homogeneous and the conditional eigenvalue problem $Tx = \lambda x$, $\lambda \ge 0$, $\|x\| = 1$ ($\|\cdot\|$ the sum-norm) has a unique solution $\lambda^* > 0, x^* > 0$.
   (b) The limit $Sx = \lim\limits_{k \to \infty} \frac{T^k x}{\lambda^{*k}}$ exists and $Sx = 0$ for all $x \in \mathbb{R}^2_+$.

(c) Statement (ii) of Theorem 2.3.1 does not apply to T.

2. Let $T\colon \mathbb{R}^2_+ \longrightarrow \mathbb{R}^2_+$ be defined by

$$T(x_1,x_2) = \begin{bmatrix} \frac{1}{4} & \frac{3}{4} \\ \frac{3}{4} & \frac{1}{4} \end{bmatrix} \begin{bmatrix} x_1 \\ x_2 \end{bmatrix}.$$

(a) Compute the unique solution $\lambda^* > 0$, $x^* > 0$ ($\|x^*\| = 1$ for the sum-norm) for T.

(b) Compute $Sx = \lim_{k\to\infty} \frac{T^k x}{\lambda^{*k}}$ directly from the iterates T^k.

(c) Verify statement (ii) of Theorem 2.3.1 directly by computing the iterates T^k.

3. Let $T\colon \mathbb{R}^2_+ \longrightarrow \mathbb{R}^2_+$, $T(x_1,x_2) = (\frac{1}{3}x_1 + \frac{2}{3}x_2, \frac{2}{3}x_1 + \frac{1}{3}x_2)$ and $r\colon \mathbb{R}^2_+ \longrightarrow \mathbb{R}^2_+$, $r(x_1,x_2) = \frac{x_1+x_2}{\max\{x_1,x_2\}}$ for $x \gneq 0$, $r(0) = 0$. (See Example 2.3.7).

(a) Calculate explicitly the iterates T^k.

(b) Use (a) to calculate the iterates $R^k x = r(T^{k-1}x)\ldots r(Tx)r(x)T^k x$.

(c) Determine the limit $Sx = \lim_{k\to\infty} \frac{R^k x}{2^k}$.

4. Consider a concave operator $T\colon \mathbb{R}^n_+ \longrightarrow \mathbb{R}^n_+$ which is primitive and positively homogeneous and such that there exists some concave operator $T'\colon \mathbb{R}^n_+ \longrightarrow \mathbb{R}^n_+$ with the property that $\langle T'y, x\rangle = \langle y, Tx\rangle$ holds for all $x, y \in \mathbb{R}^n_+$. ($\langle u, v\rangle = \sum_{i=1}^n u_i v_i$ denotes the inner product in $\mathbb{R}^n$.) Show that the operator $Sx = \lim_{k\to\infty} \frac{T^k x}{\lambda^{*k}}$ can be computed as $Sx = \frac{\langle \bar{x}, x\rangle}{\langle \bar{x}, x^*\rangle} x^*$ for all $x \in \mathbb{R}^n_+$ with a fixed vector $\bar{x} \in \mathbb{R}^n_+ \setminus \{0\}$ and (x^*, λ^*) the solution according to Theorem 2.3.1.

5. For finitely many non-negative matrices $A(k) = (a_{ij}(k))_{1\le i,j\le n}$, $1 \le k \le m$ consider the operator $T\colon \mathbb{R}^n_+ \longrightarrow \mathbb{R}^n_+$ defined by

$$T_i x = \min_{1\le k\le m} \sum_{j=1}^n a_{ij}(k) x_j \text{ for } 1 \le i \le n.$$

(a) Show that T is concave and positively homogeneous.

(b) Find conditions on the matrices such that T is primitive.

(c) Apply Theorem 2.3.1 to the special case where $A(1) = \left[\begin{smallmatrix} 2 & 1 \\ 1 & 0 \end{smallmatrix}\right]$, $A(2) = \left[\begin{smallmatrix} 1 & 2 \\ 3 & 0 \end{smallmatrix}\right]$.
Calculate the solution (x^*, λ^*) (for the sum-norm) and check whether the operators are linear.

6. Find an operator $T\colon \mathbb{R}^2_+ \longrightarrow \mathbb{R}^2_+$ which is concave, ray preserving and primitive and for which, however, part (ii) of Theorem 2.3.8 does not hold.

7. Check assumptions and statements of Theorem 2.3.8 for the operator $T\colon \mathbb{R}^n \longrightarrow \mathbb{R}^n$ given by $Tx = A|x|$ where A is a (strictly) positive $n \times n$-matrix.

8. Let T be a continuous and monotone selfmapping of $\mathbb{R}^n_+$ which maps $\mathbb{R}^n_+$ into its interior and for which it holds that $T(\lambda x) > \lambda Tx$ for $0 < \lambda < 1$ and $x \in \operatorname{int} \mathbb{R}^n_+$.

(a) Show that T has a unique fixed point $x^* \in \operatorname{int} \mathbb{R}^n_+$ with $\lim_{k\to\infty} T^k x^0 = x^*$ if $x^0 \in \operatorname{int} \mathbb{R}^n_+$ with $Tx^0 \leq x^0$.

(b) Find an example which satisfies the general assumptions and for which condition and conclusion in (a) do not hold. (Cf. [25, Appendix B].)

2.4 A special case: Linear Perron–Frobenius theory

Let $A \in \mathbb{R}^{n\times n}_+$ or $A \geq 0$ denote a real $n\times n$-matrix which is *non-negative*, i.e. all entries a_{ij} of A are non-negative. The mapping $T\colon \mathbb{R}^n_+ \longrightarrow \mathbb{R}^n_+$ defined by $Tx = Ax$, where x is understood to be a column vector, is concave, positively homogeneous and continuous. Because of $T_i e_j = a_{ij}$ the operator T is indecomposable in the sense of Definition 2.2.6 iff for any index set $\emptyset \subsetneq I \subsetneq \{1,\ldots,n\}$ there exist indices $i \in I$ and $j \notin I$ such that $a_{ij} > 0$. A non-negative matrix enjoying the latter property is called *indecomposable* or irreducible. Also, T is weakly indecomposable in the sense of Definition 2.2.6 if for any two indices $1 \leq i,j \leq n$ there exists some $p = p(i,j) \in \mathbb{N}$ such that the (i,j)-entry of the matrix power A^p is strictly positive. As already remarked (see also Exercises below), this property is equivalent to the indecomposability of A. Furthermore, T is primitive in the sense of Definition 2.2.9 if there exist some $p \in \mathbb{N}$ such that $A^m > 0$ for all $m \geq p$ or, equivalently, $A^p > 0$. Such a matrix is called *primitive*. Here $B > 0$ for a matrix B means that all entries of B are strictly positive.

By specializing results of the previous section we arrive at the following statements of classical Perron–Frobenius Theory involving the **dominant eigenvalue λ^*** *of a non-negative matrix.*

Theorem 2.4.1 (Classical Perron–Frobenius Theorem). *Let A be a non-negative $n \times n$-matrix.*

(i) *A has a maximal non-negative eigenvalue λ^* and there exists $x^* \gneqq 0$ such that $Ax^* = \lambda^* x^*$.*

(ii) *For A indecomposable the following statements hold.*
   (a) *For λ^* as in (i) one has $\lambda^* > 0$ and there exists $x^* > 0$ such that $Ax^* = \lambda^* x^*$. Furthermore, if $Ax = \lambda x$ for some $x \gneqq 0$ then $\lambda = \lambda^*$ and $x = rx^*$ with $r > 0$.*
   (b) *If $\lambda x \leq Ax$ for some $x \gneqq 0$ then $\lambda \leq \lambda^*$.*
   If $\lambda^* x \leq Ax$ for some $x \gneqq 0$ then $x = rx^*$ with $r > 0$.
   (c) *If λ is a real eigenvalue of A then $|\lambda| \leq \lambda^*$.*

(iii) *For A primitive the following statements hold.*
   (a) *If λ is a real eigenvalue of A different from λ^* then $|\lambda| < \lambda^*$.*

(b) *The dominant eigenvalue λ^* and its eigenvector x^* with $\|x^*\| = 1$ can be obtained as limits of the iterates by*

$$\lim_{k\to\infty} \frac{A^k x}{\|A^k x\|} = x^* \text{ and } \lim_{k\to\infty} \frac{\|A^{k+1}x\|}{\|A^k x\|} = \lim_{k\to\infty} \|A^k x\|^{\frac{1}{k}} = \lambda^*$$

for arbitrary $x \gneqq 0$ and any monotone norm of $\mathbb{R}^n$.

(c) *Let B be the matrix with entries $b_{ij} = x_i^* \bar{x}_j$ where x^* and $\bar{x}$ are positive vectors such that $Ax^* = \lambda^* x^*$, $A'\bar{x} = \lambda^* \bar{x}$ and $\sum_{i=1}^n x_i^* \bar{x}_i = 1$ (A' being the transposed matrix of A). Then there holds*

$$\lim_{k\to\infty} \frac{A^k}{(\lambda^*)^k} = B.$$

Proof. (i) follows from Theorem 2.1.14. Statements (a) and (b) of (ii) follow from part (ii) of Theorem 2.3.8; statement (c) of (ii) follows from part (iii) of Theorem 2.3.8. The latter also yields statement (a) of part (iii) because $A^p > 0$ for some p by assumption and, hence, $A^p x < A^p|x|$ for all $x \in \mathbb{R}^n$ with $x \neq |x|$. Statement (b) of part (iii) follows from Theorem 2.3.1 part (ii). Part (i) of Theorem 2.3.1 yields $\lim_{k\to\infty} \frac{A^k}{\lambda^{*k}} = B$ with a non-negative matrix B such that $BA = AB = \lambda^* B$ and $BB = B$. From $AB = \lambda^* B$ it follows that the j-th column of B must be $r_j x^*$ for some $r_j > 0$. For the column vector $\bar{x}$ with components r_j equation $BA = \lambda^* B$ implies that $A'\bar{x} = \lambda^* \bar{x}$ and $BB = B$ implies that $\sum_{i=1}^n x_i^* \bar{x}_i = 1$. This proves statement (c) in part (iii) of Theorem 2.4.1. □

In addition to the statements given by Theorem 2.4.1 there are many more results considering eigenvalue problems for non-negative matrices. In particular, interesting results are available concerning complex eigenvalues and the decomposability structure of the matrix (cf. [2, 19, 41, 66]). Since these issues, however, are less closely related to our concave framework we will not consider it here. Instead we like to point out some features of Linear Perron–Frobenius Theory connected with Theorem 2.4.1 which seem to have received less attention in the linterature.

Usually the assumption of indecomposability for a non-negative matrix is made to guarantee that $Ax = \lambda x$ has a unique (up to a scalar) strictly positive solution $x^* > 0$, $\lambda^* > 0$. To get this property, however, indecomposability is by no means necessary (see Exercise 2 below) . Actually, many properties derived usually for indecomposable matrices can be otained under weaker conditions (Exercise 1). By employing the *Gantmacher* or *Frobenius normal form* of a non-negative matrix (cf. [1, 19]) it can be shown that the eigenvalue problem has a unique strictly positive solution iff the matrix is what is called a *Sraffa matrix* (cf. [36]).

Furthermore, primitivity is commonly assumed to obtain the results on limits of the iterates as in part (iii) of Theorem 2.4.1. Again, for these results primitivity is a sufficient but not a necessary condition (see Exercises 1 and 2). In case A is assumed to be indecomposable, however, it can be easily shown that $\lim_{k\to\infty} \frac{A^k}{\lambda^{*k}}$ exists iff A is primitive.

Linear Perron–Frobenius Theory is commonly thought to be a theory about matrices which are non-negative. There are, however, many matrices having some entries negative to which many statements of this theory also apply. For example, to the matrix $A = \left[\begin{smallmatrix} 2 & 2 \\ 2 & -1 \end{smallmatrix}\right]$ all the statements of Theorem 2.4.1 do apply (Exercise 3; see also [14]). That this matrix has a negative entry means that the linear mapping defined by it does not map the convex cone $\mathbb{R}^2_+$ into itself. There are, however, other convex cones as, e.g., $K = \{(x_1, x_2) \in \mathbb{R}^2_+ \mid x_2 \le 2x_1\}$, which this mapping leaves invariant. By developing a linear Perron–Frobenius Theory for mappings leaving other cones than $\mathbb{R}^n_+$ invariant the realm of applicability of this theory can be considerably extended (cf. [1, 14, 59]). This point of view we will adopt in the chapters which follow where non-linear mappings will be considered which leave invariant some convex cone within a real Banach space.

Concerning our method of proof, the main idea was to show for the operator in question that it is contractive or a contraction with respect to Hilbert's projective metric on the convex cone $\mathbb{R}^n_+$. This idea we shall pursue also in the following chapters with respect to certain non-linear operators on a convex cone within a Banach space. It seems remarkable that this goal cannot be achieved by employing instead of Hilbert's projective metric a metric induced by a norm on the vector space under consideration. This is already the case in very simple situations. Consider, e.g., the linear operator $T\colon \mathbb{R}^2_+ \longrightarrow \mathbb{R}^2_+$ defined by the matrix $A = \left[\begin{smallmatrix} 1 & 1 \\ 1 & 6 \end{smallmatrix}\right]$ (cf. [16]). As it is clear from Remark 2.1.12 the operator given by $\tilde{T}x = \frac{Ax}{\|Ax\|}$ for $x \gneqq 0$ and $\|\cdot\|$ being the sum norm is a contraction for Hilbert's metric d, i.e., $d(\tilde{T}x, \tilde{T}y) \le cd(x,y)$, with a contraction factor $c = 1 - \min\{\frac{T_j e^i}{T_j e} \mid \le i, j \le n\} = 1 - \min\{\frac{1}{2}, \frac{1}{7}, \frac{6}{7}\} = \frac{6}{7} < 1$. Considering, however, any (semi-) norm $|||\cdot|||$ on $\mathbb{R}^2$ and the metric m induced by it one finds for the points $x = (1,0)$ and $y = (0.99, 0.01)$ in $\mathbb{R}^2_+$ that

$$m(\tilde{T}x, \tilde{T}y) = |||\tilde{T}x - \tilde{T}y||| = \frac{5}{4.01}|||(0.01, -0.01)||| = \frac{5}{4.01} m(x,y) > m(x,y).$$

Thus, there is no norm induced metric on $\mathbb{R}^2_+$ for which the operator is $\tilde{T}$ is a contraction.

In concluding this section we shall shortly discuss what seems to be the first extension of the Perron–Frobenius theory to non-linear mappings (cf. Remark 2.1.17). Concerning balanced growth of an economy the economists P. A. Samuelson and M. Solow prove in [62] relative stability for a selfmapping T of $K = \mathbb{R}^n_+$ which is continuous, positively homogeneous and strictly increasing, i.e., $x \lneqq y$ implies $Tx < Ty$. Relative stability means that $\lim\limits_{t\to\infty} \frac{x_i(t)}{u_i(t)} = c(x(0))$ for all $1 \le i \le n$ where $u(t) = (\lambda^*)^t x^*$ is the balanced growth path and $x(t) = T^t x(0)$ is any actual growth path. This nice result has been generalized by a "Japanese School" of economists. M. Morishima [43, Theorem 10, p. 206]and H. Nikaido [47, Theorem 10.7, p. 160] show that relative stability holds for T continuous, positively homogeneous, monotone, MN-indecomposable and primitive in 0 and x^*. If, moreover, T is globally primitive then $x(0) \lneqq x'(0)$ implies $c(x(0)) < c(x'(0))$ ([43, Theorem 11, p. 207]). (See also Section 2.2 and Exercise 8 there.) Further interesting contributions in this direction are by T. Fujimoto and Y. Oshime [13, 44, 50]. Relative stability in the sense of $\lim_{t\to\infty} \frac{x(t)}{\|x(t)\|} = x^*$ for all $x(0) \in K \setminus \{0\}$ is shown in [14] (see also [28, 29, 32]) for T strictly increasing and satisfying some weak from of homogeneity and in [51] for T continuous, positively homogeneous, monotone, primitive and (power) real analytic. All these results do not employ concavity but continuity (to apply Brouwer's fixed point theorem) and homogeneity which excludes affin-linear maps (which are covered by the first concave Perron–Frobenius theorem). The methods used to develop concave Perron–Frobenius theory we will extend later on in Chapter 5 to ascending selfmappings in infinite dimensions which need not be concave and not even monotone.

Exercises

1. Examine the non-negative matrix $A = \left[\begin{smallmatrix} a & 0 \\ a & 1 \end{smallmatrix}\right]$, where $0 \le a < 1$, with respect to the statements of Theorem 2.4.1
   (a) Show that all three statements of part (ii) hold for the decomposable matrix A with the exception of $x^* > 0$.
   (b) Show that all three statements of part (iii) hold for A, though A is not primitive. (Take $\|\cdot\|$ to be the sum norm.)
   (c) Show that the linear mapping defined by A is not a contraction for Hilbert's metric.
   (d) Find a non-negative matrix P such that PAP^{-1} is a strictly positive matrix.
2. Let $A = \left[\begin{smallmatrix} a & 0 \\ b & c \end{smallmatrix}\right]$ be such that $a, c \ge 0$ and $b > 0$.
   (a) Show that $Ax = \lambda x$ possesses a unique positive solution $x^* > 0$, $\lambda^* > 0$ (up to a scalar for x^*) if and only if $c < a$.
   (b) Examine A with respect to the statements of Theorem 2.4.1 (Take $\|\cdot\|$ to be the sum norm.)
3. Consider the matrix $A = \left[\begin{smallmatrix} 2 & 2 \\ 2 & -1 \end{smallmatrix}\right]$ which is not non-negative.
   (a) Show that all three statements of part (ii) of Theorem 2.4.1 hold true for A.
   (b) Verify that $A^n = 3^n B + (-2)^n C$ with $B = \frac{1}{5}\left[\begin{smallmatrix} 4 & 2 \\ 2 & 1 \end{smallmatrix}\right]$ and $C = \frac{1}{5}\left[\begin{smallmatrix} 1 & -2 \\ -2 & 4 \end{smallmatrix}\right]$.
   (c) Use (b) to show that all three statements of part (iii) of Theorem 2.4.1 hold true for A. (Take $\|\cdot\|$ to be the sum norm.)
4. Show that for any $A \in \mathbb{R}_+^{n\times n}$ the dominant eigenvalue λ^* satisfies the inequalities
$$\min_{1\le i\le n} \sum_{j=1}^{n} a_{ij} \le \lambda^* \le \max_{1\le i\le n} \sum_{j=1}^{n} a_{ij}$$
   and that these inequalities are strict if A is indecomposable with row sums not all equal.
5. Show that a matrix $A \in \mathbb{R}^{n\times x}$ is indecomposable if and only if it is weakly indecomposable.

6. Work out a direct proof for part (iii) of Theorem 2.4.1 by employing Hilbert's projective metric similar to the proof of Theorem 2.1.11.

2.5 Applications to difference equations of concave type

Consider the difference equation

$$u(t+n) = f(u(t), u(t+1), \dots, u(t+n-1)) \tag{2.5.1}$$

of order $n \geq 1$ where $t \in \mathbb{N}$, $u(t) \in \mathbb{R}_+$ and $f\colon \mathbb{R}_+^n \longrightarrow \mathbb{R}_+$.

The function f defines in a canonical way a *particular discrete dynamical system* $T\colon \mathbb{R}_+^n \longrightarrow \mathbb{R}_+^n$ by $Tx = (x_2, \dots, x_n, f(x))$ for $x = (x_1, \dots, x_n)$. To apply results of the previous sections to this system we need to find conditions on f which guarantee that T is indecomposable and primitive, respectively.

Lemma 2.5.1. *For an arbitrary function $f\colon \mathbb{R}_+^n \longrightarrow \mathbb{R}_+, n \geq 2$, the discrete dynamical system T defined by f is indecomposable iff $f(e_1) > 0$, where e_i denotes the i-th unit vector.*

Proof. Suppose first that T is indecomposable and let $I = \{2, \dots, n\}$. By Definition 2.2.6 there exists some $i \in I$ such that $T_i e_1 > 0$, T_i being the i-th component function of T. Since $T_i e_1 = 0$ for $1 \leq i \leq n-1$ we must have that $f(e_1) = T_n e_1 > 0$. Conversely, suppose that $f(e_1) > 0$ and consider $\emptyset \subsetneq I \subsetneq \{1, \dots, n\}$. Let k be the smallest number in I. For $k = n$ we have $T_k e_1 = f(e_1) > 0$ where $k \in I$ and $1 \notin I$. If $k < n$ and there exists some $i \in I$ with $k \leq i < n$ such that $i + 1 \notin I$ then $T_i e_j = 1 > 0$ where $i \in I$ and $j = i + 1 \notin I$. As the remaining case we have to consider $k < n$ where for any $i \in I$ with $k \leq i < n$ it holds that $i + 1 \in I$. In other words, $I = \{k, k+1, k+2, \dots, n\}$ where $k \geq 2$. In this case $T_n e_1 = f(e_1) > 0$ where $n \in I$, $1 \notin I$. □

From Theorem 2.3.8 we obtain the following consequence for difference equation (2.5.1) with **characteristic equation**

$$\lambda^n = f(1, \lambda, \lambda^2, \dots, \lambda^{n-1}) \quad \text{for} \quad \lambda \in \mathbb{R}_+.$$

Theorem 2.5.2. *Assume that $f\colon \mathbb{R}_+^n \longrightarrow \mathbb{R}_+$ is concave and positively homogeneous with $f(e_1) > 0$.*

(i) *The characteristic equation has a unique strictly positive solution λ^*.*
(ii) *For any given $u(0) > 0$ the function $u(t) = \lambda^t u(0)$, $t \in \mathbb{N}$, is a solution of the difference equation (2.5.1) iff $\lambda = \lambda^*$.*

Proof. Since statements (i) and (ii) hold trivially for $n = 1$ we assume $n \geq 2$. (i) The operator $T\colon \mathbb{R}_+^n \longrightarrow \mathbb{R}_+^n$ defined by $Tx = (x_2, \dots, x_n, f(x))$ is concave, positively homogeneous and indecomposable by Lemma 2.5.1. By Theorem 2.3.8 there exist $x^* > 0$ and $\lambda^* > 0$ such that $Tx = \lambda x$ holds for $x \gneq 0$ and $\lambda \geq 0$ iff $\lambda = \lambda^*$ and $x = rx^*$ with $r > 0$.

Equation $Tx = \lambda x$ is equivalent to $x_2 = \lambda x_1, x_3 = \lambda x_2, \dots, x_n = \lambda x_{n-1}$ and $f(x) = \lambda x_n$. Positive homogeneity of f implies that $\lambda = \lambda^* > 0$ is a solution of the characteristic equation $f(1, \lambda, \lambda^2, \dots, \lambda^{n-1}) = \lambda^n$. Conversely, for any solution $\lambda \geq 0$ of this equation by setting $x = (1, \lambda, \lambda^2, \dots, \lambda^{n-1})$ we have that $Tx = \lambda x$ and, therefore, $\lambda = \lambda^*$.

(ii) By positive homogeneity of f the function $u(t) = \lambda^t u(0)$ is a solution of (2.5.1) iff $f(1, \lambda, \lambda^2, \dots, \lambda^{n-1}) = \lambda^n$ which by (i) means that $\lambda = \lambda^*$. □

Example 2.5.3. Consider the difference equation of second order given by

$$u(t + 2) = u(t) + \sqrt{u(t)u(t + 1)}.$$

The function $f\colon \mathbb{R}_+^2 \longrightarrow \mathbb{R}_+$, $f(x) = x_1 + \sqrt{x_1 x_2}$ is concave, positively homogeneous with $f(e_1) > 0$. By Theorem 2.5.2 the characteristic equation $1 + \sqrt{\lambda} = \lambda^2$ has a unique solution $\lambda^* > 0$. Furthermore, for given initial conditions $u(0), u(1) \geq 0$ the resulting solution is of type $u(t) = \lambda^t u(0)$ with $\lambda \geq 0$ if it is either constant zero or if $u(1) = \lambda^* u(0)$. Thus, non-trivial solutions with constant growth rate must have initial conditions $u(0) > 0$, $u(1) = \lambda^* u(0)$.

To handle the primitivity of T we employ the following **comparison principle** for solutions of the difference equation.

Lemma 2.5.4. *Let $f\colon \mathbb{R}_+^n \longrightarrow \mathbb{R}_+$ be monotone, i.e. $0 \leq x \leq y$ implies $f(x) \leq f(y)$, and suppose there exist $n_1, \dots, n_r \in \{1, \dots, n\}$ with $r \geq 2$, $n_1 = 1$ and $\gcd\{n - n_1 + 1, \dots, n - n_r + 1\} = 1$ such that the following strict monotonicity property holds:*

$$0 \leq x \leq y \text{ and } x_{n_i} < y_{n_i} \text{ for some } 1 \leq i \leq r \text{ implies that } f(x) < f(y) \tag{2.5.2}$$

If $u(\cdot)$ and $v(\cdot)$ are two solutions of the difference equation (2.5.1) satisfying $(u(0), \dots, u(n-1)) \lneqq (v(0), \dots, v(n-1))$ then $u(t) \leq v(t)$ for all $t \in \mathbb{N}$ and there exists some $t_0 \in \mathbb{N}$, independently of u, v, such that $u(t) < v(t)$ for all $t \geq t_0$.

Proof. Obviously, by induction equation (2.5.1) yields $u(t) \leq v(t)$ for all $t \in \mathbb{N}$.

(i) that $u(k) < v(k)$ for some $k \geq n_i - 1$ implies that $u(k + qm_i) < v(k + qm_i)$ for all $q \in \mathbb{N}$. For $k = n_i - 1 + l$, $l \in \mathbb{N}$, we have that $u(k + m_i) = u(l + n) = f(u(l), \dots, u(l + n - 1))$. For $u(l + n_i - 1) < v(l + n_i - 1)$ the strong monotonicity property (2.5.2) yields

$$f(u(l), \dots, u(l + n - 1)) < f(v(l), \dots, v(l + n - 1)),$$

that is, $u(k + m_i) < v(k + m_i)$. By iterating this argument we arrive at

$$u(k + qm_i) < v(k + qm_i) \quad \text{for all} \quad q \in \mathbb{N}.$$

(ii) By assumption there exists some $0 \leq j \leq n - 1$ with $u(j) < v(j)$. Since $j \geq 0 = n_1 - 1$ step (i) yields

$$u(j + q_1 m_1) < v(j + q_1 m_1) \quad \text{for all} \quad q_1 \in \mathbb{N}.$$

For $q_1 \geq 1$ we have $k = j + q_1 m_1 \geq m_1 = n \geq n_2 - 1$ and applying step (i) again yields

$$u(j + q_1 m_1 + q_2 m_2) < v(j + q_1 m_1 + q_2 m_2) \quad \text{for all} \quad q_1 \geq 1,\ q_2 \in \mathbb{N}.$$

By iterating step (i) in this way we obtain

$$u(j + q_1 m_1 + \cdots + q_r m_r) < v(j + q_1 m_1 + \cdots + q_r m_r) \quad \text{for all} \quad q_1 \geq 1,\ q_i \in \mathbb{N}. \ (*)$$

Now, by assumption $gcd\{m_1, \ldots, m_r\} = 1$ and, therefore, $1 = \sum_{i=1}^{r} l_i m_i$ with $l_i \in \mathbb{Z}$. Define $d = \sum_{i=1}^{r} |l_i| m_i$ and $t_0 = j + n + d^2$. For $t \geq t_0$ we have that $d^2 \leq t - j - n = qd + s$ with $q \in \mathbb{N}$, $0 \leq s < d$ and, hence, $d < q + 1$. It follows that

$$t - j - n = qd + s \cdot 1 = \sum_{i=1}^{r} (q|l_i| + s l_i) m_i$$

where $q|l_i| + s l_i \in \mathbb{N}$ because of $q > d - 1 \geq s$. Thus, $t = j + q_1 m_1 + \cdots + q_r m_r$ with $q_1 \geq 1$, $q_i \in \mathbb{N}$ and from $(*)$ we obtain $u(t) < v(t)$ for $t \geq t_0$. Since d does not depend on u and v, t_0 can be chosen independent of u, v as $n - 1 + n + d^2$. □

Remark 2.5.5. For the comparison principle to hold it is not sufficient to require f to be monotone and strictly increasing in just one component as can be seen from Example 2.5.3 (see also Exercise 1).

Using Lemma 2.5.4 from Theorem 2.3.1 the following result for difference equations follows.

Theorem 2.5.6. *Let $f\colon \mathbb{R}_+^n \longrightarrow \mathbb{R}_+$ be concave, positively homogeneous and suppose there exist $n_1, \ldots, n_r \in \{1, \ldots, n\}$ with $r \geq 2$, $n_1 = 1$ and $gcd\{n - n_1 + 1, \ldots, n - n_r + 1\} = 1$ such that $0 \leq x$ and $0 < x_{n_i}$ for some $1 \leq i \leq r$ implies $0 < f(x)$. Then the characteristic equation of difference equation (2.5.1) has a unique strictly positive root λ^* and the following statements hold.*

(i) *Every solution $u(\cdot)$ of equation (2.5.1) with initial conditions $\bar{u} = (u(0), \ldots, u(n-1))$ is relatively stable, i.e.,*

$$\lim_{t\to\infty} \frac{u(t)}{\lambda^{*t}} = s(\bar{u})$$

where the function $s\colon \mathbb{R}_+^n \longrightarrow \mathbb{R}_+$ is concave, positively homogeneous and satisfies $s(x) > 0$ for $x \gneqq 0$, $s(x_2, \ldots, x_n, f(x)) = \lambda^* s(x)$ for $x \in \mathbb{R}_+^n$ and $s(1, \lambda^*, \ldots, \lambda^{*(n-1)}) = 1$.

(ii) *Every solution $u(\cdot)$ of equation (2.5.1) which is not constant zero grows asymptotically with the same factor λ^*, i.e.,*

$$\lim_{t\to\infty} \frac{u(t+1)}{u(t)} = \lim_{t\to\infty} u(t)^{\frac{1}{t}} = \lambda^*.$$

Proof. By assumption in particular $f(e_1) > 0$ and, hence, by Theorem 2.5.2 the characteristic equation has a unique strictly positive root λ^*. Furthermore, the operator $T\colon \mathbb{R}_+^n \longrightarrow \mathbb{R}_+^n$ defined by $Tx = (x_2, \ldots, x_n, f(x))$ is concave and positively homogeneous. Concavity of f together with positive homogeneity imply for $0 \leq x \leq y$ that

$f(y) = f(\frac{1}{2}(2x) + \frac{1}{2}(2(x-y))) \geq f(x) + f(y-x)$. Thus, the positivity assumption made for f implies the strong monotonicity property (2.5.2) for f. For $x = (u(0), \ldots, u(n-1)) \geq 0$ we have by induction that $T^t x = (u(t), \ldots, u(t+n-1))$ for all $t \in \mathbb{N}$ where $u(\cdot)$ is a solution of (2.5.1). For $y = (v(0), \ldots, v(n-1))$ and $x \lneqq y$ we, therefore, obtain by Lemma 2.5.4 that there exists some $t_0 \in \mathbb{N}$ such that $T^t x < T^t y$ for all $t \geq t_0$. In particular, T is primitive and we may apply Theorem 2.3.1. For any solution $u(\cdot)$ of equation (2.5.1) with initial conditions we obtain

$$\begin{aligned} S\bar{u} &= \lim_{t\to\infty} \left(\frac{u(t)}{\lambda^{*t}}, \frac{u(t+1)}{\lambda^{*(t+1)}}\lambda^*, \ldots, \frac{u(t+n-1)}{\lambda^{*(t+n-1)}}\lambda^{*(n-1)} \right) \\ &= s(\bar{u})(1, \lambda^*, \ldots, \lambda^{*(n-1)}) \\ \text{with} \quad s(\bar{u}) &= \lim_{t\to\infty} \frac{u(t)}{\lambda^{*t}}. \end{aligned}$$

This proves part (i) since the properties stated for S in Theorem 2.3.1 imply those for $s\colon \mathbb{R}^n_+ \longrightarrow \mathbb{R}_+$. Finally, part (ii) follows immediately from $\lim_{t\to\infty} \frac{u(t)}{\lambda^{*t}} = s(\bar{u})$. □

Examples 2.5.7. (i) Consider the difference equation (2.5.1) where f is given as the minimum of finitely many linear functions by

$$f(x) = \min_{1\leq i\leq m} (a_{i1}x_1 + \cdots + a_{in}x_n),$$

where the $m \times n$-matrix $A = (a_{ij})$ is non-negative with a set J of at least two strictly positive columns including the first one and $gcd\{n-j+1 \mid j \in J\} = 1$. The assumptions of Theorem 2.5.6 being satisfied it follows for any solution $u(\cdot)$ with initial conditions $\bar{u}$ that $\lim_{t\to\infty} \frac{u(t)}{\lambda^{*t}} = s(\bar{u})$ where $\lambda^* > 0$ is the unique root of the characteristic equation $\min_{1\leq i\leq m} (a_{i1} + a_{i2}\lambda + \cdots + a_{in}\lambda^{n-1}) = \lambda^n$.

(ii) A special case of Example (i) is given by linear difference equations as, e.g., the Fibonacci difference equation

$$u(t+n) = \sum_{i\in I} u(t+i)$$

with a subset I of $\{0, \ldots, n-1\}$ containing 0 and such that the set of numbers $n-i$ for $i \in I$ is relatively prime. λ^* is given as the unique positive root of $\sum_{i\in I} \lambda^i = \lambda^n$. The common Fibonacci difference equation $u(t+2) = u(t) + u(t+1)$ represents the special case where $n = 2$, $I = \{0, 1\}$ and λ^* is the unique positive root of $1 + \lambda = \lambda^2$. In this particular case, the asymptotic statements made by Theorem 2.5.6 can be obtained also directly from Binet's formula (see Chapter 1.2).

Consider the difference equation given by the sum of arithmetic and geometric mean as follows

$$u(t+n) = \frac{1}{n}\sum_{i=1}^{n} u(t+i-1) + \left(\prod_{i=1}^{n} u(t+i-1)\right)^{\frac{1}{n}}.$$

The corresponding function f is concave and positively homogeneous with $f(x) > 0$ for $x \gneqq 0$. For the unique positive root λ^* of the characteristic equation $\frac{1}{n}\sum_{i=1}^{n} \lambda^{i-1} +$

$\lambda^{\frac{n-1}{2}} = \lambda^n$ one has $\lambda^* > 1$ which implies in particular that all solutions different from the zero-solution are unbounded.

(iii) The particular positivity assumption made in Theorem 2.5.6 for f is crucial as can be seen from Example 2.5.3 or from the simple difference equation $u(t+4) = \frac{1}{2}(u(t) + u(t+2))$. In the latter case $f(x_1, x_2, x_3, x_4) = \frac{1}{2}(x_1 + x_3)$ and $gcd\{4-1+1, 4-3+1\} = 2 \neq 1$. The positive root of the characteristic equation $\frac{1}{2}(1+\lambda^2) = \lambda^4$ is $\lambda^* = 1$. The solution for the initial conditions $\bar{u} = (1, 2, 1, 2)$ is given by

$$u(t) = \begin{cases} 1, & t \text{ even} \\ 2, & t \text{ odd} \end{cases}$$

and, hence, $\frac{u(t)}{\lambda^{*t}}$ does not converge for $t \to \infty$. Also, $\frac{u(t+1)}{u(t)}$ does not converge, whereas $\lim_{t\to\infty} u(t)^{\frac{1}{t}} = 1 = \lambda^*$.

Exercises

1. Show that the conclusion of the comparison principle (Lemma 2.5.4) does not hold for $u(t+2) = u(t) + \sqrt{u(t)u(t+1)}$, $t \in \mathbb{N}$, $u(t) \in \mathbb{R}_+$.
2. Compute for the difference equation
$$u(t+2) = \frac{u(t) + u(t+1)}{2} + \frac{27}{4}\sqrt{u(t)u(t+1)}, \; t \in \mathbb{N}, \; u(t) \in \mathbb{R}_+,$$
the unique positive root of the characteristic equation, and show that any solution $u(\cdot)$ is either identically zero or unbounded.
3. Compute for the difference equation
$$u(t+2) = \min\left\{\frac{1}{8}u(t) + \frac{1}{4}u(t+1), \frac{1}{10}u(t) + \frac{2}{5}u(t+1)\right\}, \; t \in \mathbb{N}, \; u(t) \in \mathbb{R}_+,$$
the unique positive root of the characteristic equation, and show that all solutions $u(\cdot)$ converge to 0.
4. Determine for the difference equation
$$u(t+2) = \frac{u(t) + u(t+1)}{3} + a\min\{u(t), u(t+1)\}, \quad t \in \mathbb{N}, \; u(t) \in \mathbb{R}_+,$$
the values of the parameter $a \geq 0$ for which all solutions tend to 0 and all solutions (except the zero-solution) are unbounded, respectively.
5. Check for the difference equation of Exercise 1 if any of the following statements holds true for all positive solutions (λ^* being the unique positive root of the characteristic equation):
   - $\lim\limits_{t\to\infty} \frac{u(t)}{\lambda^{*t}}$ exists;

- $\lim_{t\to\infty} \frac{u(t+1)}{u(t)} = \lambda^*$;
- $\lim_{t\to\infty} u(t)^{\frac{1}{t}} = \lambda^*$.

6. Compute for the Fibonacci difference equation $u(t+2) = u(t) + u(t+1)$, $t \in \mathbb{N}$, $u(t) \in \mathbb{R}_+$ the function $s\colon \mathbb{R}_+^2 \longrightarrow \mathbb{R}_+$ of Theorem 2.5.6.

2.6 Relative stability in the concave Leslie model

As an application of concave Perron–Frobenius Theory we consider a concave version of the density dependent Leslie model discussed in Section 1.2. According to equation (1.2.7) the model is in the autonomous case given by

$$x(t+1) = T(x(t)) \quad \text{for} \quad t \in \mathbb{N}, \tag{2.6.1}$$

where for $x \in \mathbb{R}_+^n$,

$$Tx = L(x)x \text{ and } L(x) = \begin{bmatrix} b_1(x) & b_2(x) & \dots & b_{n-1}(x) & b_n(x) \\ s_1(x) & 0 & \dots & 0 & 0 \\ 0 & s_2(x) & & 0 & 0 \\ \vdots & \vdots & \ddots & \vdots & \vdots \\ 0 & 0 & \dots 0 & s_{n-1}(x) & s_n(x) \end{bmatrix}$$

For the **concave Leslie model** we make the following assumptions:

(a) The mappings $x \longmapsto b_i(x)x_i$ and $x \longmapsto s_i(x)x_i$ of $\mathbb{R}_+^n$ into $\mathbb{R}_+$ are concave for all $1 \le i \le n$.

(b) There exists $k_1, \dots, k_r \in \{1, \dots, n\}$, $r \ge 2$, $k_r = n$ with $gcd\{k_1, \dots, k_r\} = 1$ such that for all $1 \le i \le r$
$b_{k_i}(x) > 0$ for $x \in \mathbb{R}_+^n$ with $x_{k_i} > 0$.
Furthermore, for all $1 \le i \le n-1$ suppose $s_i(x) > 0$ for $x \in \mathbb{R}_+^n$ with $x_i > 0$.

(c) For any $x \in \mathbb{R}_+^n$ and any $\lambda > 0$ there exists a number $c(x, \lambda)$ such that $b_i(\lambda x) = c(x, \lambda) b_i(x)$ and $s_i(\lambda x) = c(x, \lambda) s_i(x)$ for all $1 \le i \le n$ with $x_i > 0$.

Assumption (a) means that for each age class the number of newborn and surviving individuals, respectively, grows with a non-increasing rate due to population pressure. According to assumption (b), for non-empty age classes all survival rates are positive (with the possible exception of the last class) and birth rates are positive for some specified selection of classes. Assumption (c) requires that a uniform population pressure where each age class grows with the same factor does not affect the ratio of birth and survival rates. It is clear that this concave Leslie model contains as a special case the classical (linear) Leslie model.

From assumption (b) we will conclude later on that the mapping T must be primitive. Without primitivity results like Theorem 2.6.1 below cannot be expected, not even in the linear case as can be seen from the following example discussed already in [3]. Let $Tx = Lx$ be given by the Leslie matrix

$$L = \begin{bmatrix} 0 & 0 & 6 \\ \frac{1}{2} & 0 & 0 \\ 0 & \frac{1}{3} & 0 \end{bmatrix}.$$

Obviously, assumptions (a) and (c) are satisfied but not assumption (b). One has that $Tx^* = \lambda^* x^*$ with $x^* \gneqq 0$, $\|x^*\| = 1$, $\lambda^* \geq 0$ has a unique solution, namely $x^* = (\frac{6}{10}, \frac{3}{10}, \frac{1}{10})$ and $\lambda^* = 1$, but age structures do not approach x^* as in Theorem 2.6.1. Indeed, L^3 is the identity matrix, therefore L is not primitive, and there are population waves as already observed by H. Bernardelli, that is every population path repeats itself after three periods.

A solution $t \longmapsto x(t)$ of equation (2.6.1) is called a *population path* and it is *normalized* if $\sum_{i=1}^n x_i(t) = 1$ for all $t \in \mathbb{N}$. Any (non-zero) population path can be normalized by $\frac{x(t)}{\|x(t)\|}$ which is called the *age structure* of the population path (where $\|x\| = \sum_{i=1}^n |x_i|$ for $x \in \mathbb{R}^n$).

A population path has the *uniform growth rate g* if $\frac{x_i(t+1)-x_i(t)}{x_i(t)} = g$ for all $1 \leq i \leq n$, all $t \in \mathbb{N}$ with $x_i(t) > 0$.

The following Theorem collects our main results for the concave Leslie model. The properties 2.6.2 and 2.6.3 are sometimes referred to as *relative stability*.

Theorem 2.6.1. (i) *There exists precisely one stationary age structure $x^* > 0$ and for every population path with $x(0) \gneqq 0$ the age structure converges to x^*, that is*

$$\lim_{t\to\infty} \frac{x(t)}{\|x(t)\|} = x^*. \tag{2.6.2}$$

(ii) *Suppose for the concave Leslie model that the vital rates do not increase with the population level in the sense that for some $0 \leq d \leq 1$ it holds for each $1 \leq i \leq n$ that $b_i(\lambda x) = \frac{1}{\lambda^{1-d}} b_i(x)$ and $s_i(\lambda x) = \frac{1}{\lambda^{1-d}} s_i(x)$ for all $\lambda > 0$, all $x \geq 0$ with $x_i > 0$.*

Then there exists a uniquely (up to a positive factor) determined population path $\hat{x}(t) = \hat{x}(0)(1+g)^t$ with uniform growth rate g and each population path x with $x(0) \gneqq 0$ grows finally uniformly with g, more precisely

$$\lim_{t\to\infty} \frac{x_i(t)}{\hat{x}_i(t)} = c(x(0)) > 0 \quad \textit{for all} \quad 1 \leq i \leq n. \tag{2.6.3}$$

In particular,

$$\lim_{t\to\infty} \frac{\|x(t+1)\|}{\|x(t)\|} = \lim_{t\to\infty} \|x(t)\|^{\frac{1}{t}} = 1 + g. \tag{2.6.4}$$

For $d = 1$ the path $\hat{x}$ is determined by $g = g^*$ and $\hat{x}(0) = cx^*$ for some $c > 0$ where x^* and g^* are uniquely determined by the eigenvalue problem $Tx^* = (1+g^*)x^*$, $x^* \geq 0$, $\|x^*\| = 1$.

For $0 \leq d < 1$ the path $\hat{x}$ is determined by $g = 0$, $\hat{x}(0)$ the unique non-zero fixed point of T and in (2.6.3) holds $c(x(0)) = 1$ for all $x(0) \gneqq 0$.

Proof. From assumption (a) for the concave Leslie model it follows that the mapping $T\colon \mathbb{R}^n_+ \longrightarrow \mathbb{R}^n_+$, $Tx = L(x)x$ is concave. Assumption (c) implies $T(\lambda x) = \lambda c(x,\lambda)Tx$ for all $x \geq 0$, $\lambda > 0$ and, hence, T is ray preserving. By assumption (b) T is primitive as will be shown towards the end of this proof.

(i) A stationary age structure is defined by $\frac{x(t)}{\|x(t)\|} = x^*$ for all $t \in \mathbb{N}$ for some $x^* \gneqq 0$ with $\|x^*\| = 1$. (Assuming without loss that $x(t) \neq 0$ for all $t \in \mathbb{N}$.) Since T is ray preserving it follows

$$\|x(1)\|x^* = x(1) = Tx(0) = T(\|x(0)\|x^*) = \lambda' Tx^* \quad \text{for some} \quad \lambda' > 0.$$

Thus, $Tx^* = \lambda x^*$ for some $\lambda > 0$.

Conversely, from $Tx^* = \lambda x^*$ for some $\lambda > 0$ and some $x^* \gneqq 0$, $\|x^*\| = 1$ it follows that $x(t) = T^t x^* = \lambda(t) x^*$ for $\lambda(t) > 0$, all $t \in \mathbb{N}$. This implies $\frac{x(t)}{\|x(t)\|} = x^*$ for all $t \in \mathbb{N}$. Therefore, the stationary age structures correspond to the solutions x^* of $Tx^* = \lambda x^*$ with $x^* \gneqq 0$, $\|x^*\| = 1$ and $\lambda > 0$. Theorem 2.2.11 (and Remarks 2.2.12 (1)) yields that x^* is uniquely determined, $x^* > 0$ and $\lim\limits_{t\to\infty} \frac{x(t)}{\|x(t)\|} = x^*$ for all $x(0) \gneqq 0$.

(ii) The assumptions made imply that $c(x,\lambda) = \frac{1}{\lambda^{1-d}}$ independently of x and, hence, $T(\lambda x) = \lambda^d Tx$, that is T is homogeneous of degree d. Consider first the case $d = 1$. Then for any population path x Theorem 2.3.1 yields $\lim\limits_{t\to\infty} \frac{x(t)}{\lambda^{*t}} = c'(x(0))x^*$ with constant $c'(x(0)) > 0$ for $x(0) \gneqq 0$ and, in particular, $\lim\limits_{t\to\infty} \frac{\|x(t+1)\|}{\|x(t)\|} = \lim\limits_{t\to\infty} \|x(t)\|^{\frac{1}{t}} = \lambda^*$. Thereby, $x^* \gneqq 0$, $\|x^*\| = 1$, $\lambda^* > 0$ and $Tx^* = \lambda^* x^*$. Furthermore, $Tx = \lambda x$ for some $x \gneqq 0$ and $\lambda \geq 0$ implies $x = rx^*$ with $r > 0$ and $\lambda = \lambda^*$. Obviously, any path $\hat{x}$ with uniform growth rate g satisfies $\hat{x}(t) = \hat{x}(0)(1+g)^t$. For such a path $\lim\limits_{t\to\infty} \frac{\hat{x}(0)(1+g)^t}{\lambda^{*t}} = c'(\hat{x}(0))x^*$ and for $\hat{x}(0) \gneqq 0$ we must have $1 + g = \lambda^*$ and $\hat{x}(0) = c'(\hat{x}(0))x^*$. Putting $g^* = \lambda^* - 1$ we have that $g = g^*$ and $Tx^* = (1+g^*)x^*$. This also shows that, conversely, $\hat{x}(t) = cx^*(1+g^*)^t$ defines a path with uniform growth rate for any $c > 0$. Finally, for any path x with $x(0) \gneqq 0$ it holds that

$$\lim_{t\to\infty} \frac{x_i(t)}{\hat{x}_i(t)} = \lim_{t\to\infty} \frac{x_i(t)}{cx_i^*(1+g^*)^t} = \frac{c'(x(0))}{c} = c(x(0)).$$

Consider now the case $0 \leq d < 1$. Then for any population path x with $x(0) \gneqq 0$ Corollary 2.3.4 yields $\lim\limits_{t\to\infty} x(t) = \bar{x}$ where $\bar{x}$ is the unique non-zero fixed point of T. Thus, for a path $\hat{x}(t) = \hat{x}(0)(1+g)^t$ with uniform growth rate and $\hat{x}(0) \gneqq 0$ we must have $g = 0$ and $\hat{x}(0) = \bar{x}$. Conversely, $\hat{x}(t) = \bar{x}$ for all $t \in \mathbb{N}$ defines a path with uniform growth rate 0 because of $T\bar{x} = \bar{x}$. Furthermore, for any path x with $x(0) \gneqq 0$ it holds that

$$\lim_{t\to\infty} \frac{x_i(t)}{\hat{x}_i(t)} = \frac{1}{\bar{x}_i} \lim_{t\to\infty} x_i(t) = c(x(0)) \quad \text{for all} \quad i,$$

where $c(x(0)) = 1$. In particular, for any path x with $x(0) \gneqq 0$

$$\lim_{t\to\infty} \frac{\|x(t+1)\|}{\|x(t)\|} = \lim_{t\to\infty} \|x(t)\|^{\frac{1}{t}} = 1 = 1 + g.$$

Finally, it remains to show that T is primitive. For $x \in \mathbb{R}^n_+$ let $m(x)$ be the minimum of the numbers $b_{k_i}(x)$ with $x_{k_i} > 0$, $1 \le i \le r$ and $s_j(x)$ with $x_j > 0$, $1 \le j \le n-1$.

Let

$$L = \begin{bmatrix} a_1 & a_2 & \dots & a_n \\ 1 & 0 & \dots & 0 \\ 0 & \ddots & & \vdots \\ \vdots & & & \\ 0 & \dots & 01 & 0 \end{bmatrix},$$

where $a_j = 1$ for $j = k_i$ with $1 \le i \le r$ and $a_j = 0$ otherwise.

First we show by induction that for all $k \ge 1$, $x \gneqq 0$,

$$T^k x \ge m(T^{k-1}x)m(T^{k-2}x)\cdots m(x)L^k x \gneqq 0. \qquad (*)$$

From the definition of T it follows $Tx \ge m(x)Lx$ and $m(x)Lx \gneqq 0$ because of $m(x) > 0$. Thus, $(*)$ holds for $k = 1$. If $(*)$ holds for k then

$$T^{k+1}x = T^k(Tx) \ge m(T^k x)m(T^{k-1}x)\dots m(Tx)L^k Tx,$$

and using $Tx \ge m(x)Lx$ it follows that $(*)$ holds for $k+1$.

Next we show that the mapping induced by L is primitive which by $(*)$ implies the wanted primitivity for T. Let $\rho(x_1,\dots,x_n) = (x_n, x_{n-1},\dots,x_2,x_1)$ and M the selfmapping of $\mathbb{R}^n_+$ defined by $Mx = (\rho L \rho)x$. For $f(x) = a_n x_1 + a_{n-1}x_2 + \cdots + a_1 x_n$ we have that $Mx = (x_2,\dots,x_n,f(x))$. By assumption (b) of the concave Leslie model f satisfies the assumptions of Lemma 2.5.4 if we set $n_i = n - k_{r+1-i} + 1$ for $1 \le i \le r$. This Lemma yields that for some $t_0 \in \mathbb{N}$ and $x = (u(0),\dots,u(u-1)) \gneqq 0$ it holds that $M^t x = (u(t),\dots u(t+n-1)) > 0$ for all $t \ge t_0$. Therefore, M is primitive and because of $Lx = (\rho M \rho)x$ the mapping for L is primitive, too. □

The following example illustrates the various conclusions in Theorem 2.6.1, in particular part (ii), and makes a connection to what has been said previously in Section 1.2 about the Leslie model.

Example 2.6.2. Consider vital rates given for $1 \le i \le n$, $x \in \mathbb{R}^n_+$ by

$$b_i(x) = b_i x_i^{d-1} \quad \text{and} \quad s_i(x) = s_i x_i^{d-1} \quad \text{if} \quad x_i > 0$$

and $b_i(x) = b_i$, $s_i(x) = s_i$ if $x_i = 0$. Suppose $0 \le d \le 1$, $0 < s_i$ for all $1 \le i \le n-1$, $0 \le s_n$ and $0 \le b_i$, $1 \le i \le n$, such that there exist $k_1,\dots,k_r \in \{1,\dots,n\}$, $r \ge 2$, $k_r = n$ with $gcd\{k_1,\dots,k_r\} = 1$ and $b_{k_i} > 0$ for $1 \le i \le r$.

The functions $b_i(x)x_i = b_i x_i^d$ and $s_i(x)x_i = s_i x_i^d$ are concave in x on $\mathbb{R}^n_+$. Furthermore, $b_i(\lambda x) = c(x,\lambda)b_i(x)$ and $s_i(\lambda x) = c(x,\lambda)s_i(x)$ with $c(\lambda, x) = \frac{1}{\lambda^{1-d}}$ for all $x \in \mathbb{R}^n_+$, $1 \le i \le n$ and $\lambda > 0$. Thus, the dynamical system given by (2.6.1) satisfies assumptions (a), (b), (c) for the concave Leslie model. By Theorem 2.6.1 (i) the age structure of each population path x with $x(0) \gneqq 0$ converges to the unique stationary age structure $x^* > 0$.

Even more can be said since for this example the assumptions of Theorem 2.6.1 (ii) are satisfied.

Consider first the case $d = 1$. In that case $b_i(x) = b_i$ and $s_i(x) = s_i$ for $1 \le i \le n$ and for any x one has $L(x) = L$ where L is the classical (constant) Leslie matrix (cf. Section 1.2). By Theorem 2.6.1 (ii) the matrix L has the dominant root $\lambda^* = 1 + g^*$ with unique eigenvector $x^* > 0$, $\|x^*\| = 1$. The reference path $\hat{x}$ is (up to a positive factor) given by $\hat{x}(t) = x^*(1 + g^*)^t$ and for each population path x with $x(0) \gneqq 0$ one has

$$\lim_{t\to\infty} \frac{x_i(t)}{x_i^*(1+g^*)^t} = c((x(0)) \quad \text{for all} \quad 1 \le i \le n$$

and

$$\lim_{t\to\infty} \frac{\|x(t+1)\|}{\|x(t)\|} = \lim_{t\to\infty} \|x(t)\|^{\frac{1}{t}} = 1 + g^* .$$

A special case is given by the Fibonacci model (1.2.1) of Section 1.2 which by interchanging the indices for the age classes we may represent also by the Leslie matrix $L = \left[\begin{smallmatrix} 1 & 1 \\ 1 & 0 \end{smallmatrix}\right]$ instead of the Fibonacci matrix $F = \left[\begin{smallmatrix} 0 & 1 \\ 1 & 1 \end{smallmatrix}\right]$.

The assumptions of the concave Leslie model are satisfied for $L(x) = L$ and we obtain

$$\lambda^* = \frac{1+\sqrt{5}}{2}, \; g^* = \frac{\sqrt{5}-1}{2} \quad \text{and} \quad x^* = \left(\frac{1+\sqrt{5}}{3+\sqrt{5}}, \frac{2}{3+\sqrt{5}}\right)$$

and for any population path x

$$\lim_{t\to\infty} \frac{x_1(t)}{\left(\frac{1+\sqrt{5}}{2}\right)^{t+1}} = \lim_{t\to\infty} \frac{x_2(t)}{\left(\frac{1+\sqrt{5}}{2}\right)^{t}} = \frac{2}{3+\sqrt{5}}\, c(x(0)).$$

In particular, $\lim\limits_{t\to\infty} \frac{x_1(t)}{x_2(t)} = \frac{1+\sqrt{5}}{2}$, which we now obtain without knowing the explicit solution given by the Binet formula (1.2.3).

Consider next the case $0 \le d < 1$. From Theorem 2.6.1 (ii) we obtain that each population path must converge and $\lim\limits_{t\to\infty} x(t) = \hat{x}$ for $x(0) \gneqq 0$. Thus, one only has to find the unique non-zero fixed point $\hat{x}$ of T. The condition $Tx = L(x)x = x$ means that

$$\sum_{i=1}^{n} b_i x_i^d = x_1 \quad \text{and} \quad s_i x_i^d = x_{i+1} \quad \text{for} \quad 1 \le i \le n-2,\; s_{n-1}x_{n-1}^d + s_n x_n^d = x_n .$$

The problem to determine $\hat{x}$ can be reduced to determine $\hat{x}_1$ as the (strictly) positive solution of the equation $a_1 x_1^d + a_2 x_1^{d^2} + \cdots + a_n x_1^{d^n} = x_1$ with coefficients $a_i > 0$ computable from the given coefficients b_j and s_j. With the exception of $d = 0$ this is, however, a quite difficult task.

For $0 < d < 1$ we illustrate the procedure in the special case given by $n = 2$, $b_1 = b_2 = 1$, $s_1 = 1$, $s_2 = 0$. Then $Tx = x$ amounts to $x_1^d + x_2^d = x_1$ and $x_1^d = x_2$. Eliminating x_2 we obtain $x_1^d + x_1^{d^2} = x_1$ and $x_1^{d^2-1} + x_1^{d-1} - 1 = 0$ since we solve for a non-zero fixed point. Putting $y = x_1^{d(1-d)}$ we obtain the non-linear equation $y^{(1+\frac{1}{d})} - y - 1 = 0$. This equation has exactly one positive solution $\hat{y}$ as we know on general grounds but which also can be confirmed easily by direct computation. The dynamics $x(t+1) = Tx(t)$ in this special case is given by $x_1(t+1) = x_1(t)^d + x_2(t)^d$ and $x_2(t+1) = x_1(t)^d$ or, equivalently, by putting $p(t) = x_1(t)$, by the non-linear Fibonacci equation

$$p(t+2) = p(t+1)^d + p(t)^{d^2} \quad \text{for} \quad t \in \mathbb{N},\ p(t) \geq 0.$$

Obviously, for $p(0), p(1)$ given this equation has a unique solution $t \longmapsto p(t)$ and by the above we have that $\lim_{t\to\infty} p(t) = \hat{p}$ with $\hat{p} = \hat{x}_1 = \hat{y}^{\frac{1}{d(1-d)}}$.

Remark 2.6.3. A particular kind of population pressure has been already analyzed by P.H. Leslie in [38]; cf. also [21, 22, 24, 64]. The model amounts to $Rx = r(x)Lx$ where L is a (constant) Leslie matrix and $r(x) = \frac{U}{U+(r-1)\|x\|}$.

Thereby, U measures environmental capacity, r is the dominant eigenvalue of L and $\|x\| = \sum_{i=1}^n x_i$ for $x \in \mathbb{R}_+^n$. The mapping R is not concave but satisfies the assumptions of Corollary 2.2.14 which implies that for any non-zero path the age structure converges to the unique stationary age structure. Non-concave Leslie models of the type $Rx = r(x)Tx$, where $Tx = L(x)x$ is a concave Leslie model but the scalar $r(x)$ measures in addition some uniform population pressure, may be analyzed also with the help of Corollary 2.3.4.

Other non-linear Leslie models $Tx = L(x)x$ which are not concave have been studied in the literature, especially for vital rates of the type $b_i(x) = b_i \exp(-\sum_{j=1}^n c_j x_j)$ and $s_i(x) = s_i \exp(-\sum_{j=1}^n d_j x_j)$ with certain coefficients $c_j, d_j \geq 0$. In contrast to the concave model global properties as relative stability can no longer be expected, on the contrary, chaotic behavior may occur as shown by computer simulations. Cf. [10, 20, 39, 53].

Exercises

1. Demonstrate for the concave Leslie model given by $Tx = (\sqrt{x_1} + \sqrt{x_2}, \sqrt{x_1})$ the following properties.
   (a) The eigenvalue problem $Tx = \lambda x$ has for every $\lambda > 0$ a unique positive solution $x = x(\lambda)$.
   (b) For every $\lambda > 0$
   $$\lim_{t\to\infty} T^t x(\lambda) = \lambda^2 x(\lambda).$$
   (c) Explain why property (b) does not contradict Theorem 2.6.1 (ii) for $d = \frac{1}{2}$.
2. Consider the Leslie model given for $n = 2$ by the vital rates $b_1(x) = b_1 + (\frac{x_2}{x_1})^\alpha$ for $x_1 > 0$ and $b_1(x) = b_1$ for $x_1 = 0$, $b_2(x) = b_2$, $s_1(x) = s_1$, $s_2(x) = 0$. Thereby, $b_1, b_2, s_1 > 0$ and $0 \leq \alpha \leq 1$.
   (a) Show that by the above a concave Leslie model is defined.
   (b) Compute for $\alpha = \frac{1}{2}$ the stationary age structure x^* and the growth rate g^* according to Theorem 2.6.1.

(c) Given a non-zero population path x derive a difference equation for $y(t) = \frac{x_1(t)}{x_2(t)}$ and show that y oscillates in approaching $\frac{x_1^*}{x_2^*}$.

3. Consider for $0 < d < 1$ the non-linear equation $y^{1+\frac{1}{d}} - y - 1 = 0$ in one real variable y.
   (a) Show that the equation has a unique positive root.
   (b) Compute the root for $d = \frac{1}{2}$ by Cardano's formula.
   (c) Determine for the non-linear Fibonacci equation

$$p(t+2) = \sqrt[\frac{1}{2}]{p(t+1)} + \sqrt[\frac{1}{4}]{p(t)},\ t \in \mathbb{N},\ p(t) \geq 0$$

the limit $\hat{p} = \lim_{t\to\infty} p(t)$.

2.7 Price setting and balanced growth in a concave Leontief model

As mentioned at the end of Section 1.4, concave Perron–Frobenius theory is useful to handle price setting in a Leontief model with choice of techniques. As discussed already in Section 1.4 the price setting modeled by equation (1.4.2) specializes for time-invariant technology sets A_i and constant real wages b^i to

$$p(t+1) = k(t)Tp(t), \tag{2.7.1}$$

where T is a selfmapping of the cone $\mathbb{R}_+^n$ given by $Tp = c(p)$ with

$$c_i(p) = \inf\{p(a + lb^i) \mid (a,l) \in A_i\} \quad \text{for} \quad 1 \leq i \leq n. \tag{2.7.2}$$

Since $r_i(t) = \frac{p_i(t+1) - c_i(p(t))}{c_i(p(t))}$ is the rate of profit for producer i at time t equation 2.7.1 amounts to $p(t+1) = (1 + r(t))Tp(t)$ with $r(t) = k(t) - 1$ the *uniform rate of profit* for all producers at time t. Obviously, the cost operator T is concave and positively homogeneous and, as shown already in Section 1.4, by introducing relative prices $q(t) = p(t)\|p(t)\|^{-1}$, the positive discrete dynamical system (2.7.1) can be written as

$$q(t+1) = \tilde{T}q(t), \quad t \in \mathbb{N}, \quad \|q(0)\| = 1, \tag{2.7.3}$$

where the normalized cost operator $\tilde{T}$ is given by $\tilde{T}q = Tq\|Tq\|^{-1}$ for $Tq \neq 0$.

Using knowledge from concave Perron–Frobenius theory we obtain the following results concerning the above price setting process.

Theorem 2.7.1. *For the price setting process* (2.7.1) *with cost function* (2.7.2) *denote by* $d_{ij} = \inf\{a_i + lb_i^j \mid (a,l) \in A_j\}$ *the minimal overall expenditure of good i in the production of one unit of good j for* $1 \leq i,j \leq n$. *If the matrix* $D = (d_{ij})$ *is indecomposable then the following statements hold.*

(i) *There exists an equilibrium rate of profit* $r^* > -1$ *with equilibrium prices* $p^* > 0$, $\|p^*\| = 1$ *such that* $p^* = (1 + r^*)c(p^*)$. *The equilibrium is unique, i.e.,* $p = (1 + r)c(p)$ *for* $r \geq -1$ *and* $p \gneqq 0$ *implies* $r = r^*$ *and* $p = sp^*$ *for some* $s > 0$.
Moreover

$$r^* = \min_{p \gneqq 0} \max_{1 \leq i \leq n} \left\{ \frac{p_i - c_i(p)}{c_i(p)} | c_i(p) > 0 \right\}.$$

(ii) *Assume in addition that for at least one good its overall expenditure is strictly positive. Then for any given initial prices* $p(0) \gneqq 0$ *it holds that* $\lim_{t \to \infty} q(t) = p^*$ *for relative prices* $q(t) = p(t)\|p(t)\|^{-1}$. *Moreover,*

$$\lim_{t \to \infty} \left((1 + r(t)) \frac{\|p(t)\|}{\|p(t + 1)\|} \right) = 1 + r^*,$$

in particular, $\lim_{t \to \infty} r(t) = r^*$ *if absolute prices* $p(t)$ *have a non-zero limit.*

Proof. (i) From (2.7.2) we have that $c_i(e_j) = \inf\{a_j + lb_j^i \mid (a, l) \in A_i\} = d_{ji}$. Therefore, the mapping $Tp = c(p)$ is indecomposable and parts (i) and (ii)(c) of Theorem 2.3.8 imply existence and uniqueness of r^*, p^*. Let $s(p) = \max_{1 \leq i \leq n} \left\{ \frac{p_i - c_i(p)}{c_i(p)} \mid c_i(p) > 0 \right\}$ for $p \gneqq 0$ and $r = \inf\{s(p) \mid p \gneqq 0\}$. From $p^* = (1 + r^*)c(p^*)$ it follows that $s(p^*) = r^*$ and $r \leq r^*$. Since $(1 + s(p))c(p) \geqq p \gneqq 0$ from part (ii) (a) of Theorem 2.3.8 it follows that $(1 + s(p))^{-1} \leq (1 + r^*)^{-1}$ and, hence, $r^* \leq s(p)$. This shows $r^* \leq r$ and $r^* = r = \min\{s(p) \mid p \gneqq 0\}$ because of $s(p^*) = r^*$.

(ii) Since T is positively homogeneous, concave, weakly indecomposable by Lemma 2.2.7 and $T_h e_h = c_h(e_h) = d_{hh} > 0$ for some h, Theorem 2.2.11 yields that

$$q(t) = \tilde{T}^t q(0) = \widetilde{T^t} q(0) = T^t q(0) \|T^t q(0)\|^{-1}$$

converges to p^* for $t \to \infty$. Furthermore, $p(t + 1) = (1 + r(t))c(p(t))$ implies that $q(t + 1) = (1 + r(t)) \frac{\|p(t)\|}{\|p(t+1)\|} c(q(t))$ and, hence,

$$\lim_{t \to \infty} \left(1 + r(t) \frac{\|p(t)\|}{\|p(t + 1)\|} \right) c(p^*) = p^* = (1 + r^*)c(p^*).$$ □

Remark 2.7.2. Concave Perron–Frobenius theory is applicable also in cases where the cost function is different from the one considered in equation (2.7.2). For example, if the (unit) cost function is given by a Cobb–Douglas technology, that is

$$c_i(p) = k_i \prod_{j=1}^{n} p_j^{\alpha_{ij}} + l_i w_i, \quad 1 \leq i \leq n,$$

with constants $k_i > 0$, $\alpha_{ij} \geq 0$ with $\sum_{j=1}^{n} \alpha_{ij} = 1$, $l_i > 0$ and $w_i = pb^i$, $b^i > 0$. The self-mapping T of $\mathbb{R}_+^n$ given by $Tp = c(p)$ is concave, positively homogeneous with $Tp > 0$ for $p \gneqq 0$. Therefore, Theorem 2.2.11 applies and yields conclusions as those in Theorem 2.7.1. (See Exercises 2 and 3 below.) For models of price setting similar to the one considered the reader is referred to the references [15, 28, 29, 31, 33, 43] and the literature given therein.

Concave Perron–Frobenius theory can be applied also to analyze balanced growth in a non-linear closed model of production. (Such a model was first investigated in [62] and in full detail subsequently in [43] and [47].)

A closed model of production is given by a mapping $T \gneqq 0$ which transforms a given input of goods into an output of the same goods. If $x(t) \in \mathbb{R}^n_+$ is the input in period t then one period later output $x(t+1) \in \mathbb{R}^n_+$ is produced and

$$x(t+1) = Tx(t), \quad \text{for all} \quad t \in \mathbb{N}. \tag{2.7.4}$$

A solution of (2.7.4) is called a *balanced growth path* $x(\cdot)$ if for all goods $1 \le i, j \le n$ the ratio $\frac{x_i(t)}{x_j(t)}$ is constant over time. Equivalently, there exists a time dependent scalar $\sigma(t) > 0$ such that $x(t) = \sigma(t)x(0)$ for all $t \in \mathbb{N}$ and $\sigma(0) = 1$.

Suppose the selfmapping T of $\mathbb{R}^n_+$ satisfies the homogeneity condition $T(\lambda x) = f(\lambda)Tx$ with some function $f\colon \mathbb{R}_+ \longrightarrow \mathbb{R}_+$. For a balanced growth path $x(\cdot)$ with $x(0) \gneqq 0$ it follows that

$$\sigma(t+1)x(0) = x(t+1) = Tx(t) = T(\sigma(t)x(0)) = f(\sigma(t))Tx(0)$$

and, hence,

$$Tx(0) = \lambda^* x(0) \quad \text{and} \quad \sigma(t+1) = f(\sigma(t))\lambda^*$$

for some $\lambda^* > 0$. Conversely, these conditions determine a balanced growth path and, therefore, $x(\cdot)$ with $x(0) \gneqq 0$ is a balanced growth path iff

$$x(t) = h^t(1)x(0) \quad \text{for all} \quad t \in \mathbb{N}, \quad \text{and} \quad Tx(0) = \lambda^* x(0), \tag{2.7.5}$$

where $h(\lambda) = f(\lambda)\lambda^*$ and h^t is the t-th iterate of the mapping h.

Theorem 2.7.3. *For the growth model* (2.7.4) *let T be concave with $T(\lambda x) = f(\lambda)Tx$ for all $\lambda \in \mathbb{R}_+$, $x \in \mathbb{R}^n_+$ and some function $f\colon \mathbb{R}_+ \longrightarrow \mathbb{R}_+$. Then there exists $0 \le d \le 1$ with $f(\lambda) = \lambda^d$, i.e., T is homogeneous of degree d, and the following statements hold.*

(i) *Let T be indecomposable. For $d = 1$ there exists a unique (up to a positive scalar) non-zero balanced growth path, namely $u(t) = (\lambda^*)^t u(0)$ where $\lambda^* > 0$ and $u(0) > 0$ satisfy $Tu(0) = \lambda^* u(0)$.*

For $0 \le d < 1$ there exists a unique non-zero fixed point x^* of T, $x^* > 0$ and $u(\cdot)$ is a balanced growth path iff $u(t) = \lambda^{-\frac{d^t}{1-d}} x^*$ for some $\lambda > 0$.

(ii) *If, in addition, $T_h e_h > 0$ for some h then any non-zero solution $x(\cdot)$ of* (2.7.4) *is relatively stable, that is*

$$\lim_{t\to\infty} \frac{x_i(t)}{u_i(t)} = c(x(0)) > 0 \quad \textit{for all} \quad 1 \le i \le n$$

where $u(\cdot)$ is any non-zero balanced growth path.

Proof. From $T(\lambda\mu x) = f(\lambda\mu)Tx$ and $T(\lambda\mu x) = f(\lambda)T(\mu x) = f(\lambda)f(\mu)Tx$ it follows that $f(\lambda\mu) = f(\lambda)f(\mu)$ for all $\lambda, \mu \in \mathbb{R}_+$ because of $T \gneqq 0$. Concavity of T implies concavity and, hence, monotonicity of f. It is well-known that a monotone mapping satisfying $f(\lambda\mu) = f(\lambda)f(\mu)$ must be of the form $f(\lambda) = \lambda^d$ for some $d \geq 0$. Concavity of f implies $d \leq 1$.

(i) Let T be indecomposable and assume first that $d = 1$. T is positively homogeneous and Theorem 2.3.8 assures that $Tu(0) = \lambda^* u(0)$ has a unique solution $\lambda^* > 0$, $u(0) > 0$ (up to a positive scalar). According to (2.7.5) there exists a unique (up to a positive scalar) non-zero balanced growth path, namely $u(t) = (\lambda^*)^t u(0)$. Consider now the case $0 \leq d < 1$. By Theorem 2.3.8 there exist $\bar{\lambda} > 0$ and $\bar{x} > 0$ such that $T\bar{x} = \bar{\lambda}\bar{x}$.

For $x^* = \bar{\lambda}^{\frac{1}{1-d}}\bar{x} > 0$ one has that

$$Tx^* = \bar{\lambda}^{\frac{d}{1-d}} T\bar{x} = \bar{\lambda}^{\frac{d}{1-d}} \bar{\lambda}\bar{x} = x^*.$$

If $\hat{x} \gneq 0$ is any fixed point of T then $\hat{x} > 0$ by Theorem 2.3.8. There exist $\alpha > 0$, $\beta > 0$ with $\alpha\hat{x} \leq x^* \leq \beta\hat{x}$ and applying T^t yields $\alpha^{d^t}\hat{x} \leq x^* \leq \beta^{d^t}\hat{x}$ for all $t \in \mathbb{N}$.

For $t \to \infty$ it follows that $\hat{x} \leqq x^* \leqq \hat{x}$. This shows that x^* is the unique (non-zero) fixed point of T. By (2.7.5) any balanced growth path is given by $u(t) = h^t(1)u(0)$ where for some $\lambda > 0$ one has $Tu(0) = \lambda u(0)$ and $h(\mu) = \mu^d\lambda$. Induction over t yields $h^t(1) = \lambda^{\frac{1-d^t}{1-d}}$. From $Tu(0) = \lambda u(0)$ it follows that $u(0)\lambda^{\frac{d}{1-d}}$ is a fixed point of T and, by the uniqueness of the fixed point, $u(0)\lambda^{\frac{1}{1-d}} = x^*$. Therefore, $u(t) = h^t(1)u(0) = \lambda^{-\frac{d^t}{1-d}}x^*$. Obviously, such a path is a balanced growth path for arbitrary $\lambda > 0$.

(ii) If T is indecomposable and $T_h e_h > 0$ for some h then T is primitive by Lemmas 2.2.7 and 2.2.10. For $d = 1$ relative stability of $x(\cdot)$ follows from Theorem 2.3.1 (i). For $d < 1$ Corollary 2.3.4 yields $\lim_{t\to\infty} x(t) = x^*$. For any non-zero balanced growth path $u(\cdot)$ it follows that $\lim_{t\to\infty} \frac{x_i(t)}{u_i(t)} = 1$ for all $1 \leq i \leq n$ and arbitrary $x(0) \gneq 0$. □

Remarks 2.7.4. Theorem 2.7.3 is conceived to illustrate concave Perron–Frobenius theory. It is possible, however, to prove similar results for the growth model (2.7.4) by weakening the assumption of concavity to that of monotonicity, see [43, 47, 62].

Exercises

1. Consider three producers each equipped with two technologies. Suppose producer 1 (producing good 1) can use a technique (a, l) with $a = (0, \frac{2}{5}, \frac{1}{8})$ and $l = 1$ or with $a = (0, \frac{1}{5}, \frac{1}{10})$ and $l = 3$, producer 2 (producing good 2) can use a technique (a, l) with $a = (\frac{1}{2}, 0, \frac{1}{5})$ and $l = 2$ or with $a = (\frac{3}{4}, 0, \frac{1}{5})$ and $l = 1$; producer 3 (producing good 3) can use a technique (a, l) with $a = (\frac{1}{3}, \frac{1}{6}, 0)$ and $l = 1$ or with $a = (\frac{1}{4}, \frac{1}{5}, 0)$ with $l = 2$. Suppose further the real wage for all producers is given by $b = \frac{1}{10}(\frac{1}{4}, 1, 1)$.

(a) Compute the equilibrium (p^*, r^*) ($\|p^*\| = p_1^* + p_2^* + p_3^* = 1$) analytically and by simulation on the computer.
(b) Determine the techniques applied by the producers at equilibrium prices.

2. Analyze the price setting model (2.7.1) for a Cobb–Douglas technology, i.e., $Tp = c(p)$ with $c_i(p) = k_i \prod_{j=1}^n p_j^{\alpha_{ij}} + l_i w_i$ with constants $k_i > 0$, $\alpha_{ij} \geq 0$, $\sum_{j=1}^n \alpha_{ij} = 1$, $l_i > 0$ and $w_i = pb^i$, $b^i > 0$.
   (a) Verify that T is concave, positively homogeneous and primitive.
   (b) Prove that there exists a unique equilibrium

$$p^* = (1 + r^*)c(p^*), \quad p^* > 0, \quad \|p^*\| = \sum_{i=1}^n p_i^* = 1, \quad r^* > -1.$$

   (c) Prove that relative prices converge to p^* and find conditions on the constants such that absolute prices converge, too.

3. Let for two producers a Cobb–Douglas technology given as in Exercise 2 with $k_1 = \frac{1}{4}$, $k_2 = \frac{1}{5}$, $l_1 = l_2 = 1$ and

$$(\alpha_{ij}) = \begin{bmatrix} \frac{1}{2} & \frac{1}{2} \\ \frac{1}{3} & \frac{2}{3} \end{bmatrix}, \quad b^1 = \frac{1}{8}\begin{bmatrix} 1 \\ 1 \end{bmatrix}, \quad b^2 = \frac{1}{10}\begin{bmatrix} 2 \\ 1 \end{bmatrix}.$$

   (a) Compute the equilibrium (p^*, r^*).
   (b) Simulate on the computer the convergence of relative prices to p^*.
   (c) Check convergence of absolute prices analytically and by doing iterations on the computer.

4. Consider the growth model given by

$$T(x_1, x_2) = \left((ax_1 + bx_2)^\alpha (cx_1 + dx_2)^\beta, x_1^{\alpha+\beta}\right)$$

   where a, b, c, d are strictly positive constants and α, β are non-negative constants with $\alpha + \beta \leq 1$.
   (a) Verify that T is concave, homogeneous of degree $\alpha + \beta$ and primitive.
   (b) Discuss the dependence of balanced growth paths on α and β for $a = b = c = d = 1$.
   (c) Compute the balanced growth paths in (b) for $\alpha = \beta = \frac{1}{4}$.

Bibliography

[1] A. Berman, M. Neumann, and R.J. Stern. *Nonnegative Matrices in Dynamic Systems*. Wiley, New York, 1989.
[2] A. Berman and R.J. Plemmons. *Nonnegative Matrices in the Mathematical Sciences*. Academic Press, New York, 1979.
[3] H. Bernardelli. Population waves. *Journal of Burma Research Society*, 31:1–18, 1941.

[4] C. Bidard and U. Krause. A monotonicity law for relative prices. *Econ. Theory*, 7:51–61, 1996.

[5] G. Birkhoff. Extensions of Jentzsch's theorem. *Trans. Amer. Math. Soc.*, 85:219–227, 1957.

[6] P. J. Bushell. Hilbert's metric and positive contraction mappings in a Banach space. *Arch. Rational Mech. Anal.*, 52:330–338, 1973.

[7] H. Caswell. *Matrix Population Models*. Sinauer Associates, Sunderland, 1989.

[8] J. E. Cohen. Ergodic theorems in demography. *Bull. Am. Math. Soc.*, N.S. 1:275–295, 1979.

[9] K. Deimling. *Nonlinear Functional Analysis*. Springer, Berlin, Heidelberg, New York, Tokyo, 1985.

[10] R. A. Desharnais and J.E. Cohen. Life not lived due to disequilibrium in heterogeneous age-structured populations. *Theoret. Biol.*, 29:385–406, 1986.

[11] G. Frobenius. Über Matrizen aus positiven Elementen. *Sitzungsber. Preuss. Akad. Wiss.*, pp. 514–518, 1909.

[12] G. Frobenius. Über Matrizen aus nichtnegativen Elementen. *Sitzungsber. Preuss. Akad. Wiss.*, pp.456–477, 1912.

[13] T. Fujimoto. Nonlinear generalization of the Frobenius theorem. *J. Math. Econ.*, 6:17–21, 1979.

[14] T. Fujimoto and U. Krause. Strong ergodicity for strictly increasing non-linear operators. *Lin. Alg. Appl.*, 71:101–112, 1985.

[15] T. Fujimoto and U. Krause. Ergodic price setting with technical progress. In W. Semmler, editor, *Competition, Instability, and Nonlinear Cycles*, pp. 115–124. Springer, Berlin, etc., 1986.

[16] T. Fujimoto and U. Krause. An ergodic property for certain classes of non-linear positive operators. In R. Nagel, U. Schlotterbeck, M. P. H. Wolff, editors, *Aspects of Positivity in Functional Analysis*, pp. 191–197. North-Holland, Amsterdam etc., 1986.

[17] T. Fujimoto and U. Krause. Asymptotic properties for inhomogeneous iterations of non-linear operators. *SIAM J. Math. Anal.*, 19:841–853, 1988.

[18] S. Gaubert and J. Gunawardena. The Perron–Frobenius Theorem for homogeneous, monotone functions. *Trans.Amer. Math. Soc.*, 356:4931–4950, 2004.

[19] F. R. Gantmacher. *The Theory of Matrices*, volume I, II. Chelsea, New York, 1959.

[20] J. Guckenheimer, G. Oster, and A. Ipaktschi. Dynamics of density dependent population models. *J. Math. Biol.*, 4:101–147, 1977.

[21] P. E. Hansen. Raising Leslie matrices to power: a method and applications to demography. *J. Math. Biol.*, 18:149–161, 1983.

[22] P. E. Hansen. Leslie matrix models. *Math. Pop. Studies*, 2:37–67, 1989.

[23] D. Hilbert. Über die gerade Linie als kürzeste Verbindung zweier Punkte. *Math. Ann.*, 46:91–96, 1895.

[24] J. N. Kapur. Age-structured population models with density dependence. *Bull. Calcutta Math. Soc.*, 74:207–215, 1982.

[25] J. P. Keener. The Perron–Frobenius Theorem and the ranking of football teams. *SIAM Review*, 35:80–93, 1993.

[26] P. E. Kloeden and A.M. Rubinov. A generalization of the Perron–Frobenius theorem. *Nonlinear Analysis*, 41:97–115, 2000.

[27] V. L. Kocic and G. Ladas. *Global Behavior of Nonlinear Difference Equations of Higher Order with Applications*. Kluwer, Dordrecht, 1993.

[28] E. Kohlberg. The Perron–Frobenius theorem without additivity. *J. Math. Econ.*, 10:299–303, 1982.

[29] U. Krause. Perron's stability theorem for non-linear mappings. *J. Math. Econ.*, 15:275–282, 1986.

[30] U. Krause. Iteration of concave positive operators with examples from economy and biology. *Semesterberichte Funktionalanalysis*, 12:43–54, 1987.

[31] U. Krause. Path stability of prices in a non-linear Leontief model. *Ann. Oper. Res.*, 37:141–148, 1992.
[32] U. Krause. Relative stability for ascending and positively homogeneous operators on Banach spaces. *J. Math. Anal. Appl.*, 188:182–202, 1994.
[33] U. Krause. Positive non-linear systems in economics. In T. Maruyama and W. Takahashi, editors, *Nonlinear and Convex Analysis in Economic Theory*, pp. 181–195. Springer, Berlin, 1995.
[34] U. Krause. Positive non-linear systems: Some results and applications. In V. Lakshmikantham, editor, *World Congress of Nonlinear Analysts 92'*, pp. 1529–1539. De Gruyter, Berlin, 1996.
[35] U. Krause. Concave Perron–Frobenius theory and applications. *Nonlinear Analysis*, 47:1457–1466, 2001.
[36] H.D. Kurz and N. Salvadori. *Theory of Production*. Cambridge University Press, Cambridge, 1995.
[37] B. Lemmens and R. Nussbaum. *Nonlinear Perron–Frobenius Theory*. Cambridge University Press, Cambridge, 2012.
[38] P.H. Leslie. Some further notes on the use of matrices in population mathematics. *Biometrika*, 35:213–245, 1948.
[39] L. Liu and J. E. Cohen. Equilibrium and local stability in a logistic matrix model for age-structured populations. *J. Math. Biol.*, 25:73–88, 1987.
[40] D. G. Luenberger. *Introduction to Dynamic Systems.* Wiley & Sons, New York, 1979.
[41] H. Minc. *Nonnegative Matrices*. Wiley, New York, 1988.
[42] M. Morishima. Generalizations of the Frobenius–Wielandt theorems for non-negative square matrices. *J. London Math. Soc.*, 36:211–220, 1961.
[43] M. Morishima. *Equilibrium, Stability, and Growth*. Oxford University Press, Oxford, 1964.
[44] M. Morishima and T. Fujimoto. The Frobenius theorem, its Solow–Samuelson extension and the Kuhn–Tucker theorem. *J. Math. Econ.*, 1:199–205, 1974.
[45] J. D. Murray. *Mathematical Biology*. Springer, Berlin, 1993.
[46] H. Nikaido. Balanced growth in multi-sectoral income propagation under autonomous expenditure schemes. *Rev. Econ. Studies*, 31:25–42, 1964.
[47] H. Nikaido. *Convex Structures and Economic Theory.* Academic Press, New York, 1968.
[48] R. D. Nussbaum. Hilbert's projective metric and iterated non-linear maps. *Memoirs Amer. Math. Soc.*, 75(391):1–137, 1988.
[49] R. D. Nussbaum. Hilbert's projective metric and iterated non-linear maps II. *Memoirs Amer. Math. Soc.*, 79(401):1–118, 1989.
[50] Y. Oshime. An extension of Morishima's non-linear Perron–Frobenius theorem. *J. Math. Kyoto Univ.*, 23:803–830, 1983.
[51] Y. Oshime. Perron–Frobenius problem for weakly sublinear maps in a euclidean positive orthant. *Japan J. Indust. Appl. Math.*, 9:313–350, 1992.
[52] A. M. Ostrowski. Positive matrices and functional analysis. In H. Schneider, editor, *Recent Advances in Matrix Theory*, 81–101, The University of Wisconsin Press, Madison, 1964.
[53] C. J. Pennycuick, R.M. Compton, and L. Beckingham. A computer model for simulating the growth of a population, or of two interacting populations. *J. Theoret. Biol.*, 18:316–329, 1968.
[54] O. Perron. Zur Theorie der Matrices. *Math. Ann.*, 64:248–263, 1907.
[55] E. C. Pielou. *An Introduction to Mathematical Ecology*. Wiley Interscience, New York, 1969.
[56] J. H. Pollard. *Mathematical Models of the Growth of Human Populations*. Cambridge University Press, Cambridge, 1973.
[57] H. Samelson. On the Perron–Frobenius theorem. *Michigan Math.*, 4:57–59, 1957.
[58] H. Schneider (ed.) *Recent Advances in Matrix Theory*. The University of Wisconsin Press, Madison, 1964.

[59] H. Schneider and R. E. L. Turner. Positive eigenvectors of order preserving maps. *J. Math. Anal. Appl.*, 37:506–515, 1972.

[60] J. T. Schwartz. *Lectures on the Mathematical Method in Analytical Economics*. Gordon and Breach, New York, 1961.

[61] E. Seneta. *Non-negative Matrices and Markov Chains*. Springer, Berlin, 2nd. edition, 1980.

[62] R. M. Solow and P. A. Samuelson. Balanced growth under constant returns to scale. *Econometrica*, 21:412–424, 1953.

[63] A. C. Thompson. *Generalizations of the Perron–Frobenius theorem to operators mapping a cone into itself*. PhD thesis, University of Newcastle upon Tyne, 1963. 114 p.

[64] M. B. Usher. Development in the Leslie matrix model. In J.N.R. Jeffers, editor, *Mathematical Models in Ecology*. Blackwell, Oxford, 1972.

[65] D. Weller. *Hilbert's metric, part metric and selfmappings of a cone*. PhD thesis, Universität Bremen, 1987. 99 p.

[66] H. Wielandt. Unzerlegbare, nicht negative Matrizen. *Math. Z.*, 52:642–648, 1950.

[67] J. E. Woods. *Mathematical Economics*. Longman, London, 1978.

3 Internal metrics on convex cones

In the previous chapter we have seen how to deal with concave mappings that leave invariant the standard cone $\mathbb{R}^n_+$ (or its interior). For many applications a finite dimensional space is not sufficient because the states of the model are no longer points in the usual sense but real valued functions which in general constitute an infinite dimensional space. For positive systems this means to consider as state space a cone in an infinite dimensional space and non-linear operators leaving this cone invariant. In the present chapter we will carry out the analysis of convex cones which will be needed later to treat particular kinds of non-linear selfmappings of convex cones, including concave operators. A cornerstone for the analysis of a selfmapping T of the cone $K = \mathbb{R}^n_+$ in the previous chapter was Hilbert's projective metric on the interior $\overset{\circ}{K}$ (see Definition 2.1.8). Thus, in this chapter we will study Hilbert's projective metric on cones in infinite dimensional spaces. This can be done, actually, not only for standard cones but for quite general convex cones, which provides more flexibility for applications. Already in finite dimensions are convex cones other than the standard cone of interest, in applications as well as theoretically. The general notion of a convex cone is so important because it allows for a concept of positivity which is coordinate-free. Moreover, we will study besides Hilbert's projective metric also other metrics which are called internal metrics because they are derived from the structure of the convex cone, too. This again provides more flexibility for applications because a selfmapping of a cone might be contractive for one internal metric but not for another one.

3.1 Extraction within convex cones

The most convenient way to introduce the various kinds of internal metrics is by using extraction within convex cones, in particular by using the order function as a building block, as explained in the following.

Let V be an arbitrary vector space over $\mathbb{R}$. A (non-empty) subset $K \subset V$ is called a **cone** if K contains for every $x \in K$ the **ray** through x, i.e. $\{\lambda x \mid \lambda > 0\}$. A cone K is called **convex** if it is closed for addition, i.e. $K + K \subset K$. According to this definition, any linear subspace of V is a convex cone, e.g. V and $\{0\}$ are convex cones. The convex cones we are interested in, however, will be **pointed**, i.e., $K \cap (-K) \subset \{0\}$, or *pointed with* 0, i.e., $K \cap (-K) = \{0\}(-K = \{-x \mid x \in K\})$. Any convex cone K induces a transitive relation $\leq$ on V by

$$x \leq y \text{ iff } y - x \in K, \quad \text{for any} \quad x, y \in V.$$

This relation is a **partial order on V**, i.e., $\leq$ is reflexive, antisymmetric and transitive iff the convex cone K is pointed with 0.

Definition 3.1.1. The function $\lambda(\cdot,\cdot)\colon K \times K \longrightarrow [0,\infty]$ defined by

$$\lambda(x,y) = \sup\{\lambda \geq 0 \mid y - \lambda x \in K\}$$

is called the **order function** or the extraction grade on K. $\mu(x,y) = \min\{\lambda(x,y), \lambda(y,x)\}$ is the **symmetric order function** (extraction grade).

The mapping $e\colon K \times K \longrightarrow K$ defined by

$$e(x,y) = y - \lambda(x,y)x \quad (\text{with } \infty \cdot 0 = 0)$$

is called the **extraction function** on K.

For $x, y \in K \setminus \{0\}$ the element x is called a **component** of y if $\lambda(x,y) > 0$, and $e(x,y)$ is called the **rest** after extracting x from y.

The idea of extraction is, for any two given elements x and y to extract from y as much as possible of the element x contained in y (see Figure 3.1). The maximal amount of x contained in y is measured by the order function λ. The rest that remains after extracting x from y is given by the extraction function e.

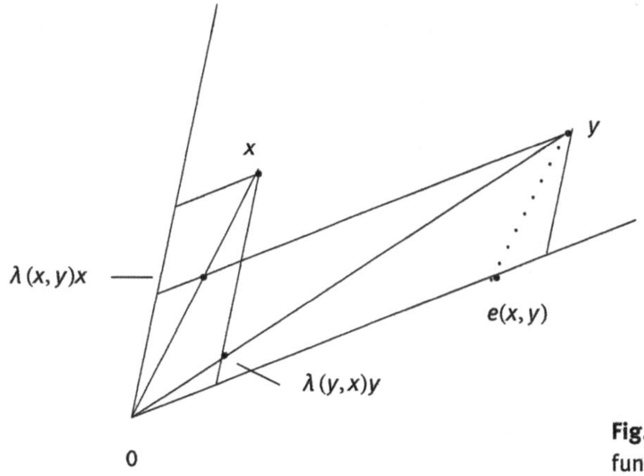

Fig. 3.1. Order function and extraction function.

Remark 3.1.2. In a way similar to the above, extraction can be defined also for arbitrary convex sets or for abstract monoids, in particular for the multiplicative monoid of an integral domain. Though we will stick here solely to convex cones, the process of extraction turns out to be a quite fundamental operation in many areas (see [24, 25, 37, 41]).

Examples 3.1.3. (1) $K = \mathbb{R}^n_+$ the standard cone in $V = \mathbb{R}^n$. Obviously, K is a convex cone pointed with 0. For $x, y \in K$, $x \neq 0$ one has $\lambda(x,y) = \min\{\frac{y_i}{x_i} \mid x_i > 0,\ 1 \leq i \leq n\} = \frac{y_{i_0}}{x_{i_0}}$ and $e(x,y)_i = \frac{1}{x_{i_0}}(y_i x_{i_0} - y_{i_0} x_i)$.

Furthermore, if $y \neq 0$ and $\frac{x_{i_1}}{y_{i_1}}$ is the minimum of all the values $\frac{x_i}{y_i}$ for $y_i > 0$ then

$$\mu(x,y) = \min\left\{\frac{y_{i_0}}{x_{i_0}}, \frac{x_{i_1}}{y_{i_1}}\right\}.$$

Obviously, one has $\lambda(x,y) = \infty$ iff $x = 0$.

(2) A little bit more involved is the computation of λ for the **ice cream cone** K, i.e., $K = \{(u,r) \mid u \in \mathbb{R}^n,\ r \in \mathbb{R}_+,\ \|u\| \leq r\}$, where $\|u\|^2 = \langle u,u\rangle$ and $\langle\cdot,\cdot\rangle$ the standard scalar product in $\mathbb{R}^n$. The double cone $K \cup (-K)$ is called the *light cone* or Lorentz cone since for $u \in \mathbb{R}^3$ the coordinates in space and r the time, the double cone is just the light cone of special relativity theory (cf. [57], [58]).

One finds (see Exercise 4) for $x = (u,r)$ and $y = (v,s)$ in K that

$$\lambda(x,y) = \frac{s^2 - \|v\|^2}{2(rs - \langle u,v\rangle)} \quad \text{for} \quad \|u\| = r$$

and

$$\lambda(x,y) = \frac{rs - \langle u,v\rangle - \sqrt{(rs - \langle u,v\rangle)^2 - (r^2 - \|u\|^2)(s^2 - \|v\|^2)}}{r^2 - \|u\|^2} \quad \text{for} \quad \|u\| < r.$$

For what follows we need various properties of the order function dependent on properties of K. Concerning the latter, K is said to be **lineless** if K does not contain an affine line. K is said to be **archimedean** for a linear subspace U with $K \subset U$ (or integrally closed in U) if $x,y \in U$ and $y + nx \in K$ for all $n \in \mathbb{N}$ imply $x \in K$.

Lemma 3.1.4. *Let K be a convex cone in the real vector space V and let $x,y,z \in K \setminus \{0\}$.*

(i) *If K is lineless then K is pointed and $\lambda(x,y) \in \mathbb{R}_+$.*

(ii) $\lambda(\alpha x, \beta y) = \frac{\beta}{\alpha}\lambda(x,y)$ *for all $\alpha, \beta > 0$.*

(iii) $\lambda(x,y) \cdot \lambda(y,z) \leq \lambda(x,z)$.

(iv) $\lambda(x,y) + \lambda(x,z) \leq \lambda(x,y+z)$.

(v) $(\lambda(x,z)^{-1} + \lambda(y,z)^{-1})^{-1} \leq \lambda(x+y,z) \leq \min\{\lambda(x,z),\ \lambda(y,z)\}$.

(vi) $\min\{\lambda(x,y),\ \lambda(y,x)\} = \sup\{\lambda > 0 \mid \lambda x \leq y \leq \frac{1}{\lambda}x\}$
($\leq$ induced by K, $\sup\emptyset = 0$).

(vii) $\lambda(x,y) \cdot \lambda(y,x) = \sup\{\lambda\mu \mid \lambda,\mu > 0,\ \lambda x \leq y \leq \frac{1}{\mu}x\}$.

(viii) $\min\{\lambda(x,y), 1\} = \sup\{\alpha \in [0,1] \mid y = \alpha x + (1-\alpha)z$ *for some $z \in K$*$\}$.

(ix) *K is pointed iff $\min\{\lambda(x,y), \lambda(y,x)\} \leq 1$ for all $x,y \in K \setminus \{0\}$ iff $\lambda(x,y) \cdot \lambda(y,x) \leq 1$ for all $x,y \in K \setminus \{0\}$.*

(x) *For K pointed and $\mu(x,y) = \min\{\lambda(x,y),\ \lambda(y,x)\}$*
$\mu(x,y) + \mu(y,z) \leq 1 + \mu(x,z)$.

(xi) *If K is pointed and archimedean in $K-K$ then K is lineless.*

(xii) *If K is archimedean in $K-K$ and $\lambda(x,y) < \infty$ then $\lambda(x,y)x \leq y$.*

(xiii) *K is lineless iff $\mu(x,y) < 1$ for all $x,y \in K \setminus \{0\}$, $x \neq y$.*

(xiv) *Suppose K is given by*

$$K = \{x \in V \mid f(x) \geq 0 \quad \text{for all} \quad f \in \mathcal{F}\},$$

where $\mathcal{F}$ is a non-empty family of linear functionals $f: V \longrightarrow \mathbb{R}$. Then K is a convex cone which is archimedean in V with $0 \in K$ and one has that

$$\lambda(x,y) = \inf\left\{\frac{f(y)}{f(x)} \mid f \in \mathcal{F},\ f(x) > 0\right\}.$$

(xv) *Let $u, v \in V$ such that $x \pm u \in K$ and $y \pm v \in K$ and let $x' = x + \alpha u \in K$, $y' = y + \beta v \in K$ with $0 < \alpha, \beta < 1$. Then*

$$\frac{1-\beta}{1+\alpha}\lambda(x,y) \le \lambda(x', y') \le \frac{1+\beta}{1-\alpha}\lambda(x,y).$$

Proof. (i) If $x \in K \cap (-K)$ then $\mathbb{R}x \subset K$. If K is lineless then the line given by $\mathbb{R}x$ must be a point that is $x = 0$, and K is pointed. Suppose now $\lambda(x,y)$ is not finite, that is $\lambda(x,y) > \lambda$ for all $\lambda > 0$. If follows that $y - \lambda x \in K$ and, because of $y + \lambda x \in K$, that $y + \lambda x \in K$ for all $\lambda \in \mathbb{R}$. Therefore the affine line determined by y and $y + x$ is contained in K.

(ii) $\beta y - \lambda(\alpha x) \in K$ is equivalent to $y - \lambda(\frac{\alpha}{\beta})x \in K$ for any $\lambda \ge 0$.

(iii) If $\lambda x \le y$ and $\lambda' y \le z$ for $\lambda, \lambda' \ge 0$ ($\le$ induced by K) then $\lambda\lambda' x \le z$.

(iv) If $\lambda x \le y$ and $\lambda' x \le z$ for $\lambda, \lambda' \ge 0$ then $\lambda x + \lambda' x \le y + z$.

(v) Consider the first inequality. It is trivial for $\lambda(x,z) = 0$ or $\lambda(y,z) = 0$. Otherwise, let $\lambda, \lambda' > 0$ and $\lambda x \le z$, $\lambda' y \le z$. It follows that $x \le \frac{z}{\lambda}, y \le \frac{z}{\lambda'}$, and, hence $x + y \le z(\frac{1}{\lambda} + \frac{1}{\lambda'})$.

This proves the first inequality. For the second inequality let $\lambda(x+z) \le z$ for $\lambda \ge 0$. Obviously, $\lambda x \le \lambda(x+y) \le z$, $\lambda y \le \lambda(x+y) \le z$ which proves the second inequality.

(vi) We may assume that $\lambda(x,y) \le \lambda(y,x)$ and $\lambda(x,y) > 0$. For $0 < \lambda < \lambda(x,y)$ we must have that $\lambda x \le y$ and, because of $\lambda < \lambda(y,x)$, also that $\lambda y \le x$.

This proves that $\min\{\lambda(x,y), \lambda(y,x)\} \le \sup\{\lambda > 0 \mid \lambda x \le y \le \frac{1}{\lambda}x\}$. Conversely, if $\lambda x \le y \le \frac{1}{\lambda}x$ then $\lambda(x,y) \ge \lambda$ and $\lambda(y,x) \ge \lambda$.

(vii) For $\lambda(x,y) = 0$ or $\lambda(y,x) = 0$ the asserted equation holds trivially ($\sup\emptyset = 0$). For $\lambda(x,y) > 0$ and $\lambda(y,x) > 0$ one has that

$$\lambda(x,y)\cdot\lambda(y,x) = \sup\{\lambda\mu \mid \lambda,\mu > 0,\ \lambda x \le y \text{ and } \mu y \le x\}.$$

(viii) Suppose $\lambda(x,y) \le 1$. Then $\lambda x \le y$ is equivalent to $y = \lambda x + (1-\lambda)\frac{z}{1-\lambda}$ for $z = y - \lambda x$, $0 \le \lambda < 1$. Suppose $\lambda(x,y) > 1$. Then $\lambda x \le y$ for some $\lambda > 1$ and, hence, for any $0 < \epsilon < 1$

$$y = (1-\epsilon)x + \epsilon z$$

with $z = \frac{y-(1-\epsilon)x}{\epsilon} \in K$.

This shows, by taking $\epsilon \longrightarrow 0$, that the right hand side of (viii) is equal to 1.

(ix) Assume first that K is not pointed, i.e., there exists $x \in K \cap (-K)$, $x \ne 0$. For $y = -x \in K$ one has that $y - 2x = -3x \in K$ and $x - 2y = 3x \in K$ and, hence,

$\lambda(x,y) \geq 2$, $\lambda(y,x) \geq 2$. Conversely, assume that for some $x,y \in K \setminus \{0\}$ one has $\min\{\lambda(x,y), \lambda(y,x)\} > 1$, which, obviously, is equivalent to $\lambda(x,y) \cdot \lambda(y,x) > 1$. By (vi) there exists $\lambda > 1$ such that $\lambda x \leq y \leq \frac{1}{\lambda}x$ and, hence, $\lambda^2 x \leq x$. This implies $x(1-\lambda^2) \in K$ and because of $1 - \lambda^2 < 0$ it follows that $0 \neq x \in K \cap -K$. Thus, K is not pointed.

(x) Let $\lambda x \leq y \leq \frac{1}{\lambda}x$ and $\lambda' y \leq z \leq \frac{1}{\lambda'}y$ for $\lambda, \lambda' > 0$. Then $\lambda\lambda' x \leq z \leq \frac{1}{\lambda\lambda'}x$ and by (vi)

$$\lambda\lambda' \leq \min\{\lambda(x,z), \lambda(z,x)\} = \mu(x,z).$$

By (ix) one has that $\lambda \leq \mu(x,y) \leq 1$, $\lambda' \leq \mu(y,z) \leq 1$ and, hence, $\lambda + \lambda' - \lambda\lambda' = \lambda(1-\lambda') + \lambda' \leq 1$.

Thus, $\lambda + \lambda' \leq 1 + \lambda\lambda' \leq 1 + \mu(x,z)$ which, by (vi), proves (x).

(xi) Suppose $x + \lambda(y-x) \in K$ for all $\lambda \in \mathbb{R}$. In particular, $x + n(y-x) \in K$ and $x + n(x-y) \in K$ for all $n \in \mathbb{N}$. Since K is archimedean in $K - K$ we must have that $y - x \in K$ and $x - y \in K$. Since K is pointed we must have that $x = y$, that is K is lineless.

(xii) Let $\lambda = \lambda(x,y) < \infty$. It follows that $(\lambda - \frac{1}{n})x \leq y$ for all $n \in \mathbb{N}$ and, hence, $x + n(y - \lambda x) \in K$ for all $n \in \mathbb{N}$. Since $y - \lambda x \in K - K$ and K is archimedean one has that $y - \lambda x \in K$.

(xiii) Suppose first that K contains an affine line, i.e., $x + \lambda(y-x) \in K$ for all $\lambda \in \mathbb{R}$, where $x,y \in K\setminus\{0\}$, $x \neq y$. For $\lambda > 1$ one obtains $\lambda y \geq (\lambda-1)x$ and, hence, $\lambda(x,y) \geq \frac{\lambda-1}{\lambda}$. For $\lambda \longrightarrow \infty$ this yields $\lambda(x,y) \geq 1$. For $\lambda < 0$ one obtains $(1-\lambda)x \geq (-\lambda)y$ and, hence, $\lambda(y,x) \geq \frac{-\lambda}{1-\lambda}$. For $\lambda \longrightarrow \infty$ this yields $\lambda(y,x) \geq 1$. Thus, $\mu(x,y) \geq 1$. Conversely, suppose that $\mu(x,y) \geq 1$ for some $x,y \in K \setminus \{0\}$, $x \neq y$. This implies for any $0 < \epsilon \leq 1$ that $(1-\epsilon)x \leq y$ and $(1-\epsilon)y \leq x$ and, hence $y - x + \epsilon x \in K$, $x - y + \epsilon y \in K$ for all $0 < \epsilon \leq 1$. Consider the affine line given by $x + \lambda(y-x)$ for $\lambda \in \mathbb{R}$. For $\lambda > 0$ we have that

$$x + \lambda(y-x) = \lambda\Big((y-x) + \epsilon x + \Big(\frac{1}{\lambda} - \epsilon\Big)x\Big) \in K,$$

provided that $0 < \epsilon \leq 1$ is small enough such that $\frac{1}{\lambda} - \epsilon \geq 0$. For $\lambda < 0$ we have that

$$x + \lambda(y-x) = y + (1-\lambda)(x-y) = (1-\lambda)\Big(x - y + \epsilon y + \Big(\frac{1}{1-\lambda} - \epsilon\Big)y\Big) \in K,$$

provided that $0 < \epsilon \leq 1$ is small enough that $\frac{1}{1-\lambda} - \epsilon \geq 0$. Thus, the affine line considered is contained in K.

(xiv) Obviously, K is a convex cone with $0 \in K$. Let $x,y \in V$ such that $y + nx \in K$ for all $n \in \mathbb{N}$. It follows that $f(y) + nf(x) = f(y + nx) \geq 0$ and, hence, $f(x) \geq -\frac{1}{n}f(y)$ for all $n \in \mathbb{N}$. Thus, $f(x) \geq 0$ and, since this holds for every $f \in \mathcal{F}$, we must have that $x \in K$. This shows that K is archimedean in V. Finally, let $\lambda x \leq y$ for $x,y \in K$ and $\lambda \geq 0$. It follows that $\lambda f(x) \leq f(y)$ and, hence, $\lambda \leq \frac{f(y)}{f(x)}$ for all $f \in \mathcal{F}$ with $f(x) > 0$. This shows that $\lambda(x,y) \leq \inf\big\{\frac{f(y)}{f(x)} \mid f \in \mathcal{F},\, f(x) > 0\big\}$. Conversely, for $x,y \in K$ and $r = \inf\big\{\frac{f(y)}{f(x)} \mid f \in \mathcal{F},\, f(x) > 0\big\}$ we have that $f(y - rx) = f(y) - rf(x) \geq 0$ for all $f \in \mathcal{F}$. By definition of K, therefore, $y - rx \in K$ which implies that $\lambda(x,y) \geq r$.

(xv) From the identities $x = \frac{1}{1+\alpha}(x + \alpha u) + \frac{\alpha}{1+\alpha}(x-u)$ and $x + \alpha u = (1-\alpha)x + \alpha(x + u)$ together with the assumptions it follows that $\lambda(x',x) \geq \frac{1}{1+\alpha}$ and $\lambda(x,x') \geq 1 - \alpha$, respectively. Similarly, $\lambda(y',y) \geq \frac{1}{1+\beta}$ and $\lambda(y,y') \geq 1 - \beta$. Together with property (iii) we obtain

$$(1-\lambda)\lambda(x',y')\frac{1}{1+\beta} \leq \lambda(x,x')\lambda(x',y')\lambda(y',y) \leq \lambda(x,y) \quad \text{and}$$
$$\frac{1}{1+\alpha}\lambda(x,y)(1-\beta) \leq \lambda(x',x)\lambda(x,y)\lambda(y,y') \leq \lambda(x',y')$$

from which the assertion follows. □

The property (xiv) in the Lemma admits a dual computation of the order function in case one has a description of the cone in terms of the dual space of V. Since $K = \mathbb{R}^n_+$ in $V = \mathbb{R}^n$ is given by the projections $f_i(x) = x_i$ from property (xiv) one gets immediately that $\lambda(x,y) = \min\{\frac{y_i}{x_i} \mid x_i > 0,\ 1 \leq i \leq n\}$ in this case (cf. Example 3.1.3 (i)). Since any convex cone in $\mathbb{R}^2$ which is closed can be described just by two linear functionals, the order function in this case can be easily obtained by the two functionals according to property (xiv). This, actually, is no longer true in $\mathbb{R}^3$, where for example ice cream cones cannot be described by finitely many but by infinitely many linear functionals (cf. Example 3.1.3 (ii)). More generally, property (xiv) applies to convex cones K in an arbitrary vector space V which are closed with respect to a locally convex Hausdorff topology on V (e.g., to closed convex cones in a Banach space). From the Hahn–Banach Theorem it follows that $K = \{x \in V \mid f(x) \geq 0,\ f \in \mathcal{F}\}$ where $\mathcal{F}$ is a set of continuous linear functionals on V.

Though the above description by linear functionals is not available in general one can, however, describe the order function of any lineless convex cone by concave functionals similar to property (xiv). Denote for a convex cone K by K^c the set of all concave functions $f\colon K \longrightarrow \mathbb{R}_+$ which are positively homogeneous. Any $f \in K^c$ must be monoton with respect to $\leq$ induced by K. (See Lemma 2.1.3 and the definitions given in Section 2.1) Obviously, a function $f\colon K \longrightarrow \mathbb{R}_+$ is in K^c iff f is *superadditive*, i.e., $f(x+y) \geq f(x) + f(y)$, and positively homogeneous. The set K^c is itself a convex cone within the vector space of real valued functions.

Two subsets A, B of K are said to satisfy a (both sided) **Harnack inequality** if there exists a constant $c > 0$ such that

$$cf(x) \leq f(y) \leq \frac{1}{c}f(x)$$

for all $f \in K^c$, all $x \in A$, all $y \in B$.

Lemma 3.1.5. *Let K be a lineless convex cone.*
(i) *For any $x \in K \setminus \{0\}$ the function $y \longmapsto \lambda(x,y)$ is concave and positively homogeneous.*
(ii) *for any $y \in K \setminus \{0\}$ the function $x \longmapsto (\lambda(x,y))^{-1}$ is convex and positively homogeneous (value $+\infty$ admitted).*

(iii) *For any $x \in K \setminus \{0\}$ it holds that*

$$\lambda(x,y) = \inf\left\{\frac{f(y)}{f(x)} \mid f \in K^c,\ f(x) > 0\right\}$$
$$= \inf\{f(y) \mid f \in K^c,\quad f(x) = 1\}.$$

(iv) *Two subsets A, B of K satisfy a Harnack inequality iff $\inf_{x\in A, y\in B} \mu(x,y)$ is not zero which, in this case, is the smallest possible constant c.*

Proof. (i) Follows from Lemma 3.1.4, properties (i), (ii), (iv).

(ii) Follows from Lemma 3.1.4, properties (ii), (v).

(iii) For $f \in K^c$ and $\lambda x \le y$ one has that $\lambda f(x) \le f(y)$ and, hence, $\lambda(x,y) \le \inf\{\frac{f(y)}{f(x)} \mid f \in K^c, f(x) > 0\}$. For $x \in K\setminus\{0\}$ the function defined by $g(y) = \lambda(x,y)$ is in K^c by (i) and, hence, $\lambda(x,y) = \frac{g(y)}{g(x)} \ge \inf\{\frac{f(y)}{f(x)} \mid f \in K^c, f(x) > 0\}$.

(iv) Suppose $cf(x) \le f(y) \le \frac{1}{c}f(x)$ for all $f \in K^c$, all $x \in A$, all $y \in B$. From (iii) it follows that

$$c \le \inf\left\{\frac{f(y)}{f(x)}\,\middle|\, f \in K^c,\ f(x) > 0\right\} = \lambda(x,y)$$

and

$$c \le \inf\left\{\frac{f(x)}{f(y)}\,\middle|\, f \in K^c,\ f(y) > 0\right\} = \lambda(y,x).$$

Therefore, $0 < c \le \min\{\lambda(x,y), \lambda(y,x)\} = \mu(x,y)$ all $x \in A$, $y \in B$. Conversely, if $c = \inf_{x\in A, y\in B} \mu(x,y) > 0$ then A, B satisfy a Harnack inequality with c. □

Remark 3.1.6. It is possible to define an order function like λ for arbitrary convex sets C by $\beta(x,y) = \sup\{\beta \in [0,1] \mid y = \beta x + (1-\beta)x' \text{ for some } x' \in C\}$. For β similar properties as those in Lemmas 3.1.4 and 3.1.5 can be proven. (Cf. [37, Sections 2 and 3]; for convex sets and an abstract version of Harnacks inequality see also [4], [5]). The order function λ or some equivalent concept appear in almost all approaches to Hilbert's projective metric and the part metric, respectively and are studied more or less explicitly. (See [57] and also the references in the next section.)

Exercises

1. Consider for $r, s \in \mathbb{R}$ the convex cone in $\mathbb{R}^2$ given by

$$K = \{x \in \mathbb{R}^2 \mid rx_1 \le x_2, sx_1 \le x_2\}.$$

   (a) Show that K is archimedean in $\mathbb{R}^2$.
   (b) For which values of r and s is K pointed and lineless, respectively?
   (c) Compute the order function for K.

2. Let $K = \mathbb{R}^n_+$ be the standard cone in $\mathbb{R}^n$.
   (a) Compare the order functions for the cones K and $\mathring{K}$ (interior of K).
   (b) For which $x \in K \setminus \{0\}$ is the mapping $y \longmapsto \lambda(x,y)$ linear on K?
   (c) Show that the standard representation $x = \sum_{i=1}^n x_i e_i$ of $x \in K$ by the standard basis $(e_1, \ldots, e_n)$ can be obtained by extracting successively from x the vectors of the standard basis. Does the ordering in which the e_i are extracted play any role for the representation?
3. Let $K = \{f \in \mathcal{C}[0,1] \mid f(x) \geq 0 \text{ for all } x \in [0,1]\}$ be the standard cone in the vector space of all real continuous functions on the interval $[0,1]$.
   (a) Show that K is a convex cone that is lineless and archimedean.
   (b) Compute $\lambda(f,g)$ for $f,g \in K$ and show that for $f \in \mathcal{C}[0,1]$ the supremum norm is given by $\|f\| = \lambda(|f|, 1)^{-1}$.
   (c) Show that for no $f \in K$ the mapping $g \longmapsto \lambda(f,g)$ is linear on K.
4. Consider the ice cream cone

$$K = \{x = (u,r) \in \mathbb{R}^n \times \mathbb{R}_+ \mid \|u\| \leq r\}$$

   where $\|u\|^2 = \langle u, u\rangle$, $\langle \cdot, \cdot \rangle$ standard scalar product.
   (a) Proof that for $\|u\| < r$

$$\lambda(x,y) = \frac{rs - \langle u,v\rangle - \sqrt{(rs - \langle u,v\rangle)^2 - (r^2 - \|u\|^2)(s^2 - \|v\|^2)}}{r^2 - \|u\|^2}$$

   where $x = (u,r), y = (v,s)$.
   (b) For $\cos h\phi = \frac{1}{2}(e^{\phi} + e^{-\phi})$ define $\Phi = \Phi(x,y)$ by

$$\cos h\Phi = \frac{rs - \langle u,v\rangle}{\sqrt{(r^2 - \|u\|^2)(s^2 - \|v\|^2)}}.$$

   Show that $-\log(\lambda(x,y) \cdot \lambda(y,x)) = 2\Phi$.
   (c) Demonstrate (a) and (b) for the infinite dimensional ice cream cone

$$K = \left\{ x = (u,r) \in \mathbb{R}^{\mathbb{N}} \times \mathbb{R}_+ \mid \sqrt{\sum_{i=1}^{\infty} u_i^2} \leq r \right\}.$$

5. A subset M of a normed vector space is **symmetrically bounded** if every symmetric subset of M is bounded. (A subset S of a real vector space is symmetric if there exists some $c \in S$, called a center of S, such that $2c - S \subset S$.)
   (a) Show that every symmetrically bounded set M is lineless.
   (b) Find a convex set in a normed space that is lineless but not symmetrically bounded.
   (c) Find a convex cone in a normed space that is pointed with 0 but not lineless. (By Lemma 3.1.4 every lineless convex cone is pointed.)

3.2 Internal metrics

From the order function λ and using its properties (Lemma 3.1.4) we can construct various metrics on a convex cone. Because $\lambda(x, y) = 0$ may happen some of the metrics may be extended, i.e., can have the value $+\infty$. This does not happen on parts of the cone defined as follows.

Definition 3.2.1. Let K be a convex cone in some real vector space with order function λ. For $x, y \in K \setminus \{0\}$ let xCy if x is a component of y, i.e., $\lambda(x, y) > 0$, and let $x \sim y$ if xCy and yCx. A non-empty subset $0 \notin P \subset K$ is called a **part** of K if $y \sim x$ for $x \in P$ is equivalent to $y \in P$. If $0 \in K$ then $\{0\}$ is called the **zero-part**.

Lemma 3.2.2. *Let K be a convex cone.*
(i) *C is a reflexive and transitive relation and $\sim$ is an equivalence relation on $K \setminus \{0\}$.*
(ii) *The parts of K are convex cones.*
(iii) *K is the disjunctive union of its parts.*
(iv) *If $P \neq \{0\}$ is a part of K, λ_K and λ_P are the order function of K and P, respectively, then λ_P and λ_K coincide on $P \times P$.*
(v) *For $x, y \in K \setminus \{0\}$ it holds that $x \sim y$ iff $x + r(y - x) \in K$ for some $r < 0$ and some $1 < r$.*

Proof. (i) C is reflexive and transitive by Lemma 3.1.4 (iii). Thus, $\sim$ is an equivalence relation.

(ii) Obviously, the zero-part is a convex cone. Let P be a part, $P = [x]$ the equivalence class for some $0 \neq x \in P$. If $y \in P$, i.e., $\lambda(x, y) > 0$ and $\lambda(y, x) > 0$, then by property (ii) of Lemma 3.1.4 it holds that $\lambda(x, \beta y) > 0$ and $\lambda(\beta y, x) > 0$ for all $\beta > 0$. Therefore, $\beta y \in P$ and P is cone. Furthermore, if $y \in P$ and $z \in P$ then $\lambda(x, y + z) \geq \lambda(x, y) + \lambda(x, z) > 0$ by Lemma 3.1.4 (iv). From property (v) we obtain $\lambda(y + z, x) \geq (\lambda(y, x)^{-1} + \lambda(z, x)^{-1})^{-1} > 0$ and, hence, $y + z \in P$. Thus, P is a convex cone.

(iii) Obvious, since the parts are the equivalence classes for $\sim$.

(iv) For $x, y \in P$ it holds $\lambda_P(x, y) \leq \lambda_K(x, y)$ because of $P \subset K$. Without loss suppose that $\lambda_K(x, y) > 0$ and $y - \lambda x \in K$ for some $\lambda > 0$. For any $0 < \epsilon < \lambda$ one has that $z = y - \lambda x + \epsilon x \in K$ and xCz. Furthermore, $zC(y + \epsilon x)$ and, by $y + \epsilon x \in P$, $(y + \epsilon x)Cx$. Therefore, zCx which shows that $z \in P$. Thus, $y - (\lambda - \epsilon)x = z \in P$ and $\lambda_P(x, y) \geq \lambda - \epsilon$. Since $0 < \epsilon < \lambda$ is arbitrary it follows that $\lambda_P(x, y) \geq \lambda$ and, hence, $\lambda_P(x, y) \geq \lambda_K(x, y)$.

(v) By definition, $x \sim y$ iff there exist $0 < \lambda, \mu$ such that $y - \lambda x \in K$ and $x - \mu y \in K$. We may assume that $\lambda, \mu < 1$. Now, $y - \lambda x \in K$ is equivalent to $x + r(y - x) \in K$ for $r = \frac{1}{1-\lambda}$ and $x - \mu y \in K$ is equivalent to $x + r(y - x) \in K$ for $r = -\frac{\mu}{1-\mu}$. □

Property (v) of Lemma 3.2.2 presents a particular simple description of the **part relation** $\sim$, as illustrated in the following figure.

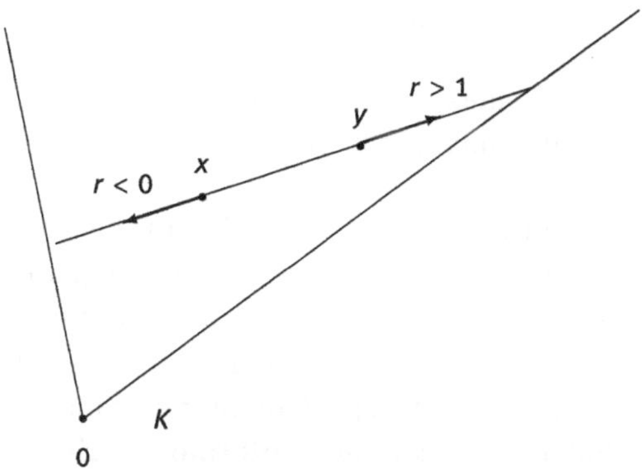

Fig. 3.2. Part relation $x \sim y$.

This description of $\sim$ by extending the segment $[x, y]$ a little bit to the right and to the left is used in [2–6] to study parts and the part metric within arbitrary convex sets.

The following figure illustrates the concept of parts for the cone $\mathbb{R}^3_+$.

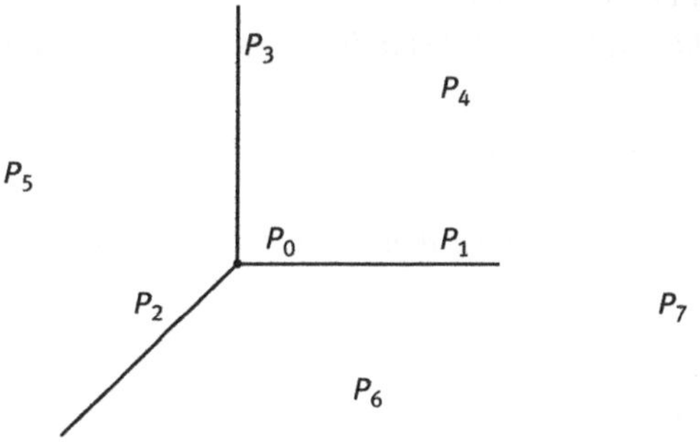

Fig. 3.3. Parts of cone $\mathbb{R}^3_+$.

P_0 is the zero part; the halflines (without 0) P_1, P_2, P_3 are the 1-dimensional parts; the cones (without boundary) P_4, P_5, P_6 are the 2-dimensional parts; the interior P_7 of $\mathbb{R}^3_+$ is the 3-dimensional part. All parts are convex cones and $\mathbb{R}^3_+$ is the disjunctive union of the P_i, $0 \leq i \leq 7$.

For an arbitrary convex cone K in some real vector space we define the following entities which will turn out to be metrics under certain conditions specified in Theorem 3.2.3.

For $x, y \in K \setminus \{0\}$, $\lambda(x, y)$ the order function and $\mu(x, y) = \min\{\lambda(x, y), \lambda(y, x)\}$ the symmetrized order function consider

- the **projective Hilbert metric:** $d(x, y) = -\log[\lambda(x, y) \cdot \lambda(y, x)]$;
- the **Thompson metric** or *part metric:* $p(x, y) = -\log \mu(x, y)$;
- the **Harnack metric:** $h(x, y) = 1 - \mu(x, y)$;
- the **Gleason metric:** $g(x, y) = 2\frac{1-\sqrt{\mu(x,y)}}{1+\sqrt{\mu(x,y)}}$;
- the **Bear metric:** $b(x, y) = \frac{1-\mu(x,y)}{1+\mu(x,y)}$;

– the **Kobayashi metric:** $k(x, y) = -\log \nu(x, y)$, where

$$\nu(x, y) = \begin{cases} \mu(x, y) & \text{if } \max\{\lambda(x, y), \lambda(y, x)\} \geq 1 \\ \lambda(x, y) \cdot \lambda(y, x) & \text{if } \max\{\lambda(x, y), \lambda(y, x)\} < 1. \end{cases}$$

The following result shows that all these "metrics" are just "ecarts" m in the sense of Bourbaki [13] which means that m is symmetric, satisfies the triangle inequality and $m(x, x) = 0$ for $x \neq 0$ (see also Remark 3.2.4). By this result also, on a lineless part of the cone the definitions of p, h, g, b, k give neat metrics, whereas d gives only a quasi-metric. In what follows we will, however, simply speak of metrics. Since all these metrics are built up from the order function we call them **internal metrics** on the cone.

Theorem 3.2.3. *Let K be a convex cone and let P be a non-zero part of K.*
(i) *On $P \times P$ the expressions for d, p, h, g, b, k are well-defined, real-valued and symmetric.*
(ii) *If P is pointed then d, p, h, g, b, k are all non-negative.*
(iii) *d, p and k satisfy the triangle inequality on $P \times P$. If P is pointed then h, g, b satisfy the triangle inequality on $P \times P$.*
(iv) *If P is lineless then p, h, g, b, k are 0 for $(x, y) \in P \times P$ iff $x = y$. Furthermore, $d(x, y) = 0$ iff $y = \lambda x$ for some $\lambda > 0$.*
(v) *If P is lineless then p, h, g, b, k are metrics on P and d is a quasi-metric (with $d(x, y) = 0$ iff x and y are on the same rag).*

Proof. (i) Obvious, because of $x \sim y$ iff $\lambda(x, y) > 0$ and $\lambda(y, x) > 0$.

(ii) This follows from Lemma 3.1.4 (ix).

(iii) From property (iii) of Lemma 3.1.4 one has that

$$[\lambda(x, y) \cdot \lambda(y, x)] \cdot [\lambda(y, z) \cdot \lambda(z, y)] \leq [\lambda(x, y) \cdot \lambda(z, x)],$$

which proves the triangle inequality for d on $P \times P$.

Also by property (iii)

$$\mu(x, y) \cdot \mu(y, z) \leq \lambda(x, y), \lambda(z, x) \text{ and, hence, } \mu(x, y) \cdot \mu(y, z) \leq \mu(x, z),$$

which proves the triangle inequality for p. Also, these inequalities imply the triangle inequality for k.

Furthermore, for h

$$\begin{aligned} h(x, z) &= 1 - \mu(x, z) \leq 1 - \mu(x, y) \cdot \mu(y, z) \\ &= (1 - \mu(x, y)) + (1 - \mu(y, z)) - (1 - \mu(x, y))(1 - \mu(y, z)) \\ &= h(x, y) + h(y, z) - h(x, y)h(y, z). \end{aligned}$$

If P is pointed then $h(x, y)h(y, z) \geq 0$ by (ii) and, hence, $h(x, z) \leq h(x, y) + h(y, z)$. To obtain the triangle inequality for b and g, respectively, observe the following inequality for non-negative numbers α, β, γ with $\alpha\beta \le \gamma$

$$\frac{1-\gamma}{1+\gamma} \le \frac{\frac{1-\alpha}{1+\alpha} + \frac{1-\beta}{1+\beta}}{1 + \frac{1-\alpha}{1+\alpha} \cdot \frac{1-\beta}{1+\beta}}. \qquad (*)$$

Put in $(*)$ $\alpha = \mu(x,y)$, $\beta = \mu(y,z)$, $\gamma = \mu(x,z)$.
The definition of b gives

$$b(x,z) \le \frac{b(x,y) + b(y,z)}{1 + b(x,y) \cdot b(y,z)} \le b(x,y) + b(y,z).$$

because $b \ge 0$ by (ii).
Similarly, putting in $(*)$ $\alpha = \sqrt{\mu(x,y)}$, $\beta = \sqrt{\mu(y,z)}$, $\gamma = \sqrt{\mu(x,z)}$ one obtains

$$g(x,z) \le \frac{g(x,y) + g(y,z)}{1 + \frac{1}{4}g(x,y) \cdot g(y,z)} \le g(x,y) + g(y,z).$$

(iv) The statement concerning p, h, g, b, k follows from Lemma 3.1.4 (xiii). Concerning d suppose that $d(x,y) = 0$, that is $\lambda(x,y) \cdot \lambda(y,x) = 1$. For $\alpha = \lambda(x,y) > 0$ property (ii) of Lemma 3.1.4 yields

$$\lambda(\alpha x, y) = \frac{1}{\alpha}\lambda(x,y) = 1 \text{ and } \lambda(y, \alpha x) = \alpha\lambda(y,x) = 1.$$

Therefore $\mu(\alpha x, y) = 1$ and, by Lemma 3.1.4 (xiii), we must have $y = \alpha x$. Conversely, if $y = \lambda x$ with $\lambda > 0$ then, by Lemma 3.1.4 (ii), we must have $d(x,y) = -\log[\lambda \cdot \frac{1}{\lambda}] = 0$. Also, k can be extended on K with $+\infty$ as a possible value.

(v) Follows from (i)–(iv) since a lineless convex cone is pointed. □

Remark 3.2.4. If K is any lineless convex cone then Theorem 3.2.3 (v) applies to all non-zero parts of K. The metrics p, h, g, b can easily be extended to a metric on the whole K since for $x, y \in K$ in different parts $\mu(x,y) = 0$ and by setting $\mu(0,0) = 1$. For p, however, it may happen that $p(x,y) = +\infty$. Since $\lambda(x,y) \cdot \lambda(y,x) = 0$ for x and y in different parts, also d can be extended to K with $d(x,y) = +\infty$ in this case and by setting $d(0,0) = 0$. Also, k can be extended on K with $+\infty$ as a possible value.

The metrics just defined play important roles in such different disciplines as potential theory, complex functions, function algebras, (non-Euclidean) geometry, and in various parts of functional analysis as positive operators, convexity, Dirichlet forms. Since these metrics also have a long and interesting history some additional comments may be in order. What today is called the projective Hilbert metric was considered by Hilbert in his investigations on the foundations of geometry for general convex bodies in $\mathbb{R}^n$ (see [23] and the discussion in the next Section 3.3). Actually, this metric has been already investigated by A. Cayley [19] and F. Klein [29] in their models of hyperbolic geometry where the metric is given as the logarithm of the cross ratio for two points in the open unit disc. For that reason, what we call the projective Hilbert metric is sometimes called the Cayley–Hilbert metric (e.g. in [18, 50]) or the Klein–Hilbert metric (e.g. in [6, 49]) or simply hyperbolic length.(G. Birkhoff in a letter from 1984 to the author remarked with respect to this metric: "... which Felix Klein says goes back to Legendre.") For the role of Hilbert's projective metric in geometry see [15, 16]. In a complete different setting, namely for positive linear operators on infinite dimensional spaces, Hilbert's projective metric was first introduced by G. Birkhoff into functional analysis. (See [10] and, with improvements, [12]. See also [38].) About the same time H. Samelson [50] used this metric in finite dimensions to give a rather elementary proof of the Perron–Frobenius Theorem (see also [8]). The first use of this metric for non-linear operators was made by P. Bushell [17], A. J. B. Potter [48], and M.A. Krasnoselski et al. [36]. A modern monography with applications to non-linear operators is by R. D. Nussbaum [44, 45] and one of the few textbooks treating this metric is V.I. Istratescu [27].

The Thompson metric was first introduced by A. C. Thompson in his Ph.D. thesis [53] (see [54] for major results) in dealing with non-linear positive operators. There Thompson refers to the use of Hilbert's projective metric made by Birkhoff and Samelson. There is another, quite interesting source, for Thompson's metric and this is the reason why it is called the part metric. For the maximal ideal space of a function algebra A.M. Gleason studied in 1957 an equivalence relation, the equivalence classes of which were later on called Gleason parts. The equivalence $x \sim y$ can be described by $G(x,y) < 2$ for a certain metric called later on Gleason metric (see [4, 5] and [32, 33] for details). Related to G two other metrics were investigated (which we called in the context of cones part metric and Bear metric) and non-trivial relations between them were established which are reflected in the definitions of part metric, Gleason metric, and Bear metric we gave here within the framework of convex cones, namely

$$p(x,y) = \log \frac{1 + b(x,y)}{1 - b(x,y)} = 2 \log \frac{2 + g(x,y)}{2 - g(x,y)}$$

(for function algebras see [33, p. 100] and [5, p. 3]). Bear, Weiss, and Bauer then made stepwise the important and beautiful finding that the concepts of parts and corresponding metrics are linked essentially, to convex sets [3, 4, 7].

What is here called the Harnack metric has been less well considered in the literature though this metric occurs already in [53]. Though almost the same as the part metric, the Harnack metric is much simpler defined and has the advantage of being finite also across parts. Metric h is called the Harnack metric because it is strongly connected to Harnack inequalities, e.g., $1 - h(x,y) = \mu(x,y)$ is the smallest constant c for which $\{x\}$ and $\{y\}$ satisfy a Harnack inequality (see Lemma 3.1.5 (iv); for the Harnack metric with respect to harmonic functions see [31]). Concerning the naming of the various metrics one has to pay attention, e.g., is the part metric also called Harnack metric in [33] and Birkhoff metric in [35] (though the metric employed by Birkhoff was the projective Hilbert metric).

The Kobayashi metric has its background in the search for invariant metrics in complex analysis. Another, very well studied invariant metric goes back to Carathéodory. Invariant means that distances do not increase when holomorphic mappings are applied. Both metrics have been considered also on convex cones as the Kobayashi-type pseudo-distance and the Carathéodory-type pseudo-distance (see [20]). Invariance here means that distances do not increase when linear selfmappings of the cone are applied or, equivalently, all linear self-mappings of the cone are non-expansive for the metric considered. Actually, the latter turns out to be equal to the part metric and the former equal to the Kobayashi metric as defined here (see also Exercise 4 (a)). This follows from two results in [20, pp. 22 and 26] which together characterize invariant metrics on a one-part lineless cone of dimension greater than one as the metrics of the form $m(x,y) = f(\log\lambda(x,y), -\log\lambda(y,x))$ where f is a so-called special function, a function $f\colon \mathbb{R}^2 \longrightarrow \mathbb{R}_+$ defined by certain properties. (See also Exercise 5.) This nice result describes invariant metrics by $\lambda(x,y)$ and, hence, as internal metrics. In later chapters it will be a major concern for which *non-linear* selfmappings of the cone the metrics under consideration are invariant, too. From a view-point completely different from the one in [20], a characterization of Hilbert's projective metric was given in [30]. Let K be a closed convex cone in $\mathbb{R}^n$ which is pointed and has non-empty interior $\mathring{K}$. Then Hilbert's projective metric is the only projective metric on $\mathring{K}$, up to a strictly increasing scaling, for which every linear mapping of $K \setminus \{0\}$ into $\mathring{K}$ is a contraction [30, p. 204].

Exercises

1. Show that for any convex lineless cone K in a real vector space V the following properties hold. (See [3]).
   (a) For any concave function $u\colon K \longrightarrow \mathbb{R}_+$ and any part P of K one has the following alternative:
   Either $u(x) > 0$ for all $x \in P$ or $u(x) = 0$ for all $x \in P$.
   (b) Parts are connected sets (with respect to the part metric).
   (c) Let V be a normed space and $x \in \operatorname{int}K$. For $y \in K$ one has that $y \sim x$ iff $y \in \operatorname{int}K$.
2. Let $K = \{(u,r) \mid u \in \mathbb{R}^n,\ r \in \mathbb{R}_+,\ \|u\| \le r\}$ be the ice cream cone (see Exercise 4 to 3.1).
   (a) Describe all parts of K.
   (b) Compute the projective Hilbert metric on K.
   (c) Compute the part metric and the Kobayashi–metric on K.
3. Let p, g, b part metric, Gleason metric and Bear metric, respectively as defined in Section 3.2.
   (a) Prove for any part $\neq \{0\}$ the equations
   $$p(x,y) = \frac{1+b(x,y)}{1-b(x,y)} = 2\frac{1+g(x,y)}{1-g(x,y)}.$$
   (b) Compute explicitly p, g, b for $K = \operatorname{int}\mathbb{R}^2_+$.

(c) Prove that every linear selfmapping f of a lineless convex cone is non-expansive for the metrics $m = p, g, b$, i.e., $m(f(x), f(x)) \leq m(x, y)$ for all x, y.

4. Let k be the Kobayashi metric on K.
   (a) Show that $k(x, y) = \max\{p(x, y), d(x, y)\}$.
   (b) By using (a) show that k is a metric on each non-zero part of K.
   (c) Compute explicitly the Kobayashi metric on $K = \operatorname{int}\mathbb{R}^2_+$.

5. Let K be a convex cone in some real vector space V. A metric m on K is called **special** if all linear mappings $f : V \longrightarrow V$ with $f(K) \subset K$ are non-expansive for m. A function $g : \mathbb{R}^2 \longrightarrow \mathbb{R}_+$ is called *special* if it stems from a special metric in the sense that
   $g(a, b) = m((1, 1), (e^a, e^b))$ for all $(a, b) \in \mathbb{R}^2$ for some special metric m on $\operatorname{int}\mathbb{R}^2_+$. (See [20].)
   (a) Prove that for $\{0\} \neq K$ lineless and consisting of one part only and for any special function g
   $$m_g(x, y) = g(\log \lambda(x, y), -\log \lambda(y, x))$$
   defines a special metric on K.
   (b) Show that for any special metric m on $\operatorname{int}\mathbb{R}^2_+$ one has that $m = m_g$ where $g(a, b) = m((1, 1), (e^a, e^b))$.
   (c) Consider for $x, y \in \operatorname{int}\mathbb{R}^2_+$ and any special function h
   $$m(x, y) = m_h(x, y) + \left|\arctan \frac{x_2}{x_1} - \arctan \frac{y_2}{y_1}\right|.$$
   Show that m is a metric on $\operatorname{int}\mathbb{R}^2_+$ which is not special.

3.3 Geometrical properties

It is often convenient to visualize or analyse a convex cone by a cone base, in the case the latter exists.

Definition 3.3.1. Let K be a convex cone in a real vector space V and $f : V \longrightarrow \mathbb{R}$ a linear functional with $f(x) > 0$ for $x \in K \setminus \{0\}$. The set $B = \{x \in K \mid f(x) = 1\}$ is called a **base of K**. For $x \in K \setminus \{0\}$ the point $\bar{x} = \frac{x}{f(x)} \in B$ is called **base point of x**.

Obviously, a base need not exist and, if it exists, it is a convex set that is not uniquely determined. Furthermore, every $x \in K \setminus \{0\}$ has a unique representation $x = \lambda \bar{x}$ with $\lambda > 0$ and $\bar{x} \in B$; thereby, $\bar{x}$ is the base point of x and $\lambda = f(x)$. (See Exercise 1.)

Any convex set C can be viewed as a base of some convex cone. Namely, define for a (non-empty) convex subset of a real vector space W the cone $K = \mathbb{R}_+(C \times \{1\}) = \{(\lambda u, \lambda) \mid \lambda \geq 0, u \in C\}$ in the real vector space $V = W \times \mathbb{R}$.

Obviously, K is a convex cone in V and $f(\lambda u, \lambda) = \lambda$ defines a linear functional $f\colon W \longrightarrow \mathbb{R}$ with $f(\lambda u, \lambda) > 0$ for $(\lambda u, \lambda) \in K \setminus \{0\}$. Furthermore, $C \times \{1\} = \{(\lambda u, \lambda) \in K \mid f(\lambda u, \lambda) = 1\}$ is a base of K. Therefore, though convex cones are special convex sets, there are "no more" convex sets than there are convex cones because the former can serve as basis of the latter.

The above correspondence between convex cones and general convex sets can be used to relate the order functions, where for an arbitrary convex set C in a real vector space its order function is given for $x, y \in C$ by

$$\beta(x,y) = \sup\{\alpha \in [0,1] \mid y = \alpha x + (1-\alpha)z,\ z \in C\}$$

(cf. [25, 37]).

Employing the function $\alpha(x,y) = \min\{\beta(x,y), \beta(y,x)\}$ instead of $\mu(x,y)$ one can define the metrics considered for convex cones in the same manner for general convex sets.

Lemma 3.3.2. *Let K be a convex cone with base $B = \{x \in K \mid f(x) = 1\}$. For $x, y \in K \setminus \{0\}$ with base points $\bar{x}, \bar{y} \in B$ one has the following relationships*
(i) $\lambda(x,y) = \frac{f(y)}{f(x)}\beta(\bar{x},\bar{y})$.
(ii) $d(x,y) = \bar{d}(\bar{x},\bar{y})$, *where* $\bar{d}(u,v) = -\log[\beta(u,v)\cdot\beta(v,u)]$ *is the projective Hilbert metric on B.*
(iii) $p(x,y) = -\log\min\{\frac{f(y)}{f(x)}\beta(\bar{x},\bar{y}), \frac{f(x)}{f(y)}\beta(\bar{y},\bar{x})$.

Proof. (i) Let $\lambda x \le y$ for $\lambda \ge 0$, that is $y = \lambda x + z$ with $z \in K$. It follows that

$$\bar{y} = \frac{y}{f(y)} = \frac{\lambda f(x)}{f(y)}\bar{x} + \frac{1}{f(y)}z.$$

If $z = 0$ then $f(y) = \lambda f(x)$ and $\bar{y} = 1 \cdot \bar{x} + 0 \cdot \bar{x}$ which imply that $\lambda = \frac{f(y)}{f(x)} = \frac{f(y)}{f(x)}\beta(\bar{x},\bar{y})$.

If $z \neq 0$ then

$$\bar{y} = \frac{\lambda f(x)}{f(y)}\bar{x} + \frac{f(z)}{f(y)}\bar{z}$$

which implies that $\frac{\lambda f(x)}{f(y)} \le \beta(\bar{x},\bar{y})$. In both cases we obtain $\lambda(x,y) \le \frac{f(y)}{f(x)}\beta(x,y)$. Conversely, let $\bar{y} = \beta\bar{x} + (1-\beta)\bar{z}$ for some $\bar{z} \in B$. It follows that

$$y = f(y)\bar{y} = \frac{\beta f(y)}{f(x)}x + (1-\beta)f(y)\bar{z}$$

and, hence, $\lambda(x,y) \ge \frac{\beta f(y)}{f(x)}$ which implies $\lambda(x,y) \ge \frac{f(y)}{f(x)}\beta(\bar{x},\bar{y})$.

(ii) By (i)

$$\lambda(x,y)\cdot\lambda(y,x) = \frac{f(y)}{f(x)}\beta(\bar{x},\bar{y})\cdot\frac{f(x)}{f(y)}\cdot\beta(\bar{y},\bar{x}).$$

(iii) Immediate from (i). □

The above Lemma says in particular that for a convex cone with base the projective Hilbert metric for the cone can be as well computed from the corresponding metric of any base. Thus, e.g., the projective Hilbert metric for an ice cream cone in three dimensions can be computed from the corresponding metric of a circle (or ellipse) in two dimensions. Generalizing considerations of Klein [29] for ellipses, Hilbert [23] defined an internal length within arbitrary convex bodies (cf. also [15, Section 18]). The points of his general geometry Hilbert mapped into a nowhere concave body in Euclidean space. Thus, let C be a convex subset of $\mathbb{R}^n$ which is closed and bounded and consider two points $A, B \in C$ which intersect the boundary of C in X, Y as in the following figure.

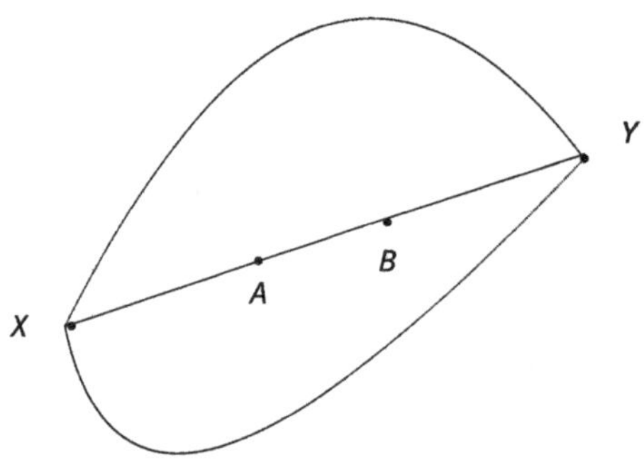

Fig. 3.4. Projective Hilbert metric in a convex set.

As distance of A and B in the general geometry Hilbert defined

$$\breve{AB} = \log\left\{\frac{\overline{YA}}{\overline{YB}}\cdot\frac{\overline{XB}}{\overline{XA}}\right\}.$$

Here $\frac{\overline{YA}}{\overline{YB}}\cdot\frac{\overline{XB}}{\overline{XA}} = \left(\frac{\overline{YA}}{\overline{YB}}\right)\left(\frac{\overline{XA}}{\overline{XB}}\right)^{-1}$ is the cross ratio of the four points A, B, X, Y where $\overline{PQ}$ is the Euclidean distance between two points P, Q given by the Euclidean norm on $\mathbb{R}^n$, that is $\overline{PQ} = \|P - Q\|$. As is obvious from Figure 3.4,

$$\begin{aligned} A &= \beta_1 B + (1-\beta_1)X, \quad & \beta_1 &= \beta(B, A) \\ B &= \beta_2 A + (1-\beta_2)Y, \quad & \beta_2 &= \beta(A, B). \end{aligned}$$

Therefore,

$$\begin{aligned} \overline{YB} &= \|B - Y\| = \beta_2\|A - Y\| = \beta_2\overline{YA} \\ \overline{XA} &= \|A - X\| = \beta_1\|B - X\| = \beta_1\overline{XB} \end{aligned}$$

and, hence,

$$\breve{AB} = \log\frac{1}{\beta_1\beta_2} = -\log[\beta(A,B)\beta(B,A)].$$

This shows that the distance defined by Hilbert for points of the general geometry is exactly what was called the projective Hilbert metric with respect to the order function of a convex body.

The following proposition collects various useful inequalities considering internal metrics and semi-norms. Thereby, $q\colon V \longrightarrow \mathbb{R}_+$ is a **semi-norm** on the real vector space V if $q(\lambda x) = |\lambda| q(x)$ for $\lambda \in \mathbb{R}$, $x \in V$ and $q(x+y) \le q(x) + q(y)$ for $x, y \in V$. (Locally convex topologies on V are defined by those semi-norms; see the next section for related results.)

Proposition 3.3.3. *Let K be a lineless convex cone and let $x, y \in K \setminus \{0\}$. The following inequalities hold.*

(i) $d(x,y) \le 2p(x,y)$.

(ii) $h(x,y) \le p(x,y) \le \frac{h(x,y)}{1-h(x,y)}$ *for* $x \sim y$.

(iii) $\frac{1}{2}h(x,y) \le g(x,y) \le 2b(x,y) \le 2h(x,y)$.

(iv) $p(x,y) \le k(x,y) \le 2p(x,y)$.

(v) *For every monotone semi-norm q on V it holds that*

$$q(x-y) \le [3 - (a + b + \max\{a,b\})] \max\{q(x), q(y)\}$$

where $a = \min\{\lambda(x,y), 1\}$, $b = \min\{\lambda(y,x), 1\}$.

(vi) *For every monotone semi-norm q on V it holds that*

$$q(x-y) \le 3h(x,y) \max\{q(x), q(y)\}$$

and for $q(x) = q(y) = 1$

$$q(x-y) \le 3(1 - \exp(-d(x,y))).$$

(vii) *For every monotone semi-norm q on V it holds that*

$$p(x,y) \le d(x,y) + |\log q(x) - \log q(y)| \text{ for } q(x), q(y) > 0$$

and for $q(x) = q(y) > 0$

$$p(x,y) \le d(x,y).$$

Proof. (i) Follows immediately from $\min\{\lambda(x,y), \lambda(y,x)\}^2 \le \lambda(x,y) \cdot \lambda(y,x)$.

(ii) For $r = p(x,y)$ one has $h(x,y) = 1 - e^{-r}$. If $x \sim y$ then $r > 0$ and the mean value theorem applied to $1 - e^{-r}$ yields

$$re^{-r} \le 1 - e^{-r} \le r,$$

that is

$$p(x,y)(1 - h(x,y)) \le h(x,y) \le p(x,y).$$

(iii) By definition of the metrics one has to show that for $0 \le \mu = \mu(x,y) \le 1$

$$\frac{1}{2}(1-\mu) \le 2\frac{1-\sqrt{\mu}}{1+\sqrt{\mu}} \le 2\frac{1-\mu}{1+\mu} \le 2(1-\mu).$$

The first inequality holds because of

$$\frac{1}{2}(1-\mu)(1+\sqrt{\mu}) = \frac{1}{2}(1-\sqrt{\mu})(1+\sqrt{\mu})^2 \le 2\frac{1-\mu}{1+\mu}2(1-\sqrt{\mu}).$$

The second inequality follows from $\mu \leq \sqrt{\mu}$ and the third inequality is obvious.

(iv) Follows from $k(x,y) = \max\{d(x,y), p(x,y)\}$ (see Exercise 4 (a) of Section 3.2) by (i).

(v) Let $x, y \in K$ with $\lambda x \leq y$ and $\mu y \leq x$ for $0 \leq \lambda,\ \mu \leq 1$.

It follows that

$$-y(1-\mu) \leq x - y \leq x(1-\lambda)$$

and, hence,

$$0 \leq x - y + y(1-\mu) \leq x(1-\lambda) + y(1-\mu).$$

Since q is a monotone semi-norm, one obtains

$$q(x-y) - q(y(1-\mu)) \leq q(x-y+y(1-\mu)) \leq q(x(1-\lambda)) + q(y(1-\mu))$$

and, hence,

$$q(x-y) \leq ((1-\lambda) + 2(1-\mu)) \max\{q(x), q(y)\}.$$

Exchanging the roles of x and y one obtains

$$q(y-x) \leq ((1-\mu) + 2(1-\lambda)) \max\{q(y), q(x)\}.$$

This yields altogether

$$q(x-y) \leq (3 - (\lambda + \mu + \max\{\lambda, \mu\})) \max\{q(x), q(y)\}.$$

Taking suprema over λ and μ this proves the required inequality.

(vi) From (v) one obtains

$$\mu(x,y) = \min\{\lambda(x,y), \lambda(y,x)\} \leq a, b$$

and

$$q(x-y) \leq [3 - 3\mu(x,y)] \max\{q(x), q(y)\},$$

which shows the first inequality.

For $q(x) = q(y) = 1$ one must have that $\lambda(x,y) \leq 1$ and $\lambda(y,x) \leq 1$ and (v) yields

$$\begin{aligned} q(x-y) &\leq [3 - (\lambda(x,y) + \lambda(y,x) + \max\{\lambda(x,y), \lambda(y,x)\})] \\ &\leq [3 - 3\lambda(x,y) \cdot \lambda(y,x)]\,. \end{aligned}$$

Since $\lambda(x,y) \cdot \lambda(y,x) = \exp(-d(x,y))$ this yields the required inequality.

(vii) $\lambda x \leq y$ implies $\lambda \leq \frac{q(y)}{q(x)}$ and, hence, $\lambda(x,y)\frac{q(x)}{q(y)} \leq 1$. Similarly, $\lambda(y,x)\frac{q(y)}{q(x)} \leq 1$ and, therefore,

$$\begin{aligned} \lambda(x,y) \cdot \lambda(y,x) = \lambda(x,y)\frac{q(x)}{q(y)} \cdot \lambda(y,x)\frac{q(y)}{q(x)} &\leq \min\left\{\lambda(x,y)\frac{q(x)}{q(y)},\ \lambda(y,x)\frac{q(y)}{q(x)}\right\}. \\ &\leq \min\{\lambda(x,y),\ \lambda(y,x)\} \cdot \max\left\{\frac{q(x)}{q(y)},\ \frac{q(y)}{q(x)}\right\} \end{aligned}$$

The assertion follows by taking the logarithm in this inequality and taking the definitions of d and p into account. □

Remark 3.3.4. The statements (i)–(iv) of Proposition 3.3.3 show that for any of the metrics p, h, g, b, k the distance between two points is small iff this holds for any of the other metrics. This does not apply to the projective Hilbert metric d. By (i) one has that $d(x,y) \leq 2p(x,y)$ but, e.g., if $y = \lambda x$ for $\lambda > 0$ arbitrary then $d(x,y) = 0$ and $p(x,y) = -\log\min\{\lambda, \frac{1}{\lambda}\} = |\log\lambda|$. Proposition 3.3.3 implies inequalities obtained in the literature. Property (vi) implies that $q(x-y) \leq 3\max\{q(x), q(y)\}(\frac{1}{\mu(x,y)} - 1)$ (see [54, pp. 438/39]). Since $e^r - 1 \geq 1 - e^{-r}$ for $r \in \mathbb{R}$ property (vi) implies that

$$q(x-y) \leq 3\max\{q(x), q(y)\}(\exp p(x,y) - 1)$$

and that

$$q(x-y) \leq 3(\exp d(x,y) - 1)$$

for $q(x) = q(y) = 1$ (see [44, pp. 14, 15]).

Employing the notation $[x,y] = \{z \in K \mid x \leq z \leq y\}$ for intervals and $B_m(x,r) = \{y \in K \mid m(x,y) \leq r\}$ for a closed ball with center x and radius r for the internal metric m we can describe balls for internal metrics as follows:

Lemma 3.3.5. *Let K be a lineless convex cone which is archimedean in $K - K$.*

For $x \in K \setminus \{0\}$ and $r > 0$ one has

(i) $B_d(x,r) = \mathbb{R}_+[x, e^r x]$, *a convex cone*

(ii) $B_p(x,r) = [e^{-r}x, e^r x]$

(iii) $B_h(x,r) = [(1-r)x, \frac{1}{1-r}x]$ *for* $r < 1$

(iv) $B_k(x,r) = \mathbb{R}_+[x, e^r x] \cap [e^{-r}x, e^r x]$.

Proof. (i) By definition

$$d(x,y) \leq r \text{ iff } e^{-r} \leq \lambda(x,y) \cdot \lambda(y,x).$$

If $y \in \mathbb{R}_+[x, e^r x]$ then $\lambda x \leq y \leq \lambda e^r x$ for some $\lambda > 0$ and, hence, $\lambda(x,y) \cdot \lambda(y,x) \geq \frac{\lambda}{\lambda e^r} = e^{-r}$. Conversely, since K is archimedean, Lemma 3.1.4 (xii) implies that

$$\lambda(x,y)x \leq y \frac{1}{\lambda(y,x)} x \leq \lambda(x,y)e^r x$$

and, hence, $\frac{y}{\lambda(x,y)} \in [x, e^r x]$.

(ii) Because K is archimedean the following equivalences hold

$$\mu(x,y) \geq \mu > 0 \Leftrightarrow \lambda(x,y) \geq \mu \quad \text{and}$$
$$\lambda(y,x) \geq \mu \Leftrightarrow \mu x \leq y \quad \text{and}$$
$$\mu y \leq x \Leftrightarrow y \in \left[\mu x, \frac{1}{\mu}x\right].$$

Since $p(x,y) \leq r \Leftrightarrow e^{-r} \leq \mu$ this proves (ii).

(iii) Follows from the equivalences in (ii).

(iv) Follows from $k(x,y) = \max\{d(x,y), p(x,y)\}$ (see Exercise 4 (a) to Section 3.2) and (i) and (ii). ☐

From property (iii) one obtains in an obvious way also a description of the balls for the Gleason metric and the Bear metric. The following figure depicts balls for some internal metrics.

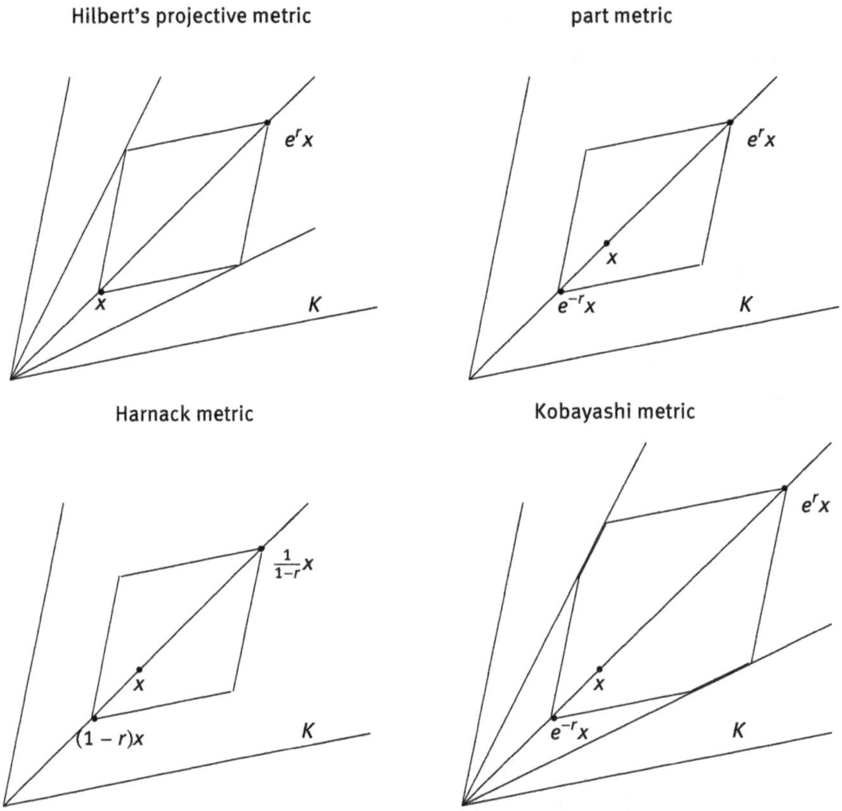

Fig. 3.5. Closed balls with center x and radius r for some internal metrics.

In case the convex cone is given by a family of linear functionals, internal metrics have a dual description in terms of these functionals. The following lemma gives a description for d, p and h; similar descriptions can be obtained for other internal metrics.

Lemma 3.3.6. *Let K be a lineless convex cone such that for some family $\mathcal{F}$ of linear functionals on V*

$$K = \{x \in V \mid f(x) \geq 0 \text{ for all } f \in \mathcal{F}\}.$$

The following formulas hold for $x, y \in K \setminus \{0\}$:

$$\begin{aligned} d(x,y) = {} & \sup\{\log f(x) - \log f(y) \mid f \in \mathcal{F}, f(x) > 0\} \\ & - \inf\{\log f(x) - \log f(y) \mid f \in \mathcal{F}, f(y) > 0\} \end{aligned}$$

$$p(x,y) = \sup\{|\log f(y) - \log f(x)| f \in \mathcal{F},\ f(x) > 0,\ f(y) > 0\}$$

$$h(x,y) = \sup\left\{\frac{|f(x) - f(y)|}{\max\{f(x), f(y)\}} | f \in \mathcal{F},\ f(x) > 0,\ f(y) > 0\right\}.$$

Proof. By Lemma 3.1.4 (xiv)

$$\lambda(x,y) = \inf\left\{\frac{f(y)}{f(x)} \mid f \in \mathcal{F},\ f(x) > 0\right\}.$$

Therefore,

$$\begin{aligned} d(x,y) &= \log \tfrac{1}{\lambda(x,y)} &&- \log \lambda(y,x) \\ &= \sup\{\log f(x) &&- \log f(y) \mid f \in \mathcal{F},\ f(x) > 0\} \\ &= \inf\{\log f(x) &&- \log f(y) \mid f \in \mathcal{F},\ f(y) > 0\}. \end{aligned}$$

Furthermore,

$$\min\{\lambda(x,y), \lambda(y,x)\} = \inf\left\{\min\left\{\frac{f(y)}{f(x)}, \frac{f(x)}{f(y)}\right\} | f \in \mathcal{F},\ f(x) > 0,\ f(y) > 0\right\}$$

and, hence,

$$\begin{aligned} p(x,y) &= -\log \min\{\lambda(x,y), \lambda(y,x)\} \\ &= \sup\left\{-\log \min\left\{\frac{f(y)}{f(x)}, \frac{f(x)}{f(y)}\right\} | f \in \mathcal{F},\ f(x) > 0,\ f(y) > 0\right\} \\ &= \sup\{|\log f(y) - \log f(x)| | f \in \mathcal{F},\ f(x) > 0,\ f(y) > 0\} \end{aligned}$$

because of

$$-\log \min\left\{\frac{f(y)}{f(x)}, \frac{f(x)}{f(y)}\right\} = \max\{\log f(x) - \log f(y),\ \log f(y) - \log f(x)\}.$$

Finally,

$$\begin{aligned} 1 - \lambda(x,y) &= 1 + \sup\left\{-\frac{f(y)}{f(x)} | f \in \mathcal{F},\ f(x) > 0\right\} \\ &= \sup\left\{\frac{f(x) - f(y)}{f(x)} | f \in \mathcal{F},\ f(x) > 0\right\} \end{aligned}$$

and, hence,

$$\begin{aligned} h(x,y) &= 1 - \min\{\lambda(x,y), \lambda(y,x)\} = \max\{1 - \lambda(x,y), 1 - \lambda(y,x)\} \\ &= \sup\left\{\max\left\{\frac{f(x) - f(y)}{f(x)}, \frac{f(y) - f(x)}{f(y)}\right\} | f \in \mathcal{F},\ f(x) > 0,\ f(y) > 0\right\} \\ &= \sup\left\{\frac{|f(x) - f(y)|}{\max\{f(x), f(y)\}} | f \in \mathcal{F},\ f(x) > 0,\ f(y) > 0\right\}. \end{aligned}$$

□

Remark 3.3.7. In case of the standard cone in $\mathbb{R}^n$ one can choose as family $\mathcal{F}$ the finitely many projections $f(x) = x_i$ for $1 \le i \le n$. Lemma 3.3.6 shows that

$$d(x,y) = \max\{\log x_i - \log y_i \mid x_i > 0\} - \min\{\log x_i - \log y_i \mid y_i > 0\},$$

which coincides with definition given in Chapter 2 (see Definition 2.1.8), and

$$\begin{aligned} p(x,y) &= \max\{|\log x_i - \log y_i| \mid x_i > 0,\ y_i > 0\} \\ h(x,y) &= \max\left\{\frac{|x_i - y_i|}{\max\{x_i, y_i\}} | x_i > 0,\ y_i > 0\right\}. \end{aligned}$$

In particular, for $n = 1$ and $x > 0, y > 0$

$$\begin{aligned} d(x,y) &= \log x - \log y - (\log x - \log y) = 0 \\ p(x,y) &= |\log x - \log y| \\ h(x,y) &= \frac{|x-y|}{\max\{x,y\}}. \end{aligned}$$

From Lemma 3.3.6 one easily obtains the following result which demonstrates that a part of a convex cone when equipped with the projective Hilbert metric and the part metric, respectively, is isomorphic as a metric space to a subset of a normed vector space.

Proposition 3.3.8. *Let K be a lineless convex cone in a real vector space V such that $K = \{x \in V \mid f(x) \geq 0,\ f \in \mathcal{F}\}$ for some family $\mathcal{F}$ of linear functionals on V.*

Define $\overset{\circ}{K} = \{x \in K \mid f(x) > 0,\ f \in \mathcal{F}\}$ and suppose there exist functions $c, d\colon \overset{\circ}{K} \longrightarrow \mathbb{R}$ such that $0 < c(x) \leq f(x) \leq d(x)$ for all $f \in \mathcal{F}$, all $x \in \overset{\circ}{K}$.

Let W be the real vector space of all bounded real valued functions on the set $\mathcal{F}$, equipped with the supremums norm $\|\cdot\|$.

(i) *Setting $\psi(x)(f) = \log f(x)$ defines an injective mapping $\psi : \overset{\circ}{K} \longrightarrow W$ with $p(x,y) = \|\psi(x) - \psi(y)\|$ for all $x, y \in \overset{\circ}{K}$.*
(ii) *Pick some $f_0 \in \mathcal{F}$ and let $W_0 = \{F \in W \mid F(f_0) = 0\}$. The mapping ψ defined in* (i) *maps $\{x \in \overset{\circ}{K} \mid f_0(x) = 1\}$ into W_0 and it holds*

$$d(x,y) = \|\,|\,\psi(x) - \psi(y)\|\,|$$

for all $x, y \in \overset{\circ}{K}$, where $\|\,|\cdot\|\,|$ is a norm on W_0 defined by $\|\,|\,F\|\,| = \sup\{F(f) \mid f \in \mathcal{F}\} - \inf\{F(f) \mid f \in \mathcal{F}\}$.

Proof. (i) For $x \in \overset{\circ}{K}$ and $f \in \mathcal{F}$ one has that $f(x) > 0$ and, hence, $\psi(x)(f)$ is defined. Furthermore, $c(x) \leq \sup\{\psi(x)(f) \mid f \in \mathcal{F}\} \leq d(x)$, for $x \in \overset{\circ}{K}$ and, hence, ψ maps $\overset{\circ}{K}$ into W. From Lemma 3.3.6 for $x, y \in \overset{\circ}{K}$ one has

$$\begin{aligned} p(x,y) &= \sup\{|\psi(x)(f) - \psi(y)(f)| \mid f \in \mathcal{F}\} \\ &= \|\psi(x) - \psi(y)\|. \end{aligned}$$

(ii) If $x \in \overset{\circ}{K}$ with $f_0(x) = 1$ then $\psi(x)(f_0) = \log f_0(x) = 0$ and, hence, ψ maps $\{x \in \overset{\circ}{K} \mid f_0(x) = 1\}$ into W_0.

From Lemma 3.3.6 for $x, y \in \overset{\circ}{K}$ one has

$$d(x,y) = \sup\{\psi(x)(f) - \psi(y)(f) \mid f \in \mathcal{F}\} - \inf\{\psi(x)(f) - \psi(y)(f) \mid f \in \mathcal{F}\}$$
$$= \|\,|\,\psi(x) - \psi(y)\|\,|$$

with $\|\,|\cdot\|\,|$ as defined in (ii).

Obviously, $\|\,|\,F\|\,| \in \mathbb{R}_+$ for all $f \in W$. If $\|\,|\,F\|\,| = 0$ for $F \in W_0$ then F must be constant on $\mathcal{F}$ and, because of $F(f_0) = 0$, one must have that $F = 0$. Finally, for $F,\ G \in \mathcal{F}$,

$$\sup\{(F+G)(f) \mid f \in \mathcal{F}\} \le \sup\{F(f) \mid f \in \mathcal{F}\} + \sup\{G(f) \mid f \in \mathcal{F}\}$$

and

$$\inf\{F(f) \mid f \in \mathcal{F}\} + \inf\{G(f) \mid f \in \mathcal{F}\} \le \inf\{(F+G)(f) \mid f \in \mathcal{F}\},$$

and, hence, $\|\,|\,F+G\|\,| \le \|\,|\,F\|\,| + \|\,|\,G\|\,|$ □

Proposition 3.3.8 applies in particular to a lineless convex cone which is closed in a normed vector space. In this case $\mathcal{F}$ can be taken to be the dual cone of K and it holds $\mathrm{int}K = \overset{\circ}{K}$, where $\mathrm{int}K$ is the interior of K with respect to the norm topology. This is illustrated in the following example for the normed space of all continuous functions on a compact space.

Example 3.3.9. (Cf. [44, pp. 20, 22], [58, pp. 29, 30].) Let T be a (non-empty) compact space, $V = \mathcal{C}(T)$ the real vector space of all real continuous functions on T and $K = \{x \in \mathcal{C}(T) \mid x(t) \ge 0, \text{ all } t \in T\}$ the standard cone in $\mathcal{C}(T)$. Obviously, $K = \{x \in V \mid \epsilon_t(x) \ge 0,\ \epsilon_t \in \mathcal{F}\}$ where $\mathcal{F}$ is the set of all evaluation functionals ϵ_t, $t \in T$, on $\mathcal{C}(t)$, i.e., $\epsilon_t(x) = x(t)$ for $x \in \mathcal{C}(T)$. It is easily seen that $\overset{\circ}{K} = \{x \in \mathcal{C}(T) \mid x(t) > 0, \text{ all } t \in T\}$ is the interior $\mathrm{int}K$ of K with respect to the supremums norm $\|\cdot\|$ on $\mathcal{C}(T)$. The functions $c, d\colon \overset{\circ}{K} \longrightarrow \mathbb{R}$ can be taken to be $c(x) = \inf\{x(t) \mid t \in T\}$ and $d(x) = \sup\{x(t) \mid t \in T\}$. By identifying $\mathcal{F}$ with T we have that $\psi(x)(t) = \log x(t)$ defines an injective mapping $\psi : \overset{\circ}{K} \longrightarrow \mathcal{C}(T)$. Moreover, ψ is surjective because of $\psi(\exp y) = y$ for $y \in \mathcal{C}(T)$. Thus we obtain that ψ yields an isometry of the metric spaces $(\mathrm{int}K, p)$ and $(C(T), \|\cdot\|)$. This isometry, however, is not an isomorphism with respect to the cone structure of $\overset{\circ}{K}$.

In a similar way, from part (ii) of Lemma 3.3.6 it follows that ψ yields for any given point $t_0 \in T$ an isometry between the metric space given by $\mathrm{int}K \cap \{x \in K \mid x(t_0) = 1\}$ equipped with Hilbert's projective metric and the vector space $\{f \in \mathcal{C}(T) \mid f(t_0) = 0\}$ equipped with the norm

$$\|\,|\,x\|\,| = \sup\{x(t) \mid t \in T\} - \inf\{x(t) \mid t \in T\}.$$

By choosing for T the discrete space $T = \{1, \dots, n\}$ the results obtained specialize to $V = \mathbb{R}^n$ and $K = \mathbb{R}^n_+$ the standard cone. It follows that $\mathrm{int}\mathbb{R}^n_+$ equipped with the part metric is isometric to $\mathbb{R}^n$ with max-norm and that $\mathrm{int}\mathbb{R}^{n-1}_+$ equipped with Hilbert's projective (pseudo-)metric is isometric to $\mathbb{R}^{n-1}$ with a pseudo-norm given by $\| \mid x\| \mid = \max_i x_i - \min_i x_i$. As already pointed out, this isometries do not respect the cone-structure.

Exercises

1. a) Let K be a convex cone with base B in some real vector space V. Show that B is a convex set such that every $x \in K \setminus \{0\}$ has a unique representation $x = \lambda \bar{x}$ with $\lambda > 0$ and $\bar{x} \in B$.
   b) Find a base of the ice cream cone
   $$K = \{x = (u, r) \in \mathbb{R}^n \times \mathbb{R}_+ \mid \|u\| \leq r\},$$
   where $\|u\|^2 = \langle u, u\rangle$; $\langle\cdot,\cdot\rangle$ standard scalar product.
   c) Find an archimedean and lineless convex cone which possesses no base.
2. Sketch the unit ball with center x for
   a) the Gleason metric on $\mathbb{R}^2_+$,
   b) the Bear metric on $\mathbb{R}^2_+$,
   c) the projective Hilbert metric on the ice cream cone
   $$\{(x, r) \in \mathbb{R}^2 \times \mathbb{R}_+ \mid x_1^2 + x_2^2 \leq r^2\}.$$
3. Let K be a lineless convex cone in a real vector space V such that $K = \{x \in V \mid f(x) \geq 0 \text{ for all } f \in \mathcal{F}\}$ for some family $\mathcal{F}$ of linear functionals on V.
   a) Describe the Gleason metric on K in terms of $\mathcal{F}$ (cf. Lemma 3.3.6).
   b) Describe the Kobayashi metric on K in terms of $\mathcal{F}$.
   c) Find for the ice cream cone $\{(x, r) \in \mathbb{R}^2 \times \mathbb{R}_+ \mid x_1^2 + x_2^2 \leq r^2\}$ a representation by a family $\mathcal{F}$ and describe the projective Hilbert metric in terms of $\mathcal{F}$.
   d) Characterize for K all its parts in terms of $\mathcal{F}$. Find the number of parts for $\mathcal{F}$ finite, in particular for $K = \mathbb{R}^n_+$.
4. Let $l_1 = \{x \in \mathbb{R}^{\mathbb{N}} \mid \sum_{i=1}^{\infty} |x_i| <+ \infty\}$ with norm $\|x\| = \sum_{i=1}^{\infty} |x_i| <+ \infty$ and K the convex cone $K = \{x \in l_1 \mid x_i \geq 0 \text{ for all } i \in \mathbb{N}\}$.
   a) Find a representation of K by a family $\mathcal{F}$ of continuous linear functionals on l_1.
   b) Compare $\overset{\circ}{K} = \{x \in K \mid f(x) > 0,\ f \in \mathcal{F}\}$ with int$K$ with respect to $\|\cdot\|$.
   c) Describe the metric space $(\overset{\circ}{K}, p)$ isometrically by a subset of a normed space.
5. Let K be a convex cone in a normed vector space V.
   a) Show that K is symmetrically bounded (cf. Exercise 5 to 3.1) iff every subset of K open for the restriction of the norm topology on K is open also for the part metric.
   b) Show that if int$K \neq \emptyset$ and K is symmetrically bounded it is **normal**, i.e., there exists a constant $c > 0$ such that $\|x\| \leq c\|x + y\|$ for all $x, y \in K$.

c) Let K be the convex cone of all sequences $x = (x_1, x_2, \ldots,) \in \mathbb{R}^{\mathbb{N}}$ with $x_i \neq 0$ only for finitely many $i \in \mathbb{N}$ and $\sum_{i=1}^{n} x_i \geq 0$ for all $n \geq 1$. Consider $V = K - K$ with the norm $\|x\| = \max_i |x_i|$. Show that $\operatorname{int} K = \emptyset$, K is symmetrically bounded but not normal.

3.4 Completeness for internal metrics

An important step in the analysis of concave Perron–Frobenius theory was the fact that the interior of the standard cone in $\mathbb{R}^n$ (or a certain subset of it) is complete for Hilbert's projective metric. This fact was easily established by an adhoc-argument (Lemma 2.1.10). To get completeness for general cones in infinite dimensions a more detailed investigation is needed. The main step will be the characterization of internal completeness, i.e., completeness of a cone with respect to any of the internal metrics, in terms of so called guided sequences of elements in the cone. This then will lead to various criteria which assure internal completeness provided the cone meets certain topological requirements. As remarked already, in general the internal metrics are just "écarts" and topological notions as well as uniform structures will be understood in the sense of Bourbaki [13].

The following lemma will allow us to concentrate without loss on cones consisting of one part only and on the internal metrics d and h only.

Lemma 3.4.1. *Let $K \neq \{0\}$ be a lineless convex cone in some real vector space.*
(i) *For m any of the internal metrics d, p, h, g, b and k on K, the cone K is complete for m if and only if every part of K is complete for m.*
(ii) *For m any of the internal metrics p, g, b and k, the completeness of K for m is equivalent to the completeness of K for h.*

Proof. (i) By Lemma 3.2.2 the order functions λ_K and λ_P of K and of a part P of K, respectively, coincide on $P \times P$. Since all internal metrics are defined by the order function it follows that the internal metrics m_K and m_P coincide on $P \times P$. Two points $x, y \in K \setminus \{0\}$ are in the same part iff $\lambda_K(x, y) > 0$ and $\lambda_K(y, x) > 0$ or, equivalently, iff $m(x, y) < \infty$ for $m = d, p, k$, $m(x, y) < 1$ for $m = h, b$ and $g(x, y) < 2$. This proves (i).

(ii) From the inequalities (ii)–(iv) of Proposition 3.3.3 it follows that K is complete for h iff K is complete for one of the metrics $m = p, g, b, k$. □

Later on we will see that completeness for d, too, is equivalent to completeness for h. The following concept of a guided sequence will be useful later on to describe more explicitly Cauchy sequences for internal metrics.

Definition 3.4.2. A sequence $(x_n) \subset K$ is a sequence guided by e, or shortly, a **guided sequence** if there exist $e \in K$ and a sequence (ϵ_n) of non-negative real numbers converging to 0 such that

$$e \le x_n \le x_{m+n} \le x_n + \epsilon_n e \quad \text{for all} \quad m, n \in \mathbb{N}.$$

($\le$ is the order relation induced by the convex cone K.)

Obviously, a sequence (x_n) in K is guided by e iff it is an increasing sequence of elements above e such that $x_{m+n} \in x_n + [0, \epsilon_n e]$ where $[a, b] = \{x \in K \mid a \le x \le b\}$ denotes an interval with respect to K. In particular, a guided sequence is increasing and order-bounded, i.e., contained in some order interval. The converse implication may hold or not as explained by the following examples.

Examples 3.4.3. (a) Consider the standard cone $K = \mathbb{R}^N_+$. For this cone one has that any increasing order-bounded sequence is also a guided sequence. If (x_n) is increasing and order-bounded then there exists $x = (x^1, \dots, x^N) \in K$ such that $x^i = \sup_n x^i_n$ for all i. Let I be the set of all $1 \le i \le N$ such that $x^i_n > 0$ for at least one n and define $e^i = \min\{x^i_n \mid x^i_n > 0\}$ for $i \in I$ and $e^i = 0$ for $i \notin I$. Obviously, $e \in K$ and $e \le x_n \le x_{m+n}$ for all $m, n \in \mathbb{N}$. By $\epsilon_n = \max\{\frac{x^i - x^i_n}{e^i} \mid i \in I\}$ and $\epsilon_n = 0$ in case of $I = \emptyset$ a sequence of non-negative numbers is defined which convergers to 0. It follows that $x^i \le x^i_n + \epsilon_n e^i$ for $i \in I$ and, trivially, for $i \notin I$ and, hence,

$$x_{m+n} \le x \le x_n + \epsilon_n e.$$

Thus (x_n) is a sequence guided by e.

(b) Consider the standard cone $K = \mathcal{C}_+([0, 1])$ in the infinite dimensional vector space of all continuous functions on the unit interval.

We construct an increasing and order-bounded sequence in the cone K which not even possesses a guided subsequence. Let $f_n \in K$ be defined by

$$f_n(t) = \begin{cases} 2^n t, & 0 \le t \le \frac{1}{2^n} \\ 1, & \frac{1}{2^n} \le t \le 1. \end{cases}$$

Obviously, $f_n \le f_{n+1}$ and $0 \le f_n \le 1$ on $[0, 1]$ for all n. One has that for the supremums norm $\|\cdot\|$

$$\|f_{m+n} - f_n\| \ge \|f_{n+1} - f_n\| \ge \frac{1}{2}$$

for all $m \ge 1$. Suppose $(f_{n'})$ is a subsequence of (f_n) that is guided by some $e \in K$. In particular, $0 \le f_{m'+n'} - f_{n'} \le \epsilon_{n'} e$ and, hence,

$$\frac{1}{2} \le \|f_{m'+n'} - f_{n'}\| \le \epsilon_{n'} \|e\|$$

with $\epsilon_{n'}$ converging to 0. This, however, is not possible.

The following lemma establishes a connection between Cauchy sequences for any of the internal metrics and guided sequences.

Lemma 3.4.4. *Let K be a lineless convex cone and m an internal metric on K.*
(i) *Every guided sequence is a Cauchy sequence for m.*
(ii) *If $(x_n)_n$ is a Cauchy sequence in K for m then there exists a subsequence $(x_{n(k)})_k$ and a sequence of real numbers $\lambda_k > 0$ such that $(z_k)_k$ for $z_k = \lambda_k x_{n(k)}$ is a guided sequence.*

Proof. Without restriction we consider sequences in $K \setminus \{0\}$.

(i) By Proposition 3.3.3 it suffices to consider $m = h$. Let (x_n) be a sequence guided by $e \in K$. Then $\lambda(x_n, x_{m+n}) \geq 1$. Furthermore,

$$x_{m+n} \leq x_n + \epsilon_n e \leq x_n + \epsilon_n x_n = (1 + \epsilon_n)x_n$$

yields that $\lambda(x_{m+n}, x_n) \geq \frac{1}{1+\epsilon_n}$.

Therefore,

$$\begin{aligned} h(x_n, x_{m+n}) &= 1 - \min\{\lambda(x_n, x_{m+n}), \lambda(x_{m+n}, x_n)\} \\ &\leq 1 - \frac{1}{1+\epsilon_n} = \frac{\epsilon_n}{1+\epsilon_n} \leq \epsilon_n \end{aligned}$$

for all m, n. Since (ϵ_n) converges to 0 for $n \to \infty$ it follows that (x_n) is a Cauchy sequence for h.

(ii) By Proposition 3.3.3 it suffices to prove the assertion for $m = d$. Let (x_n) be a Cauchy sequence for d. For $\epsilon_k = (1 + 4^{-k})^{-1}$ there exists a subsequence $y_k = x_{n(k)}$, $n(\cdot): \mathbb{N} \longrightarrow \mathbb{N}$ strictly increasing, such that $d(y_{k+1}, y_k) < -\log \epsilon_k$ for all $k \in \mathbb{N}$.

Therefore,

$$\lambda(y_{k+1}, y_k) \cdot \lambda(y_k, y_{k+1}) > \epsilon_k$$

which implies that for every $k \in \mathbb{N}$ there exist $\alpha_k > 0, \beta_k > 0$ with $\alpha_k\beta_k > \epsilon_k$ and $\alpha_k y_k \leq y_{k+1}$, $\beta_k y_{k+1} \leq y_k$.

Define recursively $\lambda_k > 0$ by $\lambda_1 = 1, \lambda_{k+1} = \lambda_k \alpha_k^{-1}$ and $z_k = \lambda_k y_k$ for $k \in \mathbb{N}$.

We shall show that (z_k) is a guided sequence with $e = z_1$.

It holds that

$$z_k = \lambda_k y_k \leq \frac{\lambda_k}{\alpha_k} y_{k+1} = \lambda_{k+1} y_{k+1} = z_{k+1}$$

and

$$z_{k+1} = \lambda_{k+1} y_{k+1} \leq \frac{\lambda_{k+1}}{\beta_k} y_k = \frac{1}{\alpha_k \beta_k} \lambda_k y_k \leq \frac{1}{\epsilon_k} z_k.$$

This yields for $n \in \mathbb{N}$

$$0 \leq z_{n+1} - z_1 = \sum_{k=1}^{n} (z_{k+1} - z_k) \leq \sum_{k=1}^{n} \left(\frac{1}{\epsilon_k} - 1\right) z_k \leq \frac{1}{3} z_n \leq \frac{1}{3} z_{n+1}$$

and, hence,

$$z_{n+1} \leq \frac{3}{2} z_1 \quad \text{for all} \quad n \in \mathbb{N}.$$

From this we obtain for $m, n \in \mathbb{N}$

$$z_{m+n} - z_n = \sum_{k=n}^{m+n-1} (z_{k+1} - z_k) \le \sum_{k=n}^{m+n-1} 4^{-k} z_k \le \frac{3}{2}\Big(\sum_{k=n}^{m+n-1} 4^{-k}\Big) z_1 \le 2 \cdot 4^{-n} z_1.$$

Therefore, for $e = z_1$, $\epsilon_n = 2 \cdot 4^{-n}$ we obtain that for all $m, n \in \mathbb{N}$

$$e \le z_n \le z_{m+n} \le z_n + \epsilon_n e.$$ □

By Lemma 3.4.4 guided sequences are a special kind of internal Cauchy sequences and, of course, the latter are not guided in general. The next lemma describes the special kind of internal convergence for guided sequences.

Since the cones we are considering are not necessarily archimedean, we consider for a lineless convex cone K in a real vector space V beside the order-relation $\le$ the following order-relation $<$, also induced by K. For $x, y \in V$ let $x < y$ if and only if $x \le ry$ for all $r > 1$.

The relation $<$ is reflexive and transitive and $x \le y$ implies $x < y$. In general, $x < y$ does not imply $x \le y$. The latter implication holds iff K is archimedean.

Lemma 3.4.5. *Let K be a lineless convex cone and m an internal metric on K. A guided sequence (x_n) in K converges for m if and only if (x_n) has a supremum in K with respect to the order relation $<$.*

Proof. (i) Let (x_n) be a guided sequence that converges for m. We shall show that (x_n) has a supremum in K for $<$. By Proposition 3.3.3 we can assume that $m = d$. If (x_n) converges for d to $x \in K$, without loss $x \neq 0$, then $\lambda(x_n, x) \cdot \lambda(x, x_n)$ converges to 1 for $n \to \infty$. We show that the limits $\lim_{n\to\infty} \lambda(x_n, x) = s$ and $\lim_{n\to\infty} \lambda(x, x_n) = S$ do exist and, hence, $s \cdot S = 1$. As a guided sequence (x_n) is increasing and, therefore, $\lambda(x_n, x) \le \lambda(x_m, x)$ for $m \le n$. Thus $(\lambda(x_n, x))$ is decreasing and, hence, $s = \lim_{n\to\infty} \lambda(x_n, x)$ exists. Similarly, $\lambda(x, x_m) \le \lambda(x, x_n)$ for $m \le n$ and the sequence $(\lambda(x, x_n))$ is increasing. Since (x_n) is guided one has that $x_n \le x_{n+1} \le (1 + \epsilon)x_1$ and, hence,

$$\lambda(x, x_n) \le \lambda(x, (1 + \epsilon_1)x_1) = (1 + \epsilon_1)\lambda(x, x_1) \quad \text{for all} \quad n.$$

By Lemma 3.1.4 (i) and $x \neq 0$ one has that $\lambda(x, x_1) < \infty$ and, therefore, $(\lambda(x, x_n))$ is bounded from above. Thus, $S = \lim_{n\to\infty} \lambda(x, x_n)$ exists.

Next we show that Sx is a supremum of (x_n) for $<$. Obviously, $\lambda(x_n, x)x_n \le rx$ for all $r > 1$ and $n \in \mathbb{N}$. Together with $s \le \lambda(x_n, x)$ we obtain $sx_n \le rx$ and, hence, $x_n < \frac{1}{s}x = Sx$ for all n. Furthermore, let $x_n < y$ for some $y \in K$ and all $n \in \mathbb{N}$. Obviously, $\lambda(x, x_n)x \le rx_n$ for all $r > 1$ and $n \in \mathbb{N}$, and we conclude that $\lambda(x, x_n)x \le r^2 y$ for all $r > 1$, all $n \in \mathbb{N}$. By definition of S, to $r > 1$ there exists $n_0 \in \mathbb{N}$ with $S \le \lambda(x, x_{n_0})r$. Altogether we obtain that

$$Sx \le \lambda(x, x_{n_0})rx \le rr^2 y.$$

Since $r > 1$ arbitrary this implies that $Sx < y$. This shows that Sx is a supremum of (x_n).

(ii) Let (x_n) be a guided sequence with a supremum x for $<$. By Proposition 3.3.3 it suffices to show that $\lim_{n\to\infty} h(x_n, x) = 0$. Since $x_n < x$ for all n one has that $x_n \le rx$ for all n and all $r > 1$ and, therefore, $\lambda(x_n, x) \ge 1$ for all n. Furthermore, because (x_n) is guided we have that $x_k \le x_n$ for $k \le n$ and $x_{m+n} \le x_n + \epsilon_n e \le (1 + \epsilon_n)e_n$ and, therefore, $x_k \le (1 + \epsilon_n)x_n$ and all $k, n \in \mathbb{N}$. Since x is a supremum of (x_n) for $<$ it follows that $x < (1 + \epsilon_n)x_n$ for all $n \in \mathbb{N}$. This implies that $\lambda(x, x_n) \ge \frac{1}{1+\epsilon_n} \ge 1 - \epsilon_n$ and, hence, $1 - \lambda(x, x_n) \le \epsilon_n$ for all n. Altogether we obtain

$$\begin{aligned} h(x, x_n) &= \max\{1 - \lambda(x_n, x),\ 1 - \lambda(x, x_n)\} \\ &\le \max\{0, \epsilon_n\} = \epsilon_n \end{aligned}$$

and, hence, $0 \le \lim_{n\to\infty} h(x, x_n) \le \lim_{n\to\infty} \epsilon_n = 0$. □

From the two lemmata we obtain immediately the following main result on internal completeness of cones.

Theorem 3.4.6 (Internal completeness theorem). *Let K be a lineless convex cone in some real vector space and let m be an internal metric on K (i.e., m is one of the metrics d, p, h, g, b, and k). K is complete for m if and only if every guided sequence in K has a supremum in K with respect to $<$.*

Proof. (i) Suppose, K is complete for m. If (x_n) is a guided sequence in K then by Lemma 3.4.4 (i) it is a Cauchy sequence for m. Therefore, (x_n) converges for m and, by Lemma 3.4.5, has a supremum in K for $<$.

(ii) Suppose, every guided sequence in K has a supremum in K for $<$. If (x_n) is a Cauchy sequence in K for m then, by Lemma 3.4.4 (ii), there exists a sequence in K given by $z_k = \lambda_k x_{n(k)}$, $\lambda_k > 0$, which is guided. By assumption, this guided sequence has a supremum in K for $<$ and, by Lemma 3.4.5, it must converge to some $z \in K$ for m. Suppose, that $m = d$. From Lemma 3.1.4 (ii) we have that

$$\begin{aligned} d(x_{n(k)}, z) &= -\log[\lambda(x_{n(k)}, z) \cdot \lambda(z, x_{n(k)})] \\ &= -\log[\lambda_k \lambda(z_k, z) \cdot \tfrac{1}{\lambda_k} \lambda(z, z_k)] \\ &= d(z_k, z). \end{aligned}$$

Therefore, $(x_{n(k)})$ converges to z and, since (x_n) is a Cauchy sequence for d, (x_n) must converge to z for d. Thus, K is complete for d.

Suppose, now, that $m \ne d$.

By Proposition 3.3.3 we may assume that $m = h$. Since (x_n) is a Cauchy sequence for h, there exists $n_0 \in \mathbb{N}$ such that $\lambda(x_{n_0}, x_n) \ge \frac{1}{2}$ for all $n \ge n_0$. Since (z_k) is a guided sequence, there exists $v \in K$ such that $z_k \le v$ for all k, and, therefore, $\lambda(x_{n(k)}, v) \ge \lambda_k$ for all k. From Lemma 3.1.4 (iii) it follows for $k_0 \in \mathbb{N}$ with $n(k_0) \ge n_0$ that

$$\lambda(x_{n_0}, v) \ge \lambda(x_{n_0}, x_{n(k)}) \cdot \lambda(x_{n(k)}, v) \ge \frac{1}{2} \cdot \lambda_k$$

for all $k \ge k_0$.

Therefore, $\sup_k \lambda_k$ is finite and there exists an increasing subsequence (λ_l) converging to some $\lambda > 0$.

Using Lemma 3.1.4 (ii) we obtain that

$$\min\left\{\lambda\left(x_{n(l)}, \frac{z}{\lambda}\right), \lambda\left(\frac{z}{\lambda}, x_{n(l)}\right)\right\} \geq \min\left\{\frac{\lambda_l}{\lambda}, \frac{\lambda}{\lambda_l}\right\} \cdot \min\{(\lambda(z_l, z), \lambda(z, z_l)\}.$$

Since $\min\{\frac{\lambda_l}{\lambda}, \frac{\lambda}{\lambda_l}\}$ and $\min\{\lambda(z_l, z), \lambda(z, z_l)\}$ converge to 1 for $l \to \infty$, we must have that $\lim_{l\to\infty} h(x_{n(l)}, \frac{z}{\lambda}) = 0$. By assumption (x_n) is a Cauchy sequence for h and, hence, (x_n) must converge to $\frac{z}{\lambda}$ for h. Thus, K is complete for h. □

The internal completeness theorem implies in particular that a cone which is complete for one internal metric must be complete for any other internal metric. Therefore, we call a lineless convex cone simply **internally complete** if it is complete for any internal metric.

Remark 3.4.7. For an earlier version of the internal completeness theorem see [38, p. 554]. Theorem 3.4.6 (together with Lemma 3.4.1) implies in particular the following criterion of [21, p. 20]:

A lineless convex cone is complete for the part metric if for every part P it holds that any increasing and order-bounded sequence (for P) has a supremum in P with respect to $<$ (for P). The conditions of this criterion are sufficient but not necessary, as can be seen from the example of all non-negative continuous functions on the unit interval (cf. Examples 3.4.3 b).

From the internal completeness theorem we can derive a characterization of internal completeness in terms of relative uniform convergence, a concept which is mainly used in vector lattices (see also Remark 3.4.10).

Definition 3.4.8. Let K be a convex cone in a real vector space V and let $\leq$ be the ordering relation induced by K. A sequence (x_n) in V converges to x in V for **relative uniform convergence (r.u. convergence)** if there exist $u \in K$ and a null-sequence (δ_n) such that

$$-\delta_n u \leq x - x_n \leq \delta_n u \quad \text{for all} \quad n \in \mathbb{N}.$$

A sequence (x_n) in V is a **Cauchy sequence for r.u. convergence,** if there exist $u \in K$ and a null-sequence (ϵ_n) such that

$$-\epsilon_n u \leq x_{m+n} - x_n \leq \epsilon_n u \quad \text{for all} \quad m, n \in \mathbb{N}.$$

A subset M of V is **complete for r.u. convergence** if every r.u. Cauchy sequence in M converges for r.u. convergence in M (with the same u).

Corollary 3.4.9. *A lineless convex cone K in the vector space V is internally complete if and only if with respect to relative uniform convergence every increasing Cauchy sequence in K converges in V (with the same u). In particular, K is internally complete if it is complete for relative uniform convergence.*

Proof. It suffices to prove the first statement.

(i) Let K be internally complete. If (x_n) is an increasing Cauchy sequence for r.u. convergence then

$$0 \le x_{m+n} - x_n \le \epsilon_n u \quad \text{for all} \quad m, n \in \mathbb{N}.$$

Let $y_n = x_n - x_1 + u \in V$. Obviously, $y_n \in K$ and

$$u \le y_n \le y_{m+n} \le y_n + \epsilon_n u,$$

that is, (y_n) is a sequence guided by u. By Theorem 3.4.6 (y_n) has a supremum y for $<$ in K. From $y_{m+n} \le y_n + \epsilon_n u$ it follows that $y_k \le y_n + \epsilon_n u$ for fixed n and all $k \in \mathbb{N}$ and, hence, $y_k < y_n + \epsilon_n u$.

The latter implies $y \le (1 + \frac{1}{n})y_n + 2\epsilon_n u$ for all $n \in \mathbb{N}$ and, hence,

$$\begin{aligned} y - y_n &\le \frac{1}{n} y_n + 2\epsilon_n u \le \frac{1}{n}(\epsilon_1 u + u) + 2\epsilon_n u \\ &\le \delta_n u \text{ with } \delta_n = \frac{1 + \epsilon_1}{n} + 2\epsilon_n. \end{aligned}$$

On the other hand, from $y_n < y$ it follows that $y_n \le (1 + \frac{1}{n})y$ and because of $y_n \le (1 + \epsilon_1)u$, we obtain

$$-\delta_n u \le -\frac{1 + \epsilon_1}{n} u \le -\frac{y_n}{n + 1} \le y - y_n.$$

Putting together, we obtain

$$-\delta_n u \le (y + x_1 - e) - x_n \le \delta_n u$$

with a null-sequence (δ_n).

Thus, the sequence (x_n) converges for r.u. convergence to $y + x_1 - e \in V$.

(ii) Suppose, every increasing Cauchy sequence in K converges for r.u. convergence in V. Let (x_n) be a guided sequence in K, i.e., $e \le x_n \le x_{m+n} \le x_n + \epsilon_n e$.

Obviously, (x_n) is an increasing Cauchy sequence for r.u. convergence and, by assumption, (x_n) is r.u. convergent to $x \in V$ with $u = e$, that is $-\delta_n e \le x - x_n \le \delta_n e$ for all n, where (δ_n) is a null-sequence.

In particular, for n big enough $0 \le (1 - \delta_n)e \le x_n - \delta_n e \le x$ and, hence, $x \in K$.

Furthermore, $x_n \le x + \delta_n e \le x + \delta_n x_n$ and, hence, $(1 - \delta_n)x_n \le x$.

For $r > 1$ given there exists $n_0 \in \mathbb{N}$ such that $\delta_n \le 1 - \frac{1}{r}$ for $n \ge n_0$, and we obtain $x_n \le \frac{1}{1-\delta_n} x \le rx$ for $n \ge n_0$.

Since (x_n) is increasing, this implies that $x_n \le rx$ for all $n \in \mathbb{N}$ and, hence, $x_n < x$ for all $n \in \mathbb{N}$.

Finally, let $x_n < y \in K$ for all $n \in \mathbb{N}$. From $x \le x_n + \delta_n e \le (1 + \delta_n)x_n$ we obtain that $x \le (1 + \delta_n)ry$ for all $r > 1$, all $n \in \mathbb{N}$. Since this implies that $x < y$ we conclude that (x_n) has for $<$ the supremum $x \in K$. By Theorem 3.4.6 the cone K is internally complete. □

Remarks 3.4.10. (i) For the notion of relative uniform convergence and related results see in particular [10, 11, 43, 47, 51, 56] (where it is called convergence with respect to a regulator).

(ii) Corollary 3.4.9 contains the following criterion due to G. Birkhoff ([11, p. 49], [12, p. 387]):

If K is a pointed and archimedean convex cone in a real vector space $V = K - K$ which is complete for r.u. convergence then K is complete with respect to Hilbert's projective metric.

For a pointed convex cone that is archimedean (in $K - K$) the internal completeness theorem can be rephrased as follows: K is internally complete if and only if every increasing Cauchy sequence for r.u. convergence has a supremum in K (for $\leq$).

If in addition $V = K - K$ is a vector lattice for $\leq$, Corollary 3.4.9 simplifies to the following characterization: K is internally complete if and only if K is complete for r.u. convergence.

Since a Banach lattice(or, more general, a complete vector lattice) is complete for r.u. convergence, it follows that the positive cone of a Banach lattice is internally complete (cf. [10, p. 227], [11, p. 50]).

Next we will derive criteria for internal completeness which employ topological assumptions. First we recall some related notions.

A **semi-norm** on a real vector space V is a mapping $q\colon V \longrightarrow \mathbb{R}_+$ such that $q(x + y) \leq q(x) + q(y)$ and $q(\lambda x) = |\lambda| q(x)$ for $x, y \in V$ and $\lambda \in \mathbb{R}$. A semi-norm q is a **norm** if in addition $q(x) = 0$ implies $x = 0$. If K is a convex cone in V with order relation $\leq$ a semi-norm or norm q is called **monotone** or increasing if $0 \leq x \leq y$ implies that $q(x) \leq q(y)$.

A **locally convex topology** on a vector space V is the coarsest topology on V for which all semi-norms q from a given family Q are continuous. A locally convex topology can be described by the following *base* for the neighborhood system of 0

$$\{x \in V \mid q(x) < \epsilon \quad \text{for all} \quad q \in F\}$$

where $\epsilon > 0$ and F is a finite subset of Q. By translation one obtains a base for any $y \in V$. The locally convex topology is separated or Hausdorff iff $q(x) = 0$ for all $q \in Q$ implies that $x = 0$. In the following a locally convex topology is always assumed to be Hausdorff.

A vector space V equipped with such a topology τ is called a **locally convex vector space**, denoted by (V, τ).

Definition 3.4.11. A convex cone K in a locally convex vector space (V, τ) is called **normal** if there exists a family Q of monotone semi-norms on V that defines τ.

For the special case of a normed space $(V, \|\cdot\|)$ a convex cone K in V is normal iff there exists a constant $c > 0$ such that $0 \leq x \leq y$ implies $\|x\| \leq c\|y\|$. (See [47] for normal cones.)

Before turning to topological criteria for internal completeness we will explore the relationship between the vector space topology τ and the *internal topologies*, that is the topologies belonging to internal metrics h, p, d, g, b, k (see also Exercises 2 to 5).

Proposition 3.4.12. *Let K be a lineless convex cone in a locally convex vector space (V,τ). Let $P \neq \{0\}$ be a part of K and let (L,τ) be the smallest subspace L of V containing P equipped with the restriction of τ on L.*

(i) *Let $x, y \in P$ and suppose $x + U \subseteq P$ and $y + U \subseteq P$ for some U from the base of the neighborhood system of 0 in (L,τ). For $x' \in x + \alpha U$, $y' \in y + \beta U$ with $0 < \alpha, \beta < 1$ it holds that*

$$\frac{1-\beta}{1+\alpha}\lambda(x,y) \leq \lambda(x',y') \leq \frac{1+\beta}{1-\alpha}\lambda(x,y). \tag{3.4.1}$$

(ii) *Let $x \in P$ with $x + U \subset K$ where $U = \{u \in L \mid q(u) \leq r,\ q \in F\}$ for a finite subset F of Q. For $x' \in x + U$ it holds that*

$$h(x',x) \leq \frac{q(x'-x)}{r} \quad \text{for all} \quad q \in F \quad \text{with} \quad q(x'-x) \neq 0. \tag{3.4.2}$$

(iii) *If P has non-empty interior $\overset{\circ}{P}$ in (L,τ) then $P = \overset{\circ}{P}$ and the order function $\lambda(\cdot,\cdot)$ and all internal metrics are τ-continuous on $P \times P$.*

(iv) *Internal open subsets of $\overset{\circ}{P}$ are open in (L,τ).*

(v) *If P is normal in (L,τ) then the topologies for the internal metrics with the exception of Hilbert's metric coincide on $\overset{\circ}{P}$ with the topology induced on $\overset{\circ}{P}$ by τ. The topology of Hilbert's metric coincides with the topology induced by τ on τ-bounded subsets S of $\overset{\circ}{P}$ with the property that for $x, y \in S$ neither $x < y$ nor $y < x$ ("$<$" the partial order defined by $\overset{\circ}{P}$).*

(vi) *Suppose τ is given by a monotone norm $\|\cdot\|$ on K such that $\overset{\circ}{K} \neq \emptyset$.*

  (a) *Let $x, y \in \overset{\circ}{K}$ with $B(x,r), B(y,r) \subseteq K$, where $B(x,r)$ is the open ball $\{z \in K \mid \|z-x\| < r\}$. Then*

$$h(x,y) \leq \frac{\|x-y\|}{r+\|x-y\|}, \quad p(x,y) \leq \frac{\|x-y\|}{r}, \quad d(x,y) \leq \frac{2\|x-y\|}{r}.$$

  (b) *For each set $M = u + K$ with $u \in \overset{\circ}{K}$ and each internal metric m there exists a positive constant K_m such that*

$$m(x,y) \leq K_m\|x-y\| \quad \text{for all} \quad x,y \in M.$$

On each norm-compact subset C of $\overset{\circ}{K}$ each internal metric $m \neq d$ is equivalent to the metric induced by the norm, that is with positive constants k_m, K_m

$$k_m\|x-y\| \leq m(x,y) \leq K_m\|x-y\| \quad \text{for all} \quad x,y \in C.$$

For $m = d$ such an equivalence holds on $C \cap \{z \in K \mid \|z\| = 1\}$.

Proof. (i) There exist $u, v \in U$ with $x' = x + \alpha u$, $y' = y + \beta v$ and $x \pm u \in P$, $y \pm v \in P$. By Lemma 3.2.2 (ii) part P is a convex cone and $\lambda(\cdot,\cdot) = \lambda_P(\cdot,\cdot)$. Applied to the convex cone P, Lemma 3.1.4 (xv) yields formula (3.4.1).

(ii) Fix a $q \in F$ with $q(x' - x) \neq 0$. For $\lambda = \frac{q(x'-x)}{r}$ and $u = r\frac{x'-x}{q(x'-x)}$ one has that

$$x = \frac{1}{1+\lambda}x' + \frac{\lambda}{1+\lambda}(x-u) \quad \text{and} \quad x' = (1-\lambda)x + \lambda(x+u).$$

Therefore, $\lambda(x', x) \geq \frac{1}{1+\lambda}$, $\lambda(x, x') \geq 1 - \lambda$ and $h(x', x) = 1 - \min\{\lambda(x', x), \lambda(x, x')\} \leq \lambda$.

(iii) Let $x_0 \in \overset{\circ}{P}$ and $x_0 + U \subseteq P$ with U from the base of 0 in (L, τ). Since P is a part, $x \sim x_0$ for $x \in P$ and, therefore, $x = \lambda x_0 + y$ with $0 < \lambda$ and $y \in P$. For $u \in U$ it follows that $x + \lambda u = \lambda x_0 + \lambda u + y = \lambda(x_0 + u) + y \in \lambda P + P \subseteq P$. Therefore, $x \in \overset{\circ}{P}$ which proves $P = \overset{\circ}{P}$.

From step (i) it follows that the order function is τ-continuous on $\overset{\circ}{P} = P$. The assertion on the internal metrics then follows according to their definitions by the order function.

(iv) From (ii) it follows that each h-open subset of $\overset{\circ}{P}$ is open in (L, τ). From Proposition 3.3.3 it follows that a subset of $\overset{\circ}{P}$ which is open for one of the internal metrics must be open for h.

(v) Because of statement (iv) and Proposition 3.3.3 it suffices to show that a τ-open subset $O \subseteq \overset{\circ}{P}$ is open for h. Let $x \in O$ and $x + U \subseteq O$ for $U = \{u \in L \mid q(u) < \epsilon,\ q \in F\}$, F finite. Let $\delta = \frac{\epsilon}{\epsilon + 3\max\{q(x) \mid q \in F\}}$ and consider $y \in P$ with $h(x, y) < \delta$. Since P is normal we may assume that the seminorms q defining τ are all monotone. Proposition 3.3.3 (vi) implies that $q(x - y) \leq 3h(x, y)\max\{q(x), q(y\}$ for $q \in F$. Furthermore, $h(x, y) < \delta$ implies that $\lambda(y, x) > 1 - \delta$ and, hence, $(1 - \delta)y \leq x$. Therefore, $(1 - \delta)q(y) \leq q(x)$ for $q \in F$ and

$$\begin{aligned}\max\left\{q(x-y) \mid q \in F\right\} &\leq h(x,y) \cdot \frac{3}{1-\delta}\max\left\{q(x) \mid q \in F\right\} \\ &< \frac{3\delta}{1-\delta}\max\left\{q(x) \mid q \in F\right\} = \epsilon\end{aligned}$$

by the definition of ϵ. This shows that $y - x \in U$ and, hence, $y = x + y - x \in O$.

Concerning the topology for Hilbert's metric d, let S be a subset of $\overset{\circ}{P}$ as in (v) and $x, y \in S$. If $\lambda(x, y) > 1$ then $y - (1 + \epsilon)x \in P$ and, hence, $y - x \in P + \epsilon x$. This implies $y > x$ which is impossible by assumption. Therefore, we must have $\lambda(x, y) \leq 1$ and, similarly, $\lambda(y, x) \leq 1$. From Proposition 3.3.3 (v) it follows for every monotone seminorm q that

$$q(x - y) \leq 3(1 - \lambda(x, y) \cdot \lambda(y, x))\max\{q(x), q(y)\}.$$

By assumption, S is τ-bounded and, therefore, for any finite set F of monotone seminorms

$$\max\{q(x-y) \mid q \in F\} \leq 3c_F(1 - e^{-d(x,y)}) \quad \text{for all} \quad x, y \in S$$

for some constant $c_F > 0$. The assertion for d follows then as for h above.

(vi) (a) Let, without loss, $x \neq y$. From $B(x, r) \subseteq K$ it follows for $\alpha = \frac{r}{\|x-y\|}$

$$\alpha x + x - \alpha y = x + \alpha(x - y) \in B(x, r) \subseteq K$$

and, hence, $\alpha y \leq (1+\alpha)x$. This gives $\lambda(y,x) \geq \frac{\alpha}{1+\alpha} = \frac{r}{r+\|x-y\|}$. In the same way $B(y,r) \subseteq K$ implies $\lambda(x,y) \geq \frac{r}{r+\|x-y\|}$. Thus, $\min\{\lambda(x,y),\lambda(y,x)\} \geq \frac{r}{r+\|x-y\|}$ and, hence,

$$h(x,y) = 1 - \min\{\lambda(x,y),\lambda(y,x)\} \leq 1 - \frac{r}{r+\|x-y\|} = \frac{\|x-y\|}{r+\|x-y\|}.$$

Furthermore, by Proposition 3.3.3 (i) and (ii)

$$p(x,y) \leq \frac{h(x,y)}{1-h(x,y)} \leq \frac{\|x-y\|}{r}$$

and $d(x,y) \leq 2p(x,y) \leq \frac{2\|x-y\|}{r}$.

(b) Let $M = u + K, u \in \mathring{K}$ and $B(u,r) \subseteq K$. If $x \in M$ and $\|x-z\| \leq r$ for $z \in L$ then $u + z - x \in B(u,r)$. Therefore, $u + z - x = v \in K$ and $z = v + x - u \in K$. This shows $B(x,r) \subseteq K$. Thus, for $x, y \in M$ from (a) it follows that

$$h(x,y) \leq K_h\|x-y\|, p(x,y) \leq K_p\|x-y\|, d(x,y) \leq K_d\|x-y\|,$$

with $K_h = K_p = \frac{1}{r}, K_d = \frac{2}{r}$. By Proposition 3.3.3 (iii), (iv) similar inequalities follow for $m = b, g, k$.

Consider now a norm-compact subset C of $\mathring{K}$. We shall show that $C \subseteq u + K$ for some $u \in \mathring{K}$. For $x \in C$, $U(x) = \frac{1}{2}x + \mathring{K}$ is an open set for $\|\cdot\|$ and $x \in U(x)$. Compactness of C for $\|\cdot\|$ implies a finite covering $C \subseteq \bigcup_{i=1}^n U(x_i)$. Since $x_i \in \mathring{K}, 1 \leq i \leq n, \lambda = \min_{1\leq i\leq n}\lambda(x_1,x_i) > 0$ and $\lambda x_1 \leq x_i$ for $1 \leq i \leq n$. Define $u = \frac{\lambda}{2}x_1$. Obviously, $u \in \mathring{K}$ and $U(x_i) = \frac{1}{2}x_i + \mathring{K} = u + (\frac{1}{2}x_i - u) + \mathring{K} \subseteq u + K$. Thus, $C \subseteq u + K$.

By the above, therefore, for each internal metric m there exist K_m such that $m(x,y) \leq K_m\|x-y\|$ for all $x, y \in C$. From Proposition 3.3.3 (vi) it follows by compactness of C that $k_h\|x-y\| \leq h(x,y)$ for some positive constant k_h and all $x, y \in C$. From Proposition 3.3.3 (ii), (iii), (iv) similar inequalities follow for all $m \neq d$. For $m = d$ such an inequality follows on $C \cap \{z \in K \mid \|z\| = 1\}$ by Proposition 3.3.3 (vii). □

Theorem 3.4.13. *Let K be a convex cone which is sequentially complete in a locally convex vector space (V,τ). Every part P of K for which the order intervals $\{x \in P \mid u - x \in P\}$ with $u \in P$ are τ-bounded is lineless and internally complete.*

Proof. Let $\leq$ be the ordering relation induced by K and $\leq_P$ the ordering relation induced by $P \subset K$. First we show that P is lineless. Suppose that $x + \lambda(y-x) \in P$ for all $\lambda \in \mathbb{R}$. Obviously, $u = 2x \in P$ and $u - (x + \lambda(y-x)) = x + (-\lambda)(y-x) \in P$ for all $\lambda \in \mathbb{R}$.

Since the order interval $[0,u] \subset P$ is τ-bounded, it follows that for every q of a defining family Q of semi-norms for τ there exists a constant $c_q > 0$ such that $q(x + \lambda(y-x)) \leq c_q$ for all $\lambda \in \mathbb{R}$.

This yields for all $\lambda \in \mathbb{R}$

$$|\lambda|q(y-x) = q(\lambda(y-x)) \leq q(x + \lambda(y-x)) + q(-x) \leq c_q + q(-x)$$

and, hence,

$$q(y - x) = 0 \quad \text{for all} \quad q \in Q, \quad \text{that is} \quad y = x.$$

This shows that P is a lineless cone.

Next, we show that any guided sequence (x_n) in P has a supremum in P for $<_P$. Theorem 3.4.6 then yields internal completeness for P. From $e \leq_P x_n \leq_P x_{m+n} \leq_P x_n + \epsilon_n e$ where $e \in P$ and (ϵ_n) a null sequence, it follows that $\frac{1}{\epsilon_n}(x_{m+n} - x_n)$ is contained in the order interval $[0, e]$ of P and, hence, τ-bounded. Therefore, (x_n) is a Cauchy sequence for τ in K and, by assumption, (x_n) converges for τ to some $x \in K$. Obviously, $e \leq x_n \leq x_{m+n} \leq x_n + \epsilon_n e$ and, therefore, $x_n \leq x \leq x_n + \epsilon_n e$ for all n. Together with $x_n \in P$, $e \in P$ this implies that $x \in P$. We show that x is a supremum of (x_n) for $<_P$. Let $r > 1$ and $y = rx - x_n$. From $0 \leq x_n \leq x$ we obtain $(r-1)x \leq y \leq rx$ and , hence, $y \in P$. This shows that $x_n \leq_P rx$ for all n, all $r > 1$, that is, $x_n <_P x$ for all n. Finally, suppose that $x_n \leq_P sz$ for some $z \in P$, all n, all $s > 1$. For any $r > 1$ and any $0 < \epsilon < r - 1$ one has that $s = r - \epsilon > 1$ and, hence, $x_n <_P (r - \epsilon)z$. Thus, $(r - \epsilon)z - x_n \in P \subset K$ for all n, which implies $(r - \epsilon)z - x \in K$. Therefore,

$$\epsilon z \leq \epsilon z + (r - \epsilon)z - x = rz - x \leq rz,$$

which shows that $rz - x \in P$ for all $r > 1$, that is $x <_P z$. We conclude that x is a supremum of (x_n) for $<_P$. □

Corollary 3.4.14. *Let K be a convex cone which is sequentially complete and normal in a locally convex vector space (V, τ). Then K is lineless and internally complete and, in particular, the interior $\overset{\circ}{K}$ with respect to τ is internally complete. Furthermore, $\{x \in \overset{\circ}{K} \mid p(x) = 1\}$ is an internally complete metric space for any functional p on V that is non-negative on K, positively homogeneous with $p(x) = 0$ only for $x = 0$.*

Proof. Since K is normal there exists for τ a defining family Q of monotone semi-norms on V. For $x, y \in K$ given and $z \in [x, y]$, that is $x \leq z \leq y$, it follows that $q(x) \leq q(z) \leq q(y)$ for all $q \in Q$. Therefore, order intervals $[x, y]$ of K are τ-bounded. In particular, K is lineless (cf. proof of Theorem 3.4.13) and for every part P of K the order intervals $[0, u]$ in P are τ-bounded. From Theorem 3.4.13 it follows that every part P of K is internally complete and, by Lemma 3.4.1, K is internally complete. Being a part of K, $P = \overset{\circ}{K}$ is internally complete. Finally, let $(x_n)_n$ be a Cauchy sequence in $\overset{\circ}{K}$ for d with $p(x_n) = 1$ for all n. This sequence converges to $y \in \overset{\circ}{P}$ and there exist $\lambda > 0, z \in K$ such that for some n it holds that $y = \lambda x_n + z \in \overset{\circ}{K} + K \subset \overset{\circ}{K}$. Therefore $p(y) > 0$ and $(x_n)_n$ converges for d to $x = \frac{y}{p(y)}$. □

Example 3.4.15. Let $(V, \|\cdot\|)$ be a Banach lattice with positive cone K. The cone K is closed and normal [51, p. 235]. By Corollary 3.4.14, K and, in particular, $\overset{\circ}{K}$ are internally complete. Furthermore, $X = \{x \in K \mid \|x\| = 1\}$ and $X = \{x \in \overset{\circ}{K} \mid \|x\| = 1\}$ are internally complete.

This applies in particular to the Banach lattice $V = \mathcal{C}(T)$ of all continuous functions on a compact space with positive cone $K = \{f \in \mathcal{C}(T) \mid f(t) \geq 0 \text{ for all } t \in T\}$. Especially, for $T = \{1, \dots, n\}$ one obtains that for $V = \mathbb{R}^n$ the standard cone $K = \mathbb{R}^n_+$, as well as its interior, is internally complete. Also, the completeness of $X = \{x \in \overset{\circ}{K} \mid \|x\| = 1\}$ with respect to Hilbert's projective metric which we proved in Lemma 2.1.10 directly is a special case. The case of $\overset{\circ}{K}$ also makes clear that in the internal completeness theorem $<$ cannot be replaced by $\leq$ in general.

Remarks 3.4.16. (i) The first result on internal completeness was established by G. Birkhoff ([10, 11]; see also [50]) in 1957 when he showed that the positive cone of a Banach lattice is complete for Hilbert's projective metric (see Remarks 3.4.10).

The first result on internal completeness for more general spaces was established by A. C. Thompson ([53, p. 69], [54]) in 1963 when he showed the statement of Corollary 3.4.14 for the part metric. From a different point of view M.A. Krasnoselskii and his group ([59]; see also [35, 36]) showed for a closed and pointed convex cone in a Banach space that a part is complete for the part metric if and only if all order intervals $[0, x]$ of the part are (norm) bounded. Thus, in the particular case of Banach spaces the condition in Theorem 3.4.13 is not only sufficient but also necessary. For results on completeness of cones for the part metric and Hilbert's projective metric, respectively, see also [17, 21, 22, 38, 44, 58].

(ii) As mentioned already (cf. Remarks 3.1.6 and Section 3.3) the part metric and Hilbert's projective metric have been studied also for general convex sets. Thompson's result for convex cones has been extended by H. Bauer [2] to (sequentially) complete convex sets in a locally convex vector space which are normal in a sense that generalizes normality of cones. H. S. Bear [5] and Bauer and Bear [3] also showed that a complete lineless convex set in a weak space is complete for the part metric. It has been observed (cf. [9], see also Section 3.3) that a convex set can be interpreted also as the base of a certain convex cone by which results on cones may be transformed into results on general convex sets. Furthermore, there is a strong relationship for convex sets between completeness for the part metric and superconvexity or σ-convexity (cf. [40]).

The results obtained, in particular the internal completeness theorem, indicate that internal completeness of a convex cone is strongly related to a principle of monotone convergence to hold for the order relation induced by the cone. Such a principle is of interest in its own and is exemplified in the context of normed spaces by the following concept of regularity due to M. A. Krasnolselskii et al. [35, 36].

Let K be a closed and pointed convex cone in a real normed space $(V, \|\cdot\|)$ and let $\leq$ be the order relation induced by K on V. The cone K is **regular** if every increasing sequence (x_n) in K with $x_n \leq y$ for all n converges for $\|\cdot\|$ to some $x \in K$. The cone is called **completely regular** if every increasing sequence in K which is norm-bounded converges in K [36, p. 47]. The following implications hold:

$$\text{completely regular} \Longrightarrow \text{regular} \Longrightarrow \text{normal},$$

but no one can be reversed in general [36, p. 48]. Furthermore, if K is normal then K is symmetrically bounded but not vice versa in general (see Exercise 5 to Section 3.1 and Exercise 5 to Section 3.3). In case the interior of K is not empty it can be shown that K is symmetrically bounded if and only if every continuous linear functional on V is the difference of two continuous linear functionals on V which are non-negative on K. This property, in turn, is equivalent to the normality of K. In case the space $(V, \|\cdot\|)$ is complete it follows from Corollary 3.4.14 that K is internally complete. Therefore, a closed, pointed convex cone K with non-empty interior in a Banach space V which is completely regular is internally complete. If V is finite dimensional then every closed, pointed convex cone K and, without loss, with non-empty interior, is completely regular (cf. [35, p. 51]) and, therefore, regular, normal, symmetrically bounded and internally complete. Moreover, it has been shown for the finite dimensional case that even *any* convex cone which is lineless possesses the properties just mentioned [42, p. 547]. Thereby it turns out that for every lineless convex cone in finite dimensions any bounded increasing sequence is guided (see Examples 3.4.3 (a) for the special case of the standard cone) and that any guided sequence converges for the norm.

Exercises

1. Consider the ice cream cone

$$K = \{(u, r) \mid u \in \mathbb{R}^n,\ r \in \mathbb{R}_+, \|u\| \leq r\}.$$

   (a) Show that K is a normal cone for $\|\cdot\|$.
   (b) Show that K is internally complete.
   (c) Examine if every increasing sequence in K which is bounded from above is a guided sequence.

2. Let K be a convex cone in a locally convex vector space (V, τ). K is called **symmetrically bounded** if every symmetric subset of K is bounded for τ. (Cf. Exercises 5 of Section 3.1)
   (a) Show that K is symmetrically bounded iff the topology generated by the part metric on K is finer than the restriction of τ to K.
   (b) Assume there exists $x_0 \in K$ such that $\lambda(x, x_0) > 0$ for all $x \in K$. Demonstrate that K is symmetrically bounded iff for every continuous functional f on (V, τ) there exists two functionals f_1 and f_2 on V which are non-negative on K and such that $f = f_1 - f_2$.
   (c) Consider the convex cone K of all sequences $x = (x_1, x_2 \ldots) \in \mathbb{R}^{\mathbb{N}}$ with $x_i \neq 0$ only for finitely many $i \in \mathbb{N}$ and $\sum_{i=1}^n x_i \geq 0$ for all $n \geq 1$. In the normed space

given by $V = K - K$, $\|x\| = \max_i |x_i|$ the cone K is symmetrically bounded but not normal (see Exercise 5 (c) to Section 3.3). Show that $f(x) = \sum_{n=1}^{\infty} \frac{(-1)^n}{n^2} x_n$, $x = (x_1, x_2, \ldots)$ defines a continuous functional on $(V, \|\cdot\|)$ which cannot have a decomposition as in (b) with f_1 and f_2 both continuous on $(V, \|\cdot\|)$.

3. (Cf. [1–3]) Let C be a convex set in a real vector space which has one part only and with $0 \in C$ Let $V = \{\lambda x \mid \lambda \geq 0, x \in C\}$ and q the Minkowski-norm for C, i.e., for $x \in V$

$$q(x) = \max\{p(x), p(-x)\}, p(x) = \inf\{\lambda \geq 0 \mid x \in \lambda C\}.$$

   (a) Show that q is a semi-norm on V which is a norm iff C is lineless.
   (b) Show that a lineless C is complete for the part metric iff (V, q) is a Banach space.
   (c) Show that every lineless and finite dimensional convex set is complete for the part metric.
   (d) Find a bounded convex subset of a Banach space which has one part only but which is not complete for the part metric.

4. (Cf. [40]). A subset C of a locally convex space (V, τ) is called **superconvex** or σ-convex if for any sequence (x_n) in C and any sequence (α_n) in $[0, 1]$ with $\sum_{n=1}^{\infty} \alpha_n = 1$ the sequence of sums $\sum_{n=1}^{N} \alpha_n x_n$ converges for $N \to \infty$ and belongs to C. Consider a convex set C, V and q as in Exercise 3 above, and let τ be a locally convex topology on V.
   (a) Show that every τ-open set is open also for q, provided that C is τ-bounded and that the converse holds, provided that C has non-empty interior for τ.
   (b) Suppose that C is superconvex in (V, τ) and has non-empty interior for τ. Show that C is complete for the part metric.
   (c) Suppose that C is complete for the part metric, τ-bounded and has non-empty interior for τ. Show that C is superconvex.

5. Prove the following characterization: A convex set is lineless, complete for the part metric and consists of one part only if it can be embedded as an open and symmetrically bounded subset into some complete locally vector space. (Compare Exercises 3 and 4 above.)

Bibliography

[1] H. Bauer. An open mapping theorem for convex sets with only one part. *Aequationes Math.*, 4:332–337, 1970.

[2] H. Bauer. Intern vollständige konvexe Mengen. In *Theory of Sets and Topology*, pp. 22–38, Berlin, 1972.

[3] H. Bauer and H. S. Bear. The part metric in convex sets. *Pacific J. Math.*, 30:15–33, 1969.

[4] H. S. Bear. A geometric characterization of Gleason parts. *Proc. Amer. Math. Soc.*, 16:407–412, 1965.

[5] H. S. Bear. *Lectures on Gleason Parts*. Springer, Berlin etc., 1970.
[6] H. S. Bear. Part metric and hyperbolic metric. *Amer. Math. Month.*, 98:109–123, 1991.
[7] H. S. Bear and H. L. Weiss. An intrinsic metric for parts. *Proc. Amer. Math. Soc.*, 18:812–817, 1967.
[8] R. Bellman and T. A. Brown. Projective metrics in dynamic programming. *Bull. Amer. Math. Soc.*, 71:773–775, 1965.
[9] R. Bellmann. Normale Mengen in Lokal–Konvexen Räumen. *Math. Ann.*, 222:49–54, 1976.
[10] G. Birkhoff. Extensions of Jentzsch's theorem. *Trans. Amer. Math. Soc.*, 85:219–227, 1957.
[11] G. Birkhoff. Uniformly semi-primitive multiplicative processes. *Trans. Amer. Math. Soc.*, 104:37–51, 1962.
[12] G. Birkhoff. *Lattice Theory*. Amer. Math. Soc., Providence R.I., 3rd edition, 1967.
[13] N. Bourbaki. *Éléments de Mathématique. Topologie Générale. Chapitre 9. Utilisation des Nombres Réels en Topologie Générale*. Hermann, Paris 1958.
[14] N. Bourbaki. *Éléments de Mathématique. Topologie Générale. Chapitre 2. Structures Uniformes*. Hermann, Paris 1961.
[15] H. Busemann. *The Geometry of Geodesics*. Academic Press, New York, 1955.
[16] H. Busemann and P.J. Kelly. *Projective Geometry and Projective Metrics*. New York, 1953.
[17] P. J. Bushell. Hilbert's metric and positive contraction mappings in a Banach space. *Arch. Rational Mech. Anal.*, 52:330–338, 1973.
[18] P. J. Bushell. The Cayley–Hilbert metric and positive operators. *Lin. Alg. Appl.*, 84:271–280, 1986.
[19] A. Cayley. On the non-Euclidean geometry. *Math. Ann. V*:630–634, 1872.
[20] G. Gentili. Distances on convex cones. In *Springer Lecture Notes in Mathematics*, volume 1022. Springer, Berlin etc., 1982.
[21] W. Gunselmann. *Interne Vollständigkeit*. PhD thesis, Universität Erlangen-Nürnberg, 1973. 44 p.
[22] D. Guo and V. Lakshmikantham. *Nonlinear Problems in Abstract Cones*. Academic Press, Boston, 1988.
[23] D. Hilbert. Über die gerade Linie als kürzeste Verbindung zweier Punkte. *Math. Ann.*, 46:91–96, 1895.
[24] D. Hinrichsen and U. Krause. Choice of techniques in joint production models. *Operations Research-Verfahren*, 34:155–171, 1979.
[25] D. Hinrichsen and U. Krause. Unique representation in convex sets by extraction of marked components. *Lin. Alg. Appl.*, 51:73–96, 1983.
[26] E. Hopf. An inequality for positive linear integral operators. *J. Math. Mech.*, 12:683–692, 1963.
[27] V. I. Istratescu. *Fixed Point Theory*. D. Reidel, Dordrecht, 1979.
[28] G. Jameson. *Ordered Linear Spaces*. Springer, Berlin etc., 1970.
[29] F. Klein. Über die sogenannte Nicht-Euklidische Geometrie. *Math. Ann.*, 4:573–625, 1871.
[30] E. Kohlberg and J. W. Pratt. The contraction mapping approach to the Perron–Frobenius theorem: Why Hilbert's metric? *Math. Oper. Res.*, 7:198–210, 1982.
[31] J. Köhn. Die Harnacksche Metrik in der Theorie der harmonischen Funktionen. *Math. Zeitschrift*, 91:50–64, 1966.
[32] H. König. Zur abstrakten Theorie der analytischen Funktionen II. *Math. Ann.*, 163:9–17, 1966.
[33] H. König. On the Gleason and Harnack metrics for uniform algebras. *Proc. Amer. Math. Soc.*, 22:100–101, 1969.
[34] M. A. Krasnoselskii. *Positive Solutions to Operator Equations*. P. Noordhoff Ltd., Groningen, 1964.
[35] M. A. Krasnoselskii, G. M. Vainikko, P. A. Zabreiko, Y. B. Rutitskii, and V. Y. Stetsenko. *Approximate Solutions to Operator Equations*. Wolters-Noordhoff, Groningen, 1972.

[36] M. A. Krasnoselskii, J. A. Lifshits, and A. V. Sobolev. *Positive Linear Systems.* Heldermann Verlag, Berlin, 1989.
[37] U. Krause. Strukturen in unendlich-dimensionalen konvexen Mengen. *Mathematik-Arbeitspapiere der Universität Bremen Nr. 1*, page 74, 1976.
[38] U. Krause. A non-linear extension of the Birkhoff–Jentzsch theorem. *J. Math. Anal.*, 114:552–568, 1986.
[39] U. Krause. Internal metrics of cones as tools for positive systems. In A. Begli, L. Finesso, and G. Picci, editors, *Mathematical Theory of Networks and Systems*, pp. 269–272. Il Poligrafo, 1998.
[40] U. Krause. Superconvexity and part-completeness. *Seminarberichte aus dem Fachbereich Mathematik der Fernuniversität Hagen*, 63(3):427–437, 1998.
[41] U. Krause. A general principle of marked extraction. In F. Colonius, U. Helmke, D. Prätzel-Wolters, and F. Wirth, editors, *Advances in Mathematical Systems Theory*. A Volume in Honor of Diederich Hinrichsen, pp. 169–183. Birkhäuser, Boston etc., 2001.
[42] H. Kriete. Internal completeness of cones. *J. Math. Anal. Appl.*, 161:545–554, 1991.
[43] W. A. J. Luxemburg and A. C. Zaanen. *Riesz Spaces*, volume 1. North Holland, Amsterdam, 1971.
[44] R. D. Nussbaum. Hilbert's projective metric and iterated non-linear maps. *Memoirs Amer. Math. Soc.*, 75(391):1–137, 1988.
[45] R. D. Nussbaum. Hilbert's projective metric and iterated non-linear maps II. *Memoirs Amer. Math. Soc.*, 79(401):1–118, 1989.
[46] A. M. Ostrowski. Positive matrices and functional analysis. In H. Schneider, editor, *Recent Advances in Matrix Theory*, pp. 81–101. The University of Wisconsin Press, Madison, 1964.
[47] A. L. Peressini. *Ordered Topological Vector Spaces*. Harper and Row, New York, 1967.
[48] A. J. B. Potter. Applications of Hilbert's projective metric to certain classes of non-homogeneous operators. *Quart. J. Math.*, 28:93–99, 1977.
[49] D. Pumplün and H. Röhrl. Convexity theories IV. Klein–Hilbert parts in convex modules. *Appl. Categorical Structures*, 3:173–200, 1995.
[50] H. Samelson. On the Perron–Frobenius theorem. *Michigan Math. J.*, 4:57–59, 1957.
[51] H. H. Schaefer. *Banach Lattices and Positive Operators*. Springer Verlag, Berlin etc., 1974.
[52] H. Schneider, editor. *Recent Advances in Matrix Theory*. The University of Wisconsin Press, Madison etc., 1964.
[53] A. C. Thompson. *Generalizations of the Perron–Frobenius theorem to operators mapping a cone into itself.* PhD thesis, University of Newcastle upon Tyne, 1963. 114 p.
[54] A. C. Thompson. On certain contraction mappings in a partially ordered vector space. *Proc. Amer. Math. Soc.*, 14:438–443, 1963.
[55] M. Turinici. A class of projective metrics on ordered linear spaces. *Bul. Stii. Tehn. Inst. Pol. Traian Vuia, Timisoara*, 26:19–26, 1981.
[56] B. Z. Vulikh. *Introduction to the Theory of Partially Ordered Spaces*. Wolters-Noordhoff, Groningen, 1967.
[57] L. J. Wallen. An abstract theorem of Bonnesen type with applications to mixed area. Typescript, no date. University of Hawaii at Manoa.
[58] D. Weller. *Hilbert's metric, part metric and selfmappings of a cone*. PhD thesis, Universität Bremen, 1987. 99 p.
[59] P. P. Zabraiko, M.A. Krasnoselskii, and Yu.V. Pokornyi. On a class of linear positive operators. *Functional Analysis and its Applications*, 5:272–279, 1972.

4 Contractive dynamics on metric spaces

The previous chapter has shown how a convex cone can be made into a complete metric space by employing one of the several internal metrics of the cone. Thus, a selfmapping of a cone becomes a selfmapping of a complete metric space. The crucial point then is that this induced selfmapping of a metric space is contractive or at least non-expansive for a great variety of non-linear selfmappings of the cone. As an example we have already seen the First Concave Perron Theorem (Theorem 2.1.11) where a concave selfmapping of the standard cone in finite dimensions became a contraction of an appropriate complete metric space for Hilbert's projective metric.

In the present chapter we now study systematically the contractive dynamics on metric spaces, that is the asymptotic behavior of the iterates of a selfmapping of a metric space which does contract in one way or another distances. In particular, we treat (ϵ, δ)-contractive mappings and contractive sequences of selfmappings and non-expansive or, more generally, power-lipschitzian selfmappings for which we prove a very useful local-global stability principle. Metric fixed point theory is a wide field and we shall concentrate on those questions which are relevant for the non-linear selfmappings of convex cones to be dealt with in the next chapter.

4.1 Iteration of contractive selfmappings

Let (X, d) be a metric space and let $f: X \to X$ be a selfmapping of X. The **forward orbit** $O(x)$ of a point $x \in X$ with respect to f is $O(x) = \{f^n(x) \mid n = 0, 1, 2, \ldots\}$, where f^n is the n-th iterate of f. The *(omega)* **limit set** $\boldsymbol{\omega(x)}$ of a point $x \in X$ with respect to f is $\omega(x) = \bigcap_{k=0}^{\infty} \overline{\{f^n(x) \mid n \geq k\}}$, where $\bar{A}$ is the closure of a subset A of (X, d). It is well known that $y \in \omega(x)$ iff $y = \lim_{i\to\infty} f^{n_i}(x)$ for a sequence $n_i \to \infty$. Obviously, limit sets are closed and invariant under f, i.e., $f(\omega(x)) \subset \omega(x)$. Limit sets may be empty and it will be an interesting point below when they are not.

Already in Section 2.1 we made use of contractivity properties of certain selfmappings of cones. Now we will study contractivity properties more systematically and in detail. The following definition collects some interesting contractivity properties, roughly in increasing generality.

Definition 4.1.1. A selfmapping f of a metric space (X, d) is called a

(i) **contraction** if there exists a constant $0 \leq c < 1$ such that $d(f(x), f(y)) \leq cd(x, y)$ for all $x, y \in X$;

(ii) $\boldsymbol{\phi}$**-contraction** if there exists a function $\phi : \mathbb{R}_+ \to \mathbb{R}_+$ with $\phi(0) = 0$ and $\phi(t) < t$ for $t > 0$ such that $d(f(x), f(y)) \leq \phi(d(x, y))$ for all $x, y \in X$;

(iii) **generalized contraction** if for any given pair $0 < \alpha \leq \beta < \infty$ there exists a constant $0 \leq L(\alpha,\beta) < 1$ such that $\alpha \leq d(x,y) \leq \beta$ implies $d(f(x), f(y)) \leq L(\alpha,\beta)d(x,y)$;
(iv) **(ϵ,δ)-contractive** if for any given $\epsilon > 0$ there exists $\delta > 0$ such that for all $x, y \in X$ $\epsilon \leq d(x,y) < \epsilon + \delta$ implies $d(f(x), f(y)) < \epsilon$;
(v) **contractive** if $d(f(x), f(y)) < d(x,y)$ for all $x, y \in X$ with $x \neq y$;
(vi) **non-expansive** if $d(f(x),f(y)) \leq d(x,y)$ for all $x, y \in X$.

All these notions make sense when restricted in an obvious way to points $x, y \in A$ for a non-empty subset A of X.

Having stated these properties we like to add some comments, in particular with respect to the widespread literature on contractivity and fixed points. Property (i), of course, is the one which is used in Banach's fixed point theorem or the contraction mapping principle to guarantee, together with completeness of (X, d), for arbitrary initial $x \in X$ the convergence of the iterates $f^k(x)$ to the unique fixed point x^* of f. A contraction is, of course, a ϕ-contraction. The reverse implication, however, is not true. ϕ-contractions have been considered by several authors, as, e.g., in [20] for $\phi(t) = \alpha(t)t$ with $\alpha(t)$ decreasing, in [2] for ϕ increasing and continuous from the right, in [1] for ϕ upper semicontinuous from the right, in [16] for ϕ increasing with $\lim_{n\to\infty} \phi^n(t) = 0$. In all these cases it was proved by the authors for (X, d) complete that for every initial value the iterates of a ϕ-contraction f converge to the unique fixed point of f. This conclusion, however, is not possible for an arbitrary ϕ-contraction (see Exercise 2). For $\phi(t) = \alpha(t)t$ with $\alpha(t)$ decreasing or for ϕ upper semicontinuous from the right, a ϕ-contraction needs to be a generalized contraction (see Exercise 1 a), b)). A generalized contraction is necessarily (ϵ,δ)-contractive but the reverse implication does not hold (see Exercises 3 and 4). Obviously, property (iv) implies property (v), which in turn implies property (vi) and none of these implications can be reversed. It is easily be seen that the conclusion of Banach's fixed point theorem no longer holds in general if f is assumed to be contractive only instead of being a contraction. Below we will see, however, that for (X, d) complete the conclusion of Banach's fixed point theorem holds already for (ϵ,δ)-contractive selfmappings and, a fortiori, for selfmappings which are a ϕ-contraction (with ϕ satisfying additional properties as above) or a generalized contraction. (See [12] for generalized contractions and [17] for (ϵ,δ)-contractive mappings which are called there weakly uniformly strict contractions.) Further generalizations of Banach's fixed point theorem can be found in [5] and [9]. In the following we shall show that contractivity together with non-empty limit sets is sufficient for the conclusion of Banach's fixed point theorem to hold (cf. [6, 14]).

Lemma 4.1.2. *Let (X, d) be a metric space, let $f: X \longrightarrow X$ be non-expansive and let $x \in X$.*

(a) *If $\omega(x)$ is not a singleton then for each $y \in \omega(x)$ there exists $c(y) > 0$ such that*

$$d(f^m(y), f^{m+1}(y)) = c(y) \text{ for all } m \geq 0.$$

(b) *If $\omega(x) \neq \emptyset$ and f is contractive on $\omega(x)$ then f has a fixed point x^* with $\lim_{n\to\infty} f^n(x) = x^*$.*

Proof. Since f is non-expansive the two sequences defined by

$$a_n = d(f^n(x), f^{n+1}(x)) \text{ and } b_n = d(f^n(x), f^n(y)) \text{ for } y \in X$$

are monotone decreasing. Therefore, the limits $a = \lim a_n$ and $b = \lim b_n$ exist.

(a) If $y \in \omega(x), y = \lim_{k\to\infty} f^{n_k}(x)$, then for $c(y) = d(y, f(y))$ by the above

$$\begin{aligned} c(y) &= \lim_{k\to\infty} d(f^{n_k}(x), f^{n_k+1}(x)) = \lim_{k\to\infty} a_{n_k} = a = \lim_{k\to\infty} a_{n_k+m} \\ &= \lim_{k\to\infty} d(f^{n_k+m}(x), f^{n_k+m+1}(x)) = d(f^m(y), f^{m+1}(y) \end{aligned}$$

for all $m \geq 0$. Suppose $c(y) = 0$, that is $f(y) = y$. Therefore, for the second sequence $(b_n)_n$

$$\lim_{n\to\infty} d(f^n(x), y) = \lim_{n\to\infty} b_n = b = \lim_{k\to\infty} b_{n_k} = \lim_{k\to\infty} d(f^{n_k}(x), y) = 0.$$

This yields $\lim_{n\to\infty} f^n(x) = y$. Thus $\omega(x) \subseteq \{y\}$ and $\omega(x) = \{y\}$ is a singleton.

(b) Since f is contractive on $\omega(x)$ it is impossible for $y \in \omega(x)$ to have that $0 < d(y, f(y)) = d(f(y), f^2(y))$. Therefore, (a) implies $\omega(x) = \{x^*\}$. It follows $f(x^*) = x^*$ and, as in (a), for $x^* = \lim_{k\to\infty} f^{n_k}(x)$

$$\lim_{n\to\infty} d(f^n(x), x^*) = \lim_{k\to\infty} (f^{n_k}(x), x^*) = 0.$$

Thus, $\lim_{n\to\infty} f^n(x) = x^*$. □

As an immediate consequence of this lemma we obtain the following characterization in case of contractive selfmappings.

Theorem 4.1.3. *Let (X, d) be a metric space and let $f: X \to X$ be contractive. There exists a unique fixed point x^* of f with $\lim_{n\to\infty} f^n(x) = x^*$ for all $x \in X$ if and only if all limit sets of points of X are non-empty.*

Proof. Suppose $\omega(x) \neq \emptyset$ for all $x \in X$. Lemma 4.1.2 (b) yields that $\lim_{n\to\infty} f^n(x) = x^*$ where x^*, by contractivity of f on X, is the unique fixed point of f. The latter, obviously, implies $\omega(x) \neq \emptyset$ for all $x \in X$. □

As a useful criterion we obtain the following generalization of Banach's fixed point theorem.

Theorem 4.1.4. *Let (X, d) be a complete metric space and let f be a selfmapping of X for which some iterate f^k is (ϵ, δ)-contractive. Then f has a unique fixed point x^* and $\lim_{n\to\infty} f^n(x) = x^*$ for all $x \in X$.*

Proof. For $x \in X$ and $x_n = f^n(x)$ for $n \geq 1$ we will show that for $n_i = ki$ the subsequence (x_{n_i}) of (x_n) is a Cauchy sequence. This then implies $\omega(x) \neq \emptyset$ and Theorem 4.1.4 follows from Theorem 4.1.3

To simplify notation let $g = f^k$ and $y_n = g^n(x) = x_{nk}$. Since g is, in particular, contractive it follows that the sequence given by $a_n = d(y_n, y_{n+1})$ is non-increasing and, hence, $a = \lim_{n\to\infty} a_n \geq 0$ exists. Suppose that $a > 0$.

For $\epsilon = a$ choose $\delta > 0$ according to the (ϵ, δ)-contractivity of g. Since there exists $m \in \mathbb{N}$ with $\epsilon \leq a_m < \epsilon + \delta$ we obtain that $d(g(y_m), g(y_{m+1})) < \epsilon$. This yields $a_{m+1} = d(y_{m+1}, y_{m+2}) < a$, which is a contradiction. Therefore, we must have that $a = 0$, that is $\lim_{n\to\infty} a_n = 0$.

Now, let $\epsilon > 0$ be arbtrarily given and choose $\delta > 0$ according to the (ϵ, δ)-contractivity of g, where we may assume that $\delta \leq \epsilon$. Choose $N \in \mathbb{N}$ such that $a_N < \delta$. Consider $j \in \mathbb{N}$ for which it holds that $d(y_N, y_j) < \epsilon + \delta$. Obviously,

$$d(y_N, y_{j+1}) \leq d(y_N, y_{N+1}) + d(g(y_N), g(y_j)).$$

If $d(y_N, y_j) < \epsilon$ then

$$d(g(y_N), g(y_j)) \leq d(y_N, y_j) < \epsilon$$

and, hence,

$$d(y_N, y_{j+1}) < \delta + \epsilon.$$

If, on the other hand, $d(y_N, y_j) \geq \epsilon$ then $\epsilon \leq d(y_N, y_j) < \epsilon + \delta$ and by the (ϵ, δ)-contractivity of g we obtain $d(y_N, y_{j+1}) < \delta + \epsilon$.

Thus, in any case we have that $d(y_N, y_{j+1}) < \epsilon + \delta$. By choice of N we have that $d(y_N, y_{N+1}) < \delta < 2\epsilon$ and, therefore, we obtain that

$d(y_N, y_{N+p}) < 2\epsilon$ for all $p \in \mathbb{N}$.

This proves that the sequence given by $y_n = x_{nk}$ for $n \geq 1$ is a Cauchy sequence. □

Remarks 4.1.5. (i) If (X, d) is complete and f is a contractive selfmapping of X for which all orbits are relatively compact then Theorem 4.1.3 yields $\lim_{n\to\infty} f^n(x) = x^*, x^*$ being the unique fixed point of f. (Cf. [6] and [13]). In particular, the latter conclusion holds for any contractive selfmapping of a compact metric space.

(ii) For $k = 1$ Theorem 4.1.4 is obtained in [17]. The proof given here is a simplified version of the one given in [17]. By the comments made following Definition 4.1.1, various generalizations of Banach's fixed point theorem as the one for generalized contractions in [12] and the ones for ϕ-contractions in [1, 2, 16, 20] follow from Theorem 4.1.4.

(iii) Since the selfmappings f of (X, d) considered above are all non-expansive, a fixed point x^* of f is automatically **stable** in the sense that for $\epsilon > 0$ given there exists $\delta > 0$ such that for all $x \in X$ from $d(x, x^*) < \delta$ it follows that $d(f^n(x), x^*) < \epsilon$ for all $n \geq 0$. Hence, $\lim_{n\to\infty} f^n(x) = x^*$ for all $x \in X$ means that x^* is **globally asymptotically stable.**

Exercises

1. Let f be a selfmapping of a metric space (X, d) which is a ϕ-contraction (see Definition 4.1.1 (ii)).
   (a) Show that f is a generalized contraction if ϕ is upper semicontinuous from the right.
   (b) Show that f is a generalized contraction if $\phi(t) = \alpha(t)t$ with $\alpha(t)$ decreasing.
2. [17] Let $X = \{\sum_{k=1}^{n}(1 + \frac{1}{k}) \mid n \geq 1\}$ be equipped with the Euclidean distance d.
   (a) Show that (X, d) is complete.
   (b) Let $f: X \to X$ be defined by
   $$f\left(\sum_{k=1}^{n}(1 + \frac{1}{k})\right) = \sum_{k=1}^{n+1}(1 + \frac{1}{k}).$$
   Show that f is a ϕ-contraction for an appropriate ϕ.
   (c) Show that f has no fixed point.
3. Prove that a generalized contraction is always (ϵ, δ)-contractive.
4. [17] Let
   $$X = [0, 1] \cup \{3n \mid n \geq 1\} \cup \{3n + 1 \mid n \geq 1\}$$
   be equipped with Euclidean distance d and let $f: X \to X$ be defined by
   $$f(x) = \begin{cases} \frac{x}{2}, & 0 \leq x \leq 1 \\ 0, & x = 3n \\ 1 - \frac{1}{n+2}, & x = 3n + 1. \end{cases}$$
   (a) Show that f is (ϵ, δ)-contractive.
   (b) Show that f is not a generalized contraction.

4.2 Non-autonomous discrete systems

Whereas the previous section was about just one single selfmapping of a metric space (X, d) we now consider a whole sequence $(f_n)_n$ of those mappings. In terms of dynamical systems, whereas f defines an autonomous system by $x_{n+1} = f(x_n)$, a sequence defines a non-autonomous system by $x_{n+1} = f_n(x_n)$, $x_1 = x \in X$. Sometimes the term *inhomogenous iteration* is used for such a system.

Definition 4.2.1. A sequence $(f_n)_n$ of selfmappings of a metric space (X, d) is called an (asymptotically) **contractive sequence on a subset** $A \subset X$ if there exists a continuous mapping $c: A \times A \to \mathbb{R}$ such that the following two conditions are satisfied
(i) $c(x, y) < d(x, y)$ for all $x, y \in A$ with $x \neq y$
(ii) To every $\epsilon > 0$ there exists a $N(\epsilon) \in \mathbb{N}$ such that
$$d(f_n(x), f_n(y)) \leq c(x, y) + \epsilon \quad \text{for all} \quad n \geq N(\epsilon), \quad \text{all} \quad x, y \in A.$$

For an autonomous system, that is $f_n = f$ for all n, this definition reduces to that of a contractive mapping f as studied in the previous section.

Let (X, d) be a metric space and let $f_n\colon X \to X$ for $n \geq 1$ define a sequence of selfmappings of X.

The *(forward)* **orbit $O_s(s)$** of a point $x \in X$ with respect to this sequence is

$$O_s(x) = \{f_n \circ f_{n-1} \circ \ldots \circ f_1(x) \mid n = 0, 1, 2, \ldots\},$$

where the inhomogeneous iteration $f_n \circ \ldots \circ f_1$ for $n = 0$ is the identity. For a given point $x_1 = x \in X$ we will also consider the sequence defined by $x_{n+1} = f_n(x_n)$ for $n \geq 1$, that is $x_{n+1} = f_n \circ \ldots \circ f_1(x)$ and, hence, $O_s(x) = \{x_{n+1} \mid n \geq O\}$.

The (omega) **limit set $\omega_s(x)$** of a point $x \in X$ with respect to $(f_n)_n$ is

$$\omega_s(x) = \bigcap_{k=0}^{\infty} \overline{\{f_n \circ f_{n-1} \circ \ldots \circ f_1(x) \mid n \geq k\}},$$

or, equivalently, $\omega_s(x)$ consists of all points $y = \lim_{i\to\infty} x_{n_i}$ for a sequence $n_i \to \infty$.

Occasionally we will consider the **joint limit set $\omega_s(x, y)$** for two points $x, y \in X$, which consists of all pairs $(\lim_{i\to\infty} x_{n_i}, \lim_{i\to\infty} y_{n_i})$ for the same sequence $n_i \to \infty$.

In what follows we will also consider for a sequence $(f_n)_n$ of selfmappings of X the sequence of **lumped mappings** $(F_m)_m$, defined for a given $r \geq 1$ by

$$F_m = f_{m+r-1} \circ \ldots \circ f_{m+1} \circ f_m \quad \text{for} \quad m \geq 1.$$

Even under strong contractivity assumptions one cannot expect orbits of a non-autonomous system to converge. The following theorem shows, however, that under assumptions similar to those for the autonomous case in Theorem 4.1.3, orbits become independent of their starting point, a property sometimes called *path stability* or *weak ergodicity*.

Theorem 4.2.2. *Let $(f_n)_n$ be a sequence of selfmappings of the metric space (X, d) such that the sequence $(F_m)_m$ of lumped mappings for some $r \geq 1$ is a contractive sequence consisting of non-expansive mappings. For any two orbits $x_{n+1} = f_n(x_n)$ and $y_{n+1} = f_n(y_n)$ with $x_1, y_1 \in X$, respectively, the following statements hold.*

(i) $\lim_{n\to\infty} d(x_n, y_n)$ *and* $\lim_{n\to\infty} c(x_n, y_n)$ *exist and coincide.*

(ii) *If $\omega_s(x_1, y_1) \neq \emptyset$ then $\lim_{n\to\infty} d(x_n, y_n) = 0$.*

Proof. (i) Let for $0 \leq 1 < r$ fixed $g_m = F_{(m-1)r+i}$ for $m \geq 1$. First, we show by induction over m that

$$g_m \circ \cdots \circ g_1(x_i) = x_{mr+i}. \tag{$*$}$$

For $m = 1$

$$g_1(x_i) = F_i(x_i) = f_{i+r-1} \circ \cdots \circ f_i(x_i) = x_{r+i}.$$

If $(*)$ is true for some $m \geq 1$, then

$$g_{m+1}(g_m \circ \cdots g_1(x_i)) = F_{mr+i}(x_{mr+i}) = f_{mr+i+r-1} \circ \cdots \circ f_{mr+i}(x_{mr+i}) = x_{(m+1)r+i},$$

which proves formula $(*)$.

Obviously, the sequence $(g_m)_m$ is a contractive sequence consisting of non-expansive mappings. In particular, for the orbits given by $(g_m)_m$, that is $\bar{x}_{m+1} = g_m \circ \cdots \circ g_1(x_i)$ and $\bar{y}_{m+1} = g_m \circ \cdots \circ g_1(y_i)$, and $a_m = d(\bar{x}_m, \bar{y}_m)$ we have that $a_{m+1} \le a_m$. Therefore, $a = \lim_{m\to\infty} a_m$ exists. Furthermore, for $\epsilon > 0$ there exists $M(\epsilon)$ such that for all $m \ge M(\epsilon)$

$$c(\bar{x}_m, \bar{y}_m) \le d(\bar{x}_m, \bar{y}_m) \le a + \epsilon \quad \text{and}$$
$$a \le d(\bar{x}_{m+1}, \bar{y}_{m+1}) \le c(\bar{x}_m, \bar{y}_m) + \epsilon.$$

Therefore, $-\epsilon \le c(\bar{x}_m, \bar{y}_m) - a \le \epsilon$ for all $m \ge M(\epsilon)$. This shows $\lim_{m\to\infty} c(\bar{x}_m, \bar{y}_m) = a = \lim_{m\to\infty} d(\bar{x}_m, \bar{y}_m)$.

Next we show that this equality holds also for sequences $(x_n)_n$ and $(y_n)_n$. By property $(*)$

$$\bar{x}_{m+1} = x_{mr+i} \quad \text{and, similarly,} \quad \bar{y}_{m+1} = y_{mr+i}.$$

Therefore,

$$\lim_{m\to\infty} d(x_{mr+i}, y_{mr+i}) = \lim_{m\to\infty} c(x_{mr+i}, y_{mr+i}).$$

Let $s_{m,i} = d(x_{mr+i}, y_{mr+i}) - c(x_{mr+i}, y_{mr+i}) \ge 0$. Since $0 \le i < r$ was arbitrary we have that $\lim_{m\to\infty} s_{m,i} = 0$ for all $0 \le i < r$. That is, for $\epsilon > 0$ given there exists $M'(\epsilon)$ such that $s_{m,i} \le \epsilon$ for all $m \ge M'(\epsilon)$. Choose $N(\epsilon) = (M'(\epsilon) + 1)r$ and let $n \ge N(\epsilon)$. Since $n = mr + i$ with $m \ge 1, 0 \le i < r$ it follows that $mr + i \ge (M'(\epsilon) + 1)r$ and, hence, $m \ge M'(\epsilon)$. Thus,

$$0 \le d(x_n, y_n) - c(x_n, y_n) = s_{m,i} \le \epsilon \quad \text{for} \quad n \ge N(\epsilon),$$

which proves $\lim_{n\to\infty} d(x_n, y_n) = \lim_{n\to\infty} c(x_n, y_n)$.

(ii) By assumption there exists $(x^*, y^*) \in \omega_s(x_1, y_1)$, that is $x^* = \lim_{k\to\infty} x_{n_k}, y^* = \lim_{k\to\infty} y_{n_k}$. From step (i) it follows

$$d(x^*, y^*) = \lim_{k\to\infty} d(x_{n_k}, y_{n_k}) = \lim_{n\to\infty} d(x_n, y_n) \quad \text{and}$$
$$c(x^*, y^*) = \lim_{k\to\infty} c(x_{n_k}, y_{n_k}) = \lim_{n\to\infty} c(x_n, y_n).$$

Therefore, $d(x^*, y^*) = c(x^*, y^*)$ and since, by assumption, $c(u, v) < d(u, v)$ for $u \neq v$ we must have $x^* = y^*$.

Thus, $\lim_{n\to\infty} d(x_n, y_n) = d(x^*, y^*) = 0$. □

Later in Section 7.2, Theorem 4.2.2 will prove to be useful in the analysis of weak ergodicity for ascending operators. Then the metric setting is specialized to the part metric and the Hilbert metric, respectively. Originally, the notion of weak ergodicity stems from the theory of inhomogeneous Markov chains and its applications to demography (see Section 7.1 for this background). There also the notion of strong ergodicity arose which says that, different from weak ergodicity, the sequence given by $x_{n+1} = f_n(x_n)$ does converge in case the mappings f_n converge to some f. This question will be pursued in the following. The first result, Theorem 4.2.3, states that strong ergodicity does hold in two cases, (i) and (ii). Whereas case (i) requires some compactness and is useful in finite dimensions, case (ii) assumes instead a stronger contraction property making it useful in infinite dimensions. This tradeoff between compactness and contractiveness is already visible for the autonomous systems of the previous section.

Theorem 4.2.3. *Let (X, d) be a metric space and let $(f_n)_n$ be a sequence of selfmappings which converges to some selfmapping f. For the orbit given by $x_{n+1} = f_n(x_n), x_1 = x$ suppose that $(f_n)_n$ converges uniformly to f on the orbit. Then $(x_n)_n$ converges to the unique fixed point x^* of f in each of the following cases.*

Case (i). f is contractive and the orbit is relatively compact.
Case (ii). f is a generalized contraction, (X, d) is complete and the orbit is bounded.

Proof. Case (i). We show for the limit set $\omega(x)$ of the sequence $(x_n)_n$ that $\omega(x)$ consists of the unique fixed point x^* of f. This then proves case (i). Pick $y \in \omega(x)$, that is $y = \lim_{i\to\infty} x_{n_i}$ for some sequence $n_i \to \infty$. Since $(x_n)_n$ is relatively compact there exist subsequences $n_j \to \infty$ and $n_k \to \infty$ of $n_i \to \infty$ such that $\lim_{j\to\infty} x_{n_j-1} = u$ and $\lim_{k\to\infty} x_{n_k+1} = v$ exist. Obviously,

$$d(y, f(u)) \le d(y, x_{n_j}) + d(f_{n_j-1}(x_{n_j-1}), f(x_{n_j-1})) + d(f(x_{n_j-1}), f(u))$$

and

$$d(f(y), v) \le d(f(y), f(x_{n_k})) + d(f(x_{n_k}), f_{n_k}(x_{n_k})) + d(x_{n_k+1}, v)$$

for all j and k, respectively.

Since $(f_n)_n$ converges uniformly to f on the orbit, we obtain

$$d(y, f(u)) = 0 \quad \text{and} \quad d(f(y), v) = 0,$$

respectively.

Therefore, we must have that $y = f(u)$ and $f(y) = v$, where $u, v \in \omega(x)$. Thus, we have shown that $f(\omega(x)) = \omega(x)$. In particular, f is a contractive selfmapping of the compact metric space $(\omega(x), d)$. Furthermore, by iteration we obtain for every $n \ge 1$ and given $y \in \omega(x)$ an element $u_n \in \omega(x)$ such that $y = f^n(u_n)$. Since $\omega(x)$ is compact there exists some sequence $n_l \to \infty$ such that $\lim_{l\to\infty} u_{n_l} = u^* \in \omega(x)$ exists.

By a well-known version of Banach's fixed point theorem (see Remarks 4.1.5(i)) it follows that $\lim_{n\to\infty} f^n(u^*) = x^*$ is the unique fixed point of f in $\omega(x)$ and, hence in X. Finally from

$$\begin{aligned} d(y, x^*) &\le d(f^{n_l}(u_{n_l}), f^{n_l}(u^*)) + d(f^{n_l}(u^*), x^*) \\ &\le d(u_{n_l}, u^*) + d(f^{n_l}(u^*), x^*) \quad \text{for all} \quad l \end{aligned}$$

it follows that $d(y, x^*) = 0$, i.e., $y = x^*$. Since $y \in \omega(x)$ was arbitrary this shows that $\omega(x) = x^*$ as required.

Case (ii). Since a generalized contraction is in particular (ϵ, δ)-contractive, from Theorem 4.1.4 it follows that f has a unique fixed point x^* for which $\lim_{n\to\infty} f^n(x) = x^*$ holds for all $x \in X$. Let $\epsilon > 0$ be given and $r_n = d(x_n, x^*)$. By uniform convergence of f_n

to f on the orbit there exists $N(\epsilon)$ such that $d(f(x_n), f_n(x_n)) \le \epsilon$ for $n \ge N(\epsilon)$ and, hence,

$$r_{n+1} = d(x_{n+1}, x^*) \le d(f_n(x_n), f(x_n)) + d(f(x_n), x^*) \le \epsilon + r_n.$$

By assumption, $\bar{r} = \sup_n r_n < \infty$. Let $\underline{r} = \inf_n r_n$ and assume first $\underline{r} > 0$. Since $0 < \underline{r} \le d(x_n, x^*) \le \bar{r}$ from the definition of a generalized contraction (Definition 4.1.1 (iii)) we obtain for all n

$$d(f(x^*), f(x_n)) \le \rho\, d(x^*, x_n),$$

where $\rho = L(\underline{r}, \bar{r}) < 1$.

Therefore, for $n \ge N(\epsilon)$

$$r_{n+1} \le \epsilon + d(f(x_n), f(x^*)) \le \epsilon + \rho r_n.$$

By iteration this implies for every $k \ge 1$ and $n \ge N(\epsilon)$

$$r_{n+k} \le \rho^k r_n + \epsilon \sum_{i=0}^{k-1} \rho^i \le \rho^k \bar{r} + \frac{\epsilon}{1-\rho}.$$

For $K(\epsilon)$ such that $\rho^{K(\epsilon)}\bar{r} \le \epsilon$ we arrive at $r_m \le \epsilon \frac{2-\rho}{1-\rho}$ for $m \ge K(\epsilon) + N(\epsilon)$. Since $\epsilon > 0$ was arbitrary we arrive at $\lim_{n\to\infty} r_n = 0$, provided $\underline{r} > 0$. Consider now the case that $\underline{r} = 0$. Choose for $\epsilon > 0$ given $N'(\epsilon)$ such that $d(f(x_n), f_n(x_n)) \le \frac{\epsilon}{2}$ for $n \ge N'(\epsilon)$. We show, $r_n \le \epsilon$ for some $n \ge N'(\epsilon)$ implies $r_{n+1} \le \epsilon$. For, if not,

$$\epsilon < r_{n+1} \le r_n + \frac{\epsilon}{2}$$

and, hence, $\frac{\epsilon}{2} \le r_n \le \epsilon$.

Since f is a generalized contraction, this gives

$$\epsilon < r_{n+1} \le L(\frac{\epsilon}{2}, \epsilon) r_n < r_n \le \epsilon,$$

which is a contradiction. By iteration we obtain $r_n \le \epsilon$ for some $n \ge N'(\epsilon)$ implies $r_{n+k} \le \epsilon$ for all $k \ge 1$. By assumption $\inf_n r_n = \underline{r} = 0$ and there exists $n_0 \ge N'(\epsilon)$ such that $r_{n_0} \le \epsilon$. Thus $r_{n_0+k} \le \epsilon$ for all $k \ge 1$ which proves $\lim_{n\to\infty} r_n = 0$ in case of $\underline{r} = 0$. This proves case (ii) and, hence, Theorem 4.2.3. □

The following consequence of Theorem 4.2.3 weakens assumptions made to certain assumptions on the lumped operators.

Corollary 4.2.4. *Let (X, d) be a metric space and let $(f_n)_n$ be a sequence of selfmappings such that for some $r \ge 1$ the sequence $(F_m)_m$ of lumped mappings converges uniformly on X to some selfmapping F. For $x \in X$ the orbit given by $x_{n+1} = f_n(x_n)$, $x_1 = x$ converges to the unique fixed point x^* of F in each of the following cases.*

Case (i). *F is contractive, in particular $(F_m)_m$ is a contractive sequence of non-expansive mappings, and the orbit $(x_n)_n$ is relatively compact.*

Case (ii). *F is a generalized contraction, (X, d) is complete and the orbit $(x_n)_n$ is bounded.*

Proof. Consider for i fixed, $0 \le i < r$, the selfmapping $g_m = F_{(m-1)r+i+1}$ for $m \ge 1$. The sequence $(g_m)_m$ converges uniformly on X to F. If (F_m) is a contractive sequence of non-expansive mappings this holds for (g_m), too. Therefore, by uniform convergence to $\epsilon > 0$ exists $M(\epsilon)$ such that

$$\begin{aligned} d(F(x), F(y)) &\le d(F(x), g_m(x)) + d(g_m(x), g_m(y)) + d(g_m(y), F(y)) \\ &\le \epsilon + c(x,y) + \epsilon + \epsilon \quad \text{for all} \quad m \ge M(\epsilon), \end{aligned}$$

where $c(x,y) < d(x,y)$ for $x \ne y$. Thus, in case (i) F is contractive. Let $(y_m)_m$ be a sequence defined by $y_{m+1} = g_m(y_m), y_1 = x_{i+1}$. By the definition of lumped mappings

$$g_m \circ g_{m-1} \circ \cdots \circ g_1 = f_{mr+i} \circ f_{mr+i-1} \circ \cdots \circ f_{i+1}$$

and, hence, $y_{m+1} = x_{mr+i+1}$. If $(x_n)_n$ is relatively compact or bounded, respectively, the same applies to $(y_m)_m$. Theorem 4.2.3 implies in both cases, (i) and (ii), that $\lim_{m\to\infty} y_m = x^*$ with x^* the unique fixed point of F. Let $n = mr + i$ where $m = m(n) \ge 0$ and $0 \le i = i(n) < r$. By the above for $0 \le i < r$ we have that $\lim_{m\to\infty} x_{mr+i+1} = x^*$ and, hence, $\lim_{n\to\infty} x_n = x^*$. □

For the next Corollary from Theorem 4.2.3 we need the following

Lemma 4.2.5. *Let (X, d) be a metric space and let $(f_n)_n$ be a sequence of selfmappings which converges uniformly on X to some uniformly continuous selfmapping f. Then to every $\epsilon > 0$ and every $k \ge 1$ there exists $N(\epsilon, k)$ such that*

$$d(f^k(x), fn_1 \circ fn_2 \circ \cdots \circ fn_k(x)) \le \epsilon \quad \text{for} \quad n_i \ge N(\epsilon, k) \quad \text{and all} \quad x \in X.$$

Proof. By assumption, to $\epsilon > 0$ there exist $\delta(\epsilon) > 0$, $N(\epsilon)$ such that

$$d(f(x), f_n(y)) \le d(f(x), f(y)) + d(f(y), f_n(y)) \le \frac{\epsilon}{2} + \frac{\epsilon}{2} = \epsilon$$

provided that $d(x,y) \le \delta(\epsilon)$ and $n \ge N(\epsilon)$.

This shows the above assertion for $k = 1$, $N(\epsilon, 1) = N(\epsilon)$. Suppose, the assertion holds for some $k \ge 1$. Then

$$d(f^k(x), f_{n_2} \circ \ldots \circ f_{n_{k+1}}(x)) \le \delta(\epsilon) \quad \text{for} \quad n_2, \ldots, n_{k+1} \ge N(\delta(\epsilon), k), \quad \text{all} \quad x \in X.$$

Setting $N(\epsilon, k+1) = \max\{N(\epsilon), N(\delta(\epsilon), k)\}$ we obtain

$$d(f^{k+1}(x), f_{n_1} \circ f_{n_2} \circ \ldots \circ f_{n_{k+1}}(x)) = d(f(f^k(x)), f_{n_1}(f_{n_2} \circ \ldots \circ f_{n_{k+1}}(x))) \le \epsilon$$

for $n_1, n_2 \ldots, n_{k+1} \ge N(\epsilon, k+1)$.

By induction, this proves the above assertion. □

Corollary 4.2.6. *Let $(X, d))$ be a metric space and let $(f_n)_n$ be a sequence of selfmappings which converges uniformly on X to some uniformly continuous selfmapping f. For $x \in X$ the orbit given by $x_{n+1} = f_n(x_n), x_1 = x$ converges to the unique fixed point x^* of f in each of the following cases.*

Case (i). *f^r is contractive, in particular $(F_m)_m$ is a contractive sequence of non-expansive mappings, and $(x_n)_n$ is relatively compact.*
Case (ii). *f^r is a generalized contraction, (X, d) is complete and $(x_n)_n$ is bounded.*

Proof. By Lemma 4.2.5 to $\epsilon > 0$ there exists $N(\epsilon, r)$ such that for all $x \in X$

$$d(f^r(x), F_m(x)) = d(f^r(x), f_{m+r-1} \circ \cdots \circ f_m(x)) \le \epsilon \quad \text{for} \quad m \ge N(\epsilon, r).$$

Therefore, $(F_m)_m$ converges uniformly on X to $F = f^r$.

Corollary 4.2.4 yields $\lim_{n\to\infty} x_n = x^*$, x^* the unique fixed point of f^r. Finally,

$$d(f(x^*), x^*) \le d(f(x^*), f(x_n)) + d(f(x_n), f_n(x_n)) + d(x_{n+1}, x^*).$$

For $n \to \infty$ this implies $d(f(x^*), x^*) = 0$, that is $f(x^*) = x^*$. Thus x^* is a fixed point of f and it is the unique fixed point of f because it is the unique fixed point of f^r. □

Remark 4.2.7. For the results of this section see [7, Section 2]. There, however, it is assumed that the f_n and F_m, respectively, map X into a compact subset $Y \subset X$ whereas here orbits are assumed to be relatively compact. The latter assumption is employed in [18, Section 4.1] to obtain similar results; the proofs, however, are different from the proofs given here.(See Exercises 2, 3.) For non-autonomous systems on metric spaces see also [3] and [4]. In [4, Theorem 2.2] case (ii) of Theorem 4.2.3 is shown, with a different proof, under the weaker assumption that $(f_n)_n$ converges uniformly to f on any bounded subset of X. From a different point of view sequences of selfmappings of a metric space are considered in [9, Section 7.1]. There, the main interest is, however, in the behavior of fixed points x_n^* of f_n (see Exercise 4).

Exercises

1. Let $(f_n)_n$ be a sequence of selfmappings of the metric space (X, d) which converges uniformly on X to a selfmappings f.
   (a) Show that f is a contractive mapping if $(f_n)_n$ is a contractive sequence.
   (b) Show that $(f_n)_n$ is a contractive sequence if f is a contractive mapping.
   (c) Find an example where $(f_n)_n$ is a contractive sequence but none of the mappings f_n is non-expansive.
2. [18] Let $(f_n)_n$ be a sequence of selfmappings of the metric space (X, d) and let $(x_n)_n$ and $(y_n)_n$ be defined by $x_{n+1} = f_n(x_n)$, $x_1 = x$ and $y_{n+1} = f_n(y_n)$, $y_1 = y$, respectively.
   (a) Show that $\lim_{n\to\infty} d(x_n, y_n) = 0$ implies that the limit sets of $(x_n)_n$ and $(y_n)_n$ coincide.
   (b) Find an example for which the implication in (a) cannot be reversed.
3. [18] Let (X, d) be a (non-empty) compact metric space and let f be a continuous selfmapping that is surjective and has a contractive iterate. Show that X consists solely of the fixed point of f.

4. [9] Let (X, d) be a complete metric space and let $(f_n)_n$ be a sequence of selfmappings of X which converges uniformly on X to a selfmapping f. Suppose each f_n has at least one fixed point x_n^*, $n \geq 1$.
   (a) Suppose f is a contraction. Show that $\lim_{n\to\infty} x_n^* = x^*$ where x^* is the unique fixed point of f.
   (b) Suppose some iterate of f is a contraction. Show that $\lim_{n\to\infty} x_n^* = x^*$ where x^* is the unique fixed point of f.

4.3 A local-global stability principle for power-lipschitzian mappings

In the previous two sections we were concerned with proper contraction dynamics for selfmappings in the sense that distances are strictly contracted. Under additional assumptions we obtained the existence of a **globally attractive fixed point** x^* of f, that is $\lim_{n\to\infty} f^n(x) = x^*$ for all $x \in X$. In this section we want to find out conditions under which we may infer global attractivity from **local attractivity of a fixed point** x^* of f, that is $\lim_{n\to\infty} f^n(x) = x^*$ for all x in some neighborhood U of x^*. As we will see this is quite possible for selfmappings which do not increase distances on certain metric spaces. More general, we shall show such a local-global principle for mappings which are power-lipschitzian in the following sense.

Definition 4.3.1. A selfmapping f of a metric space (X, d) is called **power-lipschitzian** if there exists a constant $c > 0$ such that for every two points $x, y \in X$ there exists $N(x, y) \in \mathbb{N}$ such that

$$d(f^n(x), f^n(y)) \leq c d(x, y) \quad \text{for all} \quad n \geq N(x, y).$$

Obviously, if f is nonexpansive (cf. Definition 4.1.1 (vi)) then f is power-lipschitzian with constant $c = 1$. There exist, however, power-lipschitzian mappings that are not non-expansive (see Exercise 1). The concept of a power-lipschitzian selfmapping is invariant with respect to equivalence among metrics, that is for two equivalent metrics d and d' it holds for d iff it holds for d' (see Exercise 2(a)). In contrast, the concept of a non-expansive selfmapping is not invariant, that is a selfmapping which is non-expansive for metric d need not be non-expansive for an equivalent metric d' (see Exercise 2(b)).

To infer for a fixed point global attractivity from local attractivity, we need beside an assumption on the selfmapping also an assumption on the underlying metric space. Obviously, if the fixed point is an isolated point then the pass from the local to the global will fail. Actually, we will characterize global attractivity of a fixed point x^* of a selfmapping f by its local attractivity together with the power-lipschitzian property for f and the condition that x^* is not isolated in the following sense.

Definition 4.3.2. Let f be a selfmapping of a metric space (X, d). A point $x_0 \in X$ is **strongly isolated for** f if there exists a neighborhood U of x_0 which as well as its non-empty complement is invariant for f and such that

$$\inf\{d(x, y) \mid x \in U, y \notin U\} > 0.$$

Obviously, if x_0 is strongly isolated for f then x_0 must be isolated in the topological sense. Therefore, if (X, d) is connected then there are no strongly isolated points whatever f may be (see Lemma 4.3.4 below).

Theorem 4.3.3 (Local-global stability principle). *Let f be a selfmapping of the metric space (X, d). A fixed point $x^* \in X$ of f is globally attractive if and only if x^* is locally attractive and not strongly isolated for f and that f is power-lipschitzian.*

Proof. (1) Let x^* be a globally attractive fixed point of f. Obviously, f is locally attractive as well as not strongly isolated for f. Furthermore, for $x, y \in X$ given, $x \neq y$, there exist $N(x), N(y) \in \mathbb{N}$ such that $d(f^n(x), x^*) \le d(x, y)$ for $n \ge N(x)$ and $d(f^n(y), x^*) \le d(x, y)$ for $n \ge N(y)$. It follows that

$$d(f^n(x), f^n(y)) \le d(f^n(x), x^*) + d(x^*, f^n(y)) \le 2d(x, y)$$

for all $n \ge N(x, y) = \max\{N(x), N(y)\}$. Therefore, f is power-lipschitzian.

(2) Suppose the fixed point x^* is locally attractive and not strongly isolated for f and that f is power-lipschitzian. To show global attractivity for x^* we show that the set

$$U = \{x \in X \mid \lim_{n\to\infty} d(f^n(x), x^*) = 0\}$$

coincides with X. Suppose that not, that is $U \neq X$. The set U as well as its complement are non-empty and invariant for f. Since x^* is locally attractive there exists $\epsilon > 0$ such that $B(x^*, \epsilon) = \{z \in X \mid d(x^*, z) < \epsilon\} \subset U$.

In particular, U is a neighborhood of x^* and $\inf\{d(x, y) \mid x \in U, y \notin U\} = 0$ because x^* is not strongly isolated for f. Select $x_0 \in U$ and $y_0 \notin U$ such that $d(x_0, y_0) < \frac{\epsilon}{2c}$ where $c > 0$ is a constant for which f is power-lipschitzian. There exists $N(x_0) \in \mathbb{N}$ such that $d(f^n(x_0), x^*) < \frac{\epsilon}{2}$ for all $n \ge N(x_0)$. Since f is power-lipschitzian we have that

$$\begin{aligned} d(f^n(y_0), x^*) &\le d(f^n(y_0), f^n(x_0)) + d(f^n(x_0), x^*) \\ &\le cd(x_0, y_0) + \tfrac{\epsilon}{2} < \epsilon \end{aligned}$$

for all $n \ge N(x_0)$. Therefore, $f^n(y_0) \in B(x^*, \epsilon) \subset U$ for n big enough and, hence, $y_0 \in U$. But this contradicts $y_0 \notin U$ and, therefore, $U = X$ which proves the theorem. □

The next lemma supplies conditions on the metric space which guarantee that there are no strongly isolated points whatever f may be.

Lemma 4.3.4. *Let (X, d) be a metric space which is*

(i) connected, *i.e., there is no subset $\emptyset \subsetneq U \subsetneq X$ which is open and closed.*
or

(ii) ϵ-chainable for every $\epsilon > 0$, *i.e., for $x, y \in X$ and $\epsilon > 0$ arbitrarily given there exist $z_i \in X$ such that $d(z_i, z_{i-1}) < \epsilon$ for $1 \le i \le n = n(x, y, \epsilon)$ and $z_0 = x$, $z_n = y$.*

or

(iii) *complete and* metrically convex, *i.e., for any two distincs points* $x, y \in X$ *there exists* $z \in X$ *distinct from* x *and* y *such that* $d(x,z) + d(z,y) = d(x,y)$.

Then for every $x_0 \in X$ *and every neighborhood* U *of* x_0 *with non-empty complement it holds that* $\inf\{d(x,y) \mid x \in U,\ y \notin U\} = 0$.

Proof. Let U be a neighborhood of $x_0 \in X$ with non-empty complement and let $\alpha = \inf\{d(x,y) \mid x \in U,\ y \notin U\} > 0$. We shall show that this leads in each case to a contradiction.

(i) It follows that $B(x,\alpha) \subset U$ for any $x \in U$ and, hence, U contains all its limit points. Therefore, U is open and closed, $\emptyset \subsetneq U \subsetneq X$, which contradicts connectedness.

(ii) Let $x \in U$, $y \notin U$. For $\epsilon = \alpha$ there exist $z_i \in X$ such that $d(z_i, z_{i-1}) < \alpha$. Since $z_0 = x \in U$ it follows that $z_1 \in U$ which in turn implies $z_2 \in U$ etc. Thus, $y = z_n \in U$ which is a contradiction.

(iii) There exist $x \in U$, $y \notin U$ such that $\alpha \le d(x,y) < 2\alpha$. If (X,d) is complete and metrically convex then by a theorem of Menger (see [19, p. 24]) there exists an isometry $\phi : [0,r] \to X$ with $\phi(0) = x$ and $\phi(r) = y$, where $r = d(x,y)$. Since $0 \le r - \alpha < \alpha$ there exists s with $r - \alpha < s < \alpha$. For $z = \phi(s)$ it follows that

$$\begin{aligned} d(x,z) &= d(\phi(0), \phi(s)) = s < \alpha \qquad \text{and} \\ d(z,y) &= d(\phi(s), \phi(r)) = r - s < \alpha. \end{aligned}$$

Since $x \in U$ we must have that $z \in U$ which in turn implies that $y \in U$ - a contradiction. □

Remark 4.3.5. For power-lipschitzian mappings in normed spaces see also [8] where they are called uniformly lipschitzian mappings. The property in Lemma 4.3.4 (ii) can be considered also for uniform spaces where it is called uniform connectedness [11]. For the property of ϵ-chainable in case of a fixed $\epsilon > 0$ see [6] and [9]. Property (i) in Lemma 4.3.4 implies property (ii) whereas the reverse implication is not true (see Exercise 3).

With the help of Lemma 4.3.4 from Theorem 4.3.3 we immediately obtain the following result.

Corollary 4.3.6. *Let* (X,d) *be a metric space which is* ϵ*-chainable for every* $\epsilon > 0$ *(in particular,* (X,d) *is connected) or which is complete and metrically convex. Then a fixed point* x^* *of a selfmapping is globally attractive if and only if it is locally attractive and* f *is power-lipschitzian.*

The following useful extension of Theorem 4.3.3 is obtained by employing almost the same proof.

Theorem 4.3.7. *Let* f *be a selfmapping of the metric space* (X,d) *which is power-lipschitzian. Let* F *be a non-empty subset of fixed points of* f *such that* F *is locally* *attractive in the sense there exists* $\epsilon > 0$ *such that*

$$\lim_{n\to\infty} f^n(x) \in F \quad \text{for all} \quad x \quad \text{with} \quad d(x,y) \le \epsilon \quad \text{for some} \quad y \in F.$$

Then F is globally attractive, that is

$$\lim_{n\to\infty} f^n(x) \in F \quad \text{for all} \quad x \in X,$$

provided F is not strongly isolated (that is, for any neighborhood U of F with $X \setminus U \neq \emptyset$ *and both U and* $X \setminus U$ *invariant for f it holds* $\inf\{d(x,y) \mid x \in U, y \notin U\} = 0$*).*

Proof. Consider $U = \{x \in X \mid \lim_n f^n(x) \in F\}$. Obviously, $U \neq \emptyset, f(U) \subseteq U$. By assumption, $F \subseteq \bigcup_{y\in F} B(y,\epsilon) \subseteq U$ where $B(y,\epsilon) = \{x \in X \mid d(x,y) < \epsilon\}$ and, hence, U is a neighborhood of F. Suppose $U \subsetneq X$. By definition $f(X \setminus U) \subseteq X \setminus U$. By assumption F is not strongly isolated and we must have $x_0 \in U, y_0 \notin U$ such that $d(x_0,y_0) < \frac{\epsilon}{2c}$, where $c > 0$ is a constant for which f is power-lipschitzian. Furthermore, $d(f^n(x_0), y) < \frac{\epsilon}{2}$ for some $y \in F, n \ge N(x_0,y)$. Since f is power-lipschitzian we have that

$$\begin{aligned} d(f^n(y_0),y) &\le d(f^n(y_0), f^n(x_0)) + d(f^n(x_0),y) \\ &\le cd(y_0,x_0) + \tfrac{\epsilon}{2} < \epsilon \quad \text{for} \quad n \ge N(x_0,y_0), N(x_0,y). \end{aligned}$$

Thus, $f^n(y_0) \in \bigcup_{y\in F} B(y,\epsilon) \subseteq U$ for $n \ge N(x_0,y_0), N(x_0,y)$. This, however, is a contradiction to $f(X \setminus U) \subseteq X \setminus U$. Therefore, $U = X$, that is $\lim_n f^n(x) \in F$ for all $x \in X$. □

To the setting of Theorem 4.3.7 the Corollary 4.3.6 applies analogously, in particular F is not strongly isolated if the topological space (X,d) is connected.

Remarks 4.3.8. (i) For the case of a connected metric space with a non-expansive selfmapping, Corollary 4.3.6 follows from [19, Lemma 2.3].

(ii) For the case of a complete and metrically convex metric space with a power-lipschitzian selfmapping, Corollary 4.3.6 is contained in [13, Theorem 2.1]. See also [15, Proposition 3.2.3].

(iii) Since a fixed point of a power-lipschitzian selfmapping is automatically stable (see Exercise 4) the term “globally attractive” in Corollary 4.3.6 can be replaced by the term “globally asymptotically stable”.

Exercises

1. Let $X = \{x \in \mathbb{R} \mid x \ge 0\}$ be equipped with the Euclidean distance. Find a selfmapping of X which is power-lipschitzian (with $c = 1$) but which is not non-expansive.
2. Let X be a non-empty set which carries two equivalent metrics d and d', i.e., there exist constants $a > 0$, $b > 0$ such that

$$ad(x,y) \le d'(x,y) \le bd(x,y) \quad \text{for all} \quad x,y \in X.$$

   (a) Show that a selfmapping f of X is power-lipschitzian for d if and only if f is power-lipschitzian for d'.

(b) Find an example such that f is a selfmapping of X which is non-expansive for d but not for d'.

3. Let (X, d) be a metric space.
   (a) Show that (X, d) is ϵ-chainable for every $\epsilon > 0$ if (X, d) is connected.
   (b) Find an example of a metric space for which the reverse statement of a) is not true.

4. Let (X, d) be a metric space and let f be a power-lipschitzian self-mapping of X with fixed point x^*. Show that x^* is *stable*, i.e., to $\epsilon > 0$ there exists $\delta > 0$ such that $d(x, x^*) < \delta$ implies $d(f^n(x), x^*) < \epsilon$ for all $n \geq 0$.

Bibliography

[1] D. W. Boyd and J. S. W. Wong. On non-linear contractions. *Proc. Amer. Math. Soc.*, 20:458–464, 1969.

[2] F. E. Browder. On the convergence of successive approximations for non-linear functional equations. *Indag. Math.*, 30:27–35, 1968.

[3] Y.-Z. Chen. Asymptotic behavior of inhomogeneous iterates for non-linear operators in ordered Banach spaces. In *Dynamic Systems and Application*, vol. 3, pp. 147–154, 1999.

[4] Y.-Z. Chen. Inhomogeneous iterates of contraction mappings and non-linear ergodic theorems. *Nonlinear Analysis*, 39:1–10, 2000.

[5] K. Deimling. *Nonlinear Functional Analysis*. Springer, Berlin, Heidelberg, New York, Tokyo, 1985.

[6] M. Edelstein. On fixed and periodic points of contractive mappings. *J. Lond. Math. Soc.*, 37:74–79, 1962.

[7] T. Fujimoto and U. Krause. Asymptotic properties for inhomogeneous iterations of non-linear operators. *SIAM J. Math. Anal.*, 19:841–853, 1988.

[8] K. Goebel and W. A. Kirk. *Topics in Metric Fixed Point Theory*. Cambridge University Press, Cambridge, 1990.

[9] V. I. Istratescu. *Fixed Point Theory*. D. Reidel, Dordrecht, 1981.

[10] J. R. Jachymski. Equivalence of some contractivity properties over metrical structures. *Proc. Amer. Math Soc.*, 125:2327–2335, 1997.

[11] I. M. James. *Introduction to Uniform Spaces*. Cambridge University Press, Cambridge, 1990.

[12] M. A. Krasnoselskii and P. P. Zabreiko. *Geometrical Methods of Nonlinear Analysis*. Springer-Verlag, Berlin, 1984.

[13] U. Krause. A local-global stability principle for discrete systems and difference equations. In B. Aulbach et al., editor, *Proceedings of the Sixth International Conference on Difference Equations*, Augsburg, 2001. CRC Press, Boca Raton, 2004.

[14] U. Krause and R. D. Nussbaum. A limit set trichotomy for self-mappings of normal cones in Banach spaces. *Nonlinear Analysis (TMA)*, 20:855–870, 1993.

[15] B. Lemmens and R. Nussbaum. *Nonlinear Perron–Frobenius Theory*. Cambridge University Press, Cambridge, 2012.

[16] J. Matkowski. Fixed point theorems for mappings with a contractive iterate at a point. *Proc. Amer. Math. Soc.*, 62:344–348, 1977.

[17] A. Meir and E. Keeler. A theorem on contraction mappings. *J. Math. Anal. Appl.*, 28:326–329, 1969.

[18] T. Nesemann. *Stability Behavior of Positive Nonlinear Systems with Applications to Economics*. PhD thesis at the University of Bremen. Wissenschaftlicher Verlag, Berlin 1999.

[19] R. D. Nussbaum. Hilbert's projective metric and iterated non-linear maps. *Memoirs Amer. Math. Soc.*, (391), 1988.

[20] E. Rakotch. A note on contractive mappings. *Proc. Amer. Math. Soc.*, 13:459–465, 1962.

[21] B. Scherenberger. On the behavior of the recursive sequence $u(t+n) = \sum_{i=1}^{n}(a_i(t)/u(t+i-1)^{r_i})$. *J. Difference Equ. Appl.*, 6:625–639, 2000.

5 Ascending dynamics in convex cones of infinite dimension

In this chapter we continue the investigation of non-linear selfmappings of cones we started in Chapter 2 for finite dimensions. Now we will consider arbitrary dimensions and for this we will make use of the instruments developed in Chapters 3 and 4. In the center of interest is the asymptotic behavior of the iterates of an ascending selfmapping of a convex cone in arbitrary dimensions. The notion of an ascending operator comprises the strictly positive and concave operators in finite dimensions of Chapter 2 as well as various well-known concepts of strong monotonicity for operators in arbitrary Banach spaces.

The advantage in considering ascending operators – which contrary to concave mappings need not be monotone – is that they lead to contractions with respect to internal metrics of the cone. Therefore, in this chapter we make use of internal metrics and their properties from Chapter 3 as well as of contraction dynamics from Chapter 4.

Historically, the first extension of the Perron–Frobenius theorem to infinite dimensions was made in 1912 by R. Jentzsch who considered instead of non-negative matrices integral operators with a positive kernel on function spaces. Since then the analysis of positive operators developed into a field of its own. Considering even only those contributions that use methods based on internal metrics of convex cones, one observes several schools as, e.g., an Anglo-American school, a Russian school and a Japanese school.

5.1 Definition and examples of ascending operators

In the following let K be a convex cone in some real vector space and let "$\leq$" the partial order defined by K (see Section 3.1 for the definitions). Throughout K is assumed to be non-empty and non-trivial, that is not $\{0\}$. But otherwise K can be arbitrary, it can be closed or open or just one part etc.

Definition 5.1.1. A mapping or operator $T\colon K \to K$ is called monotone increasing or **monotone** for short if $0 \leq x \leq y$ implies $Tx \leq Ty$.

T is called **positively homogeneous** if $T(\lambda x) = \lambda Tx$ for all $x \in K$, all scalars $\lambda \geq 0$.

The following simple lemma gives some general information about the number of non-negative eigenvalues of mappings as above.

Lemma 5.1.2. *Let K be a convex cone in some real vector space that is pointed and archimedean (in $K - K$). Every selfmapping T of K that is monotone and positively homogeneous has the following properties.*

(i) *T possesses no more than one non-negative eigenvalue for eigenvectors in the same part of K. In particular, T has finitely (including* 0*) many non-negative eigenvalues if K consists of finitely many parts.*
(ii) *For* $K = \mathbb{R}^n_+$ *the number of non-negative eigenvalues of T is at most* $2^n - 1$.

Proof. (i) Let $Tx = \lambda x$, $Ty = \mu y$ with $x, y \in K \setminus \{0\}$ and $\lambda, \mu \geq 0$. We suppose $x \sim y$ and show that $\lambda = \mu$. From Lemma 3.1.4 it follows for the order function that $0 < \lambda(x, y) < \infty$ and $\lambda(x, y)x \leq y$. The assumptions on T imply that $\lambda(x, y)Tx \leq Ty$ and, hence, $\lambda(x, y)\lambda x \leq \mu y$. By definition of $\lambda(x, y)$ it follows that $\lambda(x, y)\frac{\lambda}{\mu} \leq \lambda(x, y)$ for $\mu > 0$. Therefore, $\lambda \leq \mu$ for $\mu > 0$. If $\mu = 0$ then $\lambda = 0$ because of $\lambda(x, y) > 0$. Exchanging the roles of x and y we obtain in addition $\mu \leq \lambda$.

(ii) The parts of K have dimension k ranging from 0 to n. There are $\binom{n}{k}$ possibilities of forming a part of dimension k. Therefore, K possesses at most $\sum_{k=0}^{n} \binom{n}{k} = 2^n$ parts. Since 0 is not an eigenvector, the assertion follows from (i). □

Remark 5.1.3. In contrast to linear mappings, the maximal possible number $2^n - 1$ in part (ii) of Lemma 5.1.2 can actually occur (see Exercise 1). Lemma 5.1.2 implies in particular the result about finitely many eigenvalues in [32] (compare also Exercise 7 to Section 2.2). In [29, Theorem 5.2.3] it is shown for any solid polyhedral cone K in finite dimensions, which includes $K = \mathbb{R}^n_+$, any monotone and positively homogeneous selfmapping of K has at most $m - 1$ non-negative (distinct) eigenvalues, where m is the number of faces of K. Moreover, there exists a selfmapping as above which is continuous and has precisely $m - 1$ non-negative eigenvalues. (See also Exercise 1.)

Let us have a first look at the connection between order properties and metric properties for a selfmapping T of a convex cone K which we assume to be lineless. Obviously, if T is monotone and positively homogeneous, then for $x, y \in K$ and $\lambda \geq 0$ from $\lambda x \leq y$ it follows that $\lambda Tx \leq Ty$. Since the extraction process is monotone (see Section 3.1) it follows that $\lambda(x, y) \leq \lambda(Tx, Ty)$. Since internal metrics on the cone K are defined via the order function $\lambda(\cdot, \cdot)$ it follows, e.g., for Hilbert's projective metric $d(x, y) = -\log[\lambda(x, y) \cdot \lambda(y, x)]$ that $d(Tx, Ty) \leq d(x, y)$. Thus, every monotone and positively homogeneous mapping is non-expansive with respect to d. Furthermore, the internal completeness theorem (Theorem 3.4.6) and related results supply conditions under which K is a complete metric space. This brings us into the area of contractive dynamics on metric spaces. To apply, however, the results of the previous chapter we need to make some more assumptions on T.

For this let K be a convex cone in a locally convex vector space V and suppose that the interior int K of K, or $\mathring{K}$ for short, is non-empty. The interior of K is again a convex cone and the corresponding **strict order relation** for $x, y \in V$ is defined by

$$x < y \quad \text{if and only if} \quad y - x \in \mathring{K}.$$

The relation "$<$" is transitive and antisymmetric if $\mathring{K}$ is pointed.

A selfmapping T of K is called **strictly monotone** if $x \lneqq y$ implies that $Tx < Ty$. A selfmapping T of K (also called a positive mapping) is called **strictly positive** if $0 \lneqq x$ implies that $0 < Tx$, or equivalently, $T(K \setminus \{0\}) \subset \overset{\circ}{K}$.

Obviously, in the particular case of $K = \mathbb{R}_+^n$ the notion $x < y$ coincides with the earlier definition as $x_i < y_i$ for all $1 \leq i \leq n$ (see Section 2.1; there $x \lneqq y$ means $x \leq y$ and $x \neq y$ which carries over to arbitrary cones).

If the selfmapping T of K is positively homogeneous and strictly monotone then $\lambda x \lneqq y$ implies $\lambda Tx < Ty$ and, by the definition of an interior point, there exists some $\lambda' > \lambda$ such that $\lambda' Tx \leq Ty$. This implies for $x, y \in K \setminus \{0\}$ which are not on the same ray that $\lambda(x, y) < \lambda(Tx, Ty)$ and, hence, $d(Tx, Ty) < d(x, y)$ (assuming that K is archimedean). Thus, T is contractive (across rays) with respect to Hilbert's metric. This step to ascend from λ to $\lambda' > \lambda$ is the point of the following definition which, however, neither assumes T to be monotone nor positively homogeneous.

Definition 5.1.4. Let T be a selfmapping of the convex cone K and let D be a non-empty subset of K.

T is **ascending on** D (with ϕ) if there exists a selfmapping ϕ of the open unit interval $]0,1[$ with $\lambda < \phi(\lambda)$ and such that for every $0 < \lambda < 1$ and every $x, y \in D$

$$\lambda x \lneqq y \quad \text{implies} \quad \phi(\lambda) Tx \leq Ty. \tag{5.1.1}$$

T is **weakly ascending on** D (with ϕ) if there exists a ϕ as above such that for every $0 < \lambda < 1$ and every $x, y \in D$

$$\lambda x \lneqq y \lneqq \frac{1}{\lambda} x \quad \text{implies} \quad \phi(\lambda) Tx \leq Ty \leq \frac{1}{\phi(\lambda)} Tx. \tag{5.1.2}$$

T is a **cone mapping on** D if for every $0 < \lambda < 1$ and every $x, y \in D$

$$\lambda x \lneqq y \lneqq \frac{1}{\lambda} x \quad \text{implies} \quad \lambda Tx \leq Ty \leq \frac{1}{\lambda} Tx. \tag{5.1.3}$$

Remark 5.1.5. (i) The above notion "ascending on D" originates from the special case of a "p-ascending" operator in [22] where D is given by a functional p on V as $D = \{x \in K \mid p(x) = 1\}$ (the definition given above follows [26]; see also [24, 25]).

(ii) Obviously, if T is ascending on D it is weakly ascending on D (with the same ϕ) which latter is always a cone mapping on D. None of these implications can be reversed (see Exercise 2). For the notion of a cone mapping on $D = K$ see [26].

(iii) For K archimedean, T ascending on D implies that T is monoton on D. This implication does not hold for weakly ascending operators or cone mappings (see Exercise 2). Furthermore, that an operator T is ascending on $D \subset K$ does in general not imply that T is monotone on K (see Exercise 3).

The next lemma enables one to build up complex ascending operators from simple ones.

Lemma 5.1.6. (i) *Let K be archimedean (in $K - K$) and let D be a non-empty subset of K such that for $x, y \in D$, $0 < \lambda < 1$ with $\lambda x \lneqq y$ there exists $z \in D$ such that $\lambda x + (1-\lambda)z \le y$. If T is a concave selfmapping of K such that for some $e \in K$ and scalars $0 < r \le s$ it holds that*

$$re \le Tx \le se \quad \text{for all} \quad x \in D \tag{5.1.4}$$

then T is ascending on D with $\phi(\lambda) = \lambda + (1-\lambda)\frac{r}{s}$.

(ii) *Let S and T be selfmappings of K which are ascending on $D \subset K$. Then the pointwise sum $S + T$ and multiple cT for $c > 0$ are ascending on D, that is, those selfmappings of K which are ascending on D form a convex cone. Moreover, if the pointwise minimum $\min\{S, T\}$ or maximum $\max\{S, T\}$ of S and T exist with respect to the ordering $\le$ induced by K, then $\min\{S, T\}$ and $\max\{S, T\}$ are ascending on D. The above statements remain true if "ascending" is replaced by "weakly ascending" or by "cone mapping".*

(iii) *Let K and D be as in (i) and let $\{T_i\}_{i \in I}$ be a family of concave selfmappings of K such that for every $i \in I$ there exists $e_i \in K$ together with scalars $0 < r_i \le s_i$ with $r_i e_i \le T_i x \le s_i e_i$ for all $i \in I$ and $x \in D$. If $Tx = \inf\{T_i x \mid i \in I\}$ (or $Tx = \sup\{T_i x \mid i \in I\}$) exists for every $x \in K$ (with respect to $\le$) and if $c = \inf\{\frac{r_i}{s_i} \mid i \in I\} > 0$ then T is ascending on D for $\phi(\lambda) = \lambda + (1-\lambda)c$.*

Proof. (i) Since K is archimedean and T concave it follows as in Lemma 2.1.3 that T is monotone. Suppose $x, y \in D$ with $\lambda x \lneqq y$ for $0 < \lambda < 1$. By assumption there exists $z \in D$ such that by concavity of T

$$Ty \ge T(\lambda x + (1-\lambda)z) \ge \lambda Tx + (1-\lambda)Tz.$$

Furthermore, by (5.1.4) $Tz \ge re \ge \frac{r}{s}Tx$ and, hence, $Ty \ge \lambda Tx + (1-\lambda)\frac{r}{s}Tx = \phi(\lambda)Tx$ with $\phi(\lambda) = \lambda + (1-\lambda)\frac{r}{s}$. Obviously, $\lambda < \phi(\lambda)$ for $0 < \lambda < 1$.

(ii) By assumption, there exist selfmappings ϕ and ψ of $]0, 1[$ such that $\lambda < \phi(\lambda)$ and $\lambda < \psi(\lambda)$ and such that for $x, y \in D$, $0 < \lambda < 1$
$\lambda x \lneqq y$ implies $\phi(\lambda)Sx \le Sy$ and $\psi(\lambda)Tx \le Ty$.
Defining $\chi(\lambda) = \min\{\phi(\lambda), \psi(\lambda)\}$, χ is a selfmapping of $]0, 1[$ with $\lambda < \chi(\lambda)$. Furthermore, one has that for $\lambda x \lneqq y$

$$\chi(\lambda)(S+T)(x) \le (S+T)(x),\ \chi(\lambda)(cT) \le cTy \quad \text{and}$$
$$\chi(\lambda)\min\{S, T\}(x) \le \min\{S, T\},\ \chi(\lambda)\max\{S, T\}(x) \le \max\{S, T\}(x),$$

provided that min and max exist.

The same reasoning applies to weakly ascending operators and to cone mappings.

(iii) By (i) each mapping T_i is ascending with $\phi_i(\lambda) = \lambda + (1-\lambda)\frac{r_i}{s_i} \ge \lambda + (1-\lambda)c = \phi(\lambda)$. ϕ is a selfmapping of $]0, 1[$ with $\lambda < \phi(\lambda)$. For $x, y \in D$, $0 < \lambda < 1$
$\lambda x \lneqq y$ implies $\phi(\lambda)T_i x \le T_i y$ for all $i \in I$ which proves (iii). □

Now we relate the concept of an ascending operator to some other useful concepts applied in the literature on positive operators.

Example 5.1.7 (Uniformly positive linear operators). This concept was introduced by G. Birkhoff [2, 3] in 1957 in extending a Theorem by Jentzsch (see below Section 5.4) to linear operators on general vector spaces. Let V be a real vector space, K an archimedean convex cone in V inducing the partial order $\leq$ and let $e \in K \setminus \{0\}$, $k > 0$. A linear operator T on V which leaves K invariant is **uniformly positive** if for all $x \in K \setminus \{0\}$

$$\lambda e \leq Tx \leq k\lambda e \quad \text{with} \quad \lambda = \lambda(x) > 0. \tag{5.1.5}$$

Obviously, as a linear operator T must be monotone and condition (5.1.5) looks similar to condition (5.1.4). To make the connection with the concept of an ascending operator precise (cf. [22]) let p be a monotone norm on V. It follows from (5.1.5) that $\lambda(x)p(e) \leq p(Tx) \leq k\lambda(x)p(e)$ and for $p(Tx) = 1$ that $\frac{1}{kp(e)}e \leq Tx \leq \frac{k}{p(e)}$.

Taking $D = \{x \in K \mid p(Tx) = 1\}$ and choosing for $x, y \in D$, $0 < \lambda < 1$ with $\lambda x \lneqq y$ and the element $z = \frac{y-\lambda x}{p(Ty-\lambda Tx)}$, all assumptions of Lemma 5.1.6 (i) are satisfied. Therefore, by this lemma, T is ascending on D with $\phi(\lambda) = \lambda + (1-\lambda)\cdot\frac{1}{k^2}$. (In [3] it is allowed that $\lambda(x) = 0$ which requires a separate consideration of nilpotent operators.)

A concept very close to that of a uniformly positive operator is that of an e-positive operator [50, 51]. An arbitrary selfmapping of K is called **e-positive** with respect to some $e \in K \setminus \{0\}$ if for all $x \in K \setminus \{0\}$

$$\alpha(x)e \leq Tx \leq \beta(x)e \quad \text{with} \quad \alpha(x) > 0,\ \beta(x) > 0. \tag{5.1.6}$$

Obviously, a uniformly positive operator is the special case of an e-positive operator for which $\frac{\beta(x)}{\alpha(x)}$ is independent of x.

Example 5.1.8 (Zigzag operators). An example of an e-positive operator is given by the **affine-linear operator** $Sx = Tx + b$ where T is a linear selfmapping of K and b in the non-empty interior $\mathring{K}$ of K (with respect to a locally convex topology on V). Since $b \in \mathring{K}$ there exists for every $x \in K$ some $r(x) > 0$ such that $x \leq r(x)b$. Therefore, for all $x \in K \setminus \{0\}$

$$b \leq Sx \leq r(x)Tb \leq r(x)r(Tb)b$$

and (5.1.6) is satisfied for $e = b, \alpha(x) = 1,\ \beta(x) = r(x)r(Tb)$. In particular, if $r(\cdot)$ is bounded on a set $D \subset K$, consider, e.g., $D = \{x \in K \mid x \leq kb\}$ for some $k > 0$, then S is by Lemma 5.1.6 (i) ascending on D – but S need not be uniformly positive in the sense of (5.1.5). Furthermore, from Lemma 5.1.6 (ii) we may conclude that **zigzag-operators**, that is finitely many successive maxima or minima (taken in any order) of affine operators $Tx + b$ as above, are ascending on a set of type $D = \{x \in K \mid x \leq k^* b\}$ for some $k^* > 0$. A zigzag-operator need neither be concave nor convex nor monotone increasing nor monotone decreasing as the following example shows.

Consider the selfmapping T of $\mathbb{R}_+^2$ which is given as the pointwise maximum of the following selfmappings (see [23])

$$T^1x = \left(\min\left\{2x_1 + x_2, \frac{1}{2}x_1 + \frac{3}{2}x_2\right\},\ \min\{x_1 + 2x_2, 1\}\right)$$
$$T^2x = \left(\min\left\{2x_1 + \frac{1}{4}, x_1 + 4x_2\right\},\ \min\{x_1 + 2x_2, 1\}\right).$$

For $D = \{x \in \mathbb{R}_+^2 \mid x_1 + x_2 = 1\}, x \in D$ one finds $r_1e_1 \le T^1x \le s_1e_1$ and $r_2e_2 \le T^2x \le s_2e_1$ with $e^1 = e^2 = (1,1)$, $r_1 = \frac{1}{2}, s_1 = \frac{5}{4}$, $r_2 = \frac{1}{4}$, $s_2 = \frac{7}{4}$.

Thus, T is ascending on D by Lemma 5.1.6 (ii) but T is neither concave nor convex and neither monotone increasing nor decreasing.

Example 5.1.9 (u_0-concave operators). This concept has been introduced in 1964 by M. A. Krasnoselskii and his collaborators [17–21] for monotone selfmappings T of the positive cone K of an ordered Banach space V without assuming that T is linear on V. For $u_0 \in K \setminus \{0\}$ the monotone operator T is called **u_0-concave** if for all $x \in K \setminus \{0\}$ condition (5.1.6) is satisfied with $e = u_0$ and if the following condition holds.
For $x \in K \setminus \{0\}$ satisfying $\alpha_1(x)u_0 \le x \le \beta_1(x)u_0$ with $\alpha_1(x) > 0$, $\beta_1(x) > 0$ and for $0 < t_0 < 1$ there exists $\eta = \eta(x, t_0) > 0$ such that

$$T(t_0x) \ge (1 + \eta)t_0Tx. \tag{5.1.7}$$

Equivalently, for T monotone T is u_0-concave iff T is u_0-positive and for each x in the part of u_0 and each $0 < t_0 < 1$ there exists $\psi(t_0, x) > t_0$ such that
$t_0x \le y$ for $y \in K$ implies $\psi(t_0, x)Tx \le Ty$. This property reminds of the defining property (5.1.1) for an ascending operator. Indeed, a monotone u_0-concave operator is ascending on each subset of $K \setminus \{0\}$. Concerning the reverse implication, linear operators can be ascending on subsets of K but being positively homogeneous they cannot be u_0-concave for any $u_0 \in K \setminus \{0\}$. (See Exercises 4, 5 and Corollary 5.1.14 below.)

A weakened form of a u_0-concave operator was considered by M.A. Krasnoselskii and his collaborators in [18, 20]. A selfmapping T of the positive cone K with non-empty interior of a Banach space V is called **uniformly concave on the conic interval** $[u, v]$ if

- $u, v \in \mathring{K}$, $[u, v] = \{x \in K \mid u \le x \le v\}$ for $u \le v$,
- T is monotone on $[0, v]$ and $Tu \in \mathring{K}$,
- for any two real numbers $0 < a < b < 1$ there exists $\eta = \eta(a, b) > 0$ such that for all $x \in [u, v]$ and all $\lambda \in [a, b]$

$$(1 + \eta)\lambda Tx \le T(\lambda x). \tag{5.1.8}$$

Obviously, a uniformly concave operator T is ascending on the conic interval $[u, v]$. Conversely, let T be ascending with ϕ on the conic interval $[u, v]$ and define $\eta(a, b) = \inf\{\frac{\phi(\lambda)}{\lambda} \mid \lambda \in [a, b]\} - 1$. If ϕ is continuous then it follows that $\eta = \eta(a, b) > 0$.

Because of $\phi(\lambda) \geq (1+\eta)\lambda$ for $\lambda \in [a,b]$ it follows that

$$(1+\eta)\lambda Tx \leq \phi(\lambda)Tx \leq T(\lambda x).$$

Even for ϕ continuous, however, T need not be uniformly concave. Indeed a positively homogeneous operator can be weakly ascending but it can never be uniformly concave (compare Corollary 5.1.14). A similar but weaker concept is that of an ***e*-monocave mapping** introduced in [50] and which is defined by the following properties for a selfmapping T of a cone K:
- T maps the part K^e generated by $e \in K$ into itself.
- For $u, v \in K^e$, $0 < r < 1$ there exists $M = M(r,u,v) \geq 0$ with $M > 0$ for $ru \neq v$ such that $ru \leq v$ and $rv \leq u$ imply $(r+M)Tu \leq Tv$ and $(r+M)Tv \leq Tu$.

Obviously, an e-monocave mapping is weakly ascending on K^e with $\phi(\lambda) = \lambda + M(\lambda)$, provided $M(r) = M(r,u,v)$ is independent of u, v.

Example 5.1.10 (Subhomogeneous operators). Operators of this type were considered by various authors and first introduced by A. C. Thompson [47, 48]. All types are selfmappings of the convex cone of the form that for each $x \in K$ and $0 \leq \lambda (\leq 1)$ there exists $0 \leq \lambda'$ such that $T(\lambda x) \geq \lambda' Tx$. In the extreme case for "≤" equal to "=" and $\lambda' = \lambda$ the operator is positively homogeneous. For "≤" equal to "=" and $\lambda' = \lambda^d$, $d \geq 0$, the operator is **positively homogeneous of degree** d (see Definition 2.2.1). For $d < 1$ those operators play a role in [5]. If $\lambda' = \lambda$ for $0 \leq \lambda \leq 1$ then T is commonly called **subhomogeneous** or **sublinear** [45] or **co-radiant** [15]. For $\lambda' = \lambda^\alpha$ with $\alpha \in \mathbb{R}$ and $x \in \mathring{K}$, $0 < \lambda \leq 1$ the operator is called α**-concave** ([39], where for "≤" instead of "≥", T is called α-convex; see also the power non-linearities in [20]). For $\lambda' = \lambda^{\alpha(a,b)}$ with $0 < \alpha(a,b) < 1$, $x \in \mathring{K}$ and $\lambda \in [a,b]$ for some fixed numbers a, b the operator T is called α**-sublinear**, provided T is monotone in addition ([8]). Obviously, a monotone operator which is α-concave (including positively homogeneous operators of degree $d < 1$) is ascending on the whole cone. Also, α-sublinear operators are closely linked to ascending operators. More generally, call a selfmapping ϕ of $]0,1[$ a **root function** if there exsits another selfmapping r of $]0,1[$ such that

$$\phi(\lambda) = \lambda^{r(\lambda)} \quad \text{with} \quad \sup r(I) < 1 \quad \text{on compact intervals} \quad I \subset]0,1[. \tag{5.1.9}$$

In case of an α-sublinear operator the function $r(\lambda)$ can be obtained from the values $\alpha(a,b)$ as a piecewise constant function.

Thompson introduced the following weak form of subhomogeneity for a selfmapping T of a convex cone K [48]: There exists p with $0 \leq p < 1$ such that for all $x, y \in K$

$$x \leq \alpha y \quad \text{and} \quad y \leq \beta x \quad \text{imply} \quad Tx \leq \alpha' Ty \quad \text{and} \quad Ty \leq \beta' Tx \tag{5.1.10}$$

with $\max\{\alpha', \beta'\} \leq \max\{\alpha^p, \beta^p\}$.

If T is monotone and positively homogeneous of degree d or α-concave with $0 < d, \alpha < 1$ or α-linear, then T is ascending on K or $\mathring{K}$ with a root function ϕ. If T satisfies (5.1.10) then T is weakly ascending on K with a root function ϕ.

Example 5.1.11 (Strongly monotone operators with homogeneity properties). Let V be a locally convex vector space and K a closed convex cone with non-empty interior $\mathring{K}$. The convex cone $\mathring{K}$ induces a relation "$<$" by $x < y$ iff $y - x \in \mathring{K}$. A selfmapping T of K is

- **strongly monotone** or strictly increasing if for all $x, y \in K$, $x \lneq y$ implies $Tx < Ty$;
- **strongly subhomogeneous** [45] if for all $x \in K \setminus \{0\}$, all $0 < \lambda < 1$, $\lambda Tx < T(\lambda x)$;
- with $(k, \mathring{K})$ **property** [27] if for all $x \in \mathring{K}$, all $0 < \lambda < 1$, $\lambda T^k x < T^k(\lambda x)$ $(k \in \mathbb{N})$;
- **weakly homogeneous** [11] if for all $x \in K$, $\lambda \geq 0$, $T(\lambda x) = c(\lambda)Tx$, where $c\colon \mathbb{R}_+ \to \mathbb{R}_+$ with $c(0) = 0$ and $\frac{c(\lambda)}{\lambda}$ non-increasing.

If T is strongly monotone and subhomogeneous and $0 < \lambda < 1$, then

$$0 \leq \lambda x \lneq y \Rightarrow T(\lambda x) < Ty \Rightarrow \lambda Tx < Ty,$$

and if T is monotone and strongly subhomogeneous, then

$$0 \lneq \lambda x \leq y \Rightarrow T(\lambda x) \leq Ty \Rightarrow \lambda Tx < Ty.$$

Despite the similarity of the properties "strongly monotone and subhomogeneous" and "monotone and strongly subhomogeneous", none implies the other one (see Exercise 5). Both properties imply $\lambda Tx < Ty$ which, under additional assumptions, yields an ascending T as in the following result.

Proposition 5.1.12. *Let (V, τ) be a locally convex vector space and $K \subset V$ a convex cone with non-empty interior $\mathring{K}$ (for τ). Let T be a selfmapping of K and $\emptyset \neq D \subset K$ such that T is continuous on D and for $x, y \in D$, $\lambda \in]0, 1[$ it holds that*

$$\lambda x \leq y \quad \textit{implies} \quad \lambda Tx < Ty. \tag{5.1.11}$$

If D is compact then T is ascending on D.

If D is convex in addition then T is ascending on D with ϕ upper semicontinuous.

Proof. (i) First we show that T is ascending on D with ϕ defined for $0 < \lambda < 1$ by

$$\phi(\lambda) = \inf\{\lambda(Tx, Ty) \mid \lambda x \leq y,\ x, y \in D\}. \tag{5.1.12}$$

By Proposition 3.4.12 (iii) the order function $\lambda(\cdot, \cdot)$ is continuous on $\mathring{K} \times \mathring{K}$ in $(V, \tau) \times (V, \tau)$. From condition (5.1.11) it follows for $x \in D$, $\lambda = \frac{1}{2}$ that $\frac{1}{2}Tx < Tx$ and, hence, $Tx \in \mathring{K}$. Therefore, $\lambda(T\cdot, T\cdot)$ is continuous on $D \times D$ by the continuity of T on D. Since $C_\lambda = \{(x, y) \in D \times D \mid \lambda x \leq y\}$ is compact there exist $x_\lambda, y_\lambda \in D$ with $\lambda x_\lambda \leq y_\lambda$ such that $\phi(\lambda) = \lambda(Tx_\lambda, Ty_\lambda)$. Condition (5.1.11) yields $\lambda(Tx_\lambda, Ty_\lambda) > \lambda$ and, hence, $\phi(\lambda) > \lambda$.

By the definition of ϕ, for $x, y \in D$ and $0 < \lambda < 1$, $\lambda x \leq y$ implies $\phi(\lambda)Tx \leq Ty$ which proves that T is ascending on D with ϕ.

(ii) Next we consider any selfmapping T of an arbitrary convex cone K in some vector space which has the property that for a convex set $D \subset K$
$\lambda x \le y$ implies $\lambda Tx \le Ty$, for any $0 < \lambda < 1$, any $x, y \in D$.
Then it is an easy exercise in extraction analysis to verify the following formula for $x, y \in D$ and $0 < \gamma \le 1$

$$\lambda(Tx, T(\gamma x + (1-\gamma)y)) \cdot \lambda(x,y) \le \lambda(Tx, Ty)(\gamma + (1-\gamma)\lambda(x,y)). \tag{5.1.13}$$

Namely, $\lambda x \le y$ implies that $\lambda(\gamma x + (1-\gamma)y) \le (\gamma + (1-\gamma)\lambda)y$ and, by convexity of D, one has that

$$\lambda T(\gamma x + (1-\gamma)y) \le (\gamma + (1-\gamma)\lambda)Ty$$

and, by taking supremum over λ, $\lambda(T(\gamma x + (1-\gamma)y),\ Ty)(\gamma + (1-\gamma)\lambda(x,y)) \ge \lambda(x,y)$. Therefore,

$$\begin{aligned}
&\lambda(Tx, T(\gamma x + (1-\gamma)y)) \cdot \lambda(x,y)\\
&\quad \le \lambda(Tx, T(\gamma x + (1-\gamma)y)) \cdot \lambda(T(\gamma x + (1-\gamma)y), Ty)(\gamma + (1-\gamma)\lambda(x,y))\\
&\quad \le \lambda(Tx, Ty)(\gamma + (1-\gamma)\lambda(x,y)).
\end{aligned}$$

(iii) Now we show that the function ϕ defined by equation (5.1.12) is upper semi-continuous. Let $\phi(\lambda_0] < \alpha$ for some $\lambda_0 \in]0,1[$. We show that there exists $\epsilon > 0$ such that $\phi(\lambda) < \alpha$ for $\lambda \in]\lambda_0 - \epsilon, \lambda_0 + \epsilon[$. From the definition of ϕ it is obvious that ϕ is increasing and, hence, it satisfies to show that $\phi(\lambda_0 + \epsilon) < \alpha$. Since $\phi(\lambda_0) < \alpha$, there exist $x, y \in D$ with $\lambda_0 x \le y$ and $0 < \lambda(Tx, Ty) < \alpha$. Choose $\epsilon > 0$ such that $\epsilon < \min\{\lambda_0(\frac{\alpha}{\lambda(Tx,Ty)} - 1),\ 1 - \lambda_0, \lambda_0\}$ and $\gamma = \frac{\epsilon}{1-\lambda_0}$. Since D is convex, $x' = x$ and $y' = \gamma x + (1-\gamma)y$ are in D and $(\lambda_0 + \epsilon)x' = (\lambda_0 + \epsilon)x \le \gamma x + (1-\gamma)y = y'$ by choice of γ. Furthermore, by step (ii)

$$\lambda(Tx', Ty') = \lambda(Tx, T(\gamma x + (1-\gamma)y)) \le \lambda(Tx, Ty)\frac{\gamma + (1-\gamma)\lambda(x,y)}{\lambda(x,y)}.$$

Now, $\frac{\gamma+(1-\gamma)\lambda(x,y)}{\lambda(x,y)} \le (\frac{\gamma}{\lambda_0} + (1-\gamma)) \le 1 + \frac{\epsilon}{\lambda_0}$ and by choice of ϵ we arrive at $\lambda(Tx', Ty') < \alpha$. Since $(\lambda_0 + \epsilon)x' \le y'$ it follows that $\phi(\lambda_0 + \epsilon) < \alpha$. □

It should not be overlooked that condition (5.1.11) in Proposition 5.1.12 may hold also for linear operators on subsets which are sectional in the following sense.

Definition 5.1.13. A non-empty subset D of a convex cone is called **sectional** if each ray of K meets D in at most one point, i.e., if

$$\lambda x = y \quad \text{for} \quad x, y \in D,\ \lambda > 0 \quad \text{implies that} \quad \lambda = 1 \quad \text{and} \quad x = y.$$

An example of a sectional set is given by any base of a convex cone (see Definition 3.3.1).

Corollary 5.1.14. Let (V, τ) be a locally convex vector space and $K \subset V$ a convex cone with non-empty interior $\overset{\circ}{K}$. Let T be a concave (in particular, a linear) selfmapping of K which is strictly positive. Then the following statements hold.

(i) T is ascending on every sectional compact subset of K and it is ascending with an upper semi-continuous function ϕ on every compact subset of a base of K.
(ii) If T is a compact operator and M is a bounded subset of V mapped by T into a closed base of K then T is ascending on $D = \overline{T(M)}$ (for τ) with ϕ upper semicontinuous.

Proof. (i) Let $D \subset K$ be sectional and compact. By Proposition 5.1.12 it suffices to show condition (5.1.11). Let $\lambda x \leq y$ for $x, y \in D$ and $0 < \lambda < 1$. Since D is sectional we must have $\lambda x \lneq y$ and $y = \lambda x + z$ with $z \in K \setminus \{0\}$. Since T is concave and strictly positive (i.e., $T(K \setminus \{0\}) \subset \mathring{K}$) it follows that

$$Ty = T\left(\lambda x + (1-\lambda)\frac{z}{1-\lambda}\right) \geq \lambda Tx + (1-\lambda)T\left(\frac{z}{1-\lambda}\right) > \lambda Tx.$$

Consider an arbitrary base B of K, $B = \{x \in K \mid f(x) = 1\}$ where f is a linear functional with $f(x) > 0$ for $x \in K \setminus \{0\}$. Let $C \subset B$ be compact. By the continuity of the vector space operations it follows that the convex hull conv$C$ is compact, too. Because of conv$C \subset B$, the set conv$C$ is sectional and the remaining assertion of (i) follows from Proposition 5.1.12.

(ii) Since $D = \overline{T(M)}$ is a compact subset of a closed base of K the assertion follows from (i). □

Remark 5.1.15. For finite dimensions, condition (5.1.11) plays an essential role in [23]. For $V = \mathbb{R}^n$, $K = \mathbb{R}^n_+$ and $\|\cdot\|$ the l_1-norm on $\mathbb{R}^n$ the set $D = \{x \in\mid K \|x\| = 1\}$ is a sectional set which is compact and convex. From Corollary 5.1.14 it follows that a concave operator on K with $Tx > 0$ for $x \gneq 0$ is ascending on D with an upper semicontinuous function ϕ. Concerning the Concave Perron Theorem (Theorem 2.1.11), it was an essential step in its proof to get by direct calculation that T is ascending on D.

In [27] a property like (5.1.11) results from monotonicity and the $(k, \mathring{K})$-property for T. There a functin ϕ is constructed like the one in the proof of Proposition 5.1.12, but simpler due to finite dimensions. In [11, 24] strongly monotone self-mappings T are considered which are weakly homogeneous. If T possesses these properties on a subset D which is sectional then T must be ascending on D.

A concept close to that of an ascending operator is the property of “strong contractivity” for selfmappings of $\mathbb{R}^n_+$ in [38]. Concepts close to those of a weakly ascending operator and a cone mapping, respectively, are the various forms of order contractivity in [46]. The main difference between these concepts and the ones introduced here, is that the former are required **globally** for the whole cone (or its interior) whereas the latter are required only **locally** on a subset D of K. It is only in the local sense that the concept of an ascending operator or an operator satisfying condition (5.1.11) applies to linear operators.

A locally ascending (weakly ascending) operator need not to be globally ascending (weakly ascending) (see Exercise 4). Furthermore, global concepts, as the one in

[38], often imply monotonicity on the whole cone, which is not necessary for ascending operators (see Exercise 3, Exercise 4 (a)). Actually, it is an important feature of positive dynamical systems, in contrast to monotone dynamical systems, that non-linear operators may fail to be monotone.

Exercises

1. A selfmapping T as in Lemma 5.1.2 (ii) may have indeed the maximal number of non-negative eigenvalues.
   (a) Show that the following selfmapping T of $\mathbb{R}^2_+$ has 3 different non-negative eigenvalues,
   $$T(x_1, x_2) = (\min\{x_1 + 2x_2, 3x_1\},\ \min\{3x_2, x_1 + 2x_2\}).$$
   (b) Find a selfmapping of $\mathbb{R}^3_+$ which has 7 different non-negative eigenvalues.
   (c) Find a continuous, monotone, and positively homogeneous selfmapping of $\mathbb{R}^n_+$ which has $2^n - 1$ different non-negative eigenvalues.
2. Prove the following statements.
   (a) Every ascending mapping is weakly ascending (for the same D and ϕ).
   (b) Every weakly ascending mapping is a cone mapping.
   (c) A mapping which is ascending on a subset D of an archimedean cone is monotone on D.
   (d) The statement (c) fails if "ascending" is replaced by "weakly ascending" or by "cone mapping".
   (e) The implications in (a) and (b) cannot be reversed.
3. Let T be the selfmapping of $\mathbb{R}^2_+$ given by $T0 = 0$ and
   $$T(x_1, x_2) = \left(\sqrt{\frac{x_1^2 + x_2^2}{x_1 + x_2}},\ \sqrt{x_1 + x_2} \right) \quad \text{for} \quad (x_1, x_2) \neq (0, 0).$$
   Show that
   (a) T is neither increasing nor decreasing (with respect to the ordering given by $\mathbb{R}^n_+$).
   (b) T is ascending on $D = \{x \in \mathbb{R}^2_+ \mid x_1 + x_2 = 1\}$.
4. (cf. [22]) Consider the selfmapping T of $K = \mathbb{R}^2_+$ given by
   $$T(x_1, x_2) = \begin{cases} (x_1 + x_2, 1) & \text{for} \quad 0 \le x_1 + x_2 \le 1 \\ (1, \frac{1}{x_1 + x_2}) & \text{for} \quad x_1 + x_2 \ge 1. \end{cases}$$
   (a) Find a subset of K on which T is ascending and show that T is neither monotone nor u_0-concave for any $u_0 \in K_0 \setminus \{0\}$.
   (b) Show that T is not weakly ascending on the whole cone K.
   (c) Show that T is a cone mapping on the whole cone K.

(d) If S is the selfmapping of K defined by $S(x_1, x_2) = (x_1 + x_2, 1)$ and $\|\cdot\|$ is the max-norm on $\mathbb{R}^2$, then

$$Tx = \frac{Sx}{\|Sx\|} \quad \text{for} \quad x \in K \setminus \{0\}.$$

Show that S is neither α-concave nor α-convex.

(e) Show that neither S nor T are superadditive.

5. Find examples of selfmappings of a cone which show that of the properties "strongly monotone and subhomogeneous" and "monotone and strongly subhomogeneous" none implies the other one.

5.2 Relative stability for ascending operators by Hilbert's projective metric

As for operators in finite dimensions (Section 2.1) we need to consider for general vector spaces the normalized or rescaled operator. This will be done more general for a scale in the following sense of which a monotone norm is a special case.

Definition 5.2.1. For a convex cone K in a real vector space a **scale s on K** is a mapping $s\colon K \to \mathbb{R}_+$ that is not identically 0 and such that s is positively homogeneous and monotone, i.e., for all $x, y \in K$, $\lambda \in \mathbb{R}_+$

$$s(\lambda x) = \lambda s(x) \quad \text{and} \quad s(x) \le s(y),$$

provided $x \le y$.
The set $U = \{x \in K \mid s(x) = 1\}$ is non-empty and called the **unit set for s.**

A selfmapping T of K is **proper** for a scale s if

$$s(Tx) \neq 0 \quad \text{for} \quad s(x) = 1.$$

For an operator $T\colon D \to K$, $D \subset K$, and a scale s on K the **normalized** or rescaled operator $\tilde{T}$ is defined by $\tilde{T}x = (Tx)(s(Tx))^{-1}$ for $x \in D$ with $s(Tx) \neq 0$. In the following it will often be used that for a selfmapping T of K which is proper for a scale on K the normalized operator $\tilde{T}$ is defined and maps the unit set U into itself. Note the caveat that in general $(\tilde{T})^k$ and $(\widetilde{T^k})$ have to be distinguished (compare Figure 2.2).

As observed earlier, Hilbert's metric d gives distance zero for points on the same ray but by scaling elements we can obtain a metric space as follows.

Lemma 5.2.2. *If K is a lineless and internally complete convex cone then $(P \cap U, d)$ is a complete metric space for every non-zero part P of K.*

Proof. $(P \cap U, d)$ is a metric space by Theorem 3.2.3 (v). Let $(x_n)_n$ be a Cauchy sequence for d in $P \cap U$. Since K is internally complete there exists $x \in P$ with $\lim_{n\to\infty} d(x_n, x) = 0$. In particlar, there exists $m \in \mathbb{N}$ with $\lambda(x_m, x) > 0$ and, hence, $\lambda x_m \le x$ for some $\lambda > 0$.

For the scale s it follows that $\lambda s(x_m) \leq s(x)$ which implies $0 < \lambda = \lambda s(x_m) \leq s(x)$. Thus, $y = \frac{x}{s(x)} \in P \cap U$ and $\lim_{n\to\infty} d(x_n, y) = \lim_{n\to\infty} d(x_n, x) = 0$. □

The next definition describes, roughly speaking, the set of all points of the cone for which the selfmapping T is ascending on a tail of the $\tilde{T}$-orbit which is required to belong to a part of the cone.

Definition 5.2.3. Let K be a convex cone in a real vector space with scale s and let T be a proper selfmapping of K. The **ascending domain** $D(T)$ of T consists of all points $x \in K$ such that

- there exists a non-zero part P_x of K;
- there exists a non-empty subset $D_x \subset K$ on which T is ascending with some ϕ_x;
- $s(Tx) > 0$ and there exists for the normalization $\tilde{T}$ (with respect to s) some $n(x) \in \mathbb{N}$ such that for $M = \{\tilde{T}^n x \mid n \geq n(x)\}$ it holds that $M \subset P_x$ and $\overline{M} \subset D_x$ where $\overline{M}$ is the closure of M in the metric space $(P_x \cap U, d)$.

The main results of this section and the next section state that under certain conditions for every point in the ascending domain of T the iterates of the normalized operator $\tilde{T}$ (or of T itself) converge to an eigenvector of T with strictly positive eigenvalue. We shall refer to the convergence of the iterates of $\tilde{T}$ as **relative stability** (for this term see [32]). If, in contrast, the iterates of T itself converge to a fixed point of T we employ, for short, the term **absolute stability**. Note that by our method of non-expanding maps, a fixed point of $\tilde{T}$, or of T, is automatically stable in the usual sense (compare Remark 4.1.5 (iii)).

Theorem 5.2.4 (Relative stability for ascending operators). *Let K be a convex cone in a real vector space V with scale s and let T be a proper selfmapping of K with non-empty ascending domain $D(T)$.*

A. *Suppose that K is lineless and internally complete.*

(i) *For every $x \in D(T)$ for which ϕ_x is upper semicontinuous on $]0,1[$ the orbit $(\tilde{T}^n x)_{n\in\mathbb{N}}$ converges with respect to Hilbert's projective metric d to an eigenvector $x^* = x^*(x)$ of T with $s(x^*) = 1$ with eigenvalue $\lambda^* > 0$.*

(ii) *Let $x, y \in D(T)$ such that for some $k \geq 1$ both $\tilde{T}^n x$ and $\tilde{T}^n y$ belong for $n \geq k$ jointly to some part and their d-closures belong to a subset on which T is ascending with an upper semicontinuous function ϕ. Then it holds in (i) that $x^* = y^*$.*

(iii) *For every subset D of K, on which T is ascending with an upper semicontinuous ϕ and for every part P of K, there is at most one eigenvector of T with scale 1 contained in $D \cap P$.*

B. *Let τ be a locally convex topology on the vector space V for which K is sequentially complete and normal.*

(iv) *The above statements (i) to (iii) do hold where in (i) the orbit $(\tilde{T}^n x)_{n\in\mathbb{N}}$ converges to x^* also with respect to τ.*

(v) *If T is ray preserving then for each x as in (i)*

$$\lim_{n\to\infty} \frac{T^n x}{s(T^n x)} = x^* \quad \text{for} \quad \tau.$$

If T is positively homogeneous and monotone then $\lim_{n\to\infty} \frac{T^n x}{\lambda^{*n}} = c(x)x^*, c(x) \geq 0.$

Proof. Suppose that K is lineless and internally complete.

(i) Fix $x_0 \in D(T)$ as well as a non-zero part P of K and a non-empty subset $D \subset K$, according to Definition 5.2.3. Let X be the closure of $M = \{\tilde{T}^n x_0 \mid n \geq n(x_0)\}$ in the complete metric space $(P \cap U, d)$, where U is the unit set of s (see Lemma 5.2.2). Obviously, (X, d) is a complete metric space contained in D.

We show that $\tilde{T}$ is a selfmapping of the metric space (X, d) which is a generalized contraction in the sense of Definition 4.1.1(iii).

Let $0 < \alpha \leq \beta < \infty$ and consider $x, y \in X$ with $\alpha \leq d(x, y) \leq \beta$. Therefore, $e^{-\beta} \leq \lambda(x, y) \cdot \lambda(y, x) \leq e^{-\alpha}$ and, because of $x, y \in X \subset U$, we must have that $0 < \lambda(x, y), \lambda(y, x) \leq 1$. Considering $0 < \lambda, \mu < 1$ with $\lambda x \leq y$, $\mu y \leq x$ we have that $\lambda(x, y)$ and $\lambda(y, x)$ is the supremum of those λ and μ respectively. Because of $x, y \in U$, neither $\lambda x = y$ nor $\mu y = x$ is possible and we must have that $\lambda x \lneqq y$ and $\mu y \lneqq x$. Since T is ascending on D with ϕ we obtain $\phi(\lambda)Tx \leq Ty$ and $\phi(\mu)Ty \leq Tx$. This implies that $\lambda(Tx, Ty) \geq \phi(\lambda)$ and $\lambda(Ty, Tx) \geq \phi(\mu)$ from which it follows that

$$\lambda(\tilde{T}x, \tilde{T}y) \cdot \lambda(\tilde{T}y, \tilde{T}x) = \lambda(Tx, Ty) \cdot \lambda(Ty, Tx) \geq \phi(\lambda) \cdot \phi(\mu).$$

Thus, we obtain for $x, y \in X$ with $\alpha \leq d(x, y) \leq \beta$ that

$$d(\tilde{T}x, \tilde{T}y) \leq -\log[\phi(\lambda) \cdot \phi(\mu)]. \tag{5.2.1}$$

for all $0 < \lambda,\ \mu < 1$ with $\lambda x \lneqq y$, $\mu y \lneqq x$.

Now consider the compact set $I = \{(r_1, r_2) \mid r_1, r_2 \in [0, 1],\ e^{-\beta} \leq r_1 . r_2 \leq e^{-\alpha}\}$, in $\mathbb{R}^2$. By assumption ϕ is upper semicontinuous on $]0, 1[$ and ϕ becomes upper semicontinuous on $]0, 1]$ by setting $\phi(1) = 1$.

Since for $(r_1, r_2) \in I$ we must have that $r_1, r_2 > 0$ and that both r_1 and r_2 cannot be 1 we have that $\phi(r_1)\phi(r_2) > r_1 r_2$ for $(r_1, r_2) \in I$. Therefore, $-\log(r_1 r_2) > -\log(\phi(r_1) \cdot \phi(r_2)) \geq 0$ for $(r_1, r_2) \in I$. Thus the function f defined on I by

$$f(r_1, r_2) = \frac{\log \phi(r_1)\phi(r_2)}{\log(r_1 r_2)}$$

is upper semicontinuous on I (see Exercise 2) and attains therefore its supremum $\sigma(I)$ on the compact set I,

$$\sigma(I) = \sup\{f(r_1, r_2) \mid (r_1, r_2) \in I\} = \frac{\log \phi(r_1')\phi(r_2')}{\log(r_1' r_2')} < 1.$$

Since $(\lambda, \mu) \in I$ from (5.2.1) we obtain that

$$d(\tilde{T}x, \tilde{T}y) \leq -\log[\phi(\lambda) \cdot \phi(\mu)] \leq \sigma(I)(-\log(\lambda\mu)).$$

By taking the infimum over λ, μ and by setting $L(\alpha, \beta) = \sigma(I)$ we obtain finally that for $x, y \in X$ and $\alpha \le d(x,y) \le \beta$ it holds that

$$d(\tilde{T}x, \tilde{T}y) \le L(\alpha,\beta) d(x,y) \quad \text{with} \quad L(\alpha,\beta) < 1. \tag{5.2.2}$$

In particular, $\tilde{T}$ is d-continuous on X. Therefore, for $x \in X$, $x = \lim_{t\to\infty} x_n, x_n \in M$ (for d) it follows that $\tilde{T}x = \lim_{n\to\infty} \tilde{T}x_n$. This implies, because of $\tilde{T}x \in P \cap U$ that $\tilde{T}x \in X$. Thus we obtain that $\tilde{T}$ is a selfmapping of X which by (5.2.2) is a generalized contraction on (X, d).

Since a generalized contradiction is always (ϵ, δ)-contractive (cf. Exercise 3 to 4.1) from Theorem 4.1.4 it follows that $\tilde{T}$ has a (unique) fixed point $x_0^* \in X$ and $\lim_{m\to\infty} \tilde{T}^m x = x_0^*$ for all $x \in X$ with respect to d. For $x = \tilde{T}^n x_0$, $n = n(x_0)$ it follows in particular that $\lim_{m\to\infty} \tilde{T}^m x_0 = x_0^*$. Obviously, $s(x_0^*) = 1$ and $\tilde{T}x_0^* = x_0^*$ implies $Tx_0^* = \lambda^* x_0^*$ with $\lambda^* = s(Tx_0^*) > 0$. This proves statement (i) of part A of the theorem.

(ii) Let $x, y \in D(T)$ and $M = \{\tilde{T}^n x \mid n \ge k\} \cup \{\tilde{T}^n y \mid n \ge k\}$ with $X = \overline{M} \subset D \cap P$, where D and P are joint ascending set and joint part, respectively. As in the proof of statement (i) it follows that $\tilde{T}$ has a unique fixed point z^* in X such that $\lim_{m\to\infty} \tilde{T}^m z = z^*$ for all $z \in X$. In particular, $\lim_{m\to\infty} \tilde{T}^m x = z^* = \lim_{m\to\infty} \tilde{T}^m y$.

(iii) Let $x \in D \cap P$ be an eigenvector of T with $s(x) = 1$. If in $Tx = \lambda x$ one would have that $\lambda < 0$ then $x \in K \cap (-K) \subset \{0\}$ because K is lineless. Therefore, $\lambda \ge 0$ and $\lambda = s(\lambda x) = s(Tx) > 0$ because T is proper. It follows that $\tilde{T}x = \frac{\lambda x}{s(\lambda x)} = \frac{x}{s(x)} = x$ and, hence, $x \in D(T)$. Similarly $y \in D(T)$ for an eigenvector $y \in D \cap P$ with $s(y) = 1$. From statement (ii) it follows that $x = \lim_{n\to\infty} \tilde{T}^n x = \lim_{n\to\infty} \tilde{T}^n y = y$. Therefore, there is at most one eigenvector of T with scale 1 in $D \cap P$.

(iv) Now we turn to part B of the theorem. From Corollary 3.4.14 it follows that K is lineless and internally complete. Therefore, statements (i) to (iii) of part A do hold. Furthermore, for $x, y \in U$, that is $s(x) = s(y) = 1$ it follows that $\lambda(x,y) \le 1$ and $\lambda(y,x) \le 1$. From Proposition 3.3.3 (v) for each of the monotone semi-norms q which define τ it follows that

$$\begin{aligned} q(x-y) &\le [3 - (\lambda(x,y) + \lambda(y,x) + \max\{\lambda(x,y), \lambda(y,x)\})] \max\{q(x), q(y)\} \\ &\le 3(1 - \exp(-d(x,y))) \max\{q(x), q(y)\}. \end{aligned}$$

By statement (i), $\lim_{n\to\infty} x_n = x^*$ holds for d with $x_n = \tilde{T}^n x$. In particular, there exists N such that $d(x_n, x^*) < \frac{1}{2}$ and, because of $x_n, x^* \in U$, $e^{-\frac{1}{2}} x_n \le x^*$ for all $n \ge N$. Thus, we obtain for all $n \ge N$

$$q(x_n - x^*) \le 3(1 - \exp(-d(x_n, x^*))) \cdot e^{\frac{1}{2}} q(x^*).$$

Since $\lim_{n\to\infty} d(x_n, x^*) = 0$ it follows that $\lim_{n\to\infty} q(x_n - x^*) = 0$ for all q defining τ that is $\lim_{n\to\infty} x_n = x^*$ holds for τ.

(v) If T is ray preserving, $x \in D(T)$ then it follows that $\tilde{T}^n x = \frac{T^n x}{s(T^n x)}$ since $s(Tx) > 0$ and T is proper (cf. Lemma 2.2.4 (i)). Suppose now that T is positively homogeneous and monotone. Denote $x_n = \frac{T^n x}{\lambda^{*n}}$ and let $\lambda_n = \lambda(x^*, x_n)$, $\mu_n = \lambda(x_n, x^*)$. From $\lambda_n x^* \le x_n$

and $\mu_n x_n \leq x^*$ it follows that $\lambda_n Tx^* \leq Tx_n$ and $\mu_n Tx_n \leq Tx^*$ by the assumptions made. Since $Tx^* = \lambda^* x^*$ by (i) we get $\lambda_n x^* \leq x_{n+1}$ and $\mu_n x_{n+1} \leq x^*$. Thus, $\lambda_n \leq \lambda_{n+1}$ and $\mu_n \leq \mu_{n+1}$ for all n. Since $\lambda_n \mu_n \leq 1$ and $\mu_1 > 0$ the sequence $(\lambda_n)_n$ is bounded and, hence, converges to some $c = c(x) \geq 0$. By (i) we have that $\lim_{n\to\infty} d(\tilde{T}^n x, x^*) = 0$ which by the properties of Hilbert's metric implies that $\lim_{n\to\infty} d(x_n, x^*) = 0$, that is $\lim_{n\to\infty} \lambda_n \mu_n = 1$. Thus, we must have $c > 0$ and $\lim_{n\to\infty} \mu_n = \frac{1}{c}$. From

$$(\lambda_n - c)x^* \leq x_n - cx^* \leq \left(\frac{1}{\mu_n} - c\right)x^*$$

we obtain for each of the monotone semi-norms q defining τ that

$$|\lambda_n - c|q(x^*) \leq q(x_n - cx^*) \leq |\frac{1}{\mu_n} - c|q(x^*) \text{ for all } n.$$

Therefore, with respect to τ $\lim_{n\to\infty} \frac{T^n x}{\lambda^{*n}} = \lim_{n\to\infty} x_n = c(x)x^*$. □

From Theorem 5.2.4 we obtain the following conclusions.

Corollary 5.2.5. *Let (V, τ) be a locally convex vector space with a convex cone K which is sequentially complete and normal. Assume further that $\mathring{K} \neq \emptyset$ and that K has a base $B = \{x \in K \mid f(x) = 1\}$ with a scale f continuous for τ. Define for $0 < \lambda < 1$ the set $C_\lambda = \{(x, y) \mid x, y \in B, \lambda x \leq y\}$.*

(i) *Let T be a selfmapping of K, continuous on B and such that the set $\{(Tx, Ty) \mid (x, y) \in C_\lambda\}$ is compact for the product topology and*

$$(x, y) \in C_\lambda \quad \text{for} \quad 0 < \lambda < 1 \quad \text{implies} \quad \lambda Tx < Ty. \qquad (*)$$

Then the conditional eigenvalue problem

$$Tx = \lambda x \quad \text{with} \quad \lambda \in \mathbb{R},\ x \in K,\ f(x) = 1$$

has a unique solution $x = x^* \in \mathring{K}$, $\lambda = \lambda^* > 0$ and it holds for f as scale and with respect to τ that

$$\lim_{n\to\infty} \tilde{T}^n x = x^* \quad \text{for all} \quad x \in K \quad \text{with} \quad f(Tx) > 0.$$

If, in addition, T is positively homogeneous then $\lim_{n\to\infty} \frac{T^n x}{\lambda^{*n}} = c(x)x^*$ with respect to τ, where $c\colon \{x \in K \mid f(Tx) > 0\} \longrightarrow \mathbb{R}_+$ is positively homogeneous and strictly monotone.

(ii) *Let $V = \mathbb{R}^n$ with Euclidean topology, $K \subseteq V$ a closed convex cone with $\mathring{K} \neq \emptyset$ and base $B = \{x \in K \mid \|x\| = 1\}$ with norm $\|x\| = \sum_{i=1}^n |x_i|$ on V. Let T be a continuous selfmapping of K.*

   (a) *If T has property $(*)$ of (i) then the conclusions stated in (i) do hold for T with $f(x) = x_1 + \cdots + x_n$.*

   (b) *For $K = \mathbb{R}^n_+$, T has an eigenvector in $\mathring{K}$ with positive eigenvalue if*

$$\lambda x \leq y \quad \text{implies} \quad \lambda Tx \leq Ty \quad \text{for} \quad x, y \in \mathring{K},\ 0 \leq \lambda \leq 1 \qquad (**)$$

   and T maps $\mathring{K}$ into itself with a strong connected graph $G(T)$.

Proof. (i) Let $\phi(\lambda) = \inf\{\lambda(Tx, Ty) \mid (x,y) \in C_\lambda\}$. Since $\{(Tx, Ty) \mid (x,y) \in C_\lambda\}$ is assumed to be compact it follows from Proposition 5.1.12 and its proof that T is ascending on B with ϕ upper semicontinuous. For $x \in B$ by $(*)$ it follows from $\frac{1}{2}x \lneqq x$ that $0 \le \frac{1}{2}Tx < Tx$, that is $T(B) \subset \mathring{K}$. This shows in particular that T is proper for the scale $s = f$. Furthermore, for $\tilde{T}x = \frac{Tx}{f(Tx)}$ one has that $\tilde{T}(B) \subset \mathring{K}$. For $x \in K$ with $f(Tx) > 0$ choose $P_x = \mathring{K}$ and $D_x = B$. Since $M = \{\tilde{T}^n x \mid n \ge 2\} \subset \mathring{K}$ and $\overline{M} \subset B$ for the closure of M in the metric space $(\mathring{K} \cap B, d)$, from the definition of the ascending domain it follows that $D(T) = \{x \in K \mid f(Tx) > 0\}$. From Theorem 5.2.4 B (iv) it follows that $\lim_{n\to\infty} \tilde{T}^n x = x^*$ (for τ) where (by statements (ii) and (iii)) (x^*, λ^*) with $\lambda^* = f(Tx^*)$ is the unique solution of $Tx = \lambda x$, $x \in K$, $f(x) = 1$.

Suppose now that T is positively homogeneous. According to Theorem 5.2.4 B (v) it suffices to show that T is monotone. For $0 \lneqq x \lneqq y$ we must have $0 < f(x) < f(y)$. Setting $\lambda = \frac{f(x)}{f(y)} < 1$ we have that $\lambda \frac{x}{f(x)} \le \frac{y}{f(y)}$ and, by condition $(*)$, $\lambda T(\frac{x}{f(x)}) < T(\frac{y}{f(y)})$. Thus $Tx < Ty$. Therefore, T is monotone and $c(x) < c(y)$. Obviously, $0 < c(y)$ for $0 \lneqq y$.

(ii) Obviously, K is sequentially complete and normal for the Euclidean topology τ. Since B is compact and T is continuous the set $\{(Tx, Ty) \mid (x,y) \in C_\lambda\}$ is compact for the product topology. Thus, part (a) follows from (i). Concerning part (b) we shall apply an approximation of T as used already in the proof of Theorem 2.1.14. Namely, define $T(k)x = Tx + \frac{1}{k}e$ for $x \in K, k \ge 1, e = (1,\dots,1) \in \mathring{K}$. Obviously, $T(k)$ is a continuous selfmapping of K. Furthermore, if $\lambda x \le y$ for $x, y \in K$ and $0 < \lambda < 1$ from $(**)$ we obtain by continuity of T that

$$\lambda T(k)x = \lambda Tx + \frac{\lambda}{k}e < y + \frac{1}{k}e.$$

Thus, $T(k)$ has property $(*)$ of (i) and part (a) implies in particular the existence of $x(k) \in B \cap \mathring{K}$ and $\lambda(k) > 0$ such that $T(k)x(k) = \lambda(k)x(k)$ for all $k \ge 1$. By compactness of B we may assume without loss that $\lim_{k\to\infty} x(k) = x$. From $\lambda(k)x(k) = T(k)x(k) = Tx(k) + \frac{1}{k}e$ we obtain $\lim_{k\to\infty} \lambda(k)x(k) = Tx$ and, hence, $\|Tx\| = \lim_{k\to\infty} \lambda(k)$. This yields $Tx = \lambda x$ with $\lambda = \|Tx\|$, $\|x\| = 1$. It remains to show that $x \in \mathring{K}$. By assumption T maps $\mathring{K}$ into itself, $G(T)$ is strongly connected and from parts (b) and (c) of Exercise 9 to Chapter 2 we have that for each $c > 0$ the set $\{x \in \mathring{K} \mid \|x\| = 1, Tx \le cx\}$ is closed for $\|\cdot\|$. Since $x(k) \in B \cap \mathring{K}$ and

$$Tx(k) \le T(k)x(k) = \lambda(k)x(k) \le (\|Tx\| + 1)x(k)$$

for k big enough we arrive at $x \in \mathring{K}$. This proves part (b) of (ii). □

Remark 5.2.6. Condition $(*)$ in Corollary 5.2.5 is related to mappings T discussed under Example 5.1.10 and Example 5.1.11, respectively. For example, if T is monotone and positively homogeneous of degree $d < 1$ or α-sublinear for $0 \le \alpha < 1$ then T satisfies condition $(*)$ (provided $T(B) \subset \mathring{K}$). It is not difficult to see that adding operators satisfying $(*)$ to monotone operators which are subhomogeneous yields operators which again satisfy condition ($*$). (See [36] for this type of operators.) For mappings satisfying condition ($*$) see [11, 23, 27, 45]. As a combination of homogeneity and monotonicity condition ($*$) appears in the early approach [32] to non-linear Perron–Frobenius theory in finite dimensions by the economist M. Morishima.

A particularly interesting consequence of condition ($*$) is the existence of an eigenvector in the interior of the cone. The special case given in part (ii) (b) of Corollary 5.2.5 extends the Generalized Perron–Frobenius Theorem of Gaubert and Gunawardena (see [12, 29]) to mappings which are not homogeneous but subhomogeneous.

The next consequence of Theorem 5.2.4 concerns stochastic operators which can be viewed as infinite generalizations of column stochastic matrices. (See [41] for this and the dual concept of a Markov operator which is defined there more special for linear operators on Banach lattices.)

Definition 5.2.7. Let $(V, \|\cdot\|)$ be a normed vector space with a convex cone K on which $\|\cdot\|$ is additive. A selfmapping T of K is called a **stochastic operator** if $\|Tx\| = 1$ for all $x \in K$ with $\|x\| = 1$.

The following result can be viewed as an infinite generalization of the Basic Limit Theorem for Markov chains (see [30]).

Corollary 5.2.8 (Basic limit theorem for stochastic operators). *Let $(V, \|\cdot\|)$ be a Banach space with a closed normal cone K with $\mathring{K} \neq \emptyset$. Let T be a stochastic operator (on K) which is compact, concave, and primitive, i.e., there exists $p \in \mathbb{N}$ such that $T^m x > 0$ for all $m \geq p$, all $x \in K \setminus \{0\}$. Then T has a unique fixed point $x^* \in K$ with $\|x^*\| = 1$ and it holds that*

$$\lim_{n\to\infty} T^n x = x^* > 0 \quad \textit{for all} \quad x \in K,\ \|x\| = 1.$$

Proof. The set $B = \{x \in K \mid \|x\| = 1\}$ is a closed base for K. By Corollary 5.1.14 (ii) the operator $S = T^p$ is ascending on $D = \overline{S(B)}$ with ϕ upper semicontinuous.

Obviously, $Sx \in B \cap \mathring{K}$ for $x \in B$. For fixed $x_0 \in B$ it holds that $M = \{\tilde{S}^n x_0 \mid \geq 2\} = \{S^n x_0 \mid n \geq 2\} \subset B \cap \mathring{K}$. Let $x \in \overline{M}$, the closure $\overline{M}$ taken in $(B \cap \mathring{K}, d)$, and $x = \lim_{n\to\infty} x_n$ for d. Since $\|x\| = 1 = \|x_n\|$, from Proposition 3.3.3 (vi) it follows that $x = \lim_{n\to\infty} x_n$ for $\|\cdot\|$. Therefore, $\overline{M}(\text{for } d) \subset \overline{M}(\text{for} \|\cdot\|) \subset \overline{S(B)}(\text{for } \|\cdot\|) = D$.
Obviously, $\overline{M}$ (for d) $\subset \mathring{K}$. Therefore, by the definition of the ascending domain, we have that $x_0 \in D(T)$. Since $x_0 \in B$ was arbitrarily chosen, the corresponding set $\overline{M}$ for any $x \in B$ is contained in $D \cap \mathring{K}$. Thus, by Theorem 5.2.4, part B, it follows for some x^* with $\|x^*\| = 1$ that $\lim_{n\to\infty} \tilde{S}^n x = x^*$ (for $\|\cdot\|$) for all $x \in B$. Since $\tilde{S}x = Sx$ for $x \in B$ it follows that $\lim_{n\to\infty} T^{np} x = x^*$ for all $x \in B$. For $x \in B$ one has that $T^i x \in B$ and, hence, $\lim_{n\to\infty} T^{np+i} x = x^*$. This implies that $\lim_{n\to\infty} T^n x = x^*$ for all $x \in B$ and, by continuity of T, x^* is a fixed point of T. Obviously, x^* is the unique fixed point of T in B. □

From Theorem 5.2.4 we obtain the following extension to infinite dimensions of the First Concave Perron Theorem (Theorem 2.1.11) and the Second Concave Perron Theorem (Theorem 2.2.11).

Theorem 5.2.9 (Relative stability for concave and zigzag-operators). *Let (V, τ) be a locally convex vector space with a convex cone K which is sequentially complete and normal. Let q be an arbitrary seminorm on K with unit set U and let T be a concave selfmapping of K.*

(i) *Assume there are numbers $0 < r \leq s$ and $e \in K$ with $q(e) > 0$ such that*

$$re \leq Tx \leq se \quad \text{for all} \quad x \in U. \tag{$*$}$$

Then the conditional eigenvalue problem

$$Tx = \lambda x \quad \text{with} \quad \lambda \in \mathbb{R},\ x \in K,\ q(x) = 1$$

has a unique solution $x = x^*$, $\lambda = \lambda^* > 0$ and it holds with respect to τ that

$$\lim_{n\to\infty} \tilde{T}^n x = x^* \quad \text{for all} \quad x \in K \quad \text{with} \quad q(Tx) > 0.$$

If, in addition, T is ray-preserving or positively homogeneous, respectively, then for τ and $q(Tx) > 0$, $\lim_{n\to\infty} \frac{T^n x}{q(T^n x)} = x^*$ or $\lim_{n\to\infty} \frac{T^n x}{\lambda^{*n}} = c(x)x^*$, respectively. Thereby, $c(\cdot)$ is concave and positively homogeneous.

(ii) *Suppose $\overset{\circ}{K} \neq \emptyset$ and let q be a norm on V. Assume T is continuous, ray-preserving, primitive, and for some k the closure of $T^k(U)$ (for τ)is a compact subset of K which does not contain zero.*
Then the eigenvalue problem

$$Tx = \lambda x \quad \text{with} \quad \lambda \in \mathbb{R},\ x \in K \setminus \{0\}$$

has a solution $x^* > 0$ with $q(x^*) = 1$ and $\lambda^* > 0$. For any solution $x \in K \setminus \{0\}$, $\lambda \in \mathbb{R}$ it holds that $x = rx^*$ for some $r > 0$ and $\lambda > 0$. Moreover, with respect to τ and $x \in K \setminus \{0\}$

$$\lim_{n\to\infty} \frac{T^n x}{q(T^n x)} = x^* \quad \text{and} \quad \lim_{n\to\infty} \frac{T^n x}{\lambda^{*n}} = c(x)x^*.$$

if T is positively homogeneous.

(iii) *Let $\{Ti\}_{i\in I}$ be a family of concave operators as in (i) with $0 < r_i \leq s_i$ and $e_i \in K$ with $q(e_i) > 0$ and such that $\inf\{\frac{r_i}{s_i} \mid i \in I\} > 0$. If $Tx = \inf\{T_i x \mid i \in I\}$ (or $Tx = \sup\{T_i x \mid i \in I\}$) exists for every $x \in K$ then the conclusion of (i) applies to T.*

Proof. (i) Lemma 5.1.6 (i) yields for $D = U = \{x \in K \mid q(x) = 1\}$ that T is ascending on D with $\phi(\lambda) = \lambda + (1-\lambda)\frac{r}{s}$. From $(*)$ it follows for $x \in U$ that $q(Tx) > 0$ and, hence, T is proper for scale q. Fix $x_0 \in K$ with $q(Tx_0) > 0$ and let P_0 the part generated by $u_0 = \tilde{T}x_0$. For $u \in U$ from $(*)$ it follows that

$$Tu \leq se \leq \frac{s}{r} Tu_0 \quad \text{and} \quad Tu_0 \leq se \leq \frac{s}{r} Tu.$$

Therefore, $M = \{\tilde{T}^n x_0 \mid n \geq 2\} \subset P_0$ and the closure $\overline{M}$ taken in the complete metric space $(P_0 \cap U, d)$ (see Lemma 5.2.2) is contained in D. Thus, $D(T) = \{x \in K \mid q(Tx) > 0\}$ for the ascending domain of T and the conclusion of (i) follows from Theorem 5.2.4.

(ii) By assumption $\overline{T^k(U)}$ is compact in K (for τ) and does not contain 0. Since T is primitive and continuous, $T^p\left(\overline{T^k(U)}\right)$, and hence, $C = \overline{T^{p+k}(U)}$ is compact and contained in $\mathring{K}$. By Proposition 3.4.12 (iii), $\lambda(\cdot,\cdot)$ is τ-continuous and strictly positive on $\mathring{K}$. For $e \in C$ arbitrary but fixed this implies that

$$r = \inf\{\lambda(e,x) \mid x \in C\} > 0 \quad \text{and} \quad s = \sup\left\{\frac{1}{\lambda(x,e)} \mid x \in C\right\} < \infty.$$

Therefore, $re \leq x \leq se$ for all $x \in C$ and $re \leq T^{p+k}u \leq se$ for all $u \in U$. For $S = T^m$, $m = p + k$, from part (i) it follows that $Sx = \lambda x$, $\lambda \in \mathbb{R}$, $x \in K$, $q(x) = 1$ has a unique solution $x = x^*$, $\lambda > 0$ and $\lim_{n\to\infty} \frac{S^n u}{q(S^n u)} = x^* > 0$ for all $u \in U$. To confer these results from S to T, consider $y = T^i x$ for $x \in K \setminus \{0\}$. For $1 \leq i < p$ one has that $0 < T^p x = T^{p-i}(T^i x) = T^{p-i}y$. Since T is ray-preserving, $T0 = 0$, and, hence, $y \neq 0$. Therefore, $y = T^i x \neq 0$ for $x \in K \setminus \{0\}$, $i \in \mathbb{N}$. Applying

$$\lim_{n\to\infty} \frac{T^{nm}u}{q(T^{nm}u)} = x^* \quad \text{for all} \quad u \in U$$

to $u = \frac{T^i x}{q(T^i x)}$, $0 \leq i < m$, $x \in K \setminus \{0\}$ yields

$$\lim_{n\to\infty} \frac{T^n x}{q(T^n x)} = x^*.$$

Similarly the assertion for T positively homogeneous is obtained from part (i). Finally, from $Sx^* = \lambda x^*$ we obtain that

$$S\left(\frac{Tx^*}{q(Tx^*)}\right) = \rho T^{m+1}x^* = \rho T(Sx^*) = \rho T(\lambda x^*) = \mu \frac{Tx^*}{q(Tx^*)},$$

where $\rho > 0$, $\mu > 0$. By uniqueness, $\frac{Tx^*}{q(Tx^*)} = x^*$ and, hence, $Tx = \lambda x$ has the solution $x = x^* > 0$, $\lambda^* = q(Tx^*) > 0$. Finally, suppose $Tx = \lambda x$ with $x \in K \setminus \{0\}$, $\lambda \in \mathbb{R}$. Since $Tx \in K \setminus \{0\}$ it follows that $\lambda = \frac{q(Tx)}{q(x)} > 0$. Therefore, $S\frac{x}{q(x)} = \alpha T^p x = \beta \frac{x}{q(x)}$ with $\alpha, \beta > 0$, which implies $\frac{x}{q(x)} = x^*$, that is, $x = rx^*$ with $r = q(x) > 0$.

(iii) From Lemma 5.1.6 (iii) it follows that T is ascending on $D = U$ with $\phi(\lambda) = \lambda + (1-\lambda)c$ with $c = \inf\{\frac{r_i}{s_i} \mid i \in I\} > 0$. As in the proof of (i) it follows that the conclusion of (i) applies to T. □

A common extension of the Perron–Frobenius theorem as well as of Jentzsch's theorem on integral operators to linear operators in infinite dimensions is the well-known Krein–Rutman theorem (see [28, 51]; for an elegant proof see [44]). This extension treats existence and properties of the dominant eigenvalue but it is not concerned with the "dynamical aspect", that is the convergence of the normalized iterates. The results presented above, although mainly directed towards dynamical features of operators, yield also variants of Jentzsch's theorem as well as of the Krein Rutman theorem (see Exercises 4 and 6 and Section 5.4); for a generalization of parts of the Krein–Rutman theorem to non-linear operators see [4, 35]).

Exercises

1. Consider the selfmapping T of $\mathbb{R}^2_+$ with component mappings given by (see Example 5.1.9 and [23])
$$T_1x = \max\left\{\min\left\{2x_1 + x_2, \frac{1}{2}x_1 + \frac{3}{2}x_2\right\}, \min\left\{2x_1 + \frac{1}{4}, x_1 + 4x_2\right\}\right\}$$
$$T_2x = \min\{x_1 + 2x_2, 1\}.$$
   (a) Obtain from Corollary 5.2.5 that the conditional eigenvalue problem
$$Tx = \lambda x, \lambda \in \mathbb{R}, x \in \mathbb{R}^2_+, \|x\| = |x_1| + |x_2| = 1$$
   has a unique solution $x = x^*$, $\lambda = \lambda^* > 0$ and that for scale $\|\cdot\|$
$$\lim_{n\to\infty} \tilde{T}^n x = x^* \text{ for all } x \in K \setminus \{0\}.$$
   (b) Compute x^* and λ^*.
   (c) What can be said about the asymptotic behavior of $\frac{T^n x}{\|T^n x\|}$?
2. A function $f: X \to \mathbb{R}$ on a topological space X is called **upper semicontinuous** if for every $a \in \mathbb{R}$ and $r \in \mathbb{R}$ with $f(a) < r$ there exists a neighborhood $U(a)$ of a such that $f(x) < r$ for all $x \in U(a)$. (f is **lower semicontinuous** if $-f$ is upper semicontinuous.)
   (a) Show that every upper semicontinuous function on a compact topological space attains its supremum on X.
   (b) Show that for two upper semicontinuous functions f and g on X with non-negative values the function $(x, y) \mapsto f(x)g(y)$ is upper semicontinuous on $X \times X$ (with the product topology).
   (c) Find two upper semicontinous functions on X for which the conclusion of (b) does not hold.
3. Consider a lineless convex cone K in some real vector space and a part $P \neq \{0\}$ of K with $V = P - P$.
   (a) Show that there exists a norm on V with respect to which each point of P is an interior point in V.
   (b) Find an example of a closed, convex, normal cone K in some normed real vector space and a part $P \neq \{0\}$ of K such that P has empty interior in V.
4. Obtain as a special case of the Theorem on relative stability for concave operators (Theorem 5.2.9 (i)) the following **Birkhoff–Jentzsch Theorem** (see [2, Theorem 3, p. 224], cf. also [22] and Example 5.1.7):

Let $(V, \|\cdot\|)$ be a Banach lattice with $K = V_+$ the cone of positive elements. Let T be a linear, bounded operator on V which maps K into itself and which is uniformly positive. Then the eigenvalue problem $Tx = \lambda x$ has a unique solution (x^*, λ^*) with $x^* \in K, \|x^*\| = 1, \lambda^* > 0$ and for every $x \in K \setminus \{0\}$ it holds that $\lim_{n\to\infty} \frac{T^n x}{\|T^n x\|} = x^*$.

5. Obtain as a special case of Theorem 5.2.9 (i) **Thompson's theorem** (see [47, Theorem 4.3.4, p. 83]; cf. also Example 5.1.7):
   Let (V, τ) be a locally convex vector space, K a convex cone which is normal and sequentially complete. Let T be a linear operator on V which maps K into itself and which is uniformly positive with e such that $s(e) = 1$ for a linear functional $s: K \to \mathbb{R}_+$, $s(x) > 0$ for $x \neq 0$. Then the eigenvalue problem $Tx = \lambda x$ has a unique solution (x^*, λ^*) with $x^* \in K$, $s(x^*) = 1$, $\lambda^* > 0$ and for every $x \in K \setminus \{0\}$ it holds that $\lim_{n\to\infty} \frac{T^n x}{s(T^n x)} = x^*$.
6. Obtain from Theorem 5.2.9 (ii) the following **variation of the Krein–Rutman theorem** (see also [51, Theorem 7.C, p. 290]):
   Let $(V, \|\cdot\|)$ be a real Banach space with a closed, convex, pointed cone with $\operatorname{int} K \neq \emptyset$. Let T be a linear, compact operator on V which maps K into itself. Suppose that T is primitive and **strictly positive** in the following sense (see [4, p. 51]), for $x_n \in K$, $\lim_{n\to\infty} Tx_n = 0$ implies that $\lim_{n\to\infty} x_n = 0$.
   Then T has exactly one eigenvector $x^* \in K$ with $\|x^*\| = 1$. The corresponding eigenvalue is $\lambda^* > 0$ and $x^* > 0$. Furthermore, for all $x \in K \setminus \{0\}$ it holds that $\lim_{n\to\infty} \frac{T^n x}{\|T^n x\|} = x^*$.

5.3 Absolute stability for weakly ascending operators by the part metric

Whereas in the last section we treated the relative stability for ascending operators we now turn to the absolute stability for weakly ascending operators. The latter means that, on a subset of the cone, the iterates of the mapping itself converge to a unique fixed point in this subset. Whereas in the previous section Hilbert's projective metric was the main tool, it is now the part metric we shall employ. Despite this difference we can proceed in what follows in many respects similarly to the previous section.

Definition 5.3.1. Let K be a convex cone in a real vector space and let T be a selfmapping of K. The **weak ascending domain** $W(T)$ of T consists of all points $x \in K$ such that

- there exists a non-zero part P_x of K;
- there exists a non-empty subset $W_x \subset K$ on which T is weakly ascending with some ϕ_x;
- there exists $n(x) \in \mathbb{N}$ such that for $M = \{T^n x \mid n \geq n(x)\}$ it holds that $M \subset P_x$ and $\overline{M} \subset W_x$ where $\overline{M}$ is the closure of M in the metric space (P_x, p), p the part metric.

The following result is similar in spirit to Theorem 5.2.4 on relative stability.

Theorem 5.3.2 (Absolute stability for weakly ascending operators). *Let K be a convex cone in a real vector space V and let T be a selfmapping of K with non-empty weak ascending domain $W(T)$.*

A. *Suppose that K is lineless and internally complete.*

(i) *Let $x \in W(T)$ for which ϕ_x has the property (P) to be upper semicontinuous or lower semicontinuous from the left on $]0,1[$ or to be a root function. Then the orbit $(T^n x)_{n\in\mathbb{N}}$ converges with respect to the part metric p to a unique fixed point $x^* = x^*(x)$ of T.*

(ii) *Let $x, y \in W(T)$ such that for some $k \geq 1$ both $T^n x$ and $T^n y$ belong for $n \geq k$ jointly to some part and their p-closures belong to a subset on which T is weakly ascending with a function ϕ_x which has property (P). Then it holds in (i) that $x^* = y^*$.*

(iii) *For every subset W of K on which T is weakly ascending with a ϕ that has property (P) and for every part P of K there exists at most one fixed point of T in $W \cap P$.*

B. *Let τ be a locally convex topology on the vector space V for which K is sequentially complete and normal. Then the above statements (i) to (iii) do hold where in (i) the orbit $(T^n x)_{n\in\mathbb{N}}$ converges to x^* also with respect to τ.*

Proof. Suppose that K is lineless and internally complete.

(i) Fix $x_0 \in W(T)$ as well as a non-zero part P of K and a non-empty subset $W \subset K$ and a mapping ϕ according to Definition 5.3.1. Let X be the closure of $M = \{T^n x_0 \mid n \geq n(x_0)\}$ in the complete metric space (P, p) (Theorem 3.2.3 (v)). Obviously, (X, p) is a comlete metric space contained in W. We shall show that T is a selfmapping of the metric space (X, p) which is a generalized contraction or (ϵ, δ)-contractive in the sense of Definition 4.1.1.

Suppose first that ϕ is upper semicontinuous. In this case we proceed similarly to the proof of Theorem 5.2.4. Let $0 < \alpha \leq \beta < \infty$ and consider $x, y \in X$ with $\alpha \leq p(x,y) \leq \beta$. Therefore, $e^{-\beta} \leq \min\{\lambda(x,y), \lambda(y,x)\} \leq e^{-\alpha}$. Consider λ satisfying $e^{-\beta} \leq \lambda \leq e^{-\alpha}$. By Lemma 3.1.4 (vi) we conclude that $\min\{\lambda(x,y), \lambda(y,x)\} = \sup\{\lambda > 0 \mid \lambda x \lneqq y \lneqq \frac{1}{\lambda} x\}$. Since T is weakly ascending on W with ϕ it follows that $\phi(\lambda) Tx \leq Ty \leq \frac{1}{\phi(\lambda)} Tx$ and, hence,

$$p(Tx, Ty) \leq -\log \phi(\lambda). \tag{5.3.1}$$

From ϕ upper semicontinuous on $I = \{r \mid e^{-\beta} \leq r \leq e^{-\alpha}\} \subset]0,1[$ and $\phi(r) > r$ for $r \in I$ we obtain that

$$\sigma(I) = \sup\left\{\frac{\log \phi(r)}{\log r} \mid r \in I\right\} < 1.$$

Obviously, this is true also for a root function ϕ (see Example 5.1.10). From (5.3.1) we have that

$$p(Tx, Ty) \leq -\log \phi(\lambda) \leq (-\log \lambda)\sigma(I)$$

and, by taking the infimum over λ, we arrive with $L(\alpha,\beta) = \sigma(I)$ at

$$p(Tx, Ty) \le L(\alpha,\beta)p(x,y). \tag{5.3.2}$$

In particular, T is p-continuous on X and, hence, $T(X) \subset X$. This shows that T is a generalized contraction on (X,p).

Next, consider the case that ϕ is lower semicontinuous from the left. For $\epsilon > 0$ given and $\lambda_0 = e^{-\epsilon}$, $\phi(\lambda_0) > \lambda_0$ and, hence, there exists $\delta > 0$ such that

$$\lambda_0 e^{-\delta} < \lambda \le \lambda_0 \quad \text{implies that} \quad \phi(\lambda) > \lambda_0.$$

Consider $x, y \in X$ with $\epsilon \le p(x,y) < \epsilon + \delta$ or, equivalently,

$$e^{-\epsilon}e^{-\delta} < \sup\left\{\lambda > 0 \mid \lambda x \le y \le \frac{1}{\lambda}x\right\} \le e^{-\epsilon}.$$

Therefore, there exists λ such that

$$\lambda_0 e^{-\delta} < \lambda \le \lambda_0 \quad \text{and} \quad \lambda x \lneqq y \lneqq \frac{1}{\lambda}x.$$

Since T is weakly ascending with ϕ we obtain that

$$\phi(\lambda)Tx \le Ty \le \frac{1}{\phi(\lambda)}Tx.$$

Putting together, we get

$$\min\{\lambda(Tx, Ty), \lambda(Ty, Tx)\} \ge \phi(\lambda) > \lambda_0 = e^{-\epsilon}$$

and, hence, $p(Tx, Ty) < \epsilon$. Thus, to any $\epsilon > 0$ given there exists $\delta > 0$ such that for all $x, y \in X$

$$\epsilon \le p(x,y) < \epsilon + \delta \quad \text{implies that} \quad p(Tx, Ty) < \epsilon. \tag{5.3.3}$$

In particular, T is contractive and, hence, T is p-continuous. Therefore, T is a selfmapping of (X,p) which is (ϵ,δ)-contractive.

In any case, from Theorem 4.1.4 we obtain that T has a unique fixed point x_0^* in X and $\lim_{n\to\infty} T^n x = x_0^*$ for all $x \in X$.

(ii) Let $x, y \in W(T)$ and $M = \{T^n x \mid n \ge k\} \cup \{T^n y \mid n \ge k\}$ with $X = \overline{M} \subset W \cap P$ where W and P are joint weak ascending set and joint part, respectively. As in (i) above, T has a unique fixed point z^* in X and $\lim_{n\to\infty} T^m z = z^*$ for all $z \in X$. In particular, $\lim_{m\to\infty} T^m x = z^* = \lim_{n\to\infty} T^m y$.

(iii) This follows directly from (ii).

(iv) Part **B** follows similarly as for Theorem 5.2.4. From Corollary 3.4.14 it follows that K is lineless and internally complete. Therefore, statements (i) to (iii) of part **A** do hold. It remains to show that in (i) for a sequence given by $x_n = T^n x$ convergence to x^* for p implies convergence for τ. From Proposition 3.3.3 (vi) (together with (ii)), for each of the monotone seminorms q which define τ it follows that

$$q(x-y) \le 3p(x,y)\max\{q(x), q(y)\} \quad \text{for} \quad x, y \in K \setminus \{0\}.$$

From $\lim_{n\to\infty} p(x_n, x^*) = 0$ it follows in particular that $p(x_n, x^*) < \frac{1}{2}$ for $n \geq N$ and, hence, $e^{-\frac{1}{2}} x_n \leq x^*$. Therefore,

$$q(x_n - x^*) \leq 3p(x_n, x^*) e^{\frac{1}{2}} q(x^*),$$

which implies that $\lim_{n\to\infty} x_n = x^*$ for τ. □

From Theorem 5.3.2 we easily obtain the following bundle of consequences.

Corollary 5.3.3. *Let (V, τ) be a locally convex vector space with a convex cone K that is sequentially complete and normal and let T be a selfmapping of K.*

(i) *Let W be an internally closed subset of a part of K and let T be weakly ascending on W with a ϕ that has property (P). Then each orbit of T which stays finally within W converges for τ to the unique fixed point of T in W.*

(ii) *Let P be a non-zero part of K and let T be weakly ascending on P with a ϕ that has property (P). Then T has a fixed point in P if and only if T sends some point of P into P. Furthermore, each fixed point of T in P is absolutely stable in P. In particular, if T is weakly ascending on K with a ϕ as above then each fixed point of T is absolutely stable in the part generated by it.*

(iii) *Let $[u, v]$ be a conical interval in a non-zero part of K on which T is uniformly concave with $\eta(a, b)$ depending continuously on a, b. Then each orbit of T which stays finally within $[u, v]$ converges for τ to the unique fixed point of T in $[u, v]$.*

(iv) *Let K^e be the part generated by $e \in K \setminus \{0\}$ and let T be e-monocave on K^e with $M(r, u, v)$ independent of u, v and continuous in r. Then K^e contains exactly one fixed point of T which is absolutely stable in K^e.*

(v) *Suppose there exists $0 \leq p < 1$ such that for all $x, y \in K$*

$$x \leq \alpha y \quad \text{and} \quad y \leq \beta x \quad \text{imply} \quad Tx \leq \alpha' Ty \quad \text{and} \quad Ty \leq \beta' Tx \tag{5.3.4}$$

where $\alpha, \beta, \alpha', \beta'$ are non-negative numbers with $\max\{\alpha', \beta'\} \leq \max\{\alpha^p, \beta^p\}$. Then each non-zero part of P for which T sends some point of P into P contains exactly one fixed point of T and this fixed point is absolutely stable in P. This conclusion holds in particular for T monotone and p-concave or T concave and homogeneous of degree p on K.

Proof. (i) Consider $x_0 \in K$, $x_0 \neq 0$ without loss, such that its orbit stays finally in W, i.e., $M = \{T^n x_0 \mid n \geq n(x_0)\} \subset W$ for some $n(x_0)$. By assumption on T the assertion of (i) follows from Theorem 5.3.2.

(ii) Let $W = P$ in (i). If $x^* \in P$ is a fixed point of T and $x \in P$ then $x \sim x^*$ which yields $Tx \sim Tx^* = x^*$ since T is weakly ascending. Therefore, the orbit of x belongs to P. The same holds for a point $x \in P$ with $Tx \in P$. Therefore, the assertion of (ii) follows from (i).

(iii) Follows for $W = [u, v]$ from (i) together with Example 5.1.9.

(iv) Since an e-monocave mapping leaves K^e invariant, the assertion follows for $W = K^e$ from (i) together with Example 5.1.9.

(v) Conditions (5.3.4) imply that T is weakly ascending on K with $\phi(\lambda) = \lambda^p$. Therefore, the assertions of (v) follow from (ii). □

Remarks 5.3.4. (i) In Corollary 5.3.3, set W in (i) and part P in (ii) need not contain a fixed point of T in case no orbit stays finally within W and no point of P is sent into P, respectively (see Exercise 2 (d)) For T ascending with a particular root function and T given by a mixed monotone operator, part (ii) can be found in [7, Theorem 3.1] (see Example 5.1.11 and Exercise 3).

(ii) For (V, τ) a Banach space, K a normal cone with $\overset{\circ}{K} \neq \emptyset$ and a conical interval in $\overset{\circ}{K}$ that is mapped by T into itself, part (iii) of the above Corollary can be found in [18, Theorem 10.3] and [20, Theorem 3.7], where however, a continuity assumption on η is missing.

(iii) Part (iv) of the above Corollary can be found essentially in [50, Theorem 5.7].

(iv) For normed vector spaces and mappings with (5.3.4) part (v) of the above Corollary can be found in [49, Theorem] (compare Example 5.1.10).

The conclusion in (v) concerning concave operators generalizes what we obtained earlier in Corollary 2.3.6 for finite dimensions. It also has the following consequence: If T sends some point of P into P then for *every* $\lambda > 0$ there exists a unique eigenvalue $x_\lambda \in P$, that is $Tx_\lambda = \lambda x_\lambda$. (For such a property concerning monotone and p-concave operators see [39, Theorem 3.5].)

From Corollary 5.3.3 together with a criterion for T to be weakly ascending we obtain the following result:

Corollary 5.3.5. *Let (V, τ) be a locally convex vector space with a convex cone K that is sequentially complete, normal, and with $\overset{\circ}{K} \neq \emptyset$. Let T be a selfmapping of K that is continuous on a convex and compact subset W of $\overset{\circ}{K}$ (for τ) such that for $x, y \in W$, $0 < \lambda < 1$,*

$$\lambda x \leq y \leq \frac{1}{\lambda} x \quad \textit{implies} \quad \lambda Tx < Ty < \frac{1}{\lambda} Tx. \tag{5.3.5}$$

Then each orbit of T that stays finally within W converges for τ to the unique fixed point of T in W.

Proof. We show that T is weakly ascending on W with ϕ upper semicontinuous. Since W is internally closed by Proposition 3.4.12 (v), the result then follows from Corollary 5.3.3 (i). For the function ϕ defined for $0 < \lambda < 1$ by

$$\phi(\lambda) = \inf\{\min\{\lambda(Tx, Ty), \lambda(Ty, Tx)\} \mid \lambda x \leq y \leq \frac{1}{\lambda} x \text{ for } x, y \in W\}$$

we proceed similar to the proof of Proposition 5.1.12. By Proposition 3.4.12 (iii), the order function $\lambda(\cdot, \cdot)$ is τ-continuous on $\overset{\circ}{K} \times \overset{\circ}{K}$ in $(V, \tau) \times (V, \tau)$. By the τ-continuity of T on W the mapping $\lambda(T\cdot, T\cdot)$ is τ-continuous on $W \times W$. Since $W_\lambda = \{(x, y) \in W \times W \mid \lambda x \leq y \leq \frac{1}{\lambda} x\}$ is compact there exist $x_\lambda, y_\lambda \in W$ with $\lambda x_\lambda \leq y_\lambda \leq \frac{1}{\lambda} x_\lambda$ and $\phi(\lambda) = \min\{\lambda(Tx_\lambda, Ty_\lambda), \lambda(Ty_\lambda, Tx_\lambda)\}$. Condition (5.3.5) yields $\lambda(Tx_\lambda, Ty_\lambda) > \lambda$ and

$\lambda(Ty_\lambda, Tx_\lambda) > \lambda$ and, hence, $\phi(\lambda) > \lambda$. Obviously, $\phi(\lambda)Tx \le Ty \le \frac{1}{\phi(\lambda)}Tx$ for $x, y \in W$, $0 < \lambda < 1$ with $\lambda x \le y \le \frac{1}{\lambda}x$. Thus, T is weakly ascending on W with ϕ.

To see that ϕ is upper semicontinuous let $\phi(\lambda_0) < \alpha$ for some $\lambda_0 \in]0, 1[$. We show that there exists $\epsilon > 0$ such that $\phi(\lambda) < \alpha$ for $\lambda \in]\lambda_0 - \epsilon, \lambda_0 + \epsilon[$. Obviously, ϕ is increasing and it suffices to show that $\phi(\lambda_0 + \epsilon) < \alpha$.

Since $\phi(\lambda_0) < \alpha$, there exist $x, y \in W$ with $\lambda_0 x \le y \le \frac{1}{\lambda_0}x$ and

$$-p(Tx, Ty) = \log \min\{\lambda(Tx, Ty), \lambda(Ty, Tx)\} < \log \alpha.$$

Choose $0 < \epsilon < \min\{\lambda_0(\alpha e^{p(x,y)} - 1), 1 - \lambda_0, \lambda_0\}$ and $\gamma = \frac{\epsilon}{1-\lambda_0}$. Since W is convex, $x' = x$ and $y' = \gamma x + (1-\gamma)y$ are in W and $(\lambda_0 + \epsilon)x' \le y'$. Now, similarly as in part (ii) of the proof of Proposition 5.1.12 (see Exercise 5) one has that

$$p(Tx', Ty') \ge p(Tx, Ty) - \log(\gamma e^{p(x,y)} + (1-\gamma)).$$

Since $\lambda_0 x \le y \le \frac{1}{\lambda_0}x$ it follows that $e^{p(x,y)} \le \frac{1}{\lambda_0}$ and, hence,

$$\gamma e^{p(x,y)} + (1-\gamma) \le \frac{\gamma}{\gamma_0} + (1-\gamma) \le 1 + \frac{\epsilon}{\lambda_0}.$$

This gives

$$p(Tx', Ty') \ge p(Tx, Ty) - \log\left(1 + \frac{\epsilon}{\lambda_0}\right)$$

and, because of $1 + \frac{\epsilon}{\lambda_0} < \alpha e^{p(Tx,Ty)}$ it follows that $p(Tx', Ty') > -\log \alpha$. Since $(\lambda_0 + \epsilon)x' \le y'$ this shows that $\phi(\lambda_0 + \epsilon) < \alpha$ □

Specializing to concave and to zigzag operators (see Example 5.1.8) we obtain the following result.

Corollary 5.3.6 (Absolute stability for concave and for zigzag operators). *Let (V, τ) be a locally convex vector space with a convex cone K that is sequentially complete, normal and with $\mathring{K} \ne \emptyset$.*

(i) *Let T be a concave selfmapping of K which maps $K \setminus \{0\}$ into $\mathring{K}$ and which maps continuously a non-empty convex, sectional, and compact subset $W \subset \mathring{K}$ into itself. Then T has a unique fixed point in W that is absolutely stable in W.*

(ii) *Let T be a selfmapping of K given by $Tx = Ax + a$, where A is a cone mapping and $a \in K$. Suppose there exists $b \in K \setminus \{0\}$ such that $Ab \le rb$ for some $0 < r < 1$ and that a is contained in the part generated by $b - Ab$. Then T has a unique fixed point in the part P generated by b that is absolutely stable in P.*

(iii) *Let S be a zigzag operator on K, that is, S is obtained by taking finitely many successive maxima or minima of finitely many affine operators $Tx = A_T x + a_T$ with A_T a linear selfmapping of K, $a_T \in K$. Suppose, there exists $b \in K \setminus \{0\}$ such that $A_T b \le r_T b$ for some $0 < r_T < 1$ for all T and a_T is contained in the part generated by $b - A_T b$. Then S has a unique fixed point in the part P generated by b that is absolutely stable in P.*

Proof. (i) For $x, y \in W$ and $0 < \lambda < 1$ we have for $\lambda x \leq y$ by assumption that $\lambda x + (1-\lambda)z = y$ with $z \in K \setminus \{0\}$. Since T is concave on K and $Tz \in \overset{\circ}{K}$ it follows that $\lambda Tx < \lambda Tx + (1-\lambda)Tz \leq Ty$. Thus T satisfies condition (5.3.4) of Corollary 5.3.3 and each orbit of T that finally stays within W converges to the unique fixed point. Because of $T(W) \subset W$, T has a unique fixed point which is absolute stable in W.

(ii) By assumption, there exists $1 < s$ with $\frac{1}{s}(b - Ab) \leq a \leq s(b - Ab)$. Fix such an s and consider $W = \{x \in K \mid \frac{1}{s}b \leq x \leq sb\}$. We show that T maps W into itself and is ascending on W with $\phi(\lambda) = \alpha + (1-\alpha)\lambda$, where $\alpha = \frac{1-r}{1-r+s^2 r}$. Since A is a cone mapping we have for $x \in W$ by assumption that

$$\frac{1}{s}b \leq \frac{1}{s}Ab + a \leq Ax + a \leq sAb + a \leq sb.$$

Therefore, $Tx = Ax + a$ maps W into itself. Furthermore, from the assumptions we have that

$$\frac{1-r}{rs}Ab = \frac{1}{s}\left(\frac{1}{r}Ab - Ab\right) \leq \frac{1}{s}(b - Ab) \leq a$$

and, hence, for $x \in W$

$$Ax \leq sAb \leq \frac{rs^2}{1-r}a = \frac{1-\alpha}{\alpha}a.$$

Let $\lambda x \leq y \leq \frac{1}{\lambda}x$ for $x, y \in W$ and $0 < \lambda < 1$. Since A is a cone mapping, we obtain

$$\phi(\lambda)Ax - Ay \leq \phi(\lambda)Ax - \lambda Ax = (\phi(\lambda) - \lambda)Ax = \alpha(1-\lambda)Ax \leq (1-\alpha)(1-\lambda)a.$$

Putting together,

$$\phi(\lambda)Tx = \phi(\lambda)(Ax + a) \leq Ay + (1-\alpha)(1-\lambda)a + \phi(\lambda)a = Ay + a = Ty.$$

Changing the roles of x and y we arrive at $\phi(\lambda)Tx \leq Ty \leq \frac{1}{\phi(\lambda)}Tx$, that is, T is ascending with ϕ on W.

Now, we can apply Corollary 5.3.3 (i). Let P be the part generated by b. Obviously, W is an internally closed subset of P. As seen above, T is ascending on W with ϕ, where $\phi(\lambda) > \lambda$ and ϕ is continuous. Since $T(W) \subset W$ it follows that T has a unique fixed point x^* in W which is aboslutely stable in W. Finally, consider $z \in P$ arbitrary and choose $s' = s(z)$ big enough such that $z \in \{x \in K \mid \frac{1}{s'}b \leq x \leq s'b\} = W', x^* \in W'$ and $\frac{1}{s'}(b - Ab) \leq a \leq s'(b - Ab)$. The above, when applied to W' yields that the orbit of z converges to x^*. Thus, x^* is the (unique) absolutely stable fixed point of T in P.

(iii) By assumption, there exists $1 < s$ such that $\frac{1}{s}(b - A_T b) \leq a_T \leq s(b - A_T b)$ for all T. Fix such an s and let $W = \{x \in K \mid \frac{1}{s}b \leq x \leq sb\}$. We show, S maps W into itself and is ascending on W with $\phi(\lambda) = \alpha + (1-\alpha)\lambda$, where $\alpha = \frac{1-r}{1-r+s^2 r}$ with $r = \max_T r_T < 1$. Since A_T is linear, for $x \in W$ we obtain

$$\frac{1}{s}b \leq \frac{1}{s}A_T b + a_T \leq A_T x + a_T \leq sA_T b + a_T \leq sb.$$

Therefore, $\frac{1}{s}b \leq Tx \leq sb$ for all T and by performing maxima or minima of the T's with respect to "$\leq$" (provided, they exist) we obtain that S maps W into itself. Obviously,

$A_T b \le rb$ for all T and we may continue for T fixed with A_T, a_T as in part (ii). First, we obtain for $x \in W$ that $A_T x \le \frac{1-\alpha}{\alpha} a_T$. Furthermore, since A_T is linear, it is a cone mapping and for $\lambda x \le y \le \frac{1}{\lambda} x$ with $x, y \in W$, $0 < \lambda < 1$ we obtain

$$\phi(\lambda) A_T - A_T y \le (1-\lambda)(1-\alpha) a_T$$

and, hence,

$$\phi(\lambda) Tx = \phi(\lambda)(A_T x + a_T) \le A_T y + a_T = Ty.$$

Changing roles of x and y we arrive at $\phi(\lambda) Tx \le Ty \le \frac{1}{\phi(\lambda)} Tx$ and, taking maxima and minima, we arrive at $\phi(\lambda) Sx \le Sy \le \frac{1}{\phi(\lambda)} Sx$, that is, S is ascending with ϕ on W. For P the part generated by b, W is an integrally closed subset of P and from Corollary 5.3.3 (i) we obtain that S has a unique fixed point x^* in W. For $z \in P$ arbitrary there exists $s' = s(z)$ big enough such that $z \in \{x \in K \mid \frac{1}{s'} b \le x \le s' b\} = W'$ and $\frac{1}{s'}(b - A_T b) \le a_T \le s'(b - A_T b)$. As above for W, it follows for W', too, that the orbit of z converges to x^*. Thus, x^* is the (unique) absolutely stable fixed point of S in P. □

Remarks 5.3.7. (i) Whereas Corollary 5.3.6 (i) yields an absolutely stable fixed point, Theorem 5.2.9 (i) only yields a relatively stable eigenvector. For example, in constrast to Theorem 5.2.9 (i) which covers arbitrary strictly positive matrices, Corollary 5.3.6 (i) addresses more specifically strictly positive (column-) stochastic matrices. More generally, let (V, τ) and K as in Corollary 5.3.6 and let T be a concave selfmapping of K as in Theorem 5.2.9 (i) with, in addition, $e \in \overset{\circ}{K}$, q a norm, T continuous with $T(U) \subset U$. It follows that $T(K \setminus \{0\}) \subset \overset{\circ}{K}$. Consider now the convex and sectional set $W = \{x \in U \mid re \le x \le se\}$. T maps W into itself and if W is compact then Corollary 5.3.5 (i) implies that T has an absolutely stable fixed point in W. Obviously, W is compact in finite dimensions which case can be generalized to reflexive locally convex spaces (V, τ) by considering for W the weak topology $\sigma(V, V')$ on V and requiring T to be weakly continuous. (For reflexive spaces and weakly compact sets see [13, 41]; see also Corollary 5.2.8.)

(ii) A mapping $Tx = Ax + a$ as in part (ii) of Corollary 5.3.6 appears in finite dimensions in connection with a **non-linear Leontief model**. There, $a > 0$ and it is assumed that A is a monotone and subhomogeneous selfmapping of K which is productive in the sense that $Ax_0 < x_0$ for some $x_0 \in K$. Then T has a unique fixed point which can be obtained by iteration (see [50, Section 6.3] and [34] for Leontief models in general). Obviously, Corollary 5.3.6 (ii) applies with $b = x_0$, but A is neither required to be monotone nor subhomogeneous. Furthermore, the space may be infinite dimensional, allowing non-linear Leontief models with infinitely many commodities (see [9] for those models). Moreover, part (iii) of Corollary 5.3.6 can be used to treat the choice of techniques for non-linear Leontief models (see also Sections 1.3 and 1.4 as well as Sections 2.6 and 2.7).

In proving the main result of this section, Theorem 5.3.2, it was essential that the mapping under consideration was a generalized contraction with respect to the part metric. This property is interesting already in one dimension in that it is strongly related to reproduction functions in population dynamics. More precisely, call a selfmapping of $\mathbb{R}_+$ with $f(x) > 0$ for $x > 0$ a **cave function** if $\frac{f(x)}{x}$ is strictly decreasing and $xf(x)$ is strictly increasing for $x > 0$. The first condition means population pressure (see Section 1.1). The second condition, which holds in particular for f increasing, means that the population does not decrease too fast. Special cases of cave functions are concave functions and **quasiconcave** functions, that is increasing functions for which $\frac{f(x)}{x}$ is strictly increasing (see [17, § 3] [18, § 46]). There are, however, relevant reproduction functions which are cave but neither concave nor quasiconcave (see the example given by (1.1.11). It turns out that a reproduction function is cave on a compact interval if and only if it is a generalized contraction for the part metric on that interval. (See Exercises 7, 8, 9 for cave functions).

Exercises

1. A function $f\colon O \to \mathbb{R}$ on a non-empty open subset O of $\mathbb{R}$ is **lower semicontinuous from the left** (from the right) in $a \in O$ if for every $r \in \mathbb{R}$ with $f(a) > r$ there exists $\epsilon > 0$ such that

$$f(x) > r \quad \text{for all} \quad x_0 - \epsilon < x \le x_0 \quad (x_0 \le x < x_0 + \epsilon).$$

   (a) Find a selfmapping ϕ of the open interval $]0,1[$ with $\phi(r) > r$ which is lower semicontinuous from the left but not from the right.
   (b) Find an open set $\emptyset \neq O \subset \mathbb{R}$, a function $f\colon O \to \mathbb{R}$ lower semicontinuous from the left and a compact subset of O on which f does not attain its (finite) infimum.

2. Let F be the Banach space of all bounded real valued functions on $\mathbb{N}$ equipped with norm $\|f\| = \sup\{|f(n)| \mid n \in \mathbb{N}\}$. Let $K = \{f \in F \mid f(n) \ge 0 \text{ for all } n \in \mathbb{N}\}$ and let T be the selfmapping of K given by $(Tf)(n) = \sqrt{f(n)}$.
   (a) Describe all parts of K by subsets of $\mathbb{N}$.
   (b) Show that T is weakly ascending on the whole of $K \setminus \{0\}$ with ϕ continuous. Show that each part of K contains exactly one fixed point of T which is absolutely stable in this part.
   (c) Let $\tau : \mathbb{N} \to \mathbb{N}$, $\tau(n) = n + 1$, and $S = T \circ \tau$. Show that S is weakly ascending with the same ϕ as T and find the fixed points of S.
   (d) Show that only the parts $\{0\}$ and $\operatorname{int} K$ contain each an absolutely stable fixed point of S and that no other part contains a point which is sent to it by S.

3. [7] Let V be a real Banach space containing a closed and normal convex cone K and let P be a non-zero part of K. An operator $A\colon P \times P \to P$ is a **mixed monotone operator** if $A(x, y)$ is monotone in x an antimonotone in y, i.e., $y \le y'$ implies $A(x, y') \le A(x, y)$. Let T be given by $Tx = A(x, x)$.

(a) Suppose that for each interval $[a,b] \subset]0,1[$ there exists $\alpha(a,b) \in]0,1[$ such that for all $x \in P, t \in [a,b]$ it holds that

$$A\left(tx, \frac{1}{t}x\right) \geq t^{\alpha(a,b)} A(x,x).$$

Show that T is a selfmapping of P which is (weakly) ascending on P with a root function.

(b) Obtain from Corollary 5.3.3 (ii) **Chen's Theorem** which states that T has a unique fixed point in P that is absolutely stable in P ([7, Theorem 3.1]).

4. Let (V, τ) be a locally convex vector space with a convex cone K that is sequentially complete, normal and with $\mathring{K} \neq \emptyset$. Let T be a selfmapping of K which maps continuously a non-empty convex, sectional and compact subset $W \subset \mathring{K}$ into itself. Show for T strongly monotone (monotone) on W and subhomogeneous (strongly subhomogeneous) on $K \setminus \{0\}$ that T has a unique fixed point in W that is absolutely stable in W.

5. Let T be a selfmapping of an arbitrary convex cone K in some vector space which has the property that for a convex subset $D \subset K$ it holds that $\lambda x \leq y$ implies $\lambda Tx \leq Ty$ for any $0 < \lambda < 1$, any $x, y \in D$. Then for $x, y \in D$ and $0 < \gamma \leq 1$ the following formula holds for the part metric

$$p(Tx, T(\gamma x + (1-\gamma)y)) \geq p(Tx, Ty) - \log(\gamma e^{p(x,y)} + (1-\gamma))$$

(compare Proposition 5.1.12).

6. (a) Find a selfmapping T of K as in Corollary 5.3.5 (ii), except that a is not contained in the part generated by $b - Ab$, and show that the conclusion of Corollary 5.3.5 (ii) does not hold.
   (b) Let $Tx = cx + d$ be a selfmapping of $\mathbb{R}_+$ with $0 < c < 1$ and $0 < d$. Compute for T the set W and the function ϕ as in the proof of Corollary 5.3.5 (ii).

7. Let $K = \mathbb{R}_+$, $P = \{x \in K \mid x > 0\}$ the non-zero part of K, f a continuous selfmapping of K which maps P into itself and I a non-empty compact interval in P.
   (a) Show that the following conditions are equivalent
      - f is a generalized contraction for the part metric P on I (i.e., Definition 4.1.1 (iii) applies to points in I).
      - f is cave on I, that is on I is $\frac{f(x)}{x}$ strictly decreasing and $xf(x)$ strictly increasing.
   (b) Show for f differentiable on P that f is cave if and only if the following condition holds for all $x \in I$

$$|x \cdot \frac{f'(x)}{f(x)}| < 1.$$

   (c) Show that f is ascending on I if and only if f is cave and increasing on I.
   (d) Let f differentiable with $|x \cdot \frac{f'(x)}{f(x)}| < 1$ on P and let $x^* \in P$ be a fixed point of f. Prove $\lim_{n\to\infty} f^n(x) = x^*$ for all $x \in P$.

8. For the set $\mathcal{C}$ of all cave functions show that
   (a) $\mathcal{C}$ is a convex cone.
   (b) $\mathcal{C}$ contains with two functions their pointwise maximum and minimum.
   (c) $\mathcal{C}$ contains all zigzag operators in one dimension.
9. (a) Determine the combinations of parameters $\lambda, a, b > 0$ for which $f(x) = \lambda x(1 + ax)^{-b}$ defines a cave function that is neither concave nor quasiconcave (see also Section 1.1).
   (b) Determine the combinations of parameters $a, b \geq 0$ with $a + b > 0$ and $c, d > 0$ for which $f(x) = \frac{a+bx^r}{c+dx^s}$ with $r, s \geq 0$ given defines a cave function.
   (c) Discuss, analytically and by computer simulations, the asymptotic behavior of the iterates of cave functions that are increasing, like $f(x) = \frac{3+5x^2}{4+4x^2}$, and of those that are not, like $f(x) = \frac{5x}{(1+x)^2}$.

5.4 Applications to non-linear difference equations and to non-linear integral operators

The results obtained on ascending operators enable us to go beyond the concave operators in finite dimensions as studied in Chapter 2. On the one hand we may consider in finite dimensions operators which are not necessarily concave. This we will illustrate by an application to difference equations which are not of the concave type as considered in Section 2.5. On the other hand we may consider concave operators in infinite dimensional function spaces. This we will illustrate by an application to integral operators of concave type. This yields at the same time a sharpening and an extension of Jentzsch's theorem on linear integral operators which itself carries over some of Perron–Frobenius theory to infinite dimensions.

First consider the difference equation

$$u(t + n) = f(u(t),\ u(t + 1), \dots, u(t + n - 1))$$

of order $n \geq 1$ with $u(t) \in \mathbb{R}_+$ for $t \in \mathbb{N}$ and $f\colon \mathbb{R}_+^n \longrightarrow \mathbb{R}_+$ with the associated characteristic equation

$$\lambda^n = f(1, \lambda, \lambda^2, \dots, \lambda^{n-1}) \quad \text{for} \quad \lambda \in \mathbb{R}_+.$$

Call f **increasing in component** i if $0 \leq x \leq y$ and $x_i < y_i$ implies $f(x) < f(y)$. From Corollary 5.2.5 we obtain the following

Theorem 5.4.1. *Suppose $f\colon \mathbb{R}_+^n \longrightarrow \mathbb{R}_+$ is continuous, positively homogeneous and there exists a set J of at least two increasing components such that $1 \in J$ and the numbers $n + 1 - j$, $j \in J$, are relatively prime. Then the characteristic equation has a unique strictly positive root λ^* and every solution $u(\cdot)$ of the difference equation with initial conditions $\bar{u}(u(0), \dots, u(n - 1))$ is relatively stable, i.e.,*

$$\lim_{t\to\infty} \frac{u(t)}{\lambda^{*t}} = c(\bar{u}),$$

where the function $c\colon \mathbb{R}_+^n \longrightarrow \mathbb{R}_+$ *is positively homogeneous and satisfies* $c(x) < c(y)$ *for* $x \lneq y$, $c(x_2,\dots,x_n,f(x)) = \lambda^* c(x)$ *and* $c(1,\lambda^*,\dots,\lambda^{*(n-1)}) = 1$.

Proof. Apply Corollary 5.2.5 to the Euclidean vector space $V = \mathbb{R}^n$ with cone $K = \mathbb{R}_+^n$. Obviously, $\overset{\circ}{K} \neq \emptyset$ and taking f the l_1-norm K has the compact base $B = \{x \in K \mid f(x) = 1\}$. Define $Tx = (x_2,\dots,x_n,f(x))$. T is a continuous selfmapping of K which is positively homogeneous. For $\bar{u} = (u(0),\dots,u(n-1))$ and $u(\cdot)$ a solution of the difference equation we have that $T^t\bar{u} = (u(t),\dots,u(t+n-1))$. By Lemma 2.5.4 from $\bar{u} \lneq \bar{v}$ we obtain that $u(t) \le v(t)$ for all t and $u(t) < v(t)$ for $t \ge t_0$. Therefore, for any two vectors $0 \le x \lneq y$ we have that $T^t x < T^t y$ for all $t \ge t_0$. This shows that T^{t_0} satisfies condition $(*)$ of Corollary 5.2.5. Since T^{t_0} is positively homogeneous we get that $T^{t_0}x = \lambda x$, $\lambda \ge 0$, $f(x) = 1$ has a unique solution $x = x^* \in \overset{\circ}{K}$, $\lambda = \mu$ and $\lim_{n\to\infty} \frac{T^{t_0 n}}{\mu^n} = c(x)x^*$ for $Tx \neq 0$. Denote $\lambda^* = \mu^{\frac{1}{t_0}}$.

Applying T^i we obtain

$$\lim_{n\to\infty} \frac{T^{t_0 n+i}}{\lambda^{*(t_0 n+i)}}\lambda^{*i} = c(x)T^i x^* = c(x)\lambda^{*i}x^*,$$

and, hence, $\lim_{m\to\infty} \frac{T^m x}{\lambda^{*m}} = c(x)x^*$. For this note that

$$T^{t_0}\left(\frac{Tx^*}{f(Tx^*)}\right) = \rho T^{t_0+1}x^* = \rho T(T^{t_0}x^*) = \rho T(\mu x^*) = \sigma \frac{Tx^*}{f(Tx^*)}$$

with certain scalars ρ and σ and, by uniqueness, $\frac{Tx^*}{f(Tx^*)} = x^*$. Actually, we must have $\lambda^* = f(Tx^*)$ and λ^* is unique. Finally, $Tx = \lambda x$ is equivalent to $x_2 = \lambda x_1, x_3 = \lambda x_2,\dots,x_n = \lambda x_{n-1}$, $f(x) = \lambda x_n$ which by positive homogeneity of f implies that $f(1,\lambda,\dots,\lambda^{n-1}) = \lambda^n$. Conversely, if this equation holds then $Tx = \lambda x$ for $x = (1,\lambda,\dots,\lambda^{n-1})$. Therefore, λ^* is the positive unique solution of the characteristic equation of f. The properties of $c(\cdot)$ carry over from those of T and the definition of T. □

The following examples illustrate the theorem and, different from the earlier Examples 2.5.7 none of them needs to be concave.

Examples 5.4.2. (i) Consider a difference equation with a right hand side f given as the maximum of finitely many linear functions

$$f(x) = \max_{1\le i\le m}(a_{i1}x_1 + \cdots + a_{in}x_n),$$

where the $m \times n$-matrix $A = (a_{ij})$ is non-negative with a set J of at least two strictly positive columns including the first one and such that the numbers $n+1-j$, $j \in J$, are relatively prime. Whereas Example 2.5.7 (i), given as the minimum of finitely many linear functions, exhibits a concave f, the f above is convex. Obviously, Theorem 5.4.1 does apply to this convex f and yields $\lim_{t\to\infty} \frac{u(t)}{\lambda^{*t}} = c(\bar{u})$, where $\lambda^* > 0$ is the unique root of the characteristic equation $\max_{1\le i\le m}(a_{i1} + a_{i2}\lambda + \cdots + a_{in}\lambda^{n-1}) = \lambda^n$.

If A is row-stochastic $\lambda = 1$ is a root and, hence, we must have that $\lambda^* = 1$. Thus all solutions with initial condition $\bar{u} \neq 0$ converge to a positive value.

(ii) Consider

$$f(x_1, \dots, x_n) = \left(\sum_{i \in J} x_i^p \right)^{\frac{1}{p}},$$

for $\emptyset \neq J \subseteq \{1, \dots, n\}$ and $p \neq 0$. For $J = \{1, \dots, n\}$ the value $\frac{1}{n^{\frac{1}{p}}} f(x_1, \dots, x_n)$ is sometimes called a power mean or a Hölder mean. For the same J but $p \geq 1$ $f(x)$ coincides on $\mathbb{R}^n_+$ with l_p-norm $\|x\|_p$. Therefore, f is convex on $\mathbb{R}^n_+$. For $p \geq 1$ but $J \neq \{1, \dots, n\}$ f is still convex but corresponds only to a semi-norm. For $p > 0$, $1 \in J$, $|J| \geq 2$ and such that the numbers $n + 1 - j$, $j \in J$, are relatively prime, f satisfies the assumptions of Theorem 5.4.1. Therefore, $\lim_{t \to \infty} \frac{u(t)}{\lambda^{*t}} = c(\bar{u})$ where $\lambda^* > 0$ is the unique root of the characteristic equation $\sum_{j \in J} \lambda^{p(j-1)} = \lambda^{pn}$. Equivalently, $\sum_{j \in J} \lambda^{-p(n+1-j)} = 1$. Therefore, we cannot have that $\lambda^* < 1$ and, since $|J| \geq 2$, we cannot have that $\lambda^* = 1$. Thus, $\lambda^* > 1$ and all non-zero solutions must tend exponentially to infinity.

(iii) The above two examples are special cases of the following "zigzag"-mapping

$$f(x) = \min_{p \in P} \max_{1 \leq i \leq m} \left(\sum_{j=1}^{n} a_{ij}(p) x_j^p \right)^{\frac{1}{p}}.$$

Thereby, P is a finite set of values $p > 0$ and $A(p)$ is a non-negative $m \times n$-matrix with a set J_p of at least two strictly positive columns, including the first one and such that the numbers $n + 1 - j$, $j \in J_p$ are relatively prime.

If the intersection of the J_p, $p \in P$, contains at least two elements then f satisfies the assumptions of Theorem 5.4.1. The study of solutions for these examples seems hopeless but Theorem 5.4.1 enables one to judge the asymptotic behavior by examining the characteristic equation whether $\lambda^* = 1$, $\lambda^* > 1$, $\lambda^* < 1$. Consider the special case $P = \{1, 2\}$, $m = n = 2$ and strictly positive 2×2-matrices $A(1) = (a_{ij}(1))$ and $A(2) = (a_{ij}(2))$. The characteristic equation then becomes

$$\min(r(\lambda), s(\lambda)) = \lambda^2$$

where

$$r(\lambda) = \max(a_{11}(1) + a_{12}(1)\lambda,\ a_{21}(1) + a_{22}(1)\lambda) \quad \text{and}$$
$$s(\lambda) = \max\left((a_{11}(2) + a_{12}(2)\lambda^2)^{\frac{1}{2}},\ (a_{21}(2) + a_{22}(2)\lambda^2)^{\frac{1}{2}} \right).$$

Depending on the matrices $A(1), A(2)$ all three cases for λ^* are possible (see Exercise 1 (c)).

Qualitatively, Theorem 5.4.1 provides conditions under which the asymptotic behavior of non-zero solution of the difference equation exhibits the following trichotomy: Either all solutions tend to infinity or all solutions converge to a positive value or all solutions converge to zero. As the Examples 5.4.2 illustrate one has to check whether the unique root of the characteristic equation is 1, above 1 or below 1. The characteristic equation may be given by a combination of minima and maxima of certain polynomials. In the next chapter we will investigate the phenomenon of limit set trichotomy for positive discrete dynamical systems in a more general way.

Next we consider eigenvalue problems for non-linear integral operators. Let X be a non-empty compact subset of $\mathbb{R}^m$, let V be the vector space $\mathcal{C}(X)$ of all real continuous functions f on X, equipped with norm $\|f\| = \sup\{|f(x)| \mid x \in X\}$. $(V, \|\cdot\|)$ is a Banach space and the convex cone K of all non-negative functions in V is sequentially complete and normal. Consider the following integral operator $T\colon V \longrightarrow V$ given by

$$(Tf)(u) = \int_X k(u,v)\phi(f(v))dv, \quad \text{for} \quad f \in V;\ u,v \in X \qquad (*)$$

where $k\colon X \times X \longrightarrow \mathbb{R}_+$ is a continuous and strictly positive kernel, $\phi : \mathbb{R}_+ \longrightarrow \mathbb{R}_+$ a continuous function and dv is the normalized Lebesgue measure on X, $\int_X dv = 1$.

A classical theorem by Jentzsch [14] states that for ϕ = identity the eigenvalue problem $Tf = \lambda f$ has a unique solution $f = f^* \in K$, $\|f^*\| = 1$, $\lambda = \lambda^* \geq 0$. Furthermore, f^* and λ^* are strictly positive and $|\lambda| < \lambda^*$ for every real eigenvalue $\lambda \neq \lambda^*$. As far as these statements are considered there are various generalizations of Jentzsch's Theorem to non-linear integral operators with ϕ a non-linear function. (See, e.g., [51] and the references given there.)

Since we want to obtain a non-linear version of Jentzsch's Theorem which gives also an iterative approximation of the unique eigenvector we have to make more restrictive assumptions on the non-linearity ϕ. The following result will be an application of Theorem 5.2.9.

Theorem 5.4.3 (Concave Jentzsch Theorem). (i) *Suppose for the integral operator T given by $(*)$ the function $\phi : \mathbb{R}_+ \longrightarrow \mathbb{R}_+$ is concave and satisfies for nonnegative numbers a and b with $a + b > 0$ the inequality $ar + b \leq \phi(r)$ for all $r \in \mathbb{R}_+$. Then the following conclusions hold for the conditional eigenvalue problem*

$$Tf = \lambda f \quad \text{with} \quad \lambda \in \mathbb{R},\ f \in K,\ \int f(v)dv = 1.$$

It has a unique solution $f = f^*, \lambda = \lambda^* \geq 0$; f^* and λ^* are strictly positive. Furthermore, for $\tilde{T}f = \frac{Tf}{\int f(v)dv}, f \in K \setminus \{0\}$, it holds that

$$\lim_{n\to\infty} \tilde{T}^n f = f^*.$$

(ii) *Let $\{T_i\}_{i\in I}$ be a family of integral operators given on K by*

$$(T_i f)(u) = \int_X k(u,v)(a_i f(v) + b_i)dv$$

with a_i, b_i non-negative and $a_i + b_i > 0$. Then the conclusions of (i) hold for the operators given by $(Tf)(u) = \inf\{(T_i f)(u) \mid i \in I\}$ and, provided it exists, $(Tf)(u) = \sup\{(T_i f)(u) \mid i \in I\}$.

Proof. (i) To apply Theorem 5.2.9 we choose as seminorm on V $q(f) = \int_X |f(v)|dv$ and $e \equiv 1$ on X. To show the required inequalities observe that the concavity of ϕ implies for $r \in \mathbb{R}_+$ that $\phi(1) \geq \frac{1}{1+r}\phi(r) + (1 - \frac{1}{1+r})\phi(1) \geq \frac{1}{1+r}\phi(r)$. Therefore, $\phi(r) \leq (1 + r)\phi(1)$ and

$$\int \phi(f(v))dv \leq \phi(1) \int (1 + f(v))dv = \phi(1)(1 + q(f)).$$

From this we obtain according to the definition of T by $(*)$

$$(Tf)(u) = \int k(u,v)\phi(f(v))dv \leq \max_{u,v\in X} k(u,v) \cdot 2\phi(1) \text{ for } q(f) = 1.$$

Thus $Tf \leq se$ with $s = 2\phi(1)\max_{u,v\in X} k(u,v)$.

The other inequality follows from $\phi(r) \geq ar + b$,

$$(Tf)(u) \geq \min_{u,v\in X} k(u,v) \int \phi(f(v))dv \geq \min_{u,v\in X} k(u,v)(a+b) \quad \text{for} \quad q(f) = 1.$$

That is, for $r = \min_{u,v\in X} k(u,v)(a+b)$, we have that $re \leq Tf \leq se$ for $q(f) = 1$ with $0 < r \leq s$. Thus the conclusions follow from part (i) of Theorem 5.2.9; observe that ϕ must be continuous on the interior of $\mathbb{R}_+$ and, hence, $Tf > 0$ for $f \in K \setminus \{0\}$.

(ii) The operator $(Tf)(u) = \int k(u,v)(af(v) + b)dv$ with $a + b > 0$ satisfies the assumptions in part (i). Actually, $re \leq Tf \leq se$ for $q(f) = 1$ with $r = (a+b)\min_{u,v\in X} k(u,v)$ and $s = (a+b)\max_{u,v\in X} k(u,v)$. Setting $a = a_i, b = b_i, r = r_i, s = s_i$ it follows that

$$\inf\left\{\frac{r_i}{s_i} \mid i \in I\right\} = \frac{\min k(u,v)}{\max k(u,v)} > 0$$

and the conclusion follows from part (iii) of Theorem 5.2.9. □

The following examples illustrate the theorem and connect it to other results.

Examples 5.4.4. (i) **The classical Jentzsch Theorem.** This is the special case of part (i) of Theorem 5.4.3 where $\phi(r) = r$. Since in this case T is positively homogeneous, properties additional to those of Theorem 5.4.3 are available. According to Theorem 5.2.9 it follows that

$$\lim_{n\to\infty} \frac{T^n f}{\lambda^{*n}} = c(f)f^* \quad \text{for} \quad f \in K \setminus \{0\} \quad \text{with} \quad c(f) > 0.$$

For $f = f_+ - f_-$ with $f_+, f_- \not\equiv 0$ one obtains that

$$\lim_{n\to\infty} \frac{T^n f}{\lambda^{*n}} = \lim_{n\to\infty} \frac{T^n f_+}{\lambda^{*n}} - \lim_{n\to\infty} \frac{T^n f_-}{\lambda^{*n}} = c(f_+)f^* - c(f_-)f^* = c(f)f^*.$$

From this for any norm $|\cdot|$ on V one has that

$$\lim_{n\to\infty} \frac{T^n f}{|T^n f|} = \sigma(f) f^* \quad \text{for all} \quad f \in V \setminus \{0\},$$

where $\sigma(f) \in \{0, +1, -1\}$ depending on $c(f_+) - c(f_-) = 0$ or > 0 or < 0.

These approximation properties are commonly not stated as part of Jentzsch's Theorem. From the approximation it follows in particular for $Tf = \lambda f, \lambda \in \mathbb{R}, f \in V \setminus \{0\}$

$$\lim_{n\to\infty} \left(\frac{\lambda}{\lambda^*}\right)^n f = c(f) f^*.$$

Therefore, we must have that $\lambda = \lambda^*$, and f proportional to f^*, or $|\lambda| < \lambda^*$.

An extension of Jentzsch's Theorem to linear operators on vector spaces has been developed by G. Birkhoff ([2]; see also [22]).

A "Generalized Jentzsch's Theorem" is proven in [51, Proposition 7.2.3, p. 289] which yields for certain non-linear functions ϕ, not necessarily concave, the existence of a solution of the conditional eigenvalue problem, without, however, an approximation property as in Theorem 5.4.3 (Cf. also Exercise 7 to Section 2.1.)

(ii) A particular case of part (i) of Theorem 5.4.3 is an affine version of Jentzsch's Theorem, that is $\phi(r) = ar + b$ with $a + b > 0$. Moreover, ϕ can be taken to be an infimum of affine functions, $\phi(r) = \inf\{a_i r + b_i \mid i \in I\}$ where $a_i, b_i \geq 0$ and $\inf_{i\in I} a_i + \inf_{i\in I} b_i > 0$. Indeed, any concave function $\phi : \mathbb{R}_+ \longrightarrow \mathbb{R}_+, \phi(r) \geq ar + b$ and $a + b > 0$, can be obtained in this way. In general, the operator T defined according to part (i) for such a ϕ is different from the infimum of operators according to part (ii). (See Exercise 4.)

Thus, parts (i) and (ii) represent two different kinds of a non-linear extension of Jentzsch's Theorem.

Exercises

1. Consider the zigzag-difference equation $u(t+1) = f(u(t), u(t+1)), t \in \mathbb{N}, u(t) \in \mathbb{R}_+$, where f is given by the (pointwise) minimum of the functions

$$f_1(x_1, x_2) = \max\{a_{11}(1)x_1 + a_{12}(1)x_2, a_{21}(1)x_1 + a_{22}(1)x_2\}$$
$$f_2(x_1, x_2) = \max\{[a_{11}(2)x_1^2 + a_{12}(2)x_2^2]^{\frac{1}{2}}, [a_{21}(2)x_1^2 + a_{22}(2)x_2^2]^{\frac{1}{2}}\}$$

with all $a_{ij}(k) > 0$.
(a) Verify the assumptions of Theorem 5.4.1 for the above difference equation.
(b) Show that the following dichotomy holds:
"Either all non-zero solutions $u(\cdot)$ are unbounded"
or
"$\lim_{t\to\infty} u(t)$ exists for all solutions".

(c) Discuss the possible roots of the characteristic equation $\lambda^2 = f(1,\lambda)$. Supply a numerical example for each of the cases $\lambda^* = 1, \lambda^* > 1, \lambda^* < 1$.

2. Prove your own zigzag-version of Jentzsch's Theorem by showing that the conclusions of Theorem 5.4.3 hold for the (pointwise) maximum of the operators

$$(T_i f)(u) = \min\left\{\int_X k(u,v)(a_{11}(i)f(v) + a_{12}(i))dv, \int_X k(u,v)(a_{21}(i)f(v) + a_{22}(i))dv\right\}$$

for $i = 1, 2$ and all $a_{rs}(i) > 0$.

3. Prove for any concave function $\phi : \mathbb{R}_+ \longrightarrow \mathbb{R}_+$:
(a) ϕ is continuous on the interior of $\mathbb{R}_+$.
(b) For each $\bar{r} > 0$ it holds that $\inf\limits_{0<r\leq\bar{r}} \frac{\phi(r)}{r} > 0$.
(c) The following conditions are equivalent:
(α) There exist $a, b \in \mathbb{R}_+$ with $a + b > 0$ such that $\phi(r) \geq ar + b$ for all $r \in \mathbb{R}_+$.
(β) $\phi(0) > 0$ or there exists $\bar{r} > 0$ such that $\inf\limits_{r\geq\bar{r}} \frac{\phi(r)}{r} > 0$.

4. Find X, $k(\cdot,\cdot)$, and positive numbers a_1, a_2, b_1, b_2 such that the operator defined by

$$(Tf)(u) = \int_X k(u,v) \min\{a_1 f(v) + b_1, a_2 f(v) + b_2\}dv$$

is different from the operator defined by

$$(T'f)(u) = \min\left\{\int_X k(u,v)(a_1 f(v) + b_1)dv, \int_X k(u,v)(a_2 f(v) + b_2)dv\right\}.$$

5. Let k: $[0,1] \times [0,1] \longrightarrow \mathbb{R}_+$ be a continuous and strictly positive kernel and let $\phi : \mathbb{R}_+ \longrightarrow \mathbb{R}_+$ be a continuous function.
(a) Prove the following special case of the "Generalized Jentzsch's Theorem" (see [51, Proposition 7.23, p.289]). Suppose there exist $r > 0$ and $\mu > 0$ such that $\phi(x) \geq \mu x$ for all $0 \leq x \leq r$. Then

$$\int_0^1 k(u,v)\phi(f(v))dv = \lambda f(u) \text{ for all } u \in [0,1] \qquad (*)$$

has a solution $\lambda > 0$ and $f \in \mathcal{C}_+[0,1]$ with $\|f\| = \rho$, for any given $0 < \rho \leq r$.
(b) Show that (a) applies for ϕ concave.
(c) Find a kernel $k(\cdot,\cdot)$ and a function ϕ with $\phi(x) \geq ax + b$ for $x \in \mathbb{R}_+$ ($0 \leq a, b$ and $a + b > 0$) such that the problem $(*)$ has a solution, which, however, is not unique. (According to Theorem 5.4.3, ϕ cannot be concave.) Conclude that the approximation property in Theorem 5.4.3 does not hold in this case.
(d) Find a kernel $k(\cdot,\cdot)$ and positive numbers a_1, a_2, b_1, b_2 such that the conclusions of part (i) of Theorem 5.4.3 hold for

$$(Tf)(u) = \max\left\{\int_0^1 k(u,v)(a_1 f(v) + b_1)dv, \int_0^1 k(u,v)(a_2 f(v) + b_2)dv\right\}$$

but not for

$$(T'f)(u) = \int_0^1 k(u,v) \max\{a_1 f(v) + b_1, a_2 f(v) + b_2\} dv.$$

Bibliography

[1] C. D. Aliprantis and O. Burkinshaw. *Positive Operators*. Academic Press, Inc., Orlando etc., 1985.

[2] G. Birkhoff. Extensions of Jentzsch's theorem. *Trans. Amer. Math. Soc.*, 85:219–227, 1957.

[3] G. Birkhoff. Uniformly semi-primitive multiplicative processes. *Trans. Amer. Math. Soc.*, 104:37–51, 1962.

[4] F. F. Bonsall. *Lecture Note on some Fixed Point Theorems of Functional Analysis*. Tata Institute of Fundamental Research, Bombay, 1962.

[5] P. J. Bushell. Hilbert's metric and positive contraction mappings in a Banach space. *Arch. Rational Mech. Anal.*, 52:330–338, 1973.

[6] P. J. Bushell. The Cayley–Hilbert metric and positive operators. *Lin. Alg. Appl.*, 84:271–280, 1986.

[7] Y.-Z. Chen. Thompson's metric and mixed monotone operators. *J. Math. Anal. Appl.*, 177:31–37, 1993.

[8] Y.-Z. Chen. Continuation method for α-sublinear mappings. *Proc. Amer. Math. Soc* 129:203–210, 2001.

[9] T. Fujimoto. Non-linear Leontief models in abstract spaces. *J. Math. Econ.*, 15:151–156, 1986.

[10] T. Fujimoto. A generalization of the Perron–Frobenius theorem to non-linear positive operators in a Banach space. *The Kagawa University Economic Review*, 59:485–495, 1986.

[11] T. Fujimoto and U. Krause. Strong ergodicity for strictly increasing non-linear operators. *Lin. Alg. Appl.*, 71:101–112, 1985.

[12] S. Gaubert and J. Gunawardena. The Perron–Frobenius Theorem for homogeneous, monotone functions. *Trans. Amer. Math. Soc.*, 356:4931–4950, 2004.

[13] R. B. Holmes. *Geometric Functional Analysis and its Applications*. Springer, New York etc., 1975.

[14] R. Jentzsch. Über Integralgleichungen mit positivem Kern. *J. Reine Angew. Math.*, 141:235–244, 1912.

[15] P. E. Kloeden and A. M. Rubinov. Attracting sets for increasing co-radiant and topical operators. *Mathematische Nachrichten*, 243:134–145, 2002.

[16] E. Kohlberg. The Perron–Frobenius theorem without additivity. *J. Math. Econ.*, 10:299–303, 1982.

[17] M. A. Krasnoselskii. *Positive Solutions to Operator Equations*. P. Noordhoff Ltd., Groningen, 1964.

[18] M. A. Krasnoselskii, V. S. Burd, and Y. S. Kolesov. *Nonlinear Almost Periodic Oscillations*. Wiley, New York, 1973.

[19] M. A. Krasnoselskii, J. A. Lifshits, and A. V. Sobolev. *Positive Linear Systems. The Method of Positive Operators*. Heldermann Verlag, Berlin, 1989.

[20] M. A. Krasnoselskii, G. M. Vainikko, P. A. Zabreiko, Y. B. Rutitskii, and V. Y. Stetsenko. *Approximate Solutions to Operator Equations*. Wolters–Noordhoff, Groningen, 1972.

[21] M. A. Krasnoselskii and P. P. Zabreiko. *Geometrical Methods of Nonlinear Analysis*. Springer-Verlag, Berlin, 1984.

[22] U. Krause. A non-linear extension of the Birkhoff–Jentzsch theorem. *J. Math. Anal. Appl.*, 114:552–568, 1986.

[23] U. Krause. Perron's stability theorem for non-linear mappings. *J. Math. Econ.*, 15:275–282, 1986.

[24] U. Krause. Relative stability for ascending and positively homogeneous operators on Banach spaces. *J. Math. Anal. Appl.*, 188:182–202, 1994.

[25] U. Krause. Positive non-linear systems: Some results and applications. In V. Lakshmikantham, editor, *World Congress of Nonlinear Analysts*, pp. 1529–1539. De Gruyter, Berlin, 1996.

[26] U. Krause. A local-global stability principle for discrete systems and difference equations. In B. Aulbach et al., editor, *Proceedings of the Sixth International Conference on Difference Equations*, Augsburg, 2001, forthcoming.

[27] U. Krause and P. Ranft. A limit set trichotomy for monotone non-linear dynamical systems. *Nonlinear Anal. TMA*, 19:375–392, 1992.

[28] M. G. Krein and M. A. Rutman. Linear operators leaving invariant a cone in a Banach space. *Amer. Math. Soc. Transl.*, 26:1–128, 1950. original in Russian 1948.

[29] B. Lemmens and R. Nussbaum. *Nonlinear Perron–Frobenius Theory.* Cambridge University Press, Cambridge, 2012.

[30] D. G. Luenberger. *Introduction to Dynamic Systems. Theory, Models, and Applications.* Wiley & Sons, New York, 1979.

[31] J. Mallet-Paret and R.D. Nussbaum. Eigenvalues for a class of homogeneous cone maps arising from max-plus operators. *Discrete and Cont. Dyn. Syst.*, 8:519–562, 2002.

[32] M. Morishima. *Equilibrium, Stability, and Growth.* Oxford University Press, Oxford, 1964.

[33] T. Nesemann. *Stability Behavior of Positive Nonlinear Systems with Applications to Economics.* PhD thesis, University of Bremen, Berlin, 1999. Wissenschaftlicher Verlag.

[34] H. Nikaido. *Convex Structures and Economic Theory.* Academic Press, New York, 1968.

[35] R. D. Nussbaum. Eigenvectors of non-linear positive operators and the linear Krein–Rutman theorem. In E. Fadell and G. Fournier, editors, *Fixed Point Theory*, pp. 309–336. Springer Verlag, New York etc., 1977.

[36] R. D. Nussbaum. Hilbert's projective metric and iterated non-linear maps. *Memoirs Amer. Math. Soc.*, 75(391):1–137, 1988.

[37] T. Ogiwara. Nonlinear Perron–Frobenius problem on an orderen Banach space. *Japan J. Math.*, 21:43–103, 1995.

[38] Y. Oshime. Non-linear Perron–Frobenius problem for weakly contractive transformations. *Math. Japonica*, 29:681–704, 1984.

[39] A. J. B. Potter. Applications of Hilbert's projective metric to certain classes of non-homogeneous operators. *Quart. J. Math.*, 28:93–99, 1977.

[40] H. Samelson. On the Perron–Frobenius theorem. *Michigan Math.*, 4:57–59, 1957.

[41] H. H. Schaefer. *Banach Lattices and Positive Operators.* Springer Verlag, Berlin etc., 1974.

[42] H. L. Smith. Cooperative systems of differential equations with concave non-linearities. *Nonlinear Anal.*, 10:1037–1052, 1986.

[43] H. L. Smith. *Monotone Dynamical Systems. An Introduction to the Theory of Competitive and Cooperative Systems.* AMS, Providence, 1995.

[44] P. Takáč. A short elementary proof of the Krein–Rutman theorem. Typoscript.

[45] P. Takáč. Asymptotic behavior of discrete-time semigroups of sublinear, strongly increasing mappings with applications to biology. *Nonlinear Anal.*, 14:35–42, 1990.

[46] H. R. Thieme. *Mathematics in Population Biology*. Princeton University Press, Princeton, 2003.

[47] A. C. Thompson. *Generalizations of the Perron–Frobenius theorem to operators mapping a cone into itself.* PhD thesis, University of Newcastle upon Tyne, 1963. 114 p.

[48] A. C. Thompson. On certain contraction mappings in a partially ordered vector spaces. *Proc. Am. Math. Soc.*, 14:438–443, 1963.

[49] A. C. Thompson. On the eigenvectors of some not-necessarily linear transformations. *Proc. London Math. Soc.*, 15:577–598, 1965.

[50] D. Weller. *Hilbert's metric, part metric and selfmappings of a cone*. PhD thesis, Universität Bremen, 1987. 99 p.

[51] E. Zeidler. *Nonlinear Functional Analysis and its Applications. Fixed Point Theorems*, volume I. Springer, New York etc., 1986.

6 Limit set trichotomy

For a positive system the orbits of different starting points show in general a very different convergence behavior. For example, one orbit may tend to infinity whereas another one tends to zero or still another one converges to a point in the interior of the underlying cone. This is true even in one dimension as exemplified by the mapping $Tx = x^2$. The situation is, however, completely different if the positive system is linear. In finite dimensions, the positive system defined by a primitive matrix shows a uniform behavior for all starting points in $\mathbb{R}^n_+ \setminus \{0\}$ that is either **all** orbits tend to infinity or **all** orbits tend to zero, or **all** orbits converge to a fixed point in the interior of $\mathbb{R}^n_+$. This property is called **limit set trichotomy.** (For the precise definition see Section 6.1) For a system given by a primitive matrix this property follows from classical Perron–Frobenius Theory, where the trichotomy is due to the three cases whether the dominant eigenvalue is greater or smaller or equal to 1 (see Theorem 2.4.1 (iii) (c)). More general, from concave Perron–Frobenius Theory a limit set trichotomy follows from the trichotomy of the dominant eigenvalue as above, provided that the concave selfmapping of $\mathbb{R}^n_+$ is primitive and positively homogeneous (see Theorem 2.3.1 (i)). For positive systems in infinite dimensions a limit set trichotomy can be inferred in a similar manner for concave and positively homogeneous operators which satisfy a certain boundedness condition (see Theorem 5.2.9 (i)).

In this chapter limit set trichotomy will be investigated for more general positive systems. As it turns out it plays a role whether, with respect to the part metric, the operator of the system is contractive, or – for a weaker form of limit set trichotomy – is non-expansive. A stylized picture of limit set trichotomy gives the following illustration in one dimension:

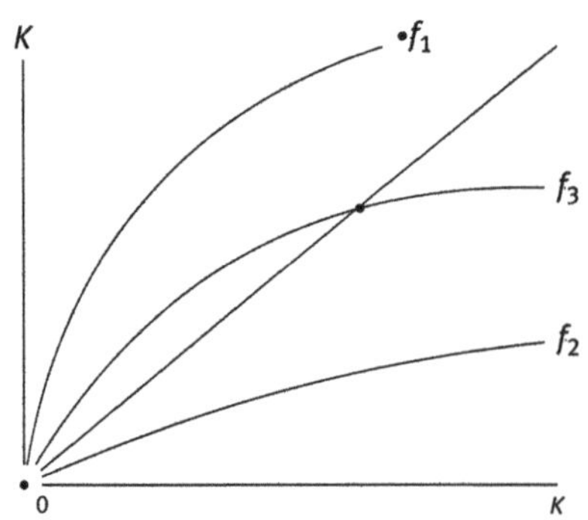

Fig. 6.1. Limit set trichotomy.

Mapping f_1 stands for the case that for all points in the interior the orbit tends to infinity; for example, $f_1(x) = x + \sqrt{x}$. For mapping f_2 all orbits converge to 0, as is the case with $f_2(x) = \frac{x}{1+x}$. For f_3 all orbits starting in the interior of K converge to a unique fixed point in the interior, as for $f_3(x) = \sqrt{x}$. It is easily seen, that in all three cases, the mapping is contractive for the part metric on the interior of $K = \mathbb{R}_+$.

The next section presents our main results on weak and strong versions of limit set trichotomy in Banach spaces. Limit set trichotomy made its first appearance in a paper by H. Smith [23] on cooperative systems of differential equations. Generalizations to non-linear positive systems were obtained in [15] for finite dimensions, in [16] for Banach spaces and in [26] for ordered topological cones. Further results including various extensions, have been obtained in [1, 20, 22]. Subsequently differentiability criteria will be developed to check whether a positive operator is contractive or non-expansive with respect to the part metric. These criteria will be useful when dealing with various applications to difference – and differential equations and models from biology.

6.1 Weak and strong forms of limit set trichotomy in Banach spaces

The first result on limit set trichotomy will be a weak form, which by strengthening the assumptions will lead us to a strong form as well as to other conclusions.

Theorem 6.1.1. *Let K be a closed convex cone in a Banach space $(E, \|\cdot\|)$ such that K is normal with non-empty interior $\mathring{K}$. Let T be a norm continuous selfmapping of K which maps $\mathring{K}$ into itself and which is non-expansive for the part metric p on $\mathring{K}$.*

A. *The following* **weak limit set trichotomy** *holds for T. Either*

(i) *for all $x \in \mathring{K}$ the orbit $O(x)$ is unbounded (for $\|\cdot\|$),*

or

(ii) *for all $x \in \mathring{K}$ the orbit $O(x)$ is bounded and the limit set $\omega(x)$ (for $\|\cdot\|$) is contained in the boundary of K,*

or

(iii) *for all $x \in \mathring{K}$ the orbit $O(x)$ is bounded and the following alternative applies: $\omega(x)$ is a singleton in $\mathring{K}$ or for each $y \in \omega(x) \cap \mathring{K}$ there exists $c(y) > 0$ such that*

$$p(T^{k+1}y, T^k y) = c(y) \text{ for all } k \geq 0. \tag{6.1.1}$$

Furthermore, $\omega(x_o) \cap \mathring{K} \neq \emptyset$ for at least one $x_o \in \mathring{K}$.

B. *If some iterate of T maps $K \setminus \{0\}$ into $\mathring{K}$ then case (i) above can be sharpened to*

(i′) *for all $x \in K \setminus \{0\}$ the orbit $O(x)$ is unbounded*
and case (ii) can be sharpened to

(ii′) *for all $x \in K \setminus \{0\}$ the orbit $O(x)$ is bounded and $\omega(x) = \emptyset$ or $\omega(x) = \{0\}$.*

If some iterate of T is contractive on $\mathring{K}$ for p then (iii) *can be sharpened to*

(iii′) *T has a fixed point $x^* \in \mathring{K}$ and $\omega(x) = \emptyset$ or $\omega(x) = \{x^*\}$ for all $x \in \mathring{K}$.*

Proof. (1) Since T is non-expansive on $(\overset{\circ}{K}, p)$, $p(T^n x, T^n y) \le p(x, y)$ for all $n \ge 0$, $x, y \in \overset{\circ}{K}$. By definition of p,

$$\lambda^{-1} T^n x \le T^n y \le \lambda T^n y \quad \text{for} \quad \lambda \ge \exp p(x, y). \tag{$*$}$$

Since K is normal we may assume $\|\cdot\|$ to be monotone which yields

$$\lambda^{-1} \|T^n x\| \le \|T^n y\| \le \lambda \|T^n x\|.$$

Therefore, either all orbits $O(x)$ for $x \in \overset{\circ}{K}$ are bounded or they are all unbounded.

(2) Assume that neither (i) nor (ii) hold. Then $O(x)$ is bounded for all $x \in \overset{\circ}{K}$ and there exists $x_0 \in \overset{\circ}{K}$ with $\omega(x_0) \cap \overset{\circ}{K} \neq \emptyset$. We shall show that for $x \in \overset{\circ}{K}$ the limit set $\omega'(x)$ of x in the metric space $(\overset{\circ}{K}, p)$ coincides with $\omega(x) \cap \overset{\circ}{K}$. For, if $y \in \omega(x) \cap \overset{\circ}{K}$ then $y = \lim_{k\to\infty} T^{n_k} x$ (for $\|\cdot\|$). Since $y \in \overset{\circ}{K}$ and by Proposition 3.4.12 the topologies induced by $\|\cdot\|$ and p do coincide on $\overset{\circ}{K}$, it follows that $y \in \omega'(x)$. For the same reason, $y \in \omega'(x)$ implies $y \in \omega(x) \cap \overset{\circ}{K}$. Since T is non-expansive on $(\overset{\circ}{K}, p)$ from Lemma 4.1.2 (a) we obtain that $\omega'(x)$ is a singleton or for each $y \in \omega'(x)$ there exists $c(y) > 0$ such that $p(T^{k+1} y, T^k y) = c(y)$ for all $k \ge 0$. This proves part A of the theorem.

(3) Considering part B, let S be an iterate of T with $S(K \setminus \{0\}) \subseteq \overset{\circ}{K}$. Case (i') is obvious. For case (ii') let $x \in K \setminus \{0\}$ and $y \in \omega(x)$, $y = \lim_{k\to\infty} T^{n_k} x$. It follows $Sy = \lim_{k\to\infty} T^{n_k} Sx \in \omega(Sx)$ and, by (ii), Sy must be contained in the boundary of K. Thus, $y \notin K \setminus \{0\}$, that is $\omega(x) = \emptyset$ or $\omega(x) = \{0\}$.

Considering case (iii') assume an iterate $S = T^m$ is contractive for p on $\overset{\circ}{K}$. First we show that for $x \in \overset{\circ}{K}$ with $\omega(x) \cap \overset{\circ}{K} \neq \emptyset$ we must have that $\omega(x)$ is a singleton in $\overset{\circ}{K}$. Otherwise, by (iii) we would have that $p(T^{k+1} y, T^k y) = c(y) > 0$ for all $y \in \omega(x) \cap \overset{\circ}{K}$ and $k \ge 0$. It follows that $Ty \in \omega(x) \cap \overset{\circ}{K}$ and

$$p(S(Ty), Sy) = p(T^{m+1} y, T^m y) = p(Ty, y) > 0.$$

But this contradicts the contractivity of S and we must have that $\omega(x)$ is singleton in $\overset{\circ}{K}$. By (iii), $\omega(x_0) \cap \overset{\circ}{K} \neq \emptyset$ for some $x_0 \in \overset{\circ}{K}$ and by the above $\omega(x) = \{x^*\}$ and x^* is a fixed point of T. For $x \in \overset{\circ}{K}$ arbitrary from $(*)$ in step (1) we have $\lambda^{-1} x^* \le T^n x \le \lambda x^*$ for some scalar λ and all n.

Therefore, $\omega(x) \subseteq \overset{\circ}{K}$. If $\omega(x) \neq \emptyset$ then $\omega(x)$ is a fixed point of T in $\overset{\circ}{K}$. By contractivity of S, x^* is the unique fixed point of T in $\overset{\circ}{K}$ and, hence, $\omega(x) = \{x^*\}$. □

This theorem yields immediately the following strong form

Corollary 6.1.2. *In the general setting of Theorem 6.1.1 assume that some iterate of T maps $K \setminus \{0\}$ into $\overset{\circ}{K}$ and some iterate of T is contractive on $\overset{\circ}{K}$ for the part metric. Suppose* *further that norm bounded orbits $O(x)$ for T and $x \in \overset{\circ}{K}$ have compact closure in the norm topology.*

Then the following **strong limit set trichotomy** *holds.*

Either

(i) *for all $x \in K \setminus \{0\}$ the orbit is unbounded (for $\|\cdot\|$),*

or

(ii) *for all $x \in K$, $\lim_{n\to\infty} T^n x = 0$ (for $\|\cdot\|$),*

or

(iii) *for all $x \in K \setminus \{0\}$, $\lim_{n\to\infty} T^n x = x^*$ (for $\|\cdot\|$), where $x^* \in \overset{\circ}{K}$ is the unique fixed point of T in $K \setminus \{0\}$.*

Proof. Since norm bounded orbits have compact closure it holds $\omega(x) \neq \emptyset$ for $x \in \overset{\circ}{K}$. Since $S(K \setminus \{0\}) \subseteq \overset{\circ}{K}$ for some iterate S of T it follows that $\omega(x) \neq \emptyset$ for all $x \in K \setminus \{0\}$. The strong limit set trichotomy then follows from the weak form together with part B of Theorem 6.1.1. □

Concerning the threefold alternative in a limit set trichotomy, in many examples and applications all three cases appear, depending on the values of parameters. (See Exercises 1, 7, 12.)

The following lemma collects some useful properties of mappings which are nonexpansive with respect to the part metric.

Lemma 6.1.3. *Let K be a lineless and archimedean convex cone in some real vector space. Denote by $\leq$ the ordering induced by K and by $N(P)$ the set of all non-expansive selfmappings on a part $P \neq \{0\}$ of K.*

(i) *$T \in N(P)$ if and only if for any $x, y \in P, \lambda \geq 1$*

$$\lambda^{-1}x \leq y \leq \lambda x \quad \textit{implies} \quad \lambda^{-1}Tx \leq Ty \leq \lambda Tx$$

or, equivalently, T maps intervals $[\lambda^{-1}x, \lambda x]$ into $[\lambda^{-1}Tx, \lambda Tx]$.

(ii) *$p(Tx, Ty) < p(x, y)$ holds for $x, y \in P, x \neq y$ if and only if for $\lambda \geq 1$ there exists $1 \leq \mu < \lambda$ such that*

$$\lambda^{-1}x \leq y \leq \lambda x \quad \textit{implies} \quad \mu^{-1}Tx \leq Ty \leq \mu Tx.$$

(iii) *$N(P)$ is a convex cone and the composition of two mappings in $N(P)$ is in $N(P)$ again. If T is a selfmapping of P, which is the restriction of a concave selfmapping of K, then $T \in N(P)$.*

(iv) *If $T \in N(P)$ then T is subhomogeneous on P, i.e., $T(\lambda x) \geq \lambda Tx$ for $x \in P$, $0 \leq \lambda \leq 1$. If T is subhomogeneous on P and T is monotone on P, then $T \in N(P)$.*

(v) *Suppose T has the following property, where $\leq_P$ denotes the ordering induced by the cone P:*
For $x, y \in P$, $\lambda^{-1}x \lneqq y \lneqq \lambda x$ implies $\lambda^{-1}Tx \leq_P Ty \leq_P \lambda Tx$. Then $T \in N(P)$ and $p(Tu, Tv) < p(u, v)$, provided $u \neq \lambda v$ for all $\lambda > 0$.

Proof. For $x, y \in P$ we have that $p(x,y) = \inf\{\log\lambda \mid \lambda^{-1}x \le y \le \lambda x, 1 \le \lambda\}$ and $p(x,y) = \log\lambda_0$ is a non-negative real number and $\lambda_0^{-1}x \le y \le \lambda_0 x$. (See Lemma 3.1.4, (vi), (xii).)

(i) Let $p(Tx, Ty) = \mu_0$. Suppose $\mu_0 \le \lambda_0$. If $\lambda^{-1}x \le y \le \lambda x$ then $\mu_0 \le \lambda_0 \le \lambda$ and $\mu_0^{-1}Tx \le Ty \le \mu_0 Tx$ implies $\lambda^{-1}Tx \le Ty \le \lambda Tx$. Conversely, $\lambda_0^{-1}x \le y \le \lambda_0 x$ implies $\lambda_0^{-1}Tx \le Ty \le \lambda_0 Tx$ and, hence, $\mu_0 \le \lambda_0$.

(ii) Similarly as above, let $\mu_0 < \lambda_0$ and $\epsilon = \lambda_0 - \mu_0$. If $\lambda^{-1}x \le y \le \lambda x$ then $\mu = \lambda_0 - \epsilon < \lambda$ and $\mu^{-1}Tx \le Ty \le \mu Tx$. Conversely, the implication in (ii) yields for $\lambda = \lambda_0$ some $1 \le \mu < \lambda$ such that $\mu^{-1}Tx \le Ty \le \mu Tx$. Therefore, $p(Tx, Ty) \le \mu < \lambda_0 = p(x, y)$.

(iii) For $S, T \in N(P)$ and $\alpha, \beta \in \mathbb{R}_+$ from $\lambda^{-1}x \le y \le \lambda x$ it follows by (i) that
$\lambda^{-1}(\alpha Sx + \beta Tx) \le \alpha Sy + \beta Ty \le \lambda(\alpha Sx + \beta Tx)$.
Therefore, by (i), $\alpha S + \beta T \in N(P)$. If $S, T \in N(P)$ then $\lambda^{-1}x \le y \le \lambda x$ implies that $\lambda^{-1}Tx \le Ty \le \lambda Tx$ which in turn implies $\lambda^{-1}S(Tx) \le S(Ty) \le \lambda S(Tx)$. Thus, $S \circ T \in N(P)$.

Finally, $\lambda^{-1}x \le y \le \lambda x$ implies for $\lambda > 1$ that $y = \lambda^{-1}x + (1 - \lambda^{-1})u$ and $x = \lambda^{-1}y + (1 - \lambda^{-1})v$ with $u, v \in K$. If T is a concave selfmapping of K then $Ty \ge \lambda^{-1}Tx$ and $Tx \ge \lambda^{-1}Ty$. This holds also for $\lambda = 1$, in which case $x = y$ and $Tx = Ty$. Thus, $T \in N(P)$.

(iv) Obviously, $\lambda^{-1}x \le y \le \lambda x$ holds for $y = \lambda^{-1}x, 1 \le \lambda$. Then for $T \in N(P), \lambda^{-1}Tx \le Ty \le T(\lambda^{-1}x)$. Therefore, T is subhomogeneous. Conversely, for T subhomogeneous and monotone, $\lambda^{-1}x \le y \le \lambda x$ implies that
$\lambda^{-1}Tx \le T(\lambda^{-1}x) \le Ty \le T(\lambda x)$ for $\lambda \ge 1$.
Since $\lambda^{-1}Tx \le T(\lambda^{-1}\lambda x) = Tx$ one obtains $\lambda^{-1}Tx \le Ty \le T(\lambda x) \le \lambda Tx$. Thus, $T \in N(P)$.

(v) From $\lambda^{-1}x \le y \le \lambda x$ it follows for $\epsilon > 0$ that $(\lambda + \epsilon)^{-1}x \lneqq y \lneqq (\lambda + \epsilon)x$ and, by assumption, $(\lambda + \epsilon)^{-1}Tx \le Ty \le (\lambda + \epsilon)Tx$. For ϵ converging to 0 this shows that $T \in N(P)$. Furthermore, let $p(u, v) = \log\lambda_0$ and $\lambda_0^{-1}u \le v \le \lambda_0 u$ for $u, v, \in P$. Assuming u and v not proportional, we must have $\lambda_0^{-1}u \lneqq v \lneqq \lambda_0 u$ and, hence, $\lambda_0^{-1}Tu \le_P Tv \le_P \lambda_0 Tu$. That is, $\lambda_0 Tu = Tv + p_1, \lambda_0 Tv = Tu + p_2$ with $p_1, p_2 \in P$. Since Tu, Tv, p_i are all in the same part P, there exists some $0 < \epsilon < \lambda_0$ such that $\epsilon Tu \le p_1$ and $\epsilon Tv \le p_2$. Therefore,

$$(\lambda_0 - \epsilon)Tu = Tv + p_1 - \epsilon Tu \ge Tv \quad \text{and} \quad (\lambda_0 - \epsilon)Tv = Tu + p_2 - \epsilon Tv \ge Tu.$$

It follows that $p(Tu, Tv) \le \log(\lambda_0 - \epsilon) < \log\lambda_0 = p(u, v)$. This proves (v) and the lemma. □

Remark 6.1.4. From Lemma 6.1.3 (i) it follows that for a selfmapping T of P that T is non-expansive if and only if T is a cone mapping on P in the sense of Definition 5.1.4. From (iii) and (iv) it follows in particular that a selfmapping of K which is concave or which is monotone and subhomogeneous is non-expansive on $(\mathring{K}, P)$ for $\mathring{K} \neq \emptyset$ and $T(\mathring{K}) \subseteq \mathring{K}$. In general, however, $T \in N(\mathring{K})$ need not possess these properties. This is true even in the most simple case of one dimension as the selfmapping $Tx = x + \frac{1}{1+x}$ of $K = \mathbb{R}_+$ shows. (See Exercise 4.) For further examples see the population model (1.1.11) in Section 1.1, the Exercises 7 and 9 to Section 5.3 and the population models in Section 7.6 including Exercise 11 to Chapter 7.

Using Lemma 6.1.3 we obtain from Theorem 6.1.1 the following form of the limit set trichotomy which is not covered by Corollary 6.1.2.

Theorem 6.1.5. *Let K be a closed convex cone in a Banach space $(E, \|\cdot\|)$ which is normal with non-empty interior $\overset{\circ}{K}$. Let T be a continuous selfmapping of K which maps $\overset{\circ}{K}$ into itself and for which bounded orbits $O(x)$, $x \in \overset{\circ}{K}$, have compact closure (for $\|\cdot\|$). Suppose T has the following property for any $x, y \in \overset{\circ}{K}$, $\lambda \geq 1$*

$$\lambda^{-1}x \leq y \leq \lambda x \quad \textit{implies} \quad \lambda^{-1}Tx \leq Ty \leq \lambda Tx \tag{6.1.2}$$

and for some iterate S of T

$$\lambda^{-1}x \lneqq y \lneqq \lambda x \quad \textit{implies} \quad \lambda^{-1}Sx < Sy < \lambda Sx$$

($\leq$, $<$ the orderings induced by K and $\overset{\circ}{K}$, respectively).

Assume further, T is monotone on rays, i.e., $T(rx) \leq T(sx)$ for $0 \leq r \leq s$, $x \in \overset{\circ}{K}$.
The following limit set trichotomy does hold (with respect to $\|\cdot\|$).
Either

(i) *for all $x \in \overset{\circ}{K}$ the orbit is unbounded,*

or

(ii) *for all $x \in \overset{\circ}{K}$ the orbit is bounded and $\omega(x) \neq \emptyset$ is contained in the boundary of K,*

or

(iii) *for all $x \in \overset{\circ}{K}$ the orbit is bounded and $\lim_{n\to\infty} T^n x = c(x)x^*$*

where $x^* \in \overset{\circ}{K}$ is a fixed point of T and $c(x) > 0$ a scalar.

If in addition $S(K \setminus \{0\}) \subseteq \overset{\circ}{K}$ then (i) and (iii) of the limit set trichotomy hold for $x \in K \setminus \{0\}$ and (ii) becomes
(ii′) *for all $x \in K$, $\lim_{n\to\infty} T^n x = 0$.*

Proof. The Theorem we obtain from Theorem 6.1.1, part **A.** To apply this theorem we need that T is non-expansive for the partmetric which follows from Lemma 6.1.3 (i) and the property (6.1.2). Thus, from Theorem 6.1.1 we have weak limit set trichotomy with cases (i) and (ii) as wanted. Concerning case (iii) of Theorem 6.1.1 suppose we have for some $y \in \overset{\circ}{K}$ that

$$p(T^{k+1}y, T^k y) = c(y) \quad \text{for all} \quad k \geq 0.$$

It follows with $S = T^m$ for all k

$$p(T^{k+1}y, T^k y) \leq p(Ty, y) = p(T^{m+k+1}y, T^{m+k}y) = p(S(T^{k+1}y), S(T^k y)) \leq p(T^{k+1}y, T^k y),$$

which implies $p(S(T^{k+1}y), S(T^k y)) = p(T^{k+1}y, T^k y)$.

Lemma 6.1.3 (v) applied to S and $P = \overset{\circ}{K}$ yields, due to property (6.1.2), that $T^{k+1}y = \lambda_k T^k y$ with $\lambda_k > 0$. It follows $T^k y = \mu_k y$ with $\mu_k > 0$ for $k \geq 0$.

In case of $\mu_k \leq \mu_{k+1}$ monotonicity of T on the ray $\mathbb{R}_+ y$ implies

$$\mu_{k+1}y = T^{k+1}y = T(T^k y) = T(\mu_k y) \leq T(\mu_{k+1}y) = T(T^{k+1}y) = \mu_{k+2}y$$

and, hence, $\mu_{k+1} \leq \mu_{k+2}$. Similarly, $\mu_k \geq \mu_{k+1}$ implies $\mu_{k+1} \geq \mu_{k+2}$. That is, the μ_k form a sequence in $\mathbb{R}_+$ which is increasing or decreasing. Since the orbit of y is bounded, the sequence $(\mu_k)_k$ must converge and, hence, $(T^k y)_k$ converges in the part topology to some $\bar{x} \in \mathring{K}$. Thus,

$$c(y) = \lim_{k\to\infty} p(T^{k+1}y, T^k y) = p(\bar{x}, \bar{x}) = 0.$$

From Theorem 6.1.1, part **A.** (iii) we obtain for $x \in \mathring{K}$ that $\omega(x)$ is a singleton or $\omega(x) \cap \mathring{K} = \emptyset$. Furthermore, $\omega(x_0) \cap \mathring{K} \neq \emptyset$ for some $x_0 \in \mathring{K}$ and, hence, $\omega(x_0) = \{x^*\}$ with $x^* \in \mathring{K}$ being a fixed point of T. For $x \in \mathring{K}$ arbitrary it holds that (see part (3) of the proof for Theorem 6.1.1) $\lambda^{-1}x^* \leq T^n x \leq \lambda x^*$ for some scalar λ and all n. Therefore, $\omega(x) \subseteq \mathring{K}$ and since $\omega(x) \neq \emptyset$ we conclude that $\omega(x)$ is a fixed point of T. If x_1, x_2 are two fixed points of T then $p(Sx_1, Sx_2) = p(x_1, x_2)$ and, by Lemma 6.1.3 (v), $x_2 = \lambda x_1$ with some scalar $\lambda > 0$. Putting together, $\omega(x) = \{c(x)x^*\}$ for $x \in \mathring{K}$ and,hence, $\lim_{n\to\infty} T^n x = c(x)x^*$ for all $x \in \mathring{K}$ with $c(x) > 0$ a scalar.

Finally, assume in addition $S(K \setminus \{0\}) \subseteq \mathring{K}$. The assertions made concerning (i) and (ii) hold trivially. Consider (iii), that is, for $x \in \mathring{K}, \omega(x) \neq \emptyset$ is contained in the boundary of K. If $0 \lneqq y \in \omega(x)$ then $Sy \in \mathring{K} \cap \omega(x)$ which is a contradiction. Therefore, $\omega(x) = \{0\}$ which implies $\lim_{n\to\infty} T^n x = \{0\}$ for $x \in \mathring{K}$. Since $T0 = 0$ and $S(K \setminus \{0\}) \subseteq \mathring{K}$ it follows that all orbits must converge to 0. □

Finally, we come back to the limit set trichotomy for concave (in particular linear) mappings as discussed earlier in the introduction and illustrated by the stylized picture. This time we do not require positive homogeneity and we do not need to assume finite dimensions. Moreover, we obtain a limit set trichotomy for monotone and subhomogeneous mappings.

Corollary 6.1.6. *With general assumption as in Theorem 6.1.5 let S be some iterate of T.*

(a) *Suppose T is monotone and subhomogeneous. The limit set trichotomy of Theorem 6.1.5 holds provided S is strictly monotone on $\mathring{K}$ (for $x, y \in \mathring{K}, x \lneqq y$ implies $Sx < Sy$) or S is strictly subhomogeneous on $\mathring{K}$ ($\alpha Sx < S(\alpha x)$ for $x \in \mathring{K}, 0 < \alpha < 1$). In the latter case, (iii) holds with $c(x) = 1$. If in addition $S(K \setminus \{0\}) \subseteq \mathring{K}$ then the following limit set trichotomy does hold.*
   Either
   (i) *for all $x \in K \setminus \{0\}$ the orbit is unbounded*
   or
   (ii) *for all $x \in K$, $\lim_{n\to\infty} T^n x = 0$*
   or

(a) *for all* $x \in K \setminus \{0\}$, $\lim_{n\to\infty} T^n x = c(x)x^*$

where $x^* \in \overset{\circ}{K}$ *is a fixed point of* T *and* $c(x) > 0$ *a scalar (with* $c(x) = 1$ *if* T *is strictly subhomogeneous).*

(b) *If* T *is concave, in particular* T *is linear, and* $S(K \setminus \{0\}) \subseteq \overset{\circ}{K}$ *then the above limit set trichotomy holds for* T.

Proof. (a) We show that property 6.1.2 is satisfied. The part for T holds since T is monotone and subhomogeneous. If S is strictly monotone on $\overset{\circ}{K}$ then $\lambda^{-1}x \lneqq y \lneqq \lambda x$ for $x, y \in \overset{\circ}{K}, \lambda \geq 1$ implies that $\lambda^{-1}Sx \leq S(\lambda^{-1}x) < Sy < S(\lambda x) \leq \lambda Sx$. If S is strictly subhomogeneous we obtain $\lambda^{-1}Sx < S(\lambda^{-1}x) \leq Sy \leq S(\lambda x) < \lambda Sx$. In both cases property (6.1.2) does hold. In the latter case, for (iii) we have $Sx^* = x^*, S(c(x)x^*) = c(x)x^*$ and for $c(x) < 1$ $S(c(x)x^* > c(x)Sx^* = c(x)x^*$ which is a contradiction. Similarly for $c(x) > 1$. Thus we must have $c(x) = 1$ if S is strictly subhomogeneous. This shows that (a) follows from Theorem 6.1.5.

(b) We show that property 6.1.2 is satisfied. The part for T holds since a concave selfmapping T of K is monotone and subhomogeneous.

If $\lambda^{-1}x \lneqq y$ for $x, y \in \overset{\circ}{K}$ and $\lambda > 1$ then $y = \lambda^{-1}x + (1-\lambda^{-1})\frac{z}{1-\lambda^{-1}}$ with $z \in K \setminus \{0\}$. Concavity of S gives

$$Sy \geq \lambda^{-1}Sx + (1-\lambda^{-1}S)\left(\frac{z}{1-\lambda^{-1}}\right) > \lambda^{-1}Sx$$

because of $S(K \setminus \{0\}) \subseteq \overset{\circ}{K}$. Similarly, from $y \lneqq \lambda x$ it follows $Sy < \lambda Sx$. The assumptions of Theorem 6.1.5 being satisfied this proves (b). □

As the stylized picture in the introduction illustrates, all three cases of a limit set trichotomy can occur even if the mapping is contractive or strongly concave. If, however, the mapping has a fixed point in the interior then only the third case survives and the fixed point is globally attractive.

This is a general feature as the following theorem will show. Furthermore, in the stylized picture the fixed point $x^* = 1$ of $f_3(x) = \sqrt{x}$ is locally attractive. By the local-global stability principle (Section 4.3) it is a general feature, too, that a locally attractive fixed point of a non-expansive mapping (for p) must be globally attractive.

Theorem 6.1.7. *Let* K *be a closed convex cone in a Banach space such that* K *is normal with non-empty interior* $\overset{\circ}{K}$. *Let* T *be a norm continuous selfmapping of* K *which maps* $\overset{\circ}{K}$ *into itself and the norm bounded orbits of which have compact closure in the norm topology. If* T *has a fixed point* $x^* \in \overset{\circ}{K}$ *then* x^* *is unique and globally attractive (within* $\overset{\circ}{K}$ *and with respect to the norm) in each of the following cases:*

(a) T *is non-expansive for the part metric on* $\overset{\circ}{K}$ *and* x^* *is locally attractive (for the norm on* $\overset{\circ}{K}$).

(b) *T is non-expansive and some iterate of T is contractive for the part metric on* $\mathring{K}$.
(c) *T is concave and some iterate of T is strongly concave on* $\mathring{K}$.

Proof. (a) By Proposition 3.4.12 the topologies induced by the norm $\|\cdot\|$ and the part metric p coincide on $\mathring{K}$. Therefore, the metric space $(\mathring{K}, p)$ is connected. Since T is non-expansive on $(\mathring{K}, p)$ it follows for this space from Corollary 4.3.6 that a locally attractive fixed point must be globally attractive. By Proposition 3.4.12 again the same is true with respect to the norm topology.

(b) Since $Tx^* = x^*$ only case (iii) in the weak limit set trichotomy of Theorem 6.1.1 survives. Since an iterate of T is contractive, case (iii') of part B of that theorem applies. Finally, $\omega(x) \neq \emptyset$ by the general assumptions made.

(c) The assertion follows from part B of Theorem 6.1.5. □

If in the setting of Theorem 6.1.5 the mapping T has a fixed point x^* in $\mathring{K}$ then only case (iii) survives and $\lim_{n\to\infty} T^n x = c(x)x^*, c(x) > 0$, for all $x \in \mathring{K}$. Also, if in the setting of Theorem 6.1.7 T has a fixed point x^* which is locally attractive then $\lim_{n\to\infty} T^n x = x^*$ for all $x \in \mathring{K}$. In what follows we shall more general place local conditions on the fixed point set of T to obtain global convergence of the iterates T^n to the fixed point set. For this we need the following lemma which uses an argument from the proof of Theorem 6.1.5.

Lemma 6.1.8. *Let K be a closed convex cone in a Banach space* $(E, \|\cdot\|)$ *which is normal with non-empty interior* $\mathring{K}$. *Let T be a non-expansive selfmapping of* $(\mathring{K}, p)$, *p the part metric. Suppose T is monotone on the ray* $R(y) = \{ry | r > 0\}$ *for a fixed point y of T. Then* $\lim_{n\to\infty} T^n x$ *exists in* $R(y)$ *for each* $x \in \mathring{K}$ *which has a non-empty and bounded limit set* $\omega(x)$ *in* $(\mathring{K}, p)$ *with* $\omega(x) \subseteq R(y)$.

Proof. For $z \in \omega(x)$ we have $T^k z \in \omega(x)$ and by assumption $T^k z = \mu_k y$ with $0 < r \le \mu_k \le s$ for all $k \ge 0$. Suppose $\mu_k \le \mu_{k+1}$ for some k. Since T is monotone on $R(y)$ it follows $\mu_{k+1} y = T(T^k z) = T(\mu_k y) \le T(\mu_{k+1} y) = T^{k+2} z = \mu_{k+2} y$ and, hence, $\mu_{k+1} \le \mu_{k+2}$. Therefore, (μ_k) is increasing in case of $\mu_0 \le \mu_1$. Similarly, (μ_k) is decreasing in case of $\mu_0 \ge \mu_1$. Since $0 < r \le \mu_k \le s$, $\mu = \lim_k \mu_k$ exists and $\mu > 0$. It follows $\lim_k T^k z = \mu y$ and, hence, $\lim_k p(T^k z, T^{k+1} z) = 0$. By Lemma 4.1.2 $\omega(x)$ must be a singleton and $\lim_n T^n x$ exists and belongs to $R(y)$. □

A fixed point of a non-expansive mapping need not be attractive as, for example, there might be points "rotating" around the fixed point. The result below provides a local condition on such rotating points which guarantees a globally attractive fixed point ray.

Definition 6.1.9. For a selfmapping f of a metric space (X, d) a point x is said to be **rotating around y** for a fixed point y of f if $d(f^m(x), y) = a > 0$ for all $m \ge 0$.

Theorem 6.1.10. *Let K be a closed convex cone in a Banach space* $(E, \|\cdot\|)$ *which is normal with non-empty interior* $\mathring{K}$. *Let T be a non-expansive selfmapping of* $(\mathring{K}, p)$ *with* *a fixed point $x^* \in \mathring{K}$ such that T is monotone on the ray $R(x^*)$. Supppose, there exists $\epsilon > 0$ such that for each fixed point y of T with $y \in R(x^*)$ and each $x \in \mathring{K}$ with $p(x, y) \le \epsilon$ the limit set $\omega(x)$ (for $\|\cdot\|$) is non-empty and has points rotating around y only in $R(y)$. Then $\lim_{n\to\infty} T^n x = c(x)x^*$ with $c(x) > 0$ for all $x \in \mathring{K}$.*

Proof. We shall show that the set F of all fixed points $y \in R(x^*)$ is locally attractive, that is

$$\lim_{n\to\infty} T^n x \in F \quad \text{for all} \quad x \in \mathring{K} \quad \text{with} \quad p(x, y) \le \epsilon \quad \text{for some} \quad y \in F. \tag{$*$}$$

Then the conclusion of the theorem follows from Theorem 4.3.7 since $(\mathring{K}, p)$ is connected and, hence, F is not strongly isolated.

To prove $(*)$ fix $\bar{y} \in F$ and $\bar{x} \in B = \{x \in \mathring{K} \mid p(x, \bar{y}) \le \epsilon\}$. Since T is non-expansive and $T\bar{y} = \bar{y}$, T maps B into itself. Therefore, $\emptyset \ne \omega(\bar{x}) \subseteq B$. In what follows we will use that the topologies for $\|\cdot\|$ and p coincide on $\mathring{K}$ (Proposition 3.4.12 (v)).

1. First we show that $p(T^m x, \bar{y}) = a$ for all $x \in \omega(\bar{x})$ and all $m \ge 0$. For $a_n = p(T^n \bar{x}, \bar{y})$ we have $a_{n+1} = p(T(T^n \bar{x}), T\bar{y}) \le p(T^n \bar{x}, \bar{y}) = a_n$ and, hence $a = \lim_k a_n$ exists. If $x \in \omega(\bar{x})$ then $x = \lim_k T^{n_k} \bar{x}$ (for $\|\cdot\|$ and p, too). Thus, for $m \ge 0$ $p(T^m x, \bar{y}) = \lim_k p(T^{m+n_k} \bar{x}, \bar{y}) = \lim_k a_{m+n_k} = a$.
2. By the above, $a = 0$ implies $x = \bar{y}$ for all $x \in \omega(\bar{x})$. If $a > 0$ then all points of $\omega(\bar{x})$ are rotating around $\bar{y}$. By the assumption made in the theorem we must then have that $\omega(\bar{x}) \subseteq R(\bar{y})$. Since T is monotone on $R(x^*)$ and, hence, on $R(\bar{y})$ it follows from Lemma 6.1.8 that $\lim_n T^n \bar{x}$ exists in $R(\bar{y})$. Since $\bar{y} \in F$ and $\bar{x} \in B$ where chosen arbitrarily this demonstrates $(*)$ and the conclusion of the theorem does follows. □

From this result we obtain the following corollary where the condition on rotating points is guaranteed by monotonicity assumptions on T.

Corollary 6.1.11. *Let K be a closed convex cone in a Banach space $(E, \|\cdot\|)$ which is normal with non-empty interior $\mathring{K}$. Let T be a selfmapping of $\mathring{K}$ which is compact for $\|\cdot\|$ and non-expansive for p with $Tx^* = \lambda^* x^*$ for some $x^* \in \mathring{K}$ and $\lambda^* > 0$. Consider for a selfmapping f of $\mathring{K}$ the following dual properties where $x, y \in \mathring{K}$, $m = m(x, y)$ and $\epsilon > 0$*

$$P_1(y, \epsilon) : p(x, y) \le \epsilon \quad \text{and} \quad x \lneqq y \quad \text{imply} \quad f^m(x) < f^m(y)$$
$$P_2(y, \epsilon) : p(x, y) \le \epsilon \quad \text{and} \quad y \lneqq x \quad \text{imply} \quad f^m(y) < f^m(x).$$

(a) *Assume T is monotone on $R(x^*)$ and subhomogeneous on $\mathring{K}$ and let $\hat{T}x = \frac{1}{\lambda^*} Tx$ on $\mathring{K}$.Suppose there exists $\epsilon > 0$ such that for all fixed points y of $\hat{T}$ in $R(x^*)$ both properties $P_1(y, \epsilon)$ and $P_2(y, \epsilon)$ hold for $f = \hat{T}$.*
Then $\lim_{n\to\infty} \hat{T}^n x = c(x)x^*$ with $c(x) > 0$ for all $x \in \mathring{K}$.

(b) *Assume T is monotone on $R(x^*)$ and positively homogeneous on $\overset{\circ}{K}$. Suppose for some $\epsilon > 0$ properties $P_1(x^*, \epsilon)$ and $P_2(x^*, \epsilon)$ hold for $f = T$. Then*

$$\lim_{n\to\infty} \frac{T^n x}{s(T^n x)} = x^* \quad \text{for all} \quad x \in \overset{\circ}{K}$$

where $s\colon \overset{\circ}{K} \to \mathbb{R}$ is any continuous, positively homogeneous mapping with $s(x) > 0$ on $\overset{\circ}{K}$ and $s(x^*) = 1$.

(c) *Assume T is monotone and positively homogeneous on $\overset{\circ}{K}$. Then the conclusion in (b) holds provided at least one of the properties $P_1(x^*, \epsilon)$ and $P_2(x^*, \epsilon)$ does hold for some $\epsilon > 0$ and $f = T$.*

Proof. (1) It is easily verified that for the part metric p and for any $\lambda > 0$ and $u, v \in \overset{\circ}{K}$ it holds that

$$p(\lambda u, v) \le -\log \min\left\{\lambda, \frac{1}{\lambda}\right\} + p(u, v). \tag{$*$}$$

This inequality implies for $p(u, v) \le \delta$, $\delta > 0$, and $e^{-\delta} \le \lambda \le e^{\delta}$ that $p(\lambda u, v) \le 2\delta$. Letting λ any of the values $\alpha = \lambda(u, v), \beta = \lambda(v, u), \mu = \min\{\alpha, \beta\}$ we obtain that $p(u, v) \le \delta$ implies $p(\lambda u, v) \le 2\delta$ and $p(\lambda^{-1} u, v) \le 2\delta$.

(2) Concerning (a) we have that $\hat{T}x^* = x^*$, $\hat{T}$ is a non-expansive selfmapping of $(\overset{\circ}{K}, p)$, $\hat{T}$ is monotone on $R(x^*)$ and $\hat{T}$ is a compact map. Let $\epsilon > 0$ according to properties $P_1(y, \epsilon), P_2(y, \epsilon)$. Fix $\bar{y} \in R(x^*)$ with $\hat{T}\bar{y} = \bar{y}$ and $\bar{x} \in \overset{\circ}{K}$ with $p(\bar{x}, \bar{y}) \le \frac{\epsilon}{2}$. Since $p(\hat{T}^n \bar{x}, \bar{y}) \le \frac{\epsilon}{2}$ for all n and $\hat{T}$ is compact it follows that $\omega(\bar{x}) \ne \emptyset$ for $\hat{T}$ keeping in mind that norm topology and part topology coincide on $\overset{\circ}{K}$. To apply Theorem 6.1.10 to $\hat{T}$ we shall show that for $x \in \omega(\bar{x})$ rotating around $\bar{y}$ we must have that $x \in R(\bar{y})$. By (1) we have for $\mu = \min\{\lambda(x, \bar{y}), \lambda(\bar{y}, x)\}$ because of $p(x, \bar{y}) \le \frac{\epsilon}{2}$ that $p(\mu x, \bar{y}) \le \epsilon$ and $p(\mu^{-1} x, \bar{y}) \le \epsilon$. Obviously, $\mu x \le \bar{y} \le \mu^{-1} x$ and suppose none of these inequalities is an equality. From properties $P_1(\bar{y}, \epsilon)$ and $P_2(\bar{y}, \epsilon)$ we obtain for $f = \hat{T}$

$f^m(\mu x) < f^m(\bar{y}) < f^m(\mu^{-1} x)$ and, since $\hat{T}$ is subhomogeneous, $\mu f^m(x) < f^m(\bar{y}) < \mu^{-1} f^m(x)$. Therefore, $\lambda(f^m(x), f^m(\bar{y})) > \mu$ and $\lambda(f^m(\bar{y}), f^m(x)) > \mu$ which implies that $p(f^m(x), f^m(\bar{y})) < p(\bar{x}, \bar{y})$. Thus, if $x \in \omega(\bar{x})$ is rotating around $\bar{y}$ (for $\hat{T}$), that is $p(\hat{T}^m x, \bar{y}) = a > 0$ for all $m \ge 0$, we must have that $\mu x = \bar{y}$ or $\bar{y} = \mu^{-1} x$ and, hence, $x \in R(\bar{y})$. From Theorem 6.1.10 we conclude that

$$\lim_n \hat{T}^n x = c(x) x^*, c(x) > 0, \quad \text{for all} \quad x \in \overset{\circ}{K}.$$

(3) Concerning (b), observe that $P_1(x^*, \epsilon), P_2(x^*, \epsilon)$ imply $P_1(y, \epsilon), P_2(y, \epsilon)$ for all $y \in R(x^*)$. Considering $P_1(x^*, \epsilon)$ note that $p(x, rx^*) \le \epsilon$ and $x \lneqq rx^*$ imply $p(\frac{1}{r}x, x^*) = p(x, rx^*) \le \epsilon$ and $\frac{1}{r}x \lneqq x^*$ as well as

$$f^m\left(\frac{1}{r}x\right) < f^m(x^*) \quad \text{implies} \quad f^m(x) < f^m(rx^*)$$

due to homogeneity of $f = T$. Similarly for $P_2(x^*, \epsilon)$. From part (a) we obtain $\lim_n \hat{T}^n x =$

$c(x)x^*, c(x) > 0$, for all $x \in \overset{\circ}{K}$. Since T is positively homogeneous we have that $\lim_n \frac{T^n x}{\lambda^{*n}} = c(x)x^*$ as well as $\lim_n s(\frac{T^n x}{\lambda^{*n}}) = c(x)$ which imply that

$$\lim_n \frac{T^n x}{s(T^n x)} = x^* \quad \text{for all} \quad x \in \overset{\circ}{K}.$$

(4) Concerning (c), we go back to the proof of Theorem 6.1.10, considering $\lambda(\cdot,\cdot)$ instead of $p(\cdot,\cdot)$. Let $\epsilon > 0$ as in property $P_1(x^*, \epsilon)$ or $P_2(x^*, \epsilon)$ with respect to $f = \hat{T}$. Fix $\bar{x} \in \overset{\circ}{K}$ with $p(\bar{x}, x^*) \le \frac{\epsilon}{2}$.

Let $a_n = \lambda(f^*(\bar{x}), x^*)$. Since T is monotone and f, too, it follows $a_{n+1} = \lambda(f(f^n(\bar{x}), x^*) \ge \lambda(f^n(\bar{x}), x^*) = a_n$. Therefore, (a_n) is increasing and because of $a_n \le e^\epsilon$ we have that $\alpha = \lim_n a_n$ exists. As in step (1) of the proof for Theorem 6.1.10 we conclude that $\lambda(f^m x, x^*) = \alpha$ for all $x \in \omega(\bar{x})$, all $m \ge 0$. Similarly, for $b_n = \lambda(x^*, f^n(\bar{x}))$ we conclude that $\lambda(x^*, f^m(x)) = \beta$ for all $x \in \omega(\bar{x}), m \ge 0$. In particular, $\alpha = \lambda(x, x^*)$ and $\alpha x \le x^*$ as well as $\beta = \lambda(x^*, x)$ and $\beta x^* \le x$. Since $p(\bar{x}, x^*) \le \frac{\epsilon}{2}$ we have $p(x, x^*) \le \frac{\epsilon}{2}$ for all $x \in \omega(\bar{x})$. From step (1) we obtain $p(\alpha x, x^*) \le \epsilon$ and $p(\beta^{-1}x, x^*) \le \epsilon$. Suppose property $P_1(x^*, \epsilon)$ applies. If $\alpha x \lneqq x^*$ then $\alpha f^m(x) = f^m(\alpha x) < x^*$ which yields that $\lambda(f^m x, x^*) > \alpha = \lambda(x, x^*)$ – a contradiction. Therefore, we must have $\alpha x = x^*$. In case property $P_2(x^*, \epsilon)$ applies and $\beta x^* \lneqq x$ it follows

$$\begin{aligned} cx^* &< f^m(\beta^{-1}x) = \beta^{-1} f^m(x) && \text{which yields that} \\ \lambda(x^*, f^m(x)) &> \beta = \lambda(x^*, x), && \text{a contradiction again,} \end{aligned}$$

and we must have $\beta x^* = x$. Thus, in any case $x \in \mathbb{R}(x^*)$ and, hence, $\omega(\bar{x}) \subseteq R(x^*)$. Since T and, hence, $f = \hat{T}$ is monotone on $\overset{\circ}{K}$ from Lemma 6.1.8 it follows $\lim_n f^n(\bar{x}) \in R(x^*)$ for $p(\bar{x}, x^*) \le \frac{\epsilon}{2}$. Finally, let $F = R(x^*)$ and $y \in F, x \in \overset{\circ}{K}$ with $p(x, y) \le \frac{\epsilon}{2}$. Since $y = rx^*$ it follows for $\bar{x} = \frac{1}{r}x$ that $p(\bar{x}, x^*) = p(x, rx^*) \le \frac{\epsilon}{2}$ and by the above $\lim_n f^n(x) = r\lim_n f^n(\bar{x}) \in R(x^*) = F$. This shows that F is locally attractive for $f = \hat{T}$. Theorem 4.3.7 yields that F is globally attractive and $\lim_n \hat{T}^n = c(x)x^*, c(x) > 0$, for all $x \in \overset{\circ}{K}$. The conclusion follows as in step (3) for (b). This proves parts (a), (b), (c) of the corollary. □

Remarks 6.1.12. The strong version of limit set trichotomy as in Corollary 6.1.2 was first obtained in [16, Theorem 3.1]. The first strong limit set trichotomy, in finite dimensions and for monotone mappings with a strongly subhomogeneous iterate (as in part B of Theorem 6.1.5), appears in [15, Corollary 1]. An extension of this result to Banach spaces can be found in [9, Theorem 5.20]. The case of a subhomogeneous mapping with a strongly monotone iterate (as in part B of Theorem 6.1.5) has been dealt with in [25, Theorem 1.1]; in a later paper the same author obtains a rather general trichotomy [26, Theorem 3.1], which admits also for 2-periodic points. This result requires an iterate to be ray-contractive, a condition which is connected to condition (6.1.1) of Theorem 6.1.1 and which, for T itself, is defined as follows: T is non-expansive on $(\overset{\circ}{K}, p)$ and $p(T^k x, T^k y) = p(x, y)$ does hold only for all $k \ge 1$ if $y = \lambda x$ for some $\lambda > 0$. See also

[3, Theorem 3.3] for a different proof of the above trichotomy. Independently, a similar result for ray-preserving mappings has been obtained in [27, Theorem 2.1]. Concerning Corollary 6.1.11 see also [17, Section 6.5, in particular Theorem 6.5.1, Theorem 6.5.6]. Various interesting extensions of a limit set trichotomy within different settings have been made in [20] to non-autonomous systems (see the next chapter), in [1] to random dynamical systems and in [22] to two-parameter semiflows on time scales.

6.2 Differentiability criteria for non-expansiveness and contractivity

For limit set trichotomy as in the last section the assumptions of non-expansiveness and contractivity, respectively were crucial. Concerning applications, as in the next two sections, it is very useful to check these assumptions by criteria in terms of differentiability. Throughout this section let the Banach space be $\mathbb{R}^n$ with a norm $\|\cdot\|$ which is monotone for the standard cone $K = \mathbb{R}^n_+$. For a self-mapping T of $\mathbb{R}^n$ denote by $\frac{\partial T_i}{\partial x_j}(x)$ the partial derivatives for $1 \le i, j \le n$ and by $J_T(x)$ the Jacobian, the $n \times n$-matrix of all partial derivatives. The following theorem describes non-expansiveness and contractivity in terms of the partial derivatives of T.

Theorem 6.2.1. *For $K = \mathbb{R}^n_+$ let $D \subseteq \overset{\circ}{K}$ be open and log-convex, that is $x^t y^{1-t} \in D$ (componentwise) for $x, y \in D, 0 \le t \le 1$. For T a continuously differentiable selfmapping of D let*

$$c(T) = \sup_{x\in D} \max_{1\le i\le n} \sum_{j=1}^{n} \frac{x_j}{T_i x} \left| \frac{\partial T_i}{\partial x_j}(x) \right| . \tag{6.2.1}$$

(i) *If $c(T)$ is finite, then it is the contraction constant of T on D for the part metric p, that is $c(T)$ is the smallest constant c such that $p(Tx, Ty) \le cp(x, y)$ for all $x, y \in D$.*
(ii) *If*

$$\sum_{j=1}^{n} \frac{x_j}{T_i x} \left| \frac{\partial T_i}{\partial x_j}(x) \right| < c$$

for all $1 \le i \le n$ and $x \in D$ then $p(Tx, Ty) < cp(x, y)$ for all $x, y \in D, x \ne y$.

Proof. (1) Fix $i \in \{1, \dots, n\}$ and consider the real function f defined by $f(u) = \log T_i(\exp u)$, where $\exp u$ is taken componentwise, $\exp u = (\exp u_1, \dots \exp u_n)$ and $u \in E = \{\log x | x \in D\}$, $\log x$ componentwise, too. f is continuously differentiable on the open set E with

$$\frac{\partial f}{\partial u_j}(u) = \frac{\exp u_j}{T_i(\exp u)} \frac{\partial T_i}{\partial x_j}(\exp u).$$

The mean value theorem yields on E which is convex since D is log-convex,

$$f(v) - f(u) = \int_0^1 \Big(\sum_{j=1}^n \frac{\partial f}{\partial u_j}(u(t))h_j \Big) dt \tag{6.2.2}$$

where $u, v \in E$, $h = v - u$, $u(t) = u + th, 0 \le t \le 1$.

(2) Since $\exp u \in D$ from the definition of $c(T)$ we have that $\sum_{j=1}^n |\frac{\partial f}{\partial u_j}(u)| \le c(T)$ and, hence,

$$|f(v) - f(u)| \le \int_0^1 \Big(\sum_{j=1}^n \Big| \frac{\partial f}{\partial u_j}(u(t)) \Big| \, |h_j| \Big) dt \le c(T) \max_j |v_j - u_j|.$$

For $x, y \in D$ given there exist $u, v \in E$ with $x = \exp u, y = \exp v$ and, hence, $f(u) = \log T_i x, f(v) = \log T_i y$. This yields $|\log T_i y - \log T_i x| \le c(T) \max_j |\log y_j - \log x_j|$. Since i was arbitrary chosen we arrive at $p(Tx, Ty) = \max_i |\log T_i x - \log T_i y| \le c(T)p(x, y)$ for all $x, y \in D$.

(3) Next we prove the existence of $x, y \in D, x \ne y$ with $p(Tx, Ty) \ge c(T)p(x, y)$. By definition of $c(T)$, to $0 < \epsilon < c(T)$ there exists some $x \in D$ and some $1 \le i \le n$ such that

$$c(T) - \epsilon < \sum_{j=1}^n \frac{x_j}{T_i x} \Big| \frac{\partial T_i}{\partial x_j}(x) \Big| .$$

Since T is continuously differentiable we may choose a neighbourhood $U = \{y \in D \mid \|\log y - \log x\| \le r\}$ of x such that the above inequality holds still in U and $\frac{\partial T_i}{\partial x_j}(z)$ does not change its sign for $z \in U$.

Define $h \in \mathbb{R}^n$ by

$$h_j = \begin{cases} r & \text{if } \frac{\partial T_i}{\partial x_j}(x) \ge 0 \\ -r & \text{if } \frac{\partial T_i}{\partial x_j}(x) < 0. \end{cases}$$

For $f(u) = \log T_i(\exp u)$ as in step (1) we have for $u = \log x$ that $\frac{\partial f}{\partial u_j}(u)h_j = \frac{x_j}{T_i x} \Big| \frac{\partial T_i}{\partial x_j}(x) \Big| \cdot r$. For $u(t) = u + th$ and $x(t) = e^{u(t)} = xe^{th}$ we have that $\|\log x(t) - \log x\| = t\|h\| \le r$ and, hence, $x(t) \in U$. Therefore,

$$\sum_{j=1}^n \frac{\partial f}{\partial u_j}(u(t))h_j = \sum_{j=1}^n \frac{x(t)_j}{T_i x(t)} \Big| \frac{\partial T_i}{\partial x_j}(x(t)) \Big| \cdot r > (c(T) - \epsilon) r$$

by the choice of U. The mean value theorem gives

$$f(u(1)) - f(u(0)) = \int_0^1 \Big(\sum_{j=1}^n \frac{\partial f}{\partial u_j}(u(t))h_j \Big) dt \ge (c(T) - \epsilon) \max_k |h_k|.$$

For $x = \exp u(0), y = \exp u(1)$ we obtain

$$\log T_i y - \log T_i x = f(u(1)) - f(u(0)) \ge (c(T) - \epsilon) \max_j |u_j(1) - u_j(0)|.$$

Since $\epsilon > 0$ was arbitrary chosen we arrive at

$$p(Tx, Ty) \geq \log T_i y - \log T_i x \geq c(T)p(x, y).$$

This proves part (i) of the theorem.

(4) For part (ii) of the theorem suppose that

$$\sum_{j=1}^{n} \frac{x_j}{T_i x} \left| \frac{\partial T_i}{\partial x_j}(x) \right| < c$$

for all i and all $x \in D$. Let for i fixed $f(u) = \log T_i(\exp u), u \in E$. Since $\frac{\partial f}{\partial u_j}(u) = \frac{\exp u_j}{T_i(\exp u)} \frac{\partial T_i}{\partial x_j}(\exp u)$ it follows that

$$\sum_{j=1}^{n} \left| \frac{\partial f}{\partial u_j}(u) \right| < c$$

for all $u \in E$. By the mean value theorem, therefore, as in step (2),

$$|f(v) - f(u)| \leq \int_0^1 \left(\sum_{j=1}^{n} \left| \frac{\partial f}{\partial u_j}(u(t)) \right| \, |h_j| \right) dt < c \max_j |v_j - u_j|$$

and, for $x = \exp u, v = \exp y$,

$$|\log T_i y - \log T_i x| < c \max_j |\log y_j - \log x_j|.$$

Thus $p(Tx, Ty) < cp(x, y)$ which proves part (ii). □

A first consequence of this Theorem is the following version of Corollary 6.1.2 in terms of differentiability.

Corollary 6.2.2. *Let $K = \mathbb{R}^n_+$ and T a continuous selfmapping of K which is a continuously differentiable selfmapping of $\overset{\circ}{K}$ such that*

$$\sum_{j=1}^{n} x_j \left| \frac{\partial T_i}{\partial x_j}(x) \right| \leq T_i x$$

for all $1 \leq i \leq n$ and $x \in \overset{\circ}{K}$.

Suppose further that some iterate of T maps $K \setminus \{0\}$ into $\overset{\circ}{K}$ and that some iterate S of T satisfies

$$\sum_{j=1}^{n} x_j \left| \frac{\partial S_i}{\partial x_j}(x) \right| < S_i x$$

for all $1 \leq i \leq n$ and $x \in \overset{\circ}{K}$.

Then strong limit set trichotomy (as in Corollary 6.1.2) holds.

Proof. Theorem 6.2.1 yields for $D = \overset{\circ}{K}$ that, both, T is non-expansive and S is contractive on $(\overset{\circ}{K}, p)$. Since in K norm bounded orbits have compact closure, the conclusion follows from Corollary 6.1.2. □

It should be noted that the assumptions in the above corollary do not imply monotonicity. This is relevant even in one dimension. The following figure illustrates for this case the conditions in the above corollary.

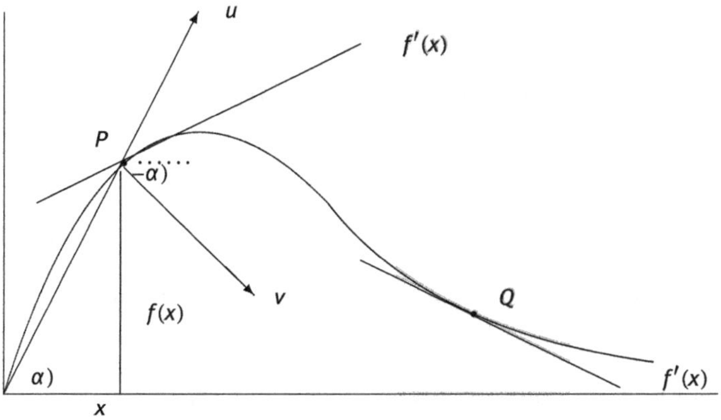

Fig. 6.2. Cave function.

At any point P the tangent must be contained in the convex cone (or its interior) spanned by u and v. At P, $f'(x) \geq 0$ and $f'(x) \leq \frac{f(x)}{x}$; at Q, $f'(x) \leq 0$ and $\frac{-f(x)}{x} \leq f'(x)$. One might say that, when increasing (decreasing), f must not increase (decrease) too much. The condition $x|f'(x)| < f(x)$ for all $x > 0$ is equivalent to f being a cave function, that is, $\frac{f(x)}{x}$ is strictly decreasing and $xf(x)$ is strictly increasing. (See the last paragraph of Section 5.3 and Exercise 7 (b). See also Figure 1.3 and the example discussed there.) The above geometric interpretation applies similarly in higher dimensions. The (positive) tangent space must be contained for $P = (x, Tx) \in \mathbb{R}_+^{2n}$ in $P + C$ where

$$C = \left\{(u, v) \in \mathbb{R}_+^n \times \mathbb{R}^n \mid \max_{1\leq i\leq n} |v_i| \leq c \sum_{i=1}^n u_i \right\}$$

is a convex polyhedral cone with $c = \max\limits_{1\leq i,j\leq n} \frac{T_i x}{x_j}$ (see Exercise 5).

Though Corollary 6.2.2 is useful in many cases, it is not applicable in many others. For example, if T is linear, $T_i x = \sum_{j=1}^n a_{ij} x_j$ with $A = (a_{ij}) \geq 0$, one hase that for each iterate $Sx = A^m x$ equality $\sum_{j=1}^n x_j |\frac{\partial S_i}{\partial x_j}(x)| = S_i x$ holds for all i. From Theorem 6.1.5 part B, however, we conclude that limit set trichotomy holds in this case if A is primitive. In what follows, therefore, we will improve the criterion given by Corollary 6.2.2. This will be achieved by examining more closely the condition (6.1.1) for the weak limit set trichotomy in Theorem 6.1.1. To do so the following lemma proves to be crucial.

Lemma 6.2.3. *For* $K = \mathbb{R}^n_+$ *let* $f : \overset{\circ}{K} \longrightarrow \operatorname{int} \mathbb{R}_+$ *be continuously differentiable and let, for* $x, y \in \overset{\circ}{K}$, $\langle x, y\rangle = \{z \in \overset{\circ}{K} \mid z_j = x_j^t \cdot y_j^{1-t},\ t \in [0,1],\ 1 \le j \le n\}$. *Assume for all* $z \in \overset{\circ}{K}$

$$\sum_{j=1}^{n} z_j \left|\frac{\partial f}{\partial x_j}(z)\right| \le f(z), \tag{6.2.3}$$

and let for $z \in \overset{\circ}{K}$

$$J_+(z) = \left\{j \,\middle|\, \frac{\partial f}{\partial x_j}(z) > 0\right\}, J_-(z) = \left\{j \,\middle|\, \frac{\partial f}{\partial x_j}(z) < 0\right\}.$$

If for $x \neq y$ *given it holds that*

$$|\log f(x) - \log f(y)| \le p(x, y), \tag{6.2.4}$$

then in (6.2.3) *equality holds for all* $z \in \langle x, y\rangle$ *and there exists some* $\lambda > 0$ *such that for each* $z \in \langle x, y\rangle$

$$x_j = \lambda y_j \text{ for } j \in J_+(z) \quad \text{and} \quad x_j = \lambda^{-1} y_j \quad \text{for} \quad j \in J_-(z)$$

or

$$x_j = \lambda^{-1} y_j \quad \text{for} \quad j \in J_+(z) \quad \text{and} \quad x_j = \lambda y_j \quad \text{for} \quad j \in J_-(z).$$

Proof. As in step (1)for the proof of Theorem 6.2.1 consider $g(u) = \log f(\exp u)$ for $D = \overset{\circ}{K}$ and $u \in \mathbb{R}^n$, where $\exp u$ is taken componentwise. For x, y there exist $v, w \in \mathbb{R}^n$ such that $x = \exp v, y = \exp w$. By the mean value theorem

$$g(v) - g(w) = \int_0^1 \left(\sum_{j=1}^{n} \frac{\partial g}{\partial u_j}(v(t))h_j\right) dt,$$

where $h = v - w, v(t) = w + th,\ t \in [0,1]$. For $z(t) = \exp v(t)$ we have that $z_j(t) = (\exp v_j)^t \cdot (\exp w_j)^{1-t} = x_j^t \cdot y_j^{1-t}$ and, hence, $z(t) \in \langle x, y\rangle$. For $a_j(t) = \frac{\partial g}{\partial u_j}(v(t))$ we have that $a_j(t) = \frac{z_j(t)}{f(z(t))}\frac{\partial f}{\partial x_j}(z(t))$. From assumption 6.2.3 if follows that $\sum_{j=1}^n |a_j(t)| \le 1$ for all $t \in [0,1]$. For $m = \max_j |h_j| > 0$ we have that

$$m = \max_j |v_j - w_j| = \max_j |\log x_j - \log y_j| = p(x, y).$$

From the assumption 6.2.4 together with the mean value theorem we obtain

$$m \le |\log f(x) - \log f(y)| = |g(v) - g(u)| \le \int_0^1 \left|\sum_j a_j(t)h_j\right| dt.$$

Since

$$\int_0^1 \Big|\sum_j a_j(t)h_j\Big| dt \le \int_0^1 \Big(\sum_j |a_j(t)| \cdot |h_j|\Big) dt \le m,$$

we conclude by the continuity of $a_j(t)$ that

$$\sum_j |a_j(t)| = 1 \text{ and } \Big|\sum_j a_j(t)h_j\Big| = m \sum_j |a_j(t)| \text{ for all } t \in [0,1].$$

The first formula means that in 6.2.3 we must have equality for all $z \in \langle x, y\rangle$. As for the second formula suppose first that $\sum_j a_j(t)h_j \ge 0$ for some $t \in [0,1]$. Then

$$\sum_j \big(|a_j(t)|m - a_j(t)h_j\big) = 0 \quad \text{and, hence,} \quad |a_j(t)|m = a_j(t)h_j \quad \text{for all} \quad j.$$

That means $h_j = m$ for $a_j(t) > 0$ and $h_j = -m$ for $a_j(t) < 0$. Equivalently, for $\lambda = \exp m$ we have that

$$x_j = \exp v_j = \exp w_j \cdot \exp h_j = \lambda y_j \quad \text{for} \quad a_j(t) > 0$$

and, similarly, $x_j = \lambda^{-1} y_j$ for $a_j(t) < 0$.

In the same manner, $\sum_j a_j(t)h_j \le 0$ for some $t \in [0,1]$ implies

$$x_j = \lambda^{-1} y_j \quad \text{for} \quad a_j(t) > 0 \quad \text{and} \quad x_j = \lambda y_j \quad \text{for} \quad a_j(t) < 0.$$

Finally, for $z \in \langle x, y\rangle$ there exists $t \in [0,1]$ with $z = z(t)$ and by definition of $a_j(t)$

$$a_j(t) > 0 (< 0) \text{ iff } \frac{\partial f}{\partial x_j}(z) > 0 (< 0) \text{ iff } j \in J_+(z) (J_-(z)).$$

This proves the lemma. □

Theorem 6.2.4. *Let $K = \mathbb{R}^n_+$ and T a continuous selfmapping of K some iterate of which maps $K \setminus \{0\}$ into $\mathring{K}$. Assume T is a continuously differentiable selfmapping of $\mathring{K}$ which satisfies for all $1 \le i \le n, x \in \mathring{K}$*

$$\sum_{j=1}^n x_j \left|\frac{\partial T_i}{\partial x_j}(x)\right| \le T_i x. \tag{6.2.5}$$

Suppose further each orbit $O(z), z \in \mathring{K}$, satisfies the following conditions:

(a) *There exists a partition $J_1 \cup J_2 = \{1, \dots, n\}$ (J_1 or J_2 may be empty) such that in case of equality in (6.2.5) for i and $x \in O(z)$ it holds that $\frac{\partial T_i}{\partial x_j}(x) > 0$ for all $j \in J_1$ and $\frac{\partial T_i}{\partial x_j}(x) < 0$ for all $j \in J_2$.*

(b) *In case of equality in (6.2.5) for i and z it holds that $\frac{\partial T_i}{\partial x_j}(u) \ge 0$ for all $j \in J_1$ and $\frac{\partial T_i}{\partial x_j}(u) \le 0$ for all $j \in J_2$ for all $u \in \mathring{K}$ with $u_j = rz_j$, $j \in J_1$ and $u_j = s^{-1}z_j$, $j \in J_2$ for some $r > 0, s > 0$.*

Then the following limit set trichotomy holds. Either

(i) *for all* $x \in K \setminus \{0\}$, $O(x)$ *is unbounded*

or

(ii) *for all* $x \in K$, $\lim_{n\to\infty} T^n x = 0$

or

(iii) *for each* $x \in K \setminus \{0\}$, $\lim_{n\to\infty} T^n x = \bar{x}$ *is a fixed point of* T *with*

$$\bar{x}_j = \begin{cases} c(x)x_j^*, & j \in I_1 \\ c(x)^{-1}x_j^*, & j \in I_2, \end{cases}$$

where $x^* \in \overset{\circ}{K}$ *is some fixed point of* T, $c(x) > 0$ *a scalar and* $I_1 \cup I_2 = \{1,\ldots,n\}$ *a partition (belonging to* x^**).*

Proof. By Theorem 6.2.1 assumption (6.2.5) implies that T is non-expansive on $(\overset{\circ}{K}, p)$. The assumptions of Theorem 6.1.1 being satisfied, part B (i′) yields (i) of the trichotomy stated in Theorem 6.2.4. Since bounded orbits have compact closure, part B (ii′) yields (ii).

(1) Considering case (iii) we shall show first that condition 6.1.1 of Theorem 6.1.1 cannot hold. For this, suppose there exists $y \in \omega(x) \cap \overset{\circ}{K}$ such that

$$p(T^{k+1}y, T^k y) = c(y) > 0 \quad \text{for all} \quad k \geq 0.$$

Then for each k there exists some $i = i(k) \in \{1,\ldots,n\}$ such that for $f(z) = T_i z, z \in \overset{\circ}{K}$ we have

$$|\log f(T^{k+1}y) - \log f(T^k y)| \geq p(T^{k+1}y, T^k y) > 0.$$

From Lemma 6.2.3, condition (6.2.4) being satisfied, we conclude that in (6.2.5) holds equality for $i = i(k)$ and $y^k = T^k y$. Applying the assumptions made in (a) on the orbits to orbit $O(y)$ and by Lemma 6.2.3 again there exists $\lambda_k > 0$ such that

$$y_j^{k+1} = \lambda_k y_j^k \quad \text{for} \quad j \in J_1 \quad \text{and} \quad y_j^{k+1} = \lambda_k^{-1} y_j^k \quad \text{for} \quad j \in J_2.$$

For any two vectors $u, v \in \overset{\circ}{K}$ and $\lambda > 0$ we shall write $u = \lambda * v$ if $u_j = \lambda v_j$ for $j \in J_1$ and $u_j = \lambda^{-1} v_j$ for $j \in J_2$. In this notation we have that $y^{k+1} = \lambda_k * y^k$ for all k and, by iteration, $y^k = \mu_k * y$ where $\mu_k = \prod_{i=0}^{k-1} \lambda_i$ for $k \geq 1$. Furthermore, since in 6.2.5 equality holds for $i = i(0)$ and $y^0 = T^0 y = y$ we may apply part (b) of the assumptions. For $0 < \alpha \leq \beta$ the mean value theorem yields

$$T_i(\beta * y) - T_i(\alpha * y) = \int_0^1 \Big(\sum_{j=1}^n \frac{\partial T_i}{\partial x_j}(u(t)) h_j \Big) dt$$

with $h = \beta * y - \alpha * y$ and $u(t) = \alpha * y + th$.

We have that $h_j = (\beta_j - \alpha_j)y_j$ for $j \in J_1$ and $h_j = (\beta_j^{-1} - \alpha_j^{-1})y_j$ for $j \in J_2$ as well as $u(t)_j = ry_j$ for $j \in J_1$ and $u(t)_j = sy_j$ for $j \in J_2$ with $r = t\beta + (1-t)\alpha$ and $s = t\beta^{-1} + (1-t)\alpha^{-1}$. By part (b) of the assumptions on orbit $O(y)$ we have for $t \in [0,1]$

$$\frac{\partial T_i}{\partial x_j}(u(t)) \geq 0 \quad \text{for} \quad j \in J_1 \quad \text{and} \quad \frac{\partial T_i}{\partial x_j}(u(t)) \leq 0 \quad \text{for} \quad j \in J_2,$$

and, hence,

$$\sum_{j=1}^{n} \frac{\partial T_i}{\partial x_j}(u(t))h_j = \sum_{j \in J_1} \frac{\partial T_i}{\partial x_j}(u(t))(\beta_j - \alpha_j)y_j + \sum_{j \in J_2} \frac{\partial T_i}{\partial x_j}(u(t))(\beta^{-1} - \alpha^{-1})y_j.$$

Since for every t each of the two terms on the right hand is non-negative, we conclude that for $i = i(0)$

$$T_i(\alpha * y) \leq T_i(\beta * y) \quad \text{for} \quad \alpha \leq \beta.$$

Suppose now, we have for some k that $\mu_k \leq \mu_{k+1}$. It follows that

$$y_i^{k+1} = T_i y^k = T_i(\mu_k * y) \leq T_i(\mu_{k+1} * y) = T_i y^{k+1} = y_i^{k+2}$$

and, hence, $\mu_{k+1}\, y_i \leq \mu_{k+2}\, y_i$, that is $\mu_{k+1} \leq \mu_{k+2}$. Similarly, $\mu_k \geq \mu_{k+1}$ implies $\mu_{k+1} \geq \mu_{k+2}$. Thus, the sequence $(\mu_k)_k$ is either increasing or decreasing. Since in case (iii) all orbits are bounded $\lim_{k\to\infty} \mu_k = \mu \geq 0$ exists. Since in case (iii) $\omega(x_0) \cap \mathring{K} \neq \emptyset$ for some $x_0 \in \mathring{K}$ we must have for some $\lambda > 0$ that $\lambda^{-1}T^n x_0 \leq T^n y \leq \lambda T^n x_0$ for n big enough as well as $\lim_{l\to\infty} T^{n_l} x_0 = u \in \mathring{K}$. Therefore,$\lambda^{-1}u \leq \mu y \leq \lambda u$ which implies $\mu > 0$. Thus, we conclude that $\lim_{k\to\infty} y^k = \mu y$ holds with respect to p. Finally, from $0 < c(y) = p(T^{k+1}y, T^k y)$ for all $k \geq 0$ we obtain $0 < c(y) = \lim_{k\to\infty} p(T^{k+1}y, T^k y) = p(\mu y, \mu y) = 0$ – which is a contradiction.

(2) By step (1), in case (iii) we must have that for each $x \in \mathring{K}$ either $\omega(x)$ is a singleton in $\mathring{K}$ or $\omega(x) \cap \mathring{K} = \emptyset$. From $\omega(x_0) \cap \mathring{K} \neq \emptyset$ for some $x_0 \in \mathring{K}$ we conclude that $\omega(x_0) = \{x^*\}$ with $x^* \in \mathring{K}$ being a fixed point of T. For $x \in \mathring{K}$ arbitrary $\lambda^{-1}T^n x_0 \leq T^n x \leq \lambda T^n x_0$ for some $\lambda > 0$ and n big enough. Since $\lim_{n\to\infty} T^n x_0 = x^*$ it follows that $\omega(x) \subseteq [\lambda^{-1}x^*, \lambda x^*] \subseteq \mathring{K}$ and $\omega(x)$ is a singleton, too. Thus, for $x \in K \setminus \{0\}$, $\lim_{n\to\infty} T^n x = \bar{x}$ and $\bar{x}$ a fixed point of T in $\mathring{K}$. Suppose $\bar{x} \neq x^*$. Then $p(T\bar{x}, Tx^*) \geq p(\bar{x}, x^*)$ and there exists i such that for $f = T_i$ we have $|\log f(\bar{x}) - \log f(x^*)| \geq p(\bar{x}, x^*) > 0$. From Lemma 6.2.3 it follows that in (6.2.5) holds equality for i and all $x \in \langle \bar{x}, x^* \rangle$ and for some $\lambda > 0$

$$\bar{x}_j = \lambda x_j^* \quad \text{for} \quad j \in J_+(x^*) \quad \text{and} \quad \bar{x}_j = \lambda^{-1} x_j^* \quad \text{for} \quad j \in J_-(x^*).$$

The orbit assumption for $O(x^*) = \{x^*\}$ yields a partition $I_1 \cup I_2 = \{1, \ldots, n\}$, belonging to x^*, such that $\frac{\partial f}{\partial x_j}(x^*) > 0$ for all $j \in I_1$ and $\frac{\partial f}{\partial x_j}(x^*) < 0$ for all $j \in I_2$, Therefore,

$$\bar{x}_j = \lambda x_j^* \quad \text{for all} \quad j \in I_1 \quad \text{and} \quad \bar{x}_j = \lambda^{-1} x_j^* \quad \text{for all} \quad j \in I_2.$$

This proves the theorem. □

In view of the applications in the next two sections the following consequences of Theorem 6.2.4 prove to be very useful.

Corollary 6.2.5. *Let $K = \mathbb{R}^n_+$ and T a continuous selfmapping of K which is a continuously differentiable selfmapping of $\mathring{K}$. For an iterate $S = T^m$ assume it maps $K \setminus \{0\}$ into $\mathring{K}$ and satisfies for all $1 \le i \le n$, $x \in \mathring{K}$*

$$\sum_{j=1}^{n} x_j \left| \frac{\partial S_i}{\partial x_j}(x) \right| \le S_i x. \tag{6.2.6}$$

A. *If for every $z \in \mathring{K}$ there exists some $k = k(z)$ such that in (6.2.6) strict inequality holds for $x = S^k z$ and all $1 \le i \le n$ then strong limit set trichotomy applies to T: Either*

(i) *for all $x \in K \setminus \{0\}$, $O(x)$ is unbounded*

or

(ii) *for all $x \in K$, $\lim_{n\to\infty} T^n x = 0$*

or

(iii) *for all $x \in K \setminus \{0\}$, $\lim_{n\to\infty} T^n x = x^*$*

where $x^* \in \mathring{K}$ is the unique fixed point of T in $K \setminus \{0\}$.

B. *Suppose there exists a partition $J_1 \cup J_2 = \{1, \ldots, n\}$ such that for all $x \in \mathring{K}$ and all $1 \le i \le n$*

$$\frac{\partial S_i}{\partial x_j}(x) > 0 \quad \text{for all} \quad j \in J_1 \quad \text{and} \quad \frac{\partial S_i}{\partial x_j}(x) < 0 \quad \text{for all} \quad j \in J_2.$$

Then a limit set trichotomy holds for T with (i) and (ii) as in A and where (iii) is replaced by

(iii′) *for all $x \in K \setminus \{0\}$ and all $0 \le i \le m-1$*

$$\lim_{n\to\infty} T^{mn+i} x = \bar{x}^i \quad \text{where} \quad \bar{x}^i_j = \begin{cases} c_i(x) x^*_j, & j \in J_1 \\ c_i(x)^{-1} x^*_j, & j \in J_2 \end{cases}$$

with $c_i(x) > 0$ and x^* a fixed point of T^m in $\mathring{K}$. In particular, if $J_1 = \emptyset$ or $J_2 = \emptyset$ and T is positively homogeneous then with a scalar $c(x) > 0$

$$\lim_{n\to\infty} T^n x = c(x) x^* \quad \text{for all} \quad x \in K \setminus \{0\}$$

where x^* is a fixed point of T in $\mathring{K}$, unique up to a positive scalar.

Proof. (1) Consider first the case $S = T$. Parts (i) and (ii) of the trichotomy follow as in the proof of Theorem 6.2.4. For parts (iii) and (iii'), respectively, conclusions A and B will be treated separately. For A we refer to arguments in the proof of Theorem 6.2.4. As there in part (1) it follows from the assumptions in A that condition (6.1.1) of Theorem 6.1.1 cannot apply. As in part (2) it follows for each $x \in K \setminus \{0\}$ that $\lim_{n\to\infty} T^n x$ is a fixed point of T in $\overset{\circ}{K}$. The assumptions applied to a fixed point yield its uniqueness. Thus, the strong limit set trichotomy does hold.

For B, Theorem 6.2.4 yields (iii′) for $m = 1$.

(2) Suppose now $S = T^m$ with $m \geq 2$. For A, part (1) above for S instead of T yields the strong limit set trichotomy for S. If an orbit is unbounded for S it is unbounded for T, too. If $\lim_{n\to\infty} S^n x = 0$ for all $x \in K$ then

$$\lim_{n\to\infty} T^{mn+i}x = \lim_{n\to\infty} S^n(T^i x) = 0 \text{ for all } x \in K, \text{all } i \in \mathbb{N},$$

and, hence, $\lim_{n\to\infty} T^n x = 0$. Considering (iii) we have $\lim_{n\to\infty} S^n x = x^*$ for $x \in K \setminus \{0\}$, x^* the unique fixed point of S. Since $S(Tx^*) = T(Sx^*) = Tx^*$ it follows that $Tx^* = x^*$ and, hence,

$$\lim_{n\to\infty} T^{mn+i}x = T^i\left(\lim_{n\to\infty} T^{mn}x\right) = T^i x^* = x^*.$$

Thus, $\lim_{n\to\infty} T^n x = x^*$ for all $x \in K \setminus \{0\}$, x^* the unique fixed point of T in $K \setminus \{0\}$.

(3) For $S = T^m$, $m \geq 2$, and B it remains to show (iii′). By part (1) above for S we have $\lim_{n\to\infty} S^n x = \bar{x}^0$ for $x \in \overset{\circ}{K}$ where, with a scalar $c_0(x) > 0$,

$$\bar{x}^0 = \begin{cases} c_0(x)x_j^*, & j \in J_1 \\ c_0(x)^{-1}x_j^*, & j \in J_2, \end{cases}$$

and x^* is a fixed point of S. Since $T(\overset{\circ}{K}) \subseteq \overset{\circ}{K}$ it follows for $0 \leq i \leq m-1$ that $\lim_{n\to\infty} S^n(T^i x) = \bar{x}^i$ where $\bar{x}^i = \overline{T^i x^0}$.

Setting $c_i(x) = c_0(T^i x)$ this yields (iii′).

Finally, suppose $J_2 = \emptyset$ and T to be positively homogeneous. In that case

$$\lim_{n\to\infty} T^{mn+i}x = T^i\left(\lim_{n\to\infty} S^n x\right) = T^i(c_0(x)x^*) = c_0(x)T^i x^*.$$

Furthermore,

$$Tx^* = \lim_{n\to\infty} S^n(Tx^*) = \overline{Tx^{*0}} = c_0(Tx^*)x^*,$$

which implies $x^* = T^m x^* = c_0(Tx^*)^m x^*$. Therefore, $c_0(Tx^*) = 1$ and, hence, $Tx^* = x^*$. Thus, we obtain for all $i \in \mathbb{N}$

$$\lim_{n\to\infty} T^{mn+i}x = c_0(x)T^i x^* = c_0(x)x^*.$$

This shows $\lim_{n\to\infty} T^n x = c(x)x^*$ with $c(x) = c_0(x)$ – in case of $J_2 = \emptyset$ – and $c(x) = c_0(x)^{-1}$ in case of $J_1 = \emptyset$. □

For all the forms of limit set trichotomy considered, the existence of a fixed point in $\overset{\circ}{K}$ implies that neither of the alternatives (i) and (ii) can hold and, hence, alternative (iii) applies. In case of the strong limit set trichotomy, (iii) means that the assumed fixed point must be globally attractive. The latter can be obtained also, by weakening the assumptions for strong limit set trichotomy to that of a non-expansive T, but assuming in addition that inequality (6.2.6) holds strictly for the fixed point. This is a consequence of the local-global stability principle, Theorem 4.3.3, as the following result demonstrates.

Theorem 6.2.6. *Let $K = \mathbb{R}^n_+$ and T a selfmapping of $\overset{\circ}{K}$ which is continuously differentiable. Suppose for an iterate $S = T^m$ it holds for all $1 \leq i \leq n, x \in \overset{\circ}{K}$*

$$\sum_{j=1}^{n} x_j \left| \frac{\partial S_i}{\partial x_j}(x) \right| \leq S_i x \text{ and } \sum_{j=1}^{n} x_j^* \left| \frac{\partial S_i}{\partial x_j}(x^*) \right| < S_i x^* \tag{6.2.7}$$

for a fixed point x^* of S. Then x^* is the unique fixed point of T in $\overset{\circ}{K}$ and it is globally attractive, that is $\lim_{n\to\infty} T^n x = x^*$ for all $x \in \overset{\circ}{K}$.

Proof. (1) It suffices to prove the theorem for $m = 1$. If the theorem holds in this case its application to $S = T^m$ yields $\lim_{n\to\infty} T^{nm} x = x^*$ for all $x \in \overset{\circ}{K}$. Therefore, $\lim_{n\to\infty} T^{mn}(T^i x) = x^*$ for all $i \in \mathbb{N}$, all $x \in \overset{\circ}{K}$, which implies $\lim_{n\to\infty} T^n x = x^*$ for all $x \in \overset{\circ}{K}$.

(2) Assume $m = 1$, $S = T$. By assumption 6.2.7 Theorem 6.2.1 implies that T is a non-expansive selfmapping of the metric space $(\overset{\circ}{K}, p)$, p the part metric. Since by Proposition 3.4.12 on $\overset{\circ}{K}$ the part topology coincides with the Euclidean topology the space $(\overset{\circ}{K}, p)$ is connected. To obtain the conclusion of the theorem from Corollary 4.3.6 it suffices to show that the fixed point x^* of T is locally attractive in $(\overset{\circ}{K}, p)$. From the second part of assumption 6.2.7 we get that

$$\sum_{j=1}^{n} z_j \left| \frac{\partial T_i}{\partial x_j}(z) \right| < T_i z \tag{$*$}$$

holds for all $1 \leq i \leq n$ and all z in some Euclidean neighbourhood U of x^*. By the coincidence of the two topologies we can assume that $U = \{z \in \overset{\circ}{K} \mid p(z, x^*) \leq \epsilon\}$ for some $\epsilon > 0$. Let for $x, y \in U$, $x \neq y$,

$$\langle x, y\rangle = \{z \in \overset{\circ}{K} \mid z_j = x_j^t \cdot y_j^{1-t},\ t \in [0,1], 1 \leq j \leq n\}.$$

For $z \in \langle x, y\rangle$

$$\begin{aligned}\log z_j - \log x_j^* &= t \log x_j + (1-t)\log y_j - \log x_j^* \\ &= t(\log x_j - \log x_j^*) + (1-t)(\log y_j - x_j^*),\end{aligned}$$

and, hence, $p(z.x^*) \leq tp(x, x^*) + (1-t)p(y, x^*)$.

Therefore, $\langle x, y\rangle \subseteq U$ and $(*)$ holds for all $z \in \langle x, y\rangle$. Lemma 6.2.3 applied to $f = T_i$ yields that $|\log T_i x - \log T_i y| < p(x, y)$. Therefore, $p(Tx, Ty) < p(x, y)$ for $x, y \in U, x \neq y$. It follows for $x \in U$,

$$p(Tx, x^*) = p(Tx, Tx^*) \leq p(x, x^*) \leq \epsilon \quad \text{and, hence,} \quad Tx \in U.$$

Thus T is a contractive selfmapping of the metric space (U, p). Since U is compact in $(\overset{\circ}{K}, p)$, we obtain that $\lim_{n\to\infty} T^n x = x^*$ in $(\overset{\circ}{K}, p)$ for all $x \in U$ (cf. Remarks 4.1.5 (i)). That is, x^* is locally attractive in $(\overset{\circ}{K}, p)$. By Proposition 3.4.12 again we finally arrive for $x \in \overset{\circ}{K}$ at $\lim_{n\to\infty} T^n x = x^*$ with respect to the Euclidean topology. □

The following example taken from the study of insect populations illustrates how the above theorem may be useful even in one dimension (cf. [14, 19]).

Example 6.2.7. For $K = \mathbb{R}_+$ let f be a selfmapping of $\overset{\circ}{K}$ given by $f(x) = \lambda x(1+x)^{-\beta}$ with parameters $\lambda > 1, \beta > 0$. f has the unique fixed point $x^* = \lambda^{\frac{1}{\beta}} - 1$ in $\overset{\circ}{K}$. One has that $x|f'(x)| \le f(x)$ for all $x \in \overset{\circ}{K}$, all $\lambda > 1$ and $\beta \le 2$. Also, for this range of parameters, $|f'(x^*)| < 1$. Therefore, by the above theorem, x^* is globally attractive. It is easily seen that x^* is globally attractive for all $\lambda > 1$ if and only if $\beta \le 2$. There are, however, values of the parameters for which x^* is locally but not globally attractive.

Indeed, for values of the parameters big enough the dynamics of this example are very complicated (see [19]; see also [6] and Exercise 7).

6.3 Applications to non-linear difference equations and cooperative systems of differential equations

Consider the difference equation

$$u(t+n) = f(u(t), u(t+1), \dots, u(t+n-1)) \tag{6.3.1}$$

of order $n \ge 1$ with $u(t) \in \mathbb{R}_+$ for $t \in \mathbb{N}$, $f \colon \mathbb{R}^n_+ \longrightarrow \mathbb{R}_+$ and with initial conditions $\bar{u} = (u(0), \dots, u(n-1)) \in \mathbb{R}^n_+$.

From results in the previous section we shall obtain the following **limit set trichotomy for difference equations:**

For solutions u of (6.3.1) either

(i) for all $0 \le \bar{u} \ne 0$, u is unbounded,

or

(ii) for all $0 \le \bar{u}$, $\lim_{t\to\infty} u(t) = 0$,

or

(iii) for all $0 \le \bar{u} \ne 0$, $\lim_{t\to\infty} u(t) = c(\bar{u})r^*$ where $c(\bar{u}) > 0$ and r^* is the unique positive solution of $f(r, \dots, r) = r$.

As in earlier Sections 2.5, 5.4 to the difference equation we associate the selfmapping T of $\mathbb{R}^n_+$ given by $T(x_1, \dots, x_n) = (x_2, \dots, x_n, f(x))$.

To apply differentiability criteria from the previous section, we have to deal with inequalities like $\sum_{j=1}^n x_j |\frac{\partial T_i}{\partial x_j}(x)| \le T_i x$ for $1 \le i \le n$. To this extent we introduce for a selfmapping T of $K = \mathbb{R}^n_+$ which maps continuously differentiable $\overset{\circ}{K}$ into itself **the**

mapping $\delta(T)$, defined by

$$\delta(T)(x) = Tx - |J(x)|x \quad \text{for} \quad x \in \overset{\circ}{K},$$

where $J(x) = J_T(x)$ is the Jacobian of T and $|J(x)|$ is the matrix of the absolute values of $J(x)$. The following lemma provides conditions under which $\delta(T^m)(x) \geq 0$ or > 0, which will be very useful in dealing with difference equations as well as with differential equations.

Lemma 6.3.1. (i) *For $m \geq 1$ it holds*

$$\delta(T^m)(x) \geq \sum_{k=0}^{m-1} |J(T^{m-1}x)| \cdot |J(T^{m-2}x)| \cdots |J(T^{m-k}x)| \delta(T)(T^{m-k-1}x). \tag{6.3.2}$$

(ii) *Suppose $\delta(T)(x) \geq 0$ for all $x \in \overset{\circ}{K}$ and let $m \geq 1$. Assume further, for $y \in \overset{\circ}{K}$ there exists some $p = p(y)$ such that there exists* **from each *i* a chain to *p*** *in the following sense: For $a_{ij}(x) = \frac{\partial T_i}{\partial x_j}(x)$ and certain indices $i = i_1, \ldots, i_k$*

$$a_{i_1 i_2}(T^{m-1}y) \neq 0, a_{i_2 i_3}(T^{m-2}y) \neq 0, \ldots, a_{i_k, p}(T^{m-k}y) \neq 0, \delta(T)(T^{m-k-1}y)_p > 0$$

(where for $k = 0$ the condition on the a_{ij} is empty).

Then $\delta(T^m)(y) > 0$ for all $y \in \overset{\circ}{K}$.

(iii) *If T is monotone and*

$$0 < J(x) \lneqq J(tx) \quad \textit{for all} \quad x \in \overset{\circ}{K}, \quad \textit{all} \quad 0 < t < 1,$$

then $\delta(T^2)(x) > 0$ for all $x \in \overset{\circ}{K}$.

Proof. (i) First, we prove

$$\delta(T \circ S)(x) \geq \delta(T)(Sx) + |J_T(Sx)| \delta(S)(x), \tag{6.3.3}$$

where S is another selfmapping of K, mapping $\overset{\circ}{K}$ continuously differentiable into itself. By definition, $\delta(T \circ S)(x) = (T \circ S)(x) - |J_{T \circ S}(x)|x$.

Using the chain rule we obtain

$$\begin{aligned}\delta(T \circ S)(x) &= (T \circ S)(x) - |J_T(Sx) \cdot J_S(x)|x \\ &\geq (T \circ S)(x) - |J_T(Sx)| \cdot |J_S(x)|x.\end{aligned}$$

Expanding the right hand side as

$$(T \circ S)(x) - |J_T(Sx)|\, Sx + |J_T(Sx)|\, Sx - |J_T(Sx)| \cdot |J_S(x)|\, x$$

and collecting the terms

$$\delta(T)(Sx) + |J_T(Sx)|\, (Sx - |J_S(x)|\, x)$$

yields inequality (6.3.3). Now we prove inequality (6.3.2) by induction on m. The case $m = 1$ is trivial. Suppose (6.3.2) holds for m. From inequality (6.3.3) we obtain for $S = T^m$

$$\delta(T^{m+1})(x) \geq \delta(T)(T^m x) + |J(T^m x)|\,\delta(T^m)(x), \quad \text{where} \quad J = J_T,$$

and by induction hypothesis for the right hand side

$$\delta(T)(T^m x) + \sum_{k=0}^{m-1} |J(T^m x)|\,|J(T^{m-1}x)| \cdots |J(T^{m-k}x)|\delta(T)(T^{m-k-1}x).$$

This expression is just the right hand side of inequality (6.3.2) with m replaced by $m + 1$. This proves inequality (6.3.2).

(ii) Let $y, p = p(y), i$ be given.

From inequality (6.3.2) and $\delta(T)(\cdot) \geq 0$, it follows

$$\delta(T^m)(y)_i \geq |a_{i_1 i_2}(T^{m-1}y)| \cdot |a_{i_2 i_3}(T^{m-2}x)| \cdots |a_{i_{k,p}}(T^{m-k}y)|\delta(T)(T^{m-k-1}y)_p,$$

and by the chain assumption $\delta(T^m)(y)_i > 0$. Since i was arbitrary this proves part (ii).

(iii) By the mean value theorem for $x \in \mathring{K}$ given

$$T_i x = \sum_{j=1}^{n} a_{ij}(t_i x)x_j + T_i 0 \geq \sum_{j=1}^{n} a_{ij}(t_i x)x_j, \quad \text{where} \quad 0 < t_i < 1.$$

For $1 > s > \max\{t_i \mid 1 \leq i \leq n\}$ by assumptions $a_{ij}(t_i x) = a_{ij}(\frac{t_i}{s} sx) \geq a_{ij}(sx)$ and, hence,

$$\delta(T)(x) = Tx - J(x)x \geq J(sx)x - J(x)x = (J(sx) - J(x))x.$$

Since $J(sx) \gneqq J(x)$ there exists for $x \in \mathring{K}$ a $p = p(x)$ such that $\delta(T)(x)_p > 0$. Since $J(Tx) > 0$, for any i we have that $a_{ip}(Tx) > 0$. Therefore, for $m = 2$ there exists a chain from i to p and from part (ii) it follows $\delta(T^2)(x) > 0$. □

With the help from Lemma 6.3.1 we obtain the following result.

Theorem 6.3.2. *Let $K = \mathbb{R}^n_+$ and T a continuous selfmapping of K some iterate of which maps $K \setminus \{0\}$ into $\mathring{K}$. Assume T to be a continuously differentiable selfmapping of $\mathring{K}$ and*

$$\sum_{j=1}^{n} x_j \left|\frac{\partial T_i}{\partial x_j}(x)\right| \leq T_i x$$

for all $1 \leq i \leq n$ and all $x \in \mathring{K}$.

(i) *Assume for each $x \in \mathring{K}$ given there exists $i = p(x)$ such that the inequality holds strictly. Suppose further there is some $m \geq 1$ such that for each $x \in \mathring{K}$ and $1 \leq i \leq n$ there exist indices $i = i_1, \ldots, i_k$ with*

$$\frac{\partial T_{i_r}}{\partial x_{i_{r+1}}}(T^{m-r}x) \neq 0, \ 1 \leq r \leq k, \ i_{k+1} = p(x).$$

Then strong limit set trichotomy holds for T.

(ii) *If T is monotone and for some iterate S of T it holds that*

$$0 < J_S(x) \lneqq J_S(tx) \quad \text{for all} \quad x \in \overset{\circ}{K}, \quad \text{all} \quad 0 < t < 1$$

then strong limit trichotomy holds for T.

Proof. (i) For $x \in \overset{\circ}{K}$ by assumption $\delta(T)(x) = Tx - \lfloor J_T(x) \rfloor x \geq 0$ and $\delta(T)(x)_p > 0$ for $p = p(x)$. The additional assumptions made imply there exists from each i a chain to p which by Lemma 6.3.1 (i) implies for the iterate $S = T^m$ that $\delta(S)(x) > 0$ for all $x \in \overset{\circ}{K}$. By Corollary 6.2.2 limit set trichotomy holds for T.

(ii) By Lemma 6.3.1 (ii), $\delta(S^2)(x) > 0$ for all $x \in \overset{\circ}{K}$. Again, Corollary 6.2.2 implies limit set trichotomy for T. □

A first consequence of Theorem 6.3.2 is the following result on difference equations.

Theorem 6.3.3. *Let $K = \mathbb{R}_+^n$ and $f : K \longrightarrow \mathbb{R}_+, f(x) > 0$ for $x > 0$.*

(i) *Assume f is monotone and subhomogeneous. If f is strongly monotone then for the difference equation (6.3.1) limit set trichotmy holds and if f is strongly subhomogeneous then limit set trichotomy holds with $c(\bar{u}) = 1$ for all $0 \leq \bar{u} \neq 0$.*

(ii) *If f is continuously differentiable on $\overset{\circ}{K}$, $f(x) > 0$ for $x \gneqq 0$ and $\sum_{j=1}^n x_j |\frac{\partial f}{\partial x_j}(x)| < f(x)$ for all $x \in \overset{\circ}{K}$ then for the difference equation (6.3.1) limit set trichotomy holds with $c(\bar{u}) = 1$ for all $0 \leq \bar{u} \neq 0$.*

(iii) *Assume f continuously differentiable on $\overset{\circ}{K}$ and $\sum_{j=1}^n x_j |\frac{\partial f}{\partial x_j}(x)| \leq f(x)$ for all $x \in \overset{\circ}{K}$ and $\sum_{j=1}^n |\frac{\partial f}{\partial x_j}(\bar{r})| < 1$ for $\bar{r} = (r, \ldots, r)$ where $r > 0$ with $f(\bar{r}) = r$. Then $\lim_{t \to \infty} u(t) = r$ for all solutions $u(\cdot)$ of the difference equation (6.3.1) with $0 \lneqq \bar{u}$. r is the unique positive solution of the equation $f(s, \ldots, s) = s$.*

Proof. Let $Tx = (x_2, \ldots, x_n, f(x)), x \in K$.

(i) T maps $\overset{\circ}{K}$ into itself, T is monotone and subhomogeneous. By iteration, $T^n x = (f(x), f(Tx), \ldots, f(T^{n-1}x))$.

If f is strongly monotone, $x \lneqq y$ implies $f(x) < f(y)$ and, hence, $Tx \lneqq Ty$. This in turn implies $f(Tx) < f(Ty)$. By iteration it follows that $T^n x < T^n y$. That is, T^n is strongly monotone. In the same way, f strongly subhomogeneous implies the same for T^n. Theorem 6.1.5 B implies limit set trichotomy, with $c(x) = 1$ in case f is strongly subhomogeneous.

From the difference equation $u(t + n) = f(u(t), \ldots, u(t + n - 1))$ it follows $T(u(t), \ldots, (u(t + n - 1)) = (u(t + 1), \ldots, u(t + n))$. By iteration $T^t \bar{u} = (u(t), \ldots, u(t + n - 1))$. Thus, the limit set trichotomy for T implies the limit set richotomy for the difference equation (6.3.1). Thereby, a fixed point $x^* \in \overset{\circ}{K}$ of T becomes $x^* = (r^*, \ldots, r^*), r^* > 0, f(r^*, \ldots, r^*) = r^*$.

(ii) T is continuous and T^n maps $K \setminus \{0\}$ into $\overset{\circ}{K}$ by step (i). Since f is continuously differentiable on $\overset{\circ}{K}$, T is a continuously differentiable selfmapping of $\overset{\circ}{K}$. Because of $Tx = (x_2, \dots, x_n, f(x))$ we have

$$\frac{\partial T_i}{\partial x_j}(x) = \delta_{i+1,j} \quad \text{for} \quad 1 \le i \le n-1 \quad \text{and} \quad \frac{\partial T_n}{\partial x_j}(x) = \frac{\partial f}{\partial x_j}(x).$$

Limit set trichotomy for the difference equation follows from Theorem 6.3.2 (i). For this take $p(x) = n$ for all $x \in \overset{\circ}{K}$. Obviously, $\sum_{j=1}^n x_j |\frac{\partial T_i}{\partial x_j}(x)| = x_{i+1} = T_i x$ if $1 \le i \le n-1$, and, by assumption

$$\sum_{j=1}^n x_j \left|\frac{\partial T_n}{\partial x_j}(x)\right| = \sum_{j=1}^n x_j \left|\frac{\partial f}{\partial x_j}(x)\right| < f(x) = T_n x.$$

Let $m = n$ and choose for $i \neq n$ as indices $i, i+1, i+2, \dots, n-1$. For $i = n$ choose just $i = i_1 = n$. For this choice the required assumptions are satisfied.

(iii) This part follows from Theorem 6.2.6. T is a continuously differentiable selfmapping of $\overset{\circ}{K}$. $x^* = \bar{r} = (r, \dots, r)$ is a fixed point of $Tx = (x_2, \dots, x_n, f(x))$. Assumption $\sum_{j=1}^n |\frac{\partial f}{\partial x_j}(\bar{r})| < 1$ implies

$$\sum_{j=1}^n x_j^* \left|\frac{\partial f}{\partial x_j}(x^*)\right| < r = f(x^*).$$

As in the proof of part (i) it follows $\sum^n j = 1 x_j |\frac{\partial T_i}{\partial x_j}(x)| \le T_i x$ for all $1 \le i \le n$, all $x \in \overset{\circ}{K}$. That is, $\delta(T)(x) \ge 0$, and, by Lemma 6.3.1 (i), $\delta(S)(x) \ge 0$, for $S = T^n$, all $x \in \overset{\circ}{K}$. Furthermore, $\sum_{j=1}^n x_j^* |\frac{\partial f}{\partial x_j}(x^*)| < f(x^*)$ by putting $y = x^*$ in Lemma 6.3.1 (ii) implies $\delta(S)(x^*) > 0$. From Theorem 6.2.6 it follows that $\lim_{t\to\infty} T^t x = x^*$ for all $x \in \overset{\circ}{K}$. Since $T^t \bar{u} = (u(t), \dots, u(t+n-1))$ for a solution $u(\cdot)$ and $f(\bar{u}) > 0$ for $\bar{u} \gneqq 0$ this proves part (iii). □

Examples 6.3.4. of difference equation (6.3.1)

$$u(t+n) = f(u(t), u(t+1), \dots, u(t+n-1)), \quad \text{for} \quad n \ge 2.$$

(i) Let $f(x_1, x_2, \dots, x_n) = \sum_{j=1}^n x_j^{p_j}, p_j \in [0,1], 1 \le j \le n$. For $n = 2, p_1 = p_2 = 1$ the difference equations generates for $\bar{u} = (1,1)$ the famous Fibonacci numbers whence one may consider the above setting a **generalized non-linear Fibonacci equation**. Obviously, f is monotone (increasing) and subhomogeneous. If $a = \min\{p_j \mid 1 \le j \le n\} > 0$ then f is strongly monotone and by Theorem 6.3.3 (i) limit set trichotomy holds for f. If, otherwise, $a = 0$ then f is strongly subhomogeneous and by Theorem 6.3.3 (i) again, limit set trichotomy holds (with $c(\bar{u}) = 1$). It is interesting to know which of the three alternatives actually applies. Since $f(1,1,\dots,1) \ge 1$ case (ii) is impossible. It is easy to verify that the equation $\sum_{j=1}^n r^{p_j} = r$ has a solution if and only if $b = \max\{p_j \mid$

$1 \le j \le n\} < 1$. This shows, case (iii) holds if and only if $b < 1$. Consequently, case (i) holds if and only if $b = 1$ (as for the Fibonacci numbers).

(ii) Consider the following "multiplicative version" of the generalized Fibonacci equation, that is,

$$f(x_1, x_2, \ldots, x_n) = \prod_{j=1}^{n} x_j^{p_j}, p_j \in [0, 1], \quad 1 \le j \le n.$$

(For $\sum_{j=1}^{n} p_j = 1$ this is the so called Cobb–Douglas production function employed in economics.)

Obviously, $f(x) > 0$ for $x > 0$ and the general assumptions of Theorem 6.3.3 are satisfied. Because of

$$\frac{\partial f}{\partial x_j}(x) = p_j x_j^{p_j - 1} \prod_{i \ne j} x_i^{p_i},$$

one has for $\sum_{j=1}^{n} p_j \le 1$ that

$$\sum_{j=1}^{n} x_j \left| \frac{\partial f}{\partial x_j}(x) \right| = \left(\sum_{j=1}^{n} p_j \right) f(x) \le f(x).$$

Assume $\sum_{j=1}^{n} p_j < 1$. Part (i) of Theorem 6.3.3 is not applicable because on $\mathbb{R}_+^n$ f is neither strictly monotone nor strictly subhomogeneous. Part (ii), too, is not applicable because $f(x) = 0$ is possible for $x \gneqq 0$. Part (iii), however, is applicable because for $r = 1, f(\bar{r}) = \bar{r}$ and

$$\sum_{j=1}^{n} \left| \frac{\partial f}{\partial x_j}(\bar{r}) \right| = \sum_{j=1}^{n} p_j < 1.$$

Thus, by Theorem 6.3.3, $\lim_{t \to \infty} u(t) = 1$ for all solutions of the difference equation (6.3.1) with $\bar{u} \gneqq 0$. In case of $\sum_{j=1}^{n} p_j = 1$, however, none of the three parts of Theorem 6.3.3 does apply. Indeed, limit set trichotomy does not hold in this case since for $\bar{u} = (1, \ldots, 1)$ the solution u is constant 1, excluding cases (i) and (ii), and for $\bar{u} = (1, 0, \ldots, 0)$ the solution u is constant 0 which excludes case (iii). On the other hand, one verifies, nevertheless, by direct calculation that $\lim_{t \to \infty} u(t) = 1$ holds true for $\bar{u} > 0$. Later on we shall examine the reasons for this phenomenon. (See the dynamics of means in Chapter 8.)

A further application of our results on limit set trichotomy is to the theory of cooperative systems of differential equations as developed by M. Hirsch and H. Smith [6, 7, 9, 23, 24]. This beautiful theory has many applications, in particular to biology. We will improve a main result in this area and illustrate it by the example of biochemical control circuits.

Consider a dynamical system in continuous time given by differential equations

$$x'(t) = F(t, x(t)), \quad x(t) \in \mathbb{R}_+^n, \quad t \in \mathbb{R}, \tag{6.3.4}$$

where $F \colon \mathbb{R} \times \mathbb{R}_+^n \longrightarrow \mathbb{R}^n$ is continuous and $x \longmapsto F(t, x)$ continuously differentiable.

(As always, for $x, y \in \mathbb{R}^n$ we write $x \le y$ if $x_i \le y_i$ for all i, $x \lneq y$ if $x \le y$ but $x \ne y$ and $x < y$ if $x_i < y_i$ for all i.)

That (6.3.4) is a dynamical system means in particular that for each $x \ge 0$ there exists a unique solution $x(t) = \phi(t, x)$ with $x(0) = \phi(t, 0) = x$. Furthermore, the mapping on int $\mathbb{R}^n_+$ given for t fixed by $x \longmapsto \phi(t, x)$, as well as its inverse, is continuously differentiable. (See [4, 5, 8] where conditions are specified for the property to hold.)

Definition 6.3.5. The system (6.3.3) is called **cooperative** if for $1 \le i, j \le n$

$$\frac{\partial F_i}{\partial x_j}(t, x) \ge 0 \text{ for } i \ne j, \text{ all } t \ge 0, \text{ all } x > 0. \tag{6.3.5}$$

Equivalently, if the Jacobian $J_F(t, x)$ is a Metzler matrix for all $t \ge 0, x > 0$. Thereby, a square matrix is a **Metzler matrix** if all non-diagonal entries are nonnegative. A Metzler matrix is **irreducible** if for some $c > 0$ the matrix $cI + M$ is irreducible as a nonnegative matrix (cf. Section 2.4).

In what follows we shall derive a limit set trichotomy for the solutions ϕ of system (6.3.3) from Theorem 6.3.2 part (ii) by setting $Tx = \phi(\tau, x)$ for some fixed τ. For this we have to translate assumptions on F into properties of T. For the required translation the following lemma will be crucial.

Lemma 6.3.6. (i) *Suppose F as in system (6.3.3) satisfies the* **Kamke condition**, *that is, for each $1 \le i \le n$, each $t \ge 0$, $0 \le x$ and $0 = x_i$ implies $0 \le F_i(t, x)$.*
Then for each solution $x(\cdot)$ of system (6.3.3) $x(0) \ge 0$ implies $x(t) \ge 0$ for all $t \ge 0$.

(ii) *Let for a solution $\phi(\cdot, x), x > 0$, of system (6.3.3) and $1 \le i, j \le n, t \ge 0$*

$$a_{ij}(t, x) = \frac{\partial F_i}{\partial x_j}(t, \phi(t, x)) \quad \text{and} \quad u_{ij}(t, x) = \frac{\partial \phi_i}{\partial x_j}(t, x). \tag{6.3.6}$$

If the matrix $A(t, x)$ of the $a_{ij}(t, x)$ is a Metzler matrix for all $t \ge 0$ then $U(t, x) \ge 0$ for all $t \ge 0$, $U(t, x)$ being the matrix of the $u_{ij}(t, x)$.

(iii) *Suppose $A(t, x)$ is a Metzler matrix for $x > 0, t > 0$ such that for $1 \le i \ne j \le n$ there exist $i_1, \dots, i_r$, pairwise different and different from i, j with*

$$a_{i,i_1}(t, x) > 0, a_{i_1,i_2}(t, x) > 0, \dots, a_{i_r,j}(t, x) > 0.$$

Then $u_{ij}(t, x) > 0$ and $u_{h,h}(t, x) > 0$ for all h.

Proof. (i) Let $x'(t) = F(t, x(t)), x(0) \ge 0$. Consider for $n \in \mathbb{N}$ the system $x'(t, n) = F(t, x(t, n)) + e(n), x(0, n) = x(0) + e(n)$, where the vector $e(n)$ has all components equal to $\frac{1}{n}$. We shall show that $x(t, n) > 0$ for all $t \ge 0$. If this does not hold then there exist i and $s > 0$ such that $x(t, n) > 0$ for $0 \le t < s$ by continuity but $x(s, n)_i = 0$. Therefore, $x'(s, n)_i \le 0$ and $0 \ge x'(s, n)_i = F_i(s, x(s, n)) + \frac{1}{n}$. Since $x(s, n) \ge 0$ by continuity and $x_i(s, n) = 0$ it follows from the Kamke condition that $F_i(s, x(s, n)) \ge 0$. This implies $0 \ge \frac{1}{n}$ – a contradiction. Thus, we must have $x(t, n) > 0$ for all $n \in \mathbb{N}$, all $t \geq 0$. For $n \longrightarrow \infty$, $x(0,n)$ approaches $x(0)$ and $x(t,n)$ approaches $x(t)$. This shows $x(t) \geq 0$ for all $t \geq 0$.

(ii) By the chain rule we obtain from (6.3.3) for $1 \leq i,j \leq n$

$$u'_{ij}(t,x) = \sum_{k=1}^{n} \frac{\partial F_i}{\partial x_k}(t,\phi(t,x))\frac{\partial \phi_k}{\partial x_j}(t,x),$$

and, hence,

$$u'_{ij}(t,x) = \sum_{k=1}^{n} a_{ik}(t,x)u_{kj}(t,x). \tag{6.3.7}$$

For j and x given let $v_i(t) = u_{ij}(t,x)$ for all i. Equation (6.3.7) gives for the vector $v(t)$

$$v'(t) = G(t,v(t)) \quad \text{with} \quad G(t,v) = A(t,x)v, \ \ t \geq 0.$$

If $v \geq 0$ with $v_i = 0$ then by assumptio- $A(t,x)$, $G_i(t,v) = a_{ii}(t,x)v_i + \sum_{k\neq i} a_{ik}(t,x)v_k \geq 0$.

Thus, G satisfies the Kamke condition and (ii) folllows from (i) because of $v_i(0) = \delta_{ij} \geq 0$.

(iii) Fix $x > 0$ and $t > 0$ and let $A = A(t,x)$, $U = U(t,x)$. First we show for any h,k,l with $k \neq l$ that

$$u_{kh} = 0 \quad \text{and} \quad a_{kl} > 0 \quad \text{imply} \quad u_{lh} = 0 \tag{$*$}$$

From (ii) we have for $U(s) = U(s,x)$ that $U(s) \geq 0$ for all $s \geq 0$. Equation (6.3.7) therefore implies $u'_{ph}(s) \geq a_{pp}(s)u_{ph}(s)$ for all p,h,s. By integration we obtain for $0 \leq s < t$
$u_{ph}(t) \geq u_{ph}\exp(\int_s^t a_{pp}(r)dr)$. Since $u_{hh}(0) = 1$ it follows that $u_{hh} > 0$ for all h. Furthermore, $u_{kh}(t) = 0$ implies $u_{kh}(s) = 0$ for $0 \leq s \leq t$ and, hence, $u'_{kh}(s) = 0$. From equation (6.3.7) it follows that $\sum_{k=1}^{n} a_{kl}(s)u_{lh}(s) = u'_{kh}(s) = 0$ for $0 \leq s \leq t$. Thus
$u_{kh} = 0$ and $a_{kl} > 0$ for $k \neq l$ implies $u_{lh} = 0$, which proves $(*)$.

Next, we show for $i \neq j$ as in (iii) and any h that

$$u_{ih} = 0 \quad \text{implies} \quad u_{jh} = 0 \tag{$**$}$$

Together with $u_{jj} > 0$ this shows $u_{ij} > 0$, the conclusion wanted in (iii). To show $(**)$ let $u_{ih} = 0$. By assumption there exists $i_1 \neq i,j$ such that $a_{i,i_1} > 0$. Property $(*)$ for $k = i, l = i_1$ yields $u_{i_1,h} = 0$. This together with $a_{i_1,i_2} > 0$ implies in the same way that $u_{i_2,h} = 0$. By iteration we arrive at $u_{i_r,h} = 0$ and $a_{i_r,j} > 0$. By $(*)$ again this gives $u_{jh} = 0$. This proves $(**)$ and, hence, part (iii) of the lemma. □

With the help of this lemma we will obtain the following limit set trichotomy for the solutions of a cooperative system of differential equations.

Theorem 6.3.7. *Assume the system (6.3.3) satisfies the following conditions where $J_F(t,x)$ denotes the Jacobian of $F(t,\cdot)$ at (t,x) for $t \geq 0, x > 0$.*

(a) *F is periodic with period $\tau > 0$, that is $F(t+\tau,x) = F(t,x)$ for all $t \geq 0, x \geq 0$.*
(b) *F satisfies the Kamke condition.*

(c) *$J_F(t,x)$ is an irreducible Metzler matrix for each $t > 0, x > 0$.*
(d) *If $0 < x < y$ and $t > 0$ then $J_F(t,y) \lneqq J_F(t,x)$.*

Then the following limit set trichotomy holds for the solutions $\phi(t,x), x \gneqq 0$.

Either

- *all solutions are unbounded,*

or

- *all solutions converge to 0,*

or

- *all solutions converge to $x^* > 0$. For $x = 0$ the solution $\phi(\cdot, 0)$ is either identically 0 or it converges to x^*, too.*

Proof. (i) Let $\phi(t,x)$ a solution with $\phi(0,x) = x \geq 0$. From Lemma 6.3.6 (i) we have that $\phi(t,x) \geq 0$ for all t. For $0 \leq x < y$ by the mean value theorem for $t > 0, 1 \leq i \leq n$,

$$\phi_i(t,y) - \phi_i(t,x) = \left[U(t,z)(y-x)\right]_i, \quad z = x + \ominus(y-x),$$

for some $0 < \ominus < 1$. Since $z > 0$, assumption (c) and Lemma 6.3.6 (iii) imply that $U(t,z) > 0$. Therefore, $0 \leq \phi(t,x) < \phi(t,y)$. In particular, $\phi(t,y) > 0$ for $y > 0$ and $0 \leq \phi(t,x) \leq \phi(t,y)$ for $0 \leq x \leq y$ by continuity.

Now define for a solution $\phi(t,x), x \geq 0$, $Tx = \phi(\tau,x)$. By the above, T is a continuous selfmapping of $\mathbb{R}^n_+$ which is monotone and maps int $\mathbb{R}^n_+$ into itself. We will obtain the limit set trichotomy wanted from Theorem 6.3.2 (ii) for $S = T$.

(ii) Next we show $0 < J_T(y) \lneqq J_T(x)$ for $0 < x < y$. Obviously, $\frac{\partial T_i}{\partial x_j}(x) = \frac{\partial \phi_i}{\partial x_j}(\tau,x)$ and T is a continuously differentiable selfmapping of int $\mathbb{R}^n_+$. From assumption (c) and Lemma 6.3.6 (iii) it follows that $\frac{\partial T_i}{\partial x_j}(y) = \frac{\partial \phi_i}{\partial x_j}(\tau,y) = u_{ij}(\tau,y) > 0$ and, hence, $J_T(y) > 0$.

Consider $Q(t) = U(t,x) - U(t,y)$. The chain rule gives

$$\begin{aligned} Q'(t) &= A(t,x)U(t,x) - A(t,y)U(t,y) \\ &= A(t,x)(U(t,x) - U(t,y)) + (A(t,x) - A(t,y))U(t,y). \end{aligned}$$

With $B(t,x,y) = (A(t,x) - A(t,y))U(t,y)$ we have that

$$Q'(t) = A(t,x)Q(t) + B(t,x,y). \tag{$*$}$$

Since $Q(0) = U(0,y) - U(0,x) = I - I = 0$, the solution of the system $(*)$ is given by

$$Q(t) = \int_0^t X(t,s)B(s,x,y)ds, \tag{$**$}$$

where $X(t,s), t \geq s$ is the fundamental solution of the homogeneous system to $(*)$. Since the latter coincides with equation (6.3.7) we have that

$$X(t,s) = U(t,x)U(s,x)^{-1} = U(t-s, \phi(s,x)).$$

By Lemma 6.3.6 (iii) $U(r,z)$ and, hence, $X(t,s)$ is strictly positive. From (i) we have for $u = \phi(t,x)$ and $v = \phi(t,y)$ that $u < v$ and, by assumption (d), $J_F(t,v) \lneqq J_F(t,u)$.

From $J_F(t,u) = A(t,x), J_F(t,v) = A(t,y)$ we conclude from the definition of $B(t,x,y)$ that $B(t,x,y) \gneqq 0$. From equation $(**)$ we obtain $Q(t) \gneqq 0$ for all $t > 0$. Finally from the definition of $Q(t)$ it follows that $J_T(x) - J_T(y) = U(\tau,x) - U(\tau,y) = Q(\tau) \gneqq 0$. This proves (ii).

(iii) To apply Theorem 6.3.2 (ii) it remains to show that $\sum_{j=1}^n x_j |\frac{\partial T_i}{\partial x_j}(x)| \le T_i x$ for $x > 0$, all i and that $Ty > 0$ for $y \gneqq 0$. By the mean value theorem and step (ii),

$$T_i x = \sum_{j=1}^n x_j \frac{\partial T_i}{\partial x_j}(\Theta x) \ge \sum_{j=1}^n x_j \frac{\partial T_i}{\partial x_j}(x) \quad \text{for} \quad x > 0,$$

which together with $J_T(x) \ge 0$ yields the first required condition. Furthermore, to $y \gneqq 0$ there exists $z > 0$ with $y \le z$. For $0 < \epsilon < 1$ arbitrary and $x = y + \epsilon z$ it holds $0 < x < 2z$. Therefore, using step (ii)

$$T_i x \ge \sum_{j=1}^n x_j \frac{\partial T_i}{\partial x_j}(x) \ge \sum_{j=1}^n x_j \frac{\partial T_i}{\partial x_j}(2z) \ge \sum_{j=1}^n y_j \frac{\partial T_i}{\partial x_j}(2z).$$

Letting $\epsilon \longrightarrow 0$ this yields, because of $J_T(2z) > 0$, $T_i y \ge \sum_{j=1}^n y_j \frac{\partial T_i}{\partial x_j}(2z) > 0$. This proves $Ty > 0$.

(iv) By (i) to (iii) all the assumptions of Theorem 6.3.2 (ii) are satisfied and the strong limit set trichotomy applies to T. Since $Tx = \phi(\tau,x)$ we have for the iterates $T^k x = \phi(k\tau,x)$. If $t \ge 0$, $t = k\tau + s$, $0 \le s < \tau$, then

$$\phi(t,x) = \phi(k\tau + s,x) = \phi(k\tau,\phi(s,x)) = T^k\phi(s,x).$$

If $(T^k x)_k$ is unbounded for all $x \gneqq 0$ then the solution $\phi(t,x)$ is unbounded for all $x \gneqq 0$. If $\lim_{k\to\infty} T^k x = 0$ for all $x \ge 0$ then $\lim_{t\to\infty} \phi(t,x) = 0$ for all $x \ge 0$. Finally, consider the case $\lim_{k\to\infty} T^k x = x^*$ for all $x \gneqq 0$ where $x^* > 0$ is the unique non-zero fixed point of T. For $x \gneqq 0, s \ge 0, k \ge 1$

$$\phi(k\tau + s,x) = \phi((k-1)\tau,\phi(s,\phi(\tau,x))) = T^{k-1}\phi(s,\phi(\tau,x)).$$

From (iii) $\phi(\tau,x) = Tx > 0$ and by (i), $\phi(s,\phi(\tau,x)) > 0$. Therefore, $\lim_{k\to\infty} \phi(k\tau + s,x) = x^*$ and for any neighborhood U of x^* there exists $k(s) \ge 1$ such that $\phi(k\tau + s,x) \in U$ for $k \ge k(s)$. By continuity of $\phi(\cdot,x)$ there exists a neighborhood V of s such that $\phi(k\tau + s',x) \in U$ for $s' \in V$ and $k \ge k(s)$. Since $[0,\tau]$ is compact there exists K such that $\phi(k\tau + s',x) \in U$ for all $k \ge K$, all $0 \le s' \le \tau$. From this we conclude that $\lim_{t\to\infty} \phi(t,x) = x^*$ for $x \gneqq 0$. Finally, let $x = 0$ and suppose $\phi(s,0) \gneqq 0$ for some $s > 0$. Then $\phi(t,0) = \phi(t-s,\phi(s,0))$ for $t \ge s$ and, hence, $\lim_{t\to\infty} \phi(t,0) = \lim_{t\to\infty} \phi(t-s,\phi(s,0)) = x^*$. This completes the limit set trichotomy for the solutions $\phi(\cdot,x)$. □

In terms of solutions we may view the limit set trichotomy of Theorem 6.3.7 as follows.

Corollary 6.3.8. *Assume in addition to conditions* (a)–(d) *of Theorem 6.3.7 that there exists at least one non-zero bounded solution. Then there exists a unique constant and global attractive solution. This solution is either the zero solution or, if* $F(\cdot, 0)$ *is not identically* 0*, a strict positive solution.*

Proof. By the additional assumption only cases (ii) and (iii) in the limit set trichotomy of Theorem 6.3.7 are possible. In case (ii) we have $\lim_{t\to\infty} \phi(t,x) = 0$ for all $x \geq 0$. By continuity, for any $s \geq 0$

$$\phi(s,0) = \phi(0, \lim_{t\to\infty} \phi(t,x)) = \lim_{t\to\infty} \phi(s, \phi(t,x)) = \lim_{t\to\infty} \phi(s+t,x) = 0.$$

That is, $\phi(\cdot, 0)$ is the zero solution and, since $\lim_{t\to\infty}(\phi(t,x) - \phi(t,0)) = \lim_{t\to\infty} \phi(t,x) = 0$, it is globally (for $x \geq 0$) attractive. There is, of course, in case (ii) no other constant solution globally attractive. Furthermore, since $\phi(\cdot, 0)$ is a solution of $x'(t) = F(t, x(t))$ we must have that $F(t, 0) = 0$ for all $t \geq 0$. For case (iii) we have $\lim_{t\to\infty} \phi(t,x) = x^*$ for all $x \gneqq 0$. By continuity, as above, $\phi(s, x^*) = x^*$ for all $s \geq 0$, that is the solution $\phi(\cdot, x^*)$ is constant to $x^* > 0$. It is globally attractive (for $x \gneqq 0$) since $\lim_{t\to\infty}(\phi(t,x) - \phi(t,x^*)) = 0$. No other constant solution is, in case (iii), globally attractive. Finally, if $F(\cdot, 0)$ is not identically zero, case (ii) is not possible by the above. Thus, we must have case (iii) and the solution under consideration must be strictly positive. □

The last Theorem as well as its Corollary will be illustrated by the following example from biology.

Example 6.3.9 (Biochemical control circuit (cf. [7, 23, 24]). A biochemical control circuit (or a single loop positive feedback system), which models for example the control of protein synthesis in the cell (cf. [24, p. 58]), is given by the system of equations

$$\begin{aligned} x_1'(t) &= f(t, x_n(t)) - \alpha_1(t)x_1(t) \\ x_i'(t) &= x_{i-1}(t) - \alpha_i(t)x_i(t) \text{ for } 2 \leq i \leq n. \end{aligned} \tag{6.3.8}$$

where $f\colon \mathbb{R} \times \mathbb{R}_+ \longrightarrow \mathbb{R}_+$ and $f(t, \cdot)$ is continuously differentiable on int $\mathbb{R}^n_+$. Assume the following concavity condition for f
$0 < v < w$ implies $0 < \frac{\partial f}{\partial u}(t, w) < \frac{\partial f}{\partial u}(t, v)$ for $t \geq 0$. Furthermore, we assume for some $\tau > 0$ that $\alpha_i(\cdot)$ and $f(\cdot, u)$ are τ-periodic in t. We write system (6.3.8) as

$$\begin{aligned} x'(t) &= F(t,(x)) \quad \text{with} \\ F_1(t,x) &= f(t, x_n) - \alpha_1(t)x_1, \quad F_i(t,x) = x_{i-1} - \alpha_i(t)x_i, \quad 2 \leq i \leq n. \end{aligned} \tag{6.3.9}$$

This system, obviously, is of the type 6.3.3 considered earlier, it is τ-periodic and it satisfies the Kamke condition. Furthermore, for $x > 0$ the Jacobian of $F(t, \cdot)$ is

$$J_F(t,x) = \begin{bmatrix} -\alpha_1(t) & 0\cdots 0 & & \frac{\partial f}{\partial u}(t,x_n) \\ 1 & -\alpha_2(t) & 0\cdots & 0 \\ 0 & 1 & -\alpha_3(t) & \cdots 0 \\ \vdots & & & \vdots \\ 0 & \cdots & 01 & -\alpha_n(t) \end{bmatrix}.$$

Therefore, $J_F(t,x)$ is a Metzler matrix which is irreducible for $t > 0, x > 0$. The concavity condition implies $J_F(t,y) \lneqq J_F(t,x)$ for $0 < x < y$. Thus, all assumptions of Theorem 6.3.7 are satisfied and the limit set trichotomy obtained holds for system (6.3.8). According to Corollary 6.3.8, if there exists a bounded orbit and $F(t,0) \neq 0$ for some t then there exists a unique strictly positive solution which is constant and attracts all solutions not beginning in 0. (This may be, however, also the case if $F(t,0) = 0$ for all t. For a particular example of a biochemical control circuit see Exercise 12.)

We shall conclude with some further remarks concerning the literature.

Remarks 6.3.10. (i) Theorem 6.3.2, part (ii) generalizes results in [15, Theorem 2] and [25, Theorem 2.1]. (See Exercise 8.) The condition $J_S(x) \lneqq J_S(tx)$ in part (ii) is a strong concavity condition which, however, is weaker than the condition $J_S(y) \lneqq J_S(x)$ for $0 < x < y$ in [23] which in turn is weaker than a similar condition in [6] (see [23, p. 1038]).

(ii) With a different proof, Theorem 6.3.3 (ii) was obtained in [14, Theorem 2(ii)]. Example 6.3.4 (ii) can be found there, too [14, Example 1].

(iii) More on the Kamke condition (Lemma 6.3.6), sometimes also called Müller–Kamke condition, can be found in [4], [9], [24]. Part (iii) of Lemma 6.3.6 refines results in [7, Theorem 1.1] and [24, Theorem 1.1], respectively, where it is proven that $U(t,x) > 0$ for $A(t,x)$ irreducible.

(iv) Though not in terms of a limit set trichotomy Theorem 6.3.7 can be found essentially in [23, Theorem 3.1]. Example 6.3.9 can be found there, too [23, p. 1049].

(v) The theory of monotone dynamical systems developed in [6, 7, 9, 23, 24] is about semiflows on a partially ordered metric space. When applied to selfmappings of a convex cone this theory requires the selfmapping to be monotone. This is different from positive dynamical systems as treated in this book. For linear selfmappings, of course, positivity is equivalent to monotonicity, for non-linear selfmappings of a convex cone, however, this need not necessarily be the case. As remarked already, cone mappings (or mappings non-expansive for the part metric) as considered for limit set trichotomy need not be monotone (see Remark 6.1.4).

Exercises

1. For Theorem 6.1.1 A find examples which show that each of the three cases of the weak limit set trichotomy is possible.
2. Let T be a selfmapping of a cone K which satisfies the general assumptions of Theorem 6.1.1 and which maps $K \setminus \{0\}$ into itself. Show that the (strong) limit set trichotomy holds for T if it holds for an iterate of T.
3. Find an example of a concave selfmapping T of $\mathbb{R}^n_+$ such that the strong limit set trichotomy holds for T but does not follow from the trichotomy for the dominant eigenvalue (in the sense of Theorem 2.3.1 (i)).
4. Show for the selfmapping of $\mathbb{R}_+$ given by $Tx = x + \frac{1}{1+x}$.
   (a) T is neither monotone nor antimonotone nor subhomogeneous.
   (b) T is non-expansive for the part metric.
5. Let K be a closed convex cone in a Banach space $(E, \|\cdot\|)$ which is normal with non-empty interior $\mathring{K}$. Let T be a selfmapping of $\mathring{K}$ which is compact for $\|\cdot\|$, non-expansive for p and monotone (for K). Suppose T has a fixed point $x^* \in \mathring{K}$ such that
$$\alpha x^* \lneqq T(\alpha x^*) \quad \text{and} \quad T(\alpha^{-1}x^*) \lneqq \alpha^{-1}x^*$$
   for all $0 < \alpha < 1$ and strict inequalities ($<$ for $\mathring{K}$) if $\alpha_0 \le \alpha < 1$ for some $0 < \alpha_0$. Obtain from Theorem 6.1.10 that $\lim_{n\to\infty} T^n x = x^*$ for all $x \in \mathring{K}$. (Cf. [10] where the above conclusion is obtained for a monotone and subhomogeneous selfmapping T of $\mathring{K}$ for $K = \mathbb{R}^n_+$ and a fixed point $x^* \in \mathring{K}$ of T which satisfies
$$\alpha^\delta x^* \le T(\alpha x) \quad \text{and} \quad T(\alpha^{-1}x^*) \le \alpha^{-\delta}x^*$$
   for some $0 < \delta < 1$ and all $0 < \alpha < 1$.)
6. Prove the following geometric interpretation of
$$\sum_{j=1}^{n} x_j \left| \frac{\partial T_i}{\partial x_j}(x) \right| \le T_i x, \quad 1 \le i \le n, \ x \in \mathring{K}$$
   where T is a selfmapping of $K = \mathbb{R}^n_+$ and a continuously differentiable selfmapping of $\mathring{K}$.
   If $A_T(x) = (x, Tx) + \{(u, J_T(x)u) \mid u \in K\}$ is the (positive) affine tangent space at $P(x) = (x, Tx)$ then
$$A_T(x) \subseteq P(x) + C(x)$$
   where
$$C(x) = \left\{ (u, v) \in K \times \mathbb{R}^n \;\middle|\; \max_{1\le i\le n} |v_i| \le c \sum_{i=1}^{n} u_i \right\}$$
   is a polyhedral convex cone with $c = \max_{1\le i,j\le n} \frac{T_i x}{x_j}$.

7. Let f be a selfmapping of $\mathbb{R}_+$ given by $f(x) = \lambda x(1+x)^{-\beta}$ with $\lambda > 1, \beta > 0$.
   (a) Prove the statements made in Example 6.2.7.
   (b) Show that the contraction constant $c(f)$ (see Theorem 6.2.1) is given by $c(f) = \max\{|1-\beta|, 1\}$.
   (c) Find for the unique fixed point $x^* = \lambda^{\frac{1}{\beta}} - 1$ a parameter value $\lambda_0 > 0$ such that x^* is globally attractive for all $1 < \lambda < \lambda_0$ and all $\beta > 0$.
   (d) Find values for parameters λ and β for which x^* is locally but not globally attractive.

8. Consider the following population models $u(t+1) = f(u(t)), u(0) \in \mathbb{R}_+, f\colon \mathbb{R}_+ \longrightarrow \mathbb{R}_+$:
   (i) The generalized Pielou equation with
   $$f(x) = \lambda x(1+ax)^{-b} + cx + d \quad \text{with} \quad \lambda > 0,\ a > 0,\ b > 0;\ c, d \geq 0.$$
   (ii) The bobwhite quail population with
   $$f(x) = \lambda x(1+x^k)^{-1} + cx + d \quad \text{with} \quad \lambda, k > 0 \quad \text{and} \quad c, d \geq 0.$$
   (a) Show that the limit set trichotomy holds for i) if $b \leq \max\{2, \frac{c}{\lambda} + 1\}$.
   (b) Show that limit set trichotomy holds for ii) if $k \leq 2 + 3\frac{c}{\lambda}$.
   (c) Find for both models parameter values for the three cases of the limit set trichotomy to hold.

9. Deduce the following result from Theorem 6.3.2 (ii) [15, Theorem 2].
   Let $K = \mathbb{R}^n_+$ and T a continuous selfmapping of K which is a continuously differentiable selfmapping of $\overset{\circ}{K}$ satisfying the following assumptions
   (a) $0 < x \leq y$ implies $0 \leq J_T(y) \leq J_T(x)$,
   (b) for some iterate S of T it holds that
   $$0 < J_S(x) \lneqq J_S(tx) \quad \text{for all} \quad x \in \overset{\circ}{K}, \quad \text{all} \quad 0 < t < 1.$$
   Then T has strong limit set trichotomy.

10. Find an example of a non-constant mapping T which satisfies the general assumptions of Theorem 6.3.2 and to which part (i) applies but not part (ii).

11. Find an example of a concave mapping $f\colon \mathbb{R}_+ \longrightarrow \mathbb{R}_+, f(x) > 0, f$ not constant to which part (iii) of Theorem 6.3.3 applies but not part (ii).

12. Prove the following superposition principle for difference equations [12, p. 333]. Let, for $K = \mathbb{R}^n_+, f\colon K \longrightarrow \mathbb{R}_+, f(x) > 0$ for $x \gneqq 0$
    and $f = \sum_{i=1}^m f^i$ where for each $1 \leq i \leq m$ the mapping $f^i\colon K \longrightarrow \mathbb{R}_+$ is continuously differentiable on $\overset{\circ}{K}$ with
    $$\sum_{j=1}^n x_j \left|\frac{\partial f^i}{\partial x_j}(x)\right| \leq f^i(x). \tag{$*$}$$
    Then for the difference equation of order n defined by f (equation (6.3.1)) the limit

set trichotomy does hold provided that for each $x \in \overset{\circ}{K}$ one of the two following conditions is met:

(i) there exists an index $k = k(x)$ for which strict inquality holds in $(*)$ for $i = k$,

or

(ii) there exist indices h, k, l (depending possibly on x) such that

$$\frac{\partial f^h}{\partial x_l}(x) \cdot \frac{\partial f^k}{\partial x_l}(x) < 0.$$

13. Consider the following biochemical control circuit with parameter $\alpha_1, \alpha_2 \geq 0$

$$\begin{aligned} x_1'(t) &= \sqrt{x_2(t)} - \alpha_1 x_1(t) \\ x_2'(t) &= x_1(t) \quad - \alpha_2 x_2(t). \end{aligned}$$

(a) Prove that the assumptions of Theorem 6.3.7 are satisfied (for any τ given).

(b) Show that each of the three cases of the limit set trichotomy in Theorem 6.3.7 can occur for appropriate values of the parameters α_1, α_2.

(c) Show that the zero solution as well as a constant strictly positive solution occurs as the unique constant globally attractive solution for certain values of the parameter α_1, α_2. Illustrate by computer simulations the behavior of the solutions for such values of the parameters.

14. (a) Obtain from the weak limit set trichotomy in case of linear operators the following result [18, Theorem 3].

Let T be a linear and continuous operator on the Banach space $(E, \|\cdot\|)$ which leaves a closed convex and normal cone K in E as well as its interior $\overset{\circ}{K}$ invariant. Suppose $\overset{\circ}{K} \neq \emptyset$, T has a bounded orbit for some $x \in \overset{\circ}{K}$ and T satisfies the following positivity condition:

$$\text{for each } x \in K \setminus \{0\} \text{ there exists } n(x) \in \mathbb{N} \text{ such that } T^{n(x)}x \in \overset{\circ}{K}. \qquad (*)$$

Then all orbits of T are bounded and either $\lim_{n\to\infty} T^n x = 0$ for all $x \in K$ with $\omega(x) \neq \emptyset$

or

$$\lim_{n\to\infty} T^n x = c(x)x^* \quad \text{for all} \quad x \in K \setminus \{0\} \quad \text{with} \quad \omega(x) \neq \emptyset,$$

where $x^* \in \overset{\circ}{K}$ is a fixed point of T and $c(x) > 0$.

(b) (Cf. [18, Example 2].) Let E be the vector space of all converging real sequences equipped with the sup-norm with K consisting of all non-negative sequences. Let T the linear operator on E defined for $x = (x_n)_n$ by $T_n x = \sum_{i=1}^{n} \frac{1}{2^i} x_i + \frac{1}{2^n} x_{n+1}$. Show that all assumptions in (a) are satisfied and that $\lim_{n\to\infty} T^n x = c(x)x^*$ where x^* is the sequence consisting of 1 and $c(x) = \sum_{n=1}^{\infty} (c_n - c_{n-1}) x_n$ with $c_n = 2^{-\frac{(n-1)n}{2}}$.

(c) Use the example in (b) to show that property $(*)$ does not necessarily imply $S(K \setminus \{0\}) \subseteq \overset{\circ}{K}$ for some iterate S of T.

Bibliography

[1] L. Arnold and I. Chueshov. A limit set trichotomy for order-preserving random systems. *Positivity*, 5: 95–114, 2001.

[2] Y.-Z. Chen. The omega limit sets of ray-contractive operators. *J. Math. Anal. Appl.*, 261: 554–561, 2001.

[3] Y.-Z. Chen. Omega limit sets in positive cones. *Bull. London Math. Soc.*, 36:72–80, 2004.

[4] W. A. Coppel. *Stability and Asymptotic Behavior of Differential Equations*. D.C. Heath and Co., Boston 1965.

[5] D. Hinrichsen and A. J. Pritchard. *Mathematical Systems Theory I. Modelling, State Space Analysis, Stability and Robustness*. Springer, Berlin etc. 2005.

[6] M. W. Hirsch. The dynamical systems approach to differential equations. *Bull. AMS*, 11:1–64, 1984.

[7] M. W. Hirsch. Systems of differential equations which are competitive or cooperative II: convergence almost everywhere. *SIAM J. Math. Anal.* 16:423–439, 1985.

[8] M. W. Hirsch and S. Smale. *Differential Equations, Dynamical Systems, and Linear Algebra*. Academic Press, New York, 1974.

[9] M. W. Hirsch and H. Smith. Monotone dynamical systems. In: *Handbook of Differential Equations*. Ordinary Differential Equations, (Eds. A. Cañada, P. Drábek, and A. Fonda), pp. 239–356. Elsevier B.V., 2005.

[10] P. E. Kloeden and A. M. Rubinov. Attracting sets for co-radiant and topical operators. *Mathematische Nachrichten*, 243:134–145, 2002.

[11] P. E. Kloeden and A. M. Rubinov. A generalization of the Perron–Frobenius theorem. *Nonlinear Anal.*, 41:97–115, 2000.

[12] U. Krause. Stability trichotomy, path stability, and relative stability for positive non-linear difference equations of higher order. *J. Differ. Equations Appl.*, 1:323–346, 1995.

[13] U. Krause. Positive non-linear difference equations. *Nonlinear Anal.*, 30:301–308, 1997.

[14] U. Krause. A local-global stability principle for discrete systems and difference equations. In B. Aulbach et al., editor, *Proceedings of the Sixth International Conference on Difference Equations*, Augsburg 2001. CRC Press, Boca Raton, 2004.

[15] U. Krause and P. Ranft. A limit set trichotomy for monotone non-linear dynamical systems. *Nonlinear Anal.*, 19:375–392, 1992.

[16] U. Krause and R. D. Nussbaum. A limit set trichotomy for self-mappings of normal cones in Banach spaces. *Nonlinear Anal.*, 20:855–870, 1993.

[17] B. Lemmens and R. Nussbaum. *Nonlinear Perron–Frobenius theory*. Cambridge University Press, Cambridge 2012.

[18] B. M. Makarov and R. M. Weber. On the asymptotic behavior of some positive semigroups. *Preprint Technische Universität Dresden*, MATH-AN-09-2000.

[19] R. M. May. Simple mathematical models with very complicated dynamics. *Nature*, 261:459–467, 1976.

[20] T. Nesemann. A limit set trichotomy for positive non-autonomous discrete dynamical systems. *J. Math. Anal. Appl.*, 237:55–73, 1999.

[21] R. D. Nussbaum. Finsler structures for the part metric and Hilbert's projective metric and applications to ordinary differential equations. *Diff. and Integral Equations*, 7:1649–1707, 1994.

[22] Ch. Pötzsche and S. Siegmund. A limit set trichotomy for order-preserving systems on time scales. *Electronic J. Diff. Equations*, 64:1–18, 2004. http://ejde.math.txstate.edu

[23] H. L. Smith. Cooperative systems of differential equations with concave non-linearities. *Nonlinear Anal.*, 10:1037–1052, 1986.

[24] H. L. Smith. *Monotone Dynamical Systems. An Introduction to the Theory of Competitive and Cooperative Systems.* AMS, Providence, 1995.
[25] P. Takáč. Asymptotic behavior of discrete-time semigroups of sublinear, strongly increasing mappings with applications in biology. *Nonlinear Anal.*, 14:35–42, 1990.
[26] P. Takáč. Convergence in the part metric for discrete dynamical systems in ordered topological cones. *Nonlinear Anal.*, 26:1753–1777, 1996.
[27] F. Tschöpe. Limesmengentrichotomien für positive nichtlineare dynamische Systeme und einige Anwendungen. Diploma thesis, Universität Bremen, 70 pages, 1996.

7 Non-autonomous positive systems

Already in Chapter 1 "How positive discrete dynamical systems do arise" we met non-autonomous systems, that is systems where the law governing the dynamics does explicitly depend on time. For the Leslie model in population dynamics as well as for the Leontief model in economic production the dynamics were modelled as

$$x(t+1) = T(t)x(t), \quad t \in \mathbb{N}, \; x(0) \in \mathbb{R}^n_+$$

(see equations (1.2.9) and (1.4.2)).

The reason for considering the "law" $T(t)$ as dependent on time t was in the case of population dynamics the dependence of birth- and survival rates on time, due possibly to changes in the environment. In the case of economic production the cost function depends on time due to changes in the technology of production. From Chapter 1 up to Chapter 6 our focus was on the non-linearity of the selfmapping T and the asymptotic behavior of its iterates. To analyse the asymptotic behavior of a non-autonomous system one has to consider the iteration of time-dependent operators $T(t)$, also called inhomogeneous or nonhomogeneous iteration. In the particular case where $T(t)$ converges for $t \to \infty$ to some operator T one has – under certain assumptions – convergence of the inhomogeneous iterates $T(t) \circ T(t-1) \circ \cdots \circ T(1) \circ T(0)$ in which case one speaks of strong ergodicity. In general, however, such a convergence cannot be expected (see Exercise 8). A new kind of stability comes into play which we call path stability. Roughly speaking, this means that a path given by $x(t+1) = T(t)x(t)$ for $t \geq 0$, when disturbed suddenly at some t_0 comes asymptotically back to its original behavior. A strongly related notion is that of weak ergodicity. (The precise definitions are given below in Section 7.1.)

The concepts of weak and strong ergodicity were developed within the fields of demography and Markov chains and do have an interesting history which we shortly sketch. (The reader can find more details in the informative articles [5, 42, 43] and in Seneta's book [44]). As early as 1931 the concept of weak ergodicity was conceived of by A. N. Kolmogoroff who called it "das Ergodenprinzip", the ergodic principle [26, p. 424]. For an **inhomogeneous Markov chain**, i.e., a sequence $(P_k)_{k\geq 1}$ of row stochastic non-negative matrices consider the forward product

$$T_{r,k} = P_{r+1}P_{r+2}\dots P_{r+k} \quad \text{with entries} \quad t^{(r,k)}_{is}.$$

Following [42, p. 507] the sequence $(P_k)_{k\geq 1}$ is **weakly ergodic in the sense of Kolmogoroff** if for all $i, j, s = 1, \dots, n$ and $r \geq 0$

$$\lim_{k\to\infty} \left(t^{(r,k)}_{is} - t^{(r,k)}_{js} \right) = 0.$$

Subsequently, conditions for weak ergodicity to hold have been obtained by W. Doeblin in 1937, S. N. Bernstein in 1946, T. A. Sarymsakov in 1956. Independently of this

"Russian School", work by J. Hajnal from 1958 on became very influential. In particular, his work was employed by A. Lopez in 1961 to prove an empirically based conjecture by the demographer A. J. Coale in 1957. This conjecture, now famous as the Coale–Lopez Theorem in demography, states that the age structure in a population is determined by the vital rates and not by the age structure years ago. Demographers had observed earlier the phenomenon that the age structure when disturbed by a war approaches the one prior to the war. One might say, equivalently, that weak ergodicity amounts to path stability with respect to the structure under consideration. Interestingly enough, considering time-continuous population dynamics, weak ergodicity has been already recognized by H. T. J. Norton in 1928, wherefore sometimes the name Norton–Coale–Lopez Theorem is used. For time-continuous population dynamics see [22, 23] where a rigorous proof of this theorem in this setting is given. Kolmogoroff in his fundamental paper [26], too, treats mainly time-continuous processes.

In what follows, Section 7.1 supplies precise definitions as well as relationships between the concepts involved.

Sections 7.2 and 7.3, the central part of this chapter, are devoted to an analysis of weak and strong ergodicity for time dependent and non-linear operators on Banach spaces which are ascending for a convex cone. In particular, a concave extension of the Coale–Lopez Theorem for Banach spaces is proven. Also, the classical results on weak and strong ergodicity of inhomogeneous Markov chains are obtained as special cases.

Section 7.4 then applies the general results to obtain a non-linear extension of a classical result of H. Poincaré on non-autonomous linear difference equations.

Sections 7.5 and 7.6 present applications to the already mentioned non-autonomous systems in population dynamics and economic production, respectively.

7.1 The concepts of path stability, asymptotic proportionality, weak and strong ergodicity

The notion of weak ergodicity for Markov chains has been extended to sequences of non-negative matrices [44, p. 85]. A sequence $(P_k)_{k\geq 1}$ of non-negative matrices is **weakly ergodic** if for the entries $t_{is}^{(r,k)}$ of the forward product $T_{r,k} = P_{r+1} \cdots P_{r+k}$ it holds for all i, j, s and r that

$$\lim_{k\to\infty} \frac{t_{is}^{(r,k)}}{t_{js}^{(r,k)}} = v_{ij}^{(r)} \tag{7.1.1}$$

exists and is independent of s. In other words, any two rows i and j of $T_{r,k}$ become propotional for k approaching infinity. In the homogeneous case, that is $P_k = P$ for all k, weak ergodicity means that any two rows of the power P^k become proportional for k approaching infinity. Actually, this is part of the classical Perron–Frobenius theory as described earlier for a primitive matrix P (Theorem 2.4.1, part (iii) (c)). In the special case of a stochastic matrix P this is the so called basic limit theorem for Markov chains. It is easily seen that equation (7.1.1) implies in the case of stochastic matrices the weak ergodicity in the sense of Kolmogoroff. Obviously, more demanding than (7.1.1) is for a sequence $(P_k)_{k\geq 1}$ the following notion of **strong ergodicity** ([44, p.92])

$$\lim_{k\to\infty} \frac{t_{ij}^{(r,k)}}{\sum_{s=1}^{n} t_{is}^{(r,k)}} = v_j^{(r)}. \tag{7.1.2}$$

In other words, the sum-normed rows of $T_{r,k}$ become equal for k approaching infinity. In the homogeneous case, $P_k = P$ for all k, the two notions of weak and strong ergodicity do coincide.

In the following we shall extend the concepts of weak and strong ergodicity, as well as the related results, to non-linear and non-autonomous systems in normed spaces. For doing so, we formalize the proportionality properties in equations (7.1.1) and (7.1.2) in a way which will allow us to make a connection to the part metric and the Hilbert metric, respectively. This in turn will allow us to apply our results on non-autonomous systems in metric spaces from Section 4.2.

The following definition distinguishes three different kinds of a generalized proportionality, where the last two are taken from [45, p. 242].

Definition 7.1.1. Let K be a convex cone in a real vector space V. Two sequences (x_n) and (y_n) in K are called
(a) **asymptotically linked** if there exist two sequences of positive real numbers (γ_n) and $(\bar{\gamma}_n)$ such that $\gamma_n x_n \leq y_n \leq \bar{\gamma}_n x_n$ for finally all n and $\lim_{n\to\infty} \frac{\gamma_n}{\bar{\gamma}_n} = 1$ ($\leq$ order relation induced by K).
(b) **asymptotically proportional** if, in addition to (a), $\lim_{n\to\infty} \gamma_n$ and $\lim_{n\to\infty} \bar{\gamma}_n$ exist and are (strictly) positive.
(c) **asymptotically equal** if, in addition to (b), $\lim_{n\to\infty} \gamma_n = 1$.

These notions are connected to part metric p and Hilbert metric d, as well as to a given norm as follows. (See also Exercise 1.)

Lemma 7.1.2. *Let K be a lineless, archimedean, convex cone in a real vector space V and let $\|\cdot\|$ be a monotone norm on V. For two sequences (x_n), (y_n) contained in $K \setminus \{0\}$ the following properties do hold.*
(i) *(x_n) and (y_n) are asymptotically equal if and only if*

$$\lim_{n\to\infty} p(x_n, y_n) = 0.$$

(ii) *(x_n) and (y_n) are asymptotically linked if and only if*

$$\lim_{n\to\infty} d(x_n, y_n) = 0.$$

(iii) *If* $\lim_{n\to\infty} p(x_n, y_n) = 0$ *and at least one of the sequences is bounded for* $\|\cdot\|$ *then*

$$\lim_{n\to\infty} \| x_n - y_n \| = 0.$$

(iv) *If* $\lim_{n\to\infty} d(x_n, y_n) = 0$ *then*

$$\lim_{n\to\infty} \left\| \frac{x_n}{\| x_n \|} - \frac{y_n}{\| y_n \|} \right\| = 0.$$

(v) *If* $\| x_n \| = \| y_n \| = 1$ *for finally all n then the properties* $\lim_{n\to\infty} p(x_n, y_n) = 0$ *and* $\lim_{n\to\infty} d(x_n, y_n) = 0$ *as well as all three types in Definition 7.1.1 are equivalent to each other.*

These equivalent statements are all equivalent to $\lim_{n\to\infty} \| x_n - y_n \| = 0$ *if, in addition, one of the sequences is contained in* $z + K$ *for some* $z \in \overset{\circ}{K}$.

Proof. (1) Let $\gamma_n x_n \le y_n \le \bar{\gamma}_n x_n$ with $\gamma_n, \bar{\gamma}_n > 0$ for, without loss, $n \ge 1$. It follows

$$\lambda(x_n, y_n) \ge \gamma_n \quad \text{and} \quad \lambda(x_n, y_n) x_n \le \bar{\gamma}_n x_n$$

and, hence, $\gamma_n \le \lambda(x_n, y_n) \le \bar{\gamma}_n$. Similarly,

$$\bar{\gamma}_n^{-1} y_n \le x_n \le \gamma_n^{-1} y_n \quad \text{implies} \quad \bar{\gamma}_n^{-1} \le \lambda(y_n, x_n) \le \gamma_n^{-1}.$$

Therefore,

$$\bar{\gamma}_n^{-1} \gamma_n \le \lambda(x_n, y_n) \cdot \lambda(y_n, x_n) \le \gamma_n \bar{\gamma}_n^{-1}.$$

If the two sequences are asymptotically linked then

$$\lim_{n\to\infty} d(x_n, y_n) = -\lim_{n\to\infty} \log[\lambda(x_n, y_n) \lambda(y_n, x_n)] = 0.$$

In case, the sequences are asymptotically equal it follows that $\lim_{n\to\infty} \lambda(x_n, y_n) = \lim_{n\to\infty} \lambda(y_n, x_n) = 1$ and, hence, $\lim_{n\to\infty} p(x_n, y_n) = -\lim_{n\to\infty} \log \min\{\lambda(x_n, y_n), \lambda(y_n, x_n)\} = 0$.

(2) Conversely, if $\lim_{n\to\infty} p(x_n, y_n) = 0$ or $\lim_{n\to\infty} d(x_n, y_n) = 0$ we may assume that $\gamma_n = \lambda(x_n, y_n) > 0$ and $\bar{\gamma}_n^{-1} = \lambda(y_n, x_n) > 0$. By definition of $\lambda(\cdot, \cdot)$ we have that $\gamma_n x_n \le y_n \le \bar{\gamma}_n x_n$. If $\lim_{n\to\infty} d(x_n, y_n) = 0$ then $\lim_{n\to\infty} \gamma_n \bar{\gamma}_n^{-1} = 1$ and the two sequences are asymptotically linked.

If $\lim_{n\to\infty} p(x_n, y_n) = 0$ then $\lim_{n\to\infty} \min\{\gamma_n, \bar{\gamma}_n^{-1}\} = 1$. To $\epsilon > 0$ there exists N such that $1 - \epsilon \le \min\{\gamma_n, \bar{\gamma}_n^{-1}\}$ and, hence, $1 - \epsilon \le \gamma_n \le \bar{\gamma}_n \le (1 - \epsilon)^{-1}$ for all $n \ge N$.

This shows $\lim_{n\to\infty} \gamma_n = \lim_{n\to\infty} \bar{\gamma}_n = 1$.

Steps (1) and (2) together proof the two equivalences (i) and (ii).

(3) Considering property (iii) let $\lim_{n\to\infty} p(x_n, y_n) = 0$ and at least one of the sequences bounded for $\|\cdot\|$. By (i) the two sequences are asymptotically equal and, hence, both sequences are bounded for $\|\cdot\|$. Proposition 3.3.3 (vi) implies $\lim_{n\to\infty} \| x_n - y_n \| = 0$. Furthermore, this proposition implies property (iv), too.

(4) As for property (v), from Proposition 3.3.3 (i) we have $d(x, y) \le 2p(x, y)$ and, hence, $\lim_{n\to\infty} p(x_n, y_n) = 0$ implies $\lim_{n\to\infty} d(x_n, y_n) = 0$. If $\| x_n \| = \| y_n \| = 1$ then

$p(x_n, y_n) \le d(x_n, y_n)$ by Proposition 3.3.3 (vii) and, hence, $\lim_{n\to\infty} p(x_n, y_n) = 0$ is equivalent to $\lim_{n\to\infty} d(x_n, y_n) = 0$. By properties (i) and (ii), parts (a) and (c) in Definition 7.1.1 are equivalent and, hence, all three parts in this definition are equivalent. Finally, let $(x_n) \subseteq z + K$ for some $z \in \mathring{K}$. By property (iii), $\lim_{n\to\infty} \| x_n - y_n \| = 0$. From Proposition 3.4.12 (ii) it follows for $P = \mathring{K}$, F consisting of $\| \cdot \|$ that

$$h(x', x'') \le \frac{\| x' - x'' \|}{r} \quad \text{for} \quad \| x' - x'' \| \le r,$$

where $h(x, y) = 1 - \min\{\lambda(x, y), \lambda(y, x)\}$ is the Harnack metric. Thus, $\lim_{n\to\infty} \| x_n - y_n \| = 0$ implies $\lim_{n\to\infty} h(x_n, y_n) = 0$ and a fortiori $\lim_{n\to\infty} p(x_n, y_n) = 0$. This proves property (v) and the lemma altogether. □

In concluding this section we give the following precise definitions for path stability and the ergodic properties.

Definition 7.1.3. Let (X, ρ) be a metric space and a non-autonomous discrete dynamical system given by a sequence $(T_n)_n$ of selfmappings of X, that is

$$x_{n+1} = T_n x_n, \quad n \ge 1, \ x_1 \in X.$$

This system has **path stability for the metric on $D \subseteq X$** if $\lim_{n\to\infty} \rho(x_n, y_n) = 0$ for all $x_1, y_1 \in D$. In particular, if $(V, \| \cdot \|)$ is a normed space with a metric given by $\| \cdot \|$ then the system has **path stability for the norm on $D \subseteq V$** if $\lim_{n\to\infty} \| x_n - y_n \| = 0$ for all $x_1, y_1 \in D$. The system has **weak ergodicity on $D \subseteq V \setminus \{0\}$** if $\lim_{n\to\infty} \| \frac{x_n}{\|x_n\|} - \frac{y_n}{\|y_n\|} \| = 0$ for all $x_1, y_1 \in D$. It has **strong ergodicity on $D \subseteq V \setminus \{0\}$** if there exists $x^* \in D$ such that $\lim_{n\to\infty} \| \frac{x_n}{\|x_n\|} - x^* \| = 0$ for all $x_1 \in D$.

In the light of Lemma 7.1.2 we can describe the above concepts also in the following way. Let $(V, \| \cdot \|)$ be a normed space with a normal, closed, pointed, convex cone K. Let $D \subseteq \mathring{K} \neq \emptyset$ and a non-autonomous system on D given by a sequence of selfmappings $(T_n)_n$ of D. For the part metric p on $\mathring{K}$ the path stability on D is equivalent to asymptotic equality of any two paths starting in D. For the Hilbert metric d on $\mathring{K}$ the path stability on D is equivalent to asymptotic linkedness of any two paths starting in D. Furthermore, in this case path statility for d implies weak ergodicity. It also follows from Lemma 7.1.2 that $\lim_{n\to\infty} d(x_n, y_n) = 0$ is equivalent to $\lim_{n\to\infty} p(x_n, y_n) = 0$ on $D \cap \{x \in V \mid \| x \| = 1\}$. Moreover, in this case these properties are equivalent to weak ergodicity provided for one sequence, say (x_n), one has $x_n \ge \| x_n \| z$ for all n and some $z \in \mathring{K}$. This yields in particular that strong ergodicity with $x^* \in \mathring{K}$, $\| x^* \| = 1$ is equivalent to $\lim_{n\to\infty} d(x_n, x^*) = 0$ or $\lim_{n\to\infty} p(\frac{x_n}{\|x_n\|}, x^*) = 0$, respectively. Strong ergodicity is, of course, stronger than weak ergodicity and, as remarked earlier, the former cannot be expected for a non-autonomous system in general. Weak ergodicity or path stability, however, play an important role in applications, especially to population dynamics. The name "path stability" is due to the following interpretation. Suppose a path (x_n) of a non-autonomous system in a metric space (X, ρ) is at a certain point of time n_0 pushed from x_{n_0} to a different value $y \in D$. If (y_n) is the path which starts in y then path stability on D gives $\lim_{n\to\infty} \rho(x_n, y_n) = 0$. Thus, path stability means that a path when disturbed (within D) comes finally back to its original behavior. Furthermore, consider two paths (x_n) and (y_n) and let $\bar{x} = \lim_{k\to\infty} x_{n_k}$ a limit point of (x_n). Then $\rho(y_{n_k}, \bar{x}) \le \rho(y_{n_k}, x_{n_k}) + \rho(x_{n_k}, \bar{x})$ and, in case of path stability, $\lim_{k\to\infty} y_{n_k} = \bar{x}$. Thus, an important consequence of path stability is that all possible paths have the same limit set.

7.2 Path stability and weak ergodicity for ascending operators

Dealing with non-autonomous systems we extend the notion of an ascending selfmapping (operator) to a sequence (T_n) of operators. (For simplicity, ϕ is assumed to be continuous.)

Definition 7.2.1. Let K be a convex cone (not $\emptyset$, $\{0\}$) in a real vector space V and let "$\le$" be the partial order defined by K. A sequence $(T_n)_{n\ge1}$ of selfmappings of K is **uniformly ascending on $D \subseteq K$** (with ϕ) if there exists a continuous selfmapping ϕ of the open interval $]0, 1[$ with $\lambda < \phi(\lambda)$ such that for every $0 < \lambda < 1$ and every $x, y \in D$

$$\lambda x \lneqq y \quad \text{implies} \quad \phi(\lambda) T_n x \le T_n y \quad \text{for all} \quad n \ge 1.$$

A sequence $(T_n)_{n\ge1}$ is called **uniformly weakly ascending on $D \subseteq K$** (with ϕ) if there exists a ϕ as above such that for every $0 < \lambda < 1$ and every $x, y \in D$

$$\lambda x \lneqq y \lneqq \frac{1}{\lambda} x \quad \text{implies} \quad \phi(\lambda) T_n x \le T_n y \le \frac{1}{\phi(\lambda)} T_n x \quad \text{for all} \quad n \ge 1.$$

For the following result which will be fundamental for what follows we draw on earlier results on non-autonomous systems on metric spaces (Section 4.2).

Theorem 7.2.2. *Let $(V, \|\cdot\|)$ be a normed real vector space and let $K \subseteq V$ be a convex cone which is closed and normal with non-empty interior $\overset{\circ}{K}$. Let $(T_n)_{n\ge1}$ be a sequence of selfmappings of $\emptyset \subsetneqq D \subseteq \overset{\circ}{K}$ and let, for some $r \ge 1$, $S_m = T_{m+r-1} \circ \cdots \circ T_m$ be a sequence of lumped operators for $m \ge 1$. Consider the system on D, defined by*

$$x_{n+1} = T_n x_n \quad \text{for} \quad n \ge 1 \quad \text{and} \quad x_1 \in D \tag{7.2.1}$$

(i) *If $(S_m)_m$ is uniformly weakly ascending on D then for the system (7.2.1) path stability holds for the part metric on D and any two sequences (x_n) and (y_n) are asymptotically equal. In particular, for (7.2.1) weak ergodicity does apply.*
If one of these sequences is norm-bounded then path stability holds for the norm on D.

(ii) *Let $U = D \cap \{x \in V \mid \|x\| = 1\}$ and assume $\frac{x}{\|x\|} \in D$ for $x \in D$. If $(T_n)_n$ is uniformly ascending on U, then for the rescaled system given by the rescaled operators $(\tilde{T}_n)_n$, path stability holds for the Hilbert metric and the part metric as well as for the norm on U. In case only $(S_m)_m$ is assumed to be uniformly ascending on U but the T_n are ray-preserving on D, then for the system (7.2.1) path stability holds for the Hilbert metric on D and weak ergodicity for the norm on D.*

Proof. (i) Let $\mu(x,y) = \sup\{\lambda > 0 \mid \lambda x \le y \le \frac{1}{\lambda}x\}$ for $x, y \in D$. Since (S_m) is uniformly weakly ascending on D with ϕ continuous we obtain for all n that

$$\lambda x \le y \le \frac{1}{\lambda}x \quad \text{implies} \quad \phi(\lambda)S_m x \le S_m y \le \frac{1}{\phi(\lambda)}S_m x \quad \text{for} \quad x, y \in D,\ 0 < \lambda < 1.$$

Therefore, $\mu(S_m x, S_m y) \ge \phi(\mu(x,y))$ and, hence, for the part metric $p(S_m x, S_m y) \le -\log[\phi(\mu(x,y))]$.

Let $c(x,y) = -\log[\phi(\mu(x,y))]$. If $\mu(x,y) = 1$ then $x = y$ because K is normal. Therefore, for $x \ne y$ we have that $c(x,y) < -\log\mu(x,y) = p(x,y)$. (D,p) is a metric space with selfmappings T_n and lumped operators S_m for which $p(S_m x, S_m y) \le c(x,y)$. Thus, (S_m) is a contractive sequence consisting of non-expansive mappings. For any two orbits given by $x_{n+1} = T_n x_n, y_{n+1} = T_n y_n$ and $x_1, y_1 \in D$ we obtain from Theorem 4.2.2 (i) that $\lim_{n\to\infty} p(x_n, y_n) = \lim_{n\to\infty} c(x_n, y_n)$. By definitions of p and c therefore, $\lim_{n\to\infty}\mu(x_n, y_n) = \lim_{n\to\infty}[\phi(\mu(x_n, y_n))]$.

By continuity of ϕ this means for $a = \lim_{n\to\infty}\mu(x_n, y_n)$ that $\phi(a) = a$. Since $\lambda < \phi(\lambda)$ for $0 < \lambda < 1$ we must have that $a = 0$ or $a = 1$. The former is impossible since $\lim_{n\to\infty} p(x_n, y_n) = -\log a$ exists.

Thus, we arrive at $\lim_{n\to\infty} p(x_n, y_n) = -\log 1 = 0$ showing that path stability holds for the part metric on D. The remaining statements follow from Lemma 7.1.2.

(ii) Let $\nu(x,y) = \sup\{\lambda\mu \mid \lambda, \mu > 0, \lambda x \le y \le \frac{1}{\mu}x\}$ for $x, y \in D$.

(1) Consider first the case that $(T_n)_n$ is uniformly ascending on U. By continuity of ϕ we have that $\lambda x \le y \le \frac{1}{\mu}x$ implies $\phi(\lambda)T_n x \le T_n y \le \frac{1}{\phi(\mu)}T_n x$ and, hence,

$$\nu(T_n x, T_n y) \ge \sup\left\{\phi(\lambda)\phi(\mu) \mid \lambda, \mu > 0, \lambda x \le y \le \frac{1}{\mu}x\right\}.$$

This shows $\nu(T_n x, T_n y) \ge \phi(\lambda(x,y)) \cdot \phi(\lambda(y,x))$ for all $n \le 1$ and, hence, for the Hilbert metric d on U

$$d(\tilde{T}_n x, \tilde{T}_n y) = d(T_n x, T_n y) \le c(x,y) \quad \text{for} \quad x, y \in U,$$

where $c(x,y) = -\log[\phi(\lambda(x,y)) \cdot \phi(\lambda(y,x))]$.

For $x, y \in U$ we have $0 < \lambda(x,y), \lambda(y,x) \le 1$. If $\lambda(x,y) = 1$ or $\lambda(y,x) = 1$ then $x = y$ because K is normal. Thus, for $x \ne y$ we must have that $\lambda(x,y)\lambda(y,x) < \phi(\lambda(x,y)) \cdot \phi(\lambda(y,x))$ and, hence, $c(x,y) < -\log[\lambda(x,y) \cdot \lambda(y,x)] = d(x,y)$. Thus, the sequence $(\tilde{T}_n)_n$ of rescaled operators is on the metric space (U,d) a contractive sequence consisting of non-expansive mappings. For any two orbits given by $\tilde{x}_{n+1} = \tilde{T}_n\tilde{x}_n, \tilde{y}_{n+1} = \tilde{T}_n\tilde{y}_n$ with

$\tilde{x}_1 = x_1, \tilde{y}_1 = y_1$ in U we obtain from Theorem 4.2.2 (i) that $\lim_{n\to\infty} d(\tilde{x}_n, \tilde{y}_n) = \lim_{n\to\infty} c(\tilde{x}_n, \tilde{y}_n)$. By definitions of d and c, therefore,

$$\lim_{n\to\infty} [\lambda(\tilde{x}_n, \tilde{y}_n)\lambda(\tilde{y}_n, \tilde{x}_n)] = \lim_{n\to\infty} [\phi(\lambda(\tilde{x}_n, \tilde{y}_n))\phi(\lambda(\tilde{y}_n, \tilde{x}_n))].$$

Since $0 \le \lambda(u, v) \le 1$ for $\| u \| = \| v \| = 1$ there exists a sequence of natural numbers $(n_k)_k$ such that $\lim_{k\to\infty} \lambda(\tilde{x}_{n_k}, \tilde{y}_{n_k}) = \lambda$, $\lim_{k\to\infty} \lambda(\tilde{y}_{n_k}, \tilde{x}_{n_k}) = \mu$. By continuity of ϕ we arrive at $\lambda\mu = \phi(\lambda)\phi(\mu)$. Because $\lim_{n\to\infty} d(\tilde{x}_n, \tilde{y}_n)$ exists we cannot have $\lambda\mu = 0$, that is $\lambda > 0$ and $\mu > 0$. If $\lambda < 1$ or $\mu < 1$ then $\lambda < \phi(\lambda)$ or $\mu < \phi(\mu)$ which implies $\lambda\mu < \phi(\lambda)\phi(\mu)$ – a contradiction. Thus, $\lambda = \mu = 1$ which yields $\lim_{n\to\infty} d(\tilde{x}_n, \tilde{y}_n) = \lim_{k\to\infty} d(\tilde{x}_{n_k}, \tilde{y}_{n_k}) = -\log(\lambda\mu) = 0$, that is, for the rescaled system path stability holds for Hilbert's metric on U. Path stability holds for the part metric as well as for the norm on U by Lemma 7.1.2, (iv) and (v).

(2) Consider now the case where (S_m) is uniformly ascending on U and the T_n are ray-preserving on D. As in step (1) it follows for $x, y \in U$ that $d(S_m x, S_m y) \le c(x, y)$, $c(x, y) < d(x, y)$ for $x \ne y$ where $c(x, y) = -\log[\phi(\lambda(x, y))\phi(\lambda(y, x))]$. Now, for the rescaled operators of ray-preserving mappings $S, T\colon D \to D$ it holds with some $\alpha > 0$

$$(\tilde{S} \circ \tilde{S})x = \frac{S(\tilde{T}x)}{\| S(\tilde{T}x) \|} = \frac{\alpha S(Tx)}{\| \alpha S(Tx) \|} = \frac{(S \circ T)x}{\| (S \circ T)x \|} = (\widetilde{S \circ T})x.$$

Therefore, by iteration $\tilde{S}_m = \tilde{T}_{mr} \circ \cdots \circ \tilde{T}_{(m-1)r+1}$ and for the lumped operators we have $d(\tilde{S}_m x, \tilde{S}_m y) = d(S_m x, S_m y) \le c(x, y)$. From Theorem 4.2.2 (i) we obtain $\lim_{n\to\infty} d(\tilde{x}_n, \tilde{y}_n) = \lim_{n\to\infty} c(\tilde{x}_n, \tilde{y}_n)$. As in step (1) this yields $\lim_{n\to\infty} d(\tilde{x}_n, \tilde{y}_n) = 0$. Since the T_n are ray-preserving, $\tilde{x}_n$ is obtained by applying the rescaled operator of $T_{n-1} \circ \cdots \circ T_1$ to x_1, that is $\tilde{x}_n = \frac{x_n}{\|x_n\|}$. Therefore, $\lim_{n\to\infty} d(x_n, y_n) = 0$ and for the original system holds path stability for the Hilbert metric on D. Finally, weak ergodicity for the norm on D follows from Lemma 7.1.2 (iv). □

Theorem 7.2.2 has interesting consequences. (For an example see Exercise 2.) It yields in particular the following extension of the Coale–Lopez theorem mentioned earlier to non-linear operators in infinite dimensions.

Corollary 7.2.3 (Concave weak ergodicity/concave Coale–Lopez theorem). *Let K be a convex cone in the normed space $(V, \| \cdot \|)$ as in Theorem 7.2.2. Let (T_n) be a sequence of concave selfmappings of $\mathring{K}$ and let $S_m = T_{m+r-1} \circ \cdots \circ T_m$, $m \ge 1$ be a sequence of lumped operators. Consider the system on $\mathring{K}$ given by $x_{n+1} = T_n x_n$ for $n \ge 1$, $x_1 \in \mathring{K}$.*

(i) *If, for some $e \in \mathring{K}$ and real numbers $0 < r \le s$,*

$$re \le T_n x \le se \quad \text{for all} \quad n \ge 1, \quad \text{all} \quad x \in U = \{x \in \mathring{K} \mid \|x\| = 1\}, \tag{7.2.2}$$

then for the rescaled system given by $(\tilde{T}_n)_n$ path stability holds on U for the Hilbert metric, the part metric as well as for the norm.

(ii) *If* $re \leq S_m x \leq se$ *for all* $m \geq 1$, *all* $x \in U$ *and the* T_n *are ray-preserving on* $\overset{\circ}{K}$ *then for the original system* $x_{n+1} = T_n x_n$ *path stability holds for the Hilbert metric and weak ergodicity for the norm on* $\overset{\circ}{K}$. *If the* T_n *are even positively homogeneous, then any two orbits of the original system are asymtotically proportional.*

Proof. (i) Let $\lambda x \lneqq y$ for $x, y \in U$ and $0 < \lambda < 1$. If $0 < \lambda' < \lambda$ then $y - \lambda' x = (y - \lambda x) + (\lambda - \lambda')x \in \overset{\circ}{K}$ and, hence, $z = \frac{y - \lambda' x}{\|y - \lambda' x\|} \in U$. Since $1 = \|y\| \leq \| y - \lambda' x \| + \| \lambda' x \|$ it follows that $\| y - \lambda' x \| \geq 1 - \lambda'$ and, hence, $y = \lambda' x + \| y - \lambda' x \| z \geq \lambda' x + (1 - \lambda')z$. Condition (7.2.2) and concavity of T_n imply that

$$T_n y \geq \lambda' T_n x + (1 - \lambda')\frac{r}{s} T_n x = (\lambda' + (1 - \lambda')\frac{r}{s}) T_n x \quad \text{for all } n.$$

Since $0 < \lambda' < \lambda$ is arbitrary it follows

$$T_n y \geq \phi(\lambda) T_n x \quad \text{for all} \quad n \quad \text{with} \quad \phi(\lambda) = \lambda + (1 - \lambda)\frac{r}{s}.$$

Thus, the sequence (T_n) is uniformly ascending on U with ϕ and (i) follows from Theorem 7.2.2 (ii).

(ii) Obviously, each S_m is concave and therefore from the assumption made on S_m it follows as under (i) that (S_m) is uniformly ascending on U with $\phi(\lambda) = \lambda + (1 - \lambda)\frac{r}{s}$. By Theorem 7.2.2 (ii) the original system has on $\overset{\circ}{K}$ path stability for the Hilbert metric and weak ergodicity for the norm. Suppose now the T_n are positively homogeneous. Being concave, T_n is monotone on $\overset{\circ}{K}$ and, hence, $\lambda(x_n, y_n) x_n \leq y_n$ implies $\lambda(x_n, y_n) T_n x_n \leq T_n y_n$ for any wo orbits $(x_n), (y_n)$ of the original system. It follows $\lambda(x_{n+1}, y_{n+1}) \geq \lambda(x_n, y_n)$ and, similarly, $\lambda(y_{n+1}, x_{n+1}) \geq \lambda(y_n, x_n)$.

Since $\lambda(x_n, y_n)\lambda(y_n, x_n) \leq 1$ for all n it follows $\lambda(x_n, y_n)\lambda(y_1, x_1) \leq 1$ for all n. Thus, $(\lambda(x_n, y_n))_n$ is a monotone bounded sequence and converges to some λ. Similarly, $\lambda(y_n, x_n)$ converges to some μ. From $\lim_{n\to\infty} d(x_n, y_n) = 0$ we have $\lim_{n\to\infty} [\lambda(x_n, y_n)\lambda(y_n, x_n)] = 1$ and, hence $\lambda\mu = 1$.

Thus, for $\gamma_n = \lambda(x_n, y_n)$ and $\bar{\gamma}_n = \lambda(y_n, x_n)^{-1}$ we have that $\gamma_n x_n \leq y_n \leq \bar{\gamma}_n x_n$ with $\lim_{n\to\infty} \gamma_n = \lambda$, $\lim_{n\to\infty} \bar{\gamma}_n = \mu^{-1}$. Because of $\lambda = \mu^{-1}$ the two orbits are asymptotically proportional which proves (ii). □

Corollary 7.2.3 applies especially to linear operators and we obtain in particular the classical weak ergodicity result or common Coale–Lopez theorem for non-negative matrices as discussed in Section 7.1.

Corollary 7.2.4 (Linear weak ergodicity/linear Coale–Lopez theorem). (i) *Let* K *be a convex cone in the normed space* $(V, \|\cdot\|)$ *as in Theorem 7.2.2. Let* (T_n) *be a sequence of linear selfmappigs of* $\overset{\circ}{K}$ *such that for some* $e \in \overset{\circ}{K}$ *and* $0 < r \leq s$ *for the lumped operators* S_m *it holds*

$$re \leq S_m x \leq se \quad \textit{for all} \quad m \geq 1, \quad \textit{all} \quad x \in U.$$

Then any two orbits of the sytem $x_{n+1} = T_n x_n$ *on* $\overset{\circ}{K}$ *are asymptotically proportional.*

(ii) *Let $(P_n)_n$ be a sequence of non-negative $d \times d$-matrices such that uniformly for all $n \geq 1$ the maximal entry of P_n is bounded from above and the minimal strictly positive entry is bounded from below. If for some $r \geq 1$ the products $P_{m+r-1} \cdots P_m$ are strictly positive for all $m \geq 1$ then for all $1 \leq i, j, k \leq d$*

$$\lim_{n\to\infty} \frac{(P_n \cdots P_1)_{ik}}{(P_n \cdots P_1)_{ij}} = v_{kj},$$

that is the columns of the matrix $P_n \cdots P_1$ tend to be proportional as n tends to ∞.

Proof. (i) is a special case of Corollary 7.2.3 (ii).
(ii) By assumption, for the entries $p_{ij}(n)$ of P_n

$$p_{ij}(n) \leq \beta, p_{ij}(n) = 0 \quad \text{or} \quad \alpha \leq p_{ij}(n)$$

for all i, j, n and some real numbers $0 < \alpha \leq \beta$. For $S_m = P_{m+r-1} \cdots P_m$ and $x \in \mathbb{R}^d_+ \setminus \{0\}$ $(S_m x)_i = \sum_{(i_1,\ldots,i_{r-1},j)} p_{ii_1}(m+r-1, j) \cdots p_{i_r-1,j}(m) x_j$ and, hence,

$$\alpha^r \sum_{j=1}^d x_j \leq (S_m x)_i \leq \beta^r \sum_{j=1}^d x_j \quad \text{for all } m, \quad \text{all } i.$$

For $K = \mathbb{R}^n_+, e = (1, \ldots, 1)' \in \mathring{K}$ and $\|x\| = \sum_{i=1}^d |x_i|$ from part (i) it follows for any two orbits $(x_n), (y_n)$ of $x_{n+1} = P_n x_n$ on $\mathring{K}$

$$\gamma_n x_n \leq y_n \leq \bar{\gamma}_n x_n \quad \text{for all } n \quad \text{and} \quad \lim_{n\to\infty} \gamma_n = \lim_{n\to\infty} \bar{\gamma}_n = \gamma > 0.$$

Since S_m is strictly positive we can choose as starting points $x_1 = e_j, y_1 = e_k$ (e_i the i-th unit vector) to obtain

$$(x_{n+1})_i = (P_n \ldots P_1)_{ij} \quad \text{and} \quad (y_{n+1})_i = (P_n \ldots P_1)_{ik} \quad \text{for all } i.$$

Therefore,

$$\lim_{n\to\infty} \frac{(P_n \ldots P_1)_{ik}}{(P_n \ldots P_1)_{ij}} = \lim_{n\to\infty} \frac{(y_{n+1})_i}{(x_{n+1})_i} = \gamma \quad \text{for all } i,$$

where γ does depend on k and j only. □

Another consequence of Theorem 7.2.2 is the following differentiability criterion in finite dimensions which will be useful later on for an application to non-autonomous population dynamics (see Section 7.4, for examples see Exercises 4, 5, 6). This criterion is complementary to Corollary 7.2.3. Whereas the latter assumes concavity and positive homogeneity (for the original system) none of these assumptions is required for this criterion. On the other hand, the criterion does not apply to positive homogeneous, especially linear, operators (see Example 7.2.7 below).

Corollary 7.2.5. *Let $V = \mathbb{R}^d, K = \mathbb{R}^d_+, \|\cdot\|$ any monotone norm on $\mathbb{R}^d$. For a log-convex subset $D \subseteq \mathring{K}$ let (T_n) be a sequence of selfmappings of D differentiable on $\mathring{K}$ and let for lumped operators S_m and $x \in \mathring{K}$*

$$A(x) = \sup_{m\geq 1} \max_{1\leq i\leq d} \sum_{j=1}^{d} \frac{x_j}{(S_m x)_i} \left| \frac{\partial (S_m)_i}{\partial x_j}(x) \right|.$$

Consider for the system $x_{n+1} = T_n x_n, x_1 \in D$, the following two cases:

Case (a). *$A(x) \leq c < 1$ for all $x \in D$.*

Then all orbits of the system are asymptotically equal and weak ergodicity does apply. If at least one orbit is norm-bounded then path stability for the norm holds on D.

Case (b). *$A(x) < 1$ for all $x \in D$.*

Then any two orbits contained in a bounded and closed subset of D are asymptotically equal. Path stability for the norm holds for all orbits contained in a bounded and closed subset of D.

Proof. For $1 \leq i \leq d$ fixed let $f_m\colon E \to \mathbb{R}$ where $E = \log D$, be defined by $f_m(u) = \log(S_m)_i(\exp u)$. The mean value theorem yields

$$\left|f_m(u) - f_m(v)\right| \leq \int_0^1 \sum_{j=1}^{d} \left|\frac{\partial f_m}{\partial u_j}(u(t))\right| \left|v_j - u_j\right| dt,$$

where $u(t) = u + t(v - u)$ and $u, v \in E$. By the chain rule

$$\frac{\partial f_m}{\partial u_j}(u) = \frac{\exp u_j}{(S_m)_i(\exp u)} \frac{\partial (S_m)_i}{\partial x_j}(\exp u).$$

For $x, y \in D$ given and $u = \log x, v = \log y$ (componentwise),

$$u(t) = \log x + t(\log y - \log x) = \log\left(x\left(\frac{y}{x}\right)^t\right) = \log(x^{1-t}y^t).$$

Since $\exp u(t) = x^{1-t}y^t \in D$ it follows by definition of $A(x)$

$$\sum_{j=1}^{d} \left|\frac{\partial f_m}{\partial u_j}(u(t))\right| \leq A(x^{1-t}y^t).$$

It follows that

in case (a): $\quad \sup_m |f_m(u) - f_m(v)| \leq c \max_j |v_j - u_j| = c \max_j |\log x_j - \log y_j| \quad (*)$

in case (b): $\quad \sup_m |f_m(u) - f_m(v)| < \max_j |\log x_j - \log y_j| \text{ for } x \neq y. \quad (**)$

Considering case (a) from $(*)$ it follows that $\lambda x \leq y \leq \frac{1}{\lambda}x$ implies $\lambda^c (S_m)_i(x) \leq (S_m)_i(y) \leq \frac{1}{\lambda^c}(S_m)_i(x)$.

Since i was arbitrary choosen this means that (S_m) is uniformly weakly ascending on D with $\phi(\lambda) = \lambda^c$.

Thus, for case (a) it follows from Theorem 7.2.2 (i) that any two orbits are asymptotically equal, weak ergodicity applies and path stability holds for the norm provided that one bounded orbit exists.

Considering case (b), $(**)$ means that for the part metric $\sup_m p(S_m x, S_m y) < p(x,y)$ for $x \neq y$.

Thus, (S_m) is a contractive sequence of non-expansive mappings on (D, p). Let $(x_n), (y_n)$ orbits contained in a bounded and closed subset of D. Since bounded subsets are relatively compact in $(\mathbb{R}^d, \|\cdot\|)$ there exist subsequences $(x_{n_k}), (y_{n_k})$ converging in norm to $x^* \in D$ and $y^* \in D$, respectively. Norm topology and part topology coincide on $\overset{\circ}{K}$ by Proposition 3.4.12 (v) and the convergence holds also for p. Therefore, the joint limit set $w_s(x_1, y_1)$ in (D, p) is non-empty and Theorem 4.2.2 (ii) implies $\lim_{n\to\infty} p(x_n, y_n) = 0$. Finally, from Lemma 7.1.2 parts (i) and (iii) we obtain for case (b) that (x_n) and (y_n) are asymptotically equal as well as $\lim_{n\to\infty} \| x_n - y_n \| = 0$. □

Remarks 7.2.6. Particular cases of Theorem 7.2.2 can be found in [16, 36]. For the concave Coale–Lopez Theorem see [16, 35]. For Corollary 7.2.4 in finite dimensions, the classical linear weak ergodicity see [44], where the dual result for "forward products" of matrices is proven. See also [21]. For other results on non-linear weak ergodicity see [2, 3, 34, 37, 45]. For a general approach to non-autonomous systems in discrete time which is based on 2-parameter semigroups see [40].

The results obtained we shall illustrate by non-linear Leslie models as considered already in Sections 1.2 and 2.6 for which we now admit the birth rates and survival rates to depend on time. (See also [16].

Example 7.2.7 (Nonlinear and non-autonomous Leslie models). (1) Consider first a concave and non-autonomous Leslie model given by $x(t+1) = T_t x(t)$ for $t = 0, 1, \dots$ and $x(0) \in \mathbb{R}^n_+$ where $T_t x = L(t,x)x$ for $x \in \mathbb{R}^n_+$ and

$$L(t,x) = \begin{bmatrix} b_1(t,x) & b_2(t,x) & \cdots & b_{n-1}(t,x) & b_n(t,x) \\ s_1(t,x) & 0 & \cdots & 0 & 0 \\ 0 & s_2(t,x) & \cdots & 0 & 0 \\ \vdots & \vdots & & \vdots & \vdots \\ 0 & 0 & \cdots 0 & s_{n-1}(t,x) & s_n(t,x) \end{bmatrix}$$

is the (generalized) Leslie matrix (see Section 1.2). On the vital rates we make similar assumptions as for the concave Leslie model in Section 2.6 taking care, however, of the time dependence:

(a) The mappings $x \mapsto b_i(t,x)x_i$ and $x \mapsto s_i(t,x)x_i$ of $\mathbb{R}^n_+$ into $\mathbb{R}_+$ are concave for each $1 \le i \le n$ and each $t = 0, 1, 2 \dots$.
(b) There exist functions μ and ν from $\mathbb{R}^n_+ \setminus \{0\}$ into $\mathbb{R}_+ \setminus \{0\}$ such that for all i, t, x

$$\mu(x) \le b_i(t,x), s_i(t,x) \quad \text{and} \quad b_i(t,x)x_i, s_i(t,x)x_i \le \nu(x).$$

Assume further for $0 \lneqq x \le y$ that $\mu(y) \le \mu(x)$ and $\nu(x) \le \nu(y)$.

(c) For any $x \in \mathbb{R}^n_+$, $t \in \mathbb{N}$ and $\lambda > 0$ there exists a number $c_t(x,\lambda)$ such that $b_i(t,\lambda x) = c_t(x,\lambda)b_i(t,x)$ and $s_i(t,\lambda x) = c_t(x,\lambda)s_i(t,x)$ for all $1 \le i \le n$ with $x_i > 0$.

As in Section 2.6, (a) models population pressure and implies that T_t is a concave self-mapping of $\mathbb{R}^n_+$ for each t.

Similarly, assumption (c) means that a population pressure uniform over age classes does not affect the ratio of birth and survival rates. By this assumption, $T_t(\lambda x) = \lambda c_t(x,\lambda)T_t x$ and, hence, the operators T_t are ray-preserving. We want to apply part (ii) of Corollary 7.2.3 for the norm $\|x\| = \sum_{i=1}^n |x_i|$ on $K = \mathbb{R}^n_+$. For this it remains to show for some $0 < \alpha \le \beta$ and $e \in \mathring{K}$

$$\alpha e \le S_m x \le \beta e \quad \text{for all} \quad m \ge 1,\ x \in \mathring{K},\ \|x\| = 1.$$

For $S_{m,r} = T_{m+r-1} \circ \cdots \circ T_m$ we show by induction over r

$$u(r,x) \le S_{m,r}x \le v^{(r)}(x) \quad \text{for} \quad x \in \mathbb{R}^n_+ \setminus \{0\} \tag{$*$}$$

where $u(r,x) = \prod_{i=0}^{r-1} \mu(v^{(i)}(x))L^r x$, $v^{(i)}(x)$ the i-th iterate of $v(x) = (n v(x), \dots, v(x)$ and

$$L = \begin{bmatrix} 1 & 1 & \cdots & 1 \\ 1 & 0 & \cdots & 0 \\ 0 & 1 & \cdots & 0 \\ \vdots & & & \\ 0 & 0 & \cdot\cdot 1 & 0 \end{bmatrix}.$$

Assertion $(*)$ for $r = 1$ means that $\mu(x)Lx \le T_m x \le v(x)$ which is true by assumption (b). If $(*)$ holds for r then

$$\mu(S_{m,r}x)LS_{m,r}x \le S_{m,r+1}x = T_{m,r}(S_{m,r}x) \le v(S_{m,r}x).$$

From $S_{m,r}x \le v^{(r)}(x)$ it follows by assumption on v that $v(S_{m,r}x) \le v(v^{(r)}(x)) = v^{(r+1)}(x)$. From $S_{m,r}x \le v^{(r)}(x)$ and the assumption on μ it follows that

$$\mu(S_{m,r}x)LS_{m,r}x \ge \mu(v^{(r)}(x))Lu(r,x) = \mu(v^{(r)}(x))\prod_{i=0}^{r-1}\mu(v^{(i)}(x))L^{r+1}x.$$

Thus

$$\prod_{i=0}^{r}\mu(v^{(i)}(x))L^{r+1}x \le S_{m,r+1}x \le v^{(r+1)}(x),$$

which proves property $(*)$. Now, since $L^n > 0$ it follows for $r = n$ and $S_m = S_{m,n}$ from $(*)$ that

$$0 < u(n,e_i) \le S_m e_i \le v^{(n)} e_i \quad \text{for} \quad 1 \le i \le n \quad \text{and all } m,$$

where e_i is the i-th unit vector.

Since S_m is concave it follows for $x = \sum_{i=1}^n x_i e_i$ with $\sum_{i=1}^n x_i = 1$ that $S_m x \geq \sum_{i=1}^n x_i S_m e_i \geq \alpha e$ where $\alpha = \min_{1 \leq i,j \leq n} u(n, e_i))_j > 0$ and e the vector with all entries equal to 1. Finally, as a concave mapping S_m is monotone and $x \leq e$ implies that $S_m x \leq S_m e \leq v^{(n)}(e)$ by property (∗). Since $v^n(e) \overset{\circ}{>} 0$ we have $v^n(e) \leq \beta e$ for some $\beta > 0$ and arrive at $\alpha e \leq S_m x \leq \beta e$ for all m and $x \in \overset{\circ}{K}, \|x\| = 1$. All assumptions being satisfied, from Corollary 7.2.3 (ii) we conclude that weak ergodicity for the norm $\| \cdot \|$ holds on $\overset{\circ}{K}$ and, hence, on $\mathbb{R}_+^n \setminus \{0\}$.

By the way, for $T_t = T$ for all t from weak ergodicity we get back earlier results. Since by the Concave Perron–Frobenius Theorem (Theorem 2.1.14) the eigenvalue problem $Tx = \lambda x$ has a solution $0 \lneqq x^*, 0 < \lambda^*$ (by assumption (b)) it follows $\frac{T^n x^*}{\|T^n x^*\|} = \frac{x^*}{\|x^*\|}$ since T is ray-preserving. Thus, weak ergodicity yields $\lim_{n\to\infty} \left\| \frac{T^n x}{\|T^n x\|} - \frac{x^*}{\|x^*\|} \right\| = 0.$ that is $\lim_{n\to\infty} \frac{T^n x}{\|T^n x\|} = \frac{x^*}{\|x^*\|}$ for all $x \gneqq 0$.

(2) Consider now the special case of a linear and non-autonomous Leslie model where $T_t x = L(t)x$ with

$$L(t) = \begin{bmatrix} b_1(t) & b_2(t) & \cdots & b_{n-1}(t) & b_n(t) \\ s_1(t) & 0 & \cdots & 0 & 0 \\ 0 & s_2(t) & \cdots & 0 & 0 \\ \vdots & \vdots & & \vdots & \vdots \\ 0 & 0 & \cdots 0 & s_{n-1}(t) & s_n(t) \end{bmatrix}.$$

Concerning the assumptions made for (1), (a) holds trivially and (c) holds with $c_t(x, \lambda) = 1$ for all t, x, λ. Assumption (b) means that uniformly for all t the maximal entry of $L(t)$ is bounded from above and the minimal strictly positive entry is bounded from below. Being a special case of (1) under these assumptions we obtain weak ergodicity. Actually, since T_t is not only ray-preserving but positively homogeneous we have that any two orbits are asymptotically proportional. This illustrates Corollary 7.2.3 part (i) as well as part (ii) since lumped products $L(m + n-1) \cdots L(m)$ are strictly positive for all m by the very structure of $L(t)$.(For non-linear and non-autonomous Leslie models see also Exercises 3, 4, 7.)

7.3 Strong ergodicity for ascending operators

In this section we extend the strong ergodicity property as it is known from inhomogeneous Markov chains, population dynamics and demography to non-linear mappings in not necessarily finite dimensional vector spaces.

Consider for a (non-empty) subset $D \subseteq V \setminus \{0\}$ of a normed vector space $(V, \| \cdot \|)$ a sequence $(T_n)_n$ of selfmappings of D defining a non-autonomous system by

$$x_{n+1} = T_n x_n, \quad n \geq 1, \quad x_1 \in D. \tag{7.3.1}$$

According to Definition 7.1.3 this system has **strong ergodicity on D** if there exists $x^* \in D$ such that $\lim_{n\to\infty} \| \frac{x_n}{\|x_n\|} - x^* \| = 0$ for all $x_1 \in D$. As in the case of inhomogeneous Markov chains we shall assume that the mappings T_n converge to a selfmapping T of D. If T, or some iterate of T, has a contraction property for an internal metric, we will be able to obtain strong ergodicity from results for non-autonomous systems on metric spaces in Section 4.2. To ensure the contraction property we assume T to be ascending – where throughout this section the function ϕ in Definition 5.1.4 is assumed to be continuous. For the particular case that $T_n = T$ for all n, we are back to the case of an autonomous system defined by an ascending operator T as it has been studied in Chapter 5.

The main result in this section is the following theorem.

Theorem 7.3.1. *Let $(V, \|\cdot\|)$ be a normed real vector space and let $K \subseteq V$ be a convex cone which is closed and normal with non-empty interior $\overset{\circ}{K}$. Let $(T_n)_n$ and T be selfmappings of $\emptyset \neq D \subseteq \overset{\circ}{K}$.*

(i) *Suppose $(T_n)_n$ converges uniformly to T and T is uniformly continuous (for the norm). Suppose, furthermore, an iterate T^k is weakly ascending on D and for some $u \in \overset{\circ}{K}$ it holds that $u \leq T^k x$ for all $x \in D$.*

   Then for each orbit $(x_n)_n$ of system (7.3.1) with compact closure (for the norm) in D it holds that

$$\lim_{n\to\infty} \|x_n - x^*\| = 0 \text{ where } x^* \text{ is the unique fixed point of } T \text{ in } D.$$

(ii) *Let $(V, \|\cdot\|)$ be a Banach space, $U = \{x \in D \mid \|x\| = 1\}$, D internally closed in $\overset{\circ}{K}$ and assume $\frac{x}{\|x\|} \in D$ for $x \in D$. Suppose on U the sequence $(T_n)_n$ converges uniformly to T, T is uniformly continuous and $a \leq \|Tx\|$ for some $a > 0$ and all $x \in U$. Suppose, further for some $k \geq 1$, T^k is ascending and norm-bounded on U and for some $u \in \overset{\circ}{K}$ it holds that $u \leq T^k x$ for all $x \in U$. Then for each orbit of the rescaled system given by*

$$\tilde{x}_{n+1} = \tilde{T}_n \tilde{x}_n, \quad n \geq 1, \ \tilde{x}_1 = x_1 \in D \tag{7.3.2}$$

   one has for $k = 1$ or T ray-preserving that $\lim_{n\to\infty} \|\tilde{x}_n - x^*\| = 0$ where x^* is the unique solution of $Tx = \lambda x$ in U with $\lambda > 0$. Moreover, in case T and $T_n, n \geq 1$, are ray-preserving it holds for k arbitrary $\lim_{n\to\infty} \| \frac{x_n}{\|x_n\|} - x^* \| = 0$ with x^* as above.

Proof. (i) The result will be obtained from Corollary 4.2.4 for the metric space (D, p), p the part metric on $\overset{\circ}{K}$. For the lumped mappings $S_m = T_{m+k-1} \circ \ldots \circ T_m$ from Lemma 4.2.5 it follows for the metric space $(D, \|\cdot\|)$ that $(S_m)_m$ converges uniformly to $S = T^k$. Next we show this holds true in (D, p), too. For $u \in \overset{\circ}{K}$ we have that for some $r > 0$ the open ball $B(\frac{1}{2}u, r)$ is contained in K. Since $\|S_m x - Sx\| \leq r$ for $m \geq N(r)$ and all $x \in D$ we have that $\frac{1}{2}u + S_m x - Sx \in B(\frac{1}{2}u, r) \subseteq K$ and, hence,

$$S_m x \geq Sx - \frac{1}{2}u \geq u - \frac{1}{2}u = \frac{1}{2}u \quad \text{for all } m \geq N(r), \quad \text{all } x \in D.$$

From Proposition 3.4.12, part (vi)(b), it follows

$$p(S_m x, Sx) \le K_p \|S_m x - Sx\| \quad \text{for all } m \ge N(r), \quad \text{all } x \in D.$$

Similarly, $p(Sx, Sy) \le K_p \|Sx - Sy\|$ for all $x, y \in D$, where $K_p > 0$ is a constant. To apply Corollary 4.2.4 (i) we show that $S = T^k$ is contractive on (D, p). By assumption, S is weakly ascending, therefore (see Proof (i) for Theorem 7.2.2)

$$\lambda x \le y \le \frac{1}{\lambda} x \quad \text{implies} \quad \phi(\lambda) Sx \le Sy \le \frac{1}{\phi(\lambda)} Sx,$$

with $\mu(x, y) = \sup\{\lambda > 0 \mid \lambda x \le y \le \frac{1}{\lambda} x\}$, therefore,

$$\mu(Sx, Sy) \ge \phi(\mu(x, y)) \quad \text{and} \quad p(Sx, Sy) \le -\log[\phi(\mu(x, y))].$$

Since $0 < \mu(x, y) < 1$, for $x \ne y$, it follows $\phi(\mu(x, y)) > \mu(x, y)$. Thus, $p(Sx, Sy) < p(x, y)$ for $x \ne y$, that is, S is contractive on (D, p). If $(x_n)_n$ has compact closure (for $\|\cdot\|$) in D then, by Proposition 3.4.12 (v), $(x_n)_n$ has compact closure in (D, p). Thus, Corollary 4.2.4 yields $\lim_{n\to\infty} p(x_n, x^*) = 0$ and, hence, $\lim_{n\to\infty} \|x_n - x^*\| = 0$. Thereby, x^* is the unique fixed point of $S = T^k$ in D. Now,

$$\|Tx^* - x^*\| \le \|Tx^* - Tx_n\| + \|Tx_n - T_n x_n\| + \|x_{n+1} - x^*\|,$$

and the assumptions made imply $Tx^* = x^*$. This proves part (i) of the theorem.

(ii) The result will be obtained from Corollary 4.2.4 for the metric space (U, d), d the Hilbert metric on U. Consider the rescaled operators $\tilde{T}_n$ and $\tilde{T}$ which are selfmappings of U. For any two $z, w \in V \setminus \{0\}$ it holds that $\|\frac{z}{\|z\|} - \frac{w}{\|w\|}\| \le \frac{2}{\|z\|}\|z - w\|$. Therefore, $\|\tilde{T}x - \tilde{T}_n x\| \le \frac{2}{\|Tx\|}\|Tx - T_n x\|$ and by assumption it follows that $\|\tilde{T}x - \tilde{T}_n x\| \le \frac{2}{a}\|Tx - T_n x\|$ for $x \in U$. Thus, $\tilde{T}_n$ converges uniformly to $\tilde{T}$ on $(U, \|\cdot\|)$. Similarly, $\tilde{T}$ is uniformly continuous on $(U, \|\cdot\|)$. From Lemma 4.2.5 it follows for the metric space $(U, \|\cdot\|)$ that the lumped operators $F_m = \tilde{T}_{m+k-1} \circ \ldots \circ \tilde{T}_m$ converge uniformly to $F = \tilde{T}^k$. From $u \le T^k x$ for $x \in U$ it follows $\frac{u}{\|T^k x\|} \le \frac{T^k x}{\|T^k x\|} = \widetilde{T^k} x$ and, since $\|T^k x\| \le b$ for some $b > 0$ and for $x \in U$, we have that $v = \frac{u}{b} \le \widetilde{T^k} x$ for $x \in U$.

Let us assume that $\widetilde{T^k} = \tilde{T}^k$. This assumption is satisfied if $k = 1$ or if T is ray-preserving. Then we have that $v \le Fx$ for all $x \in U$ where $v \in \mathring{K}$. The uniform convergence of F_m to F implies, as in step (i), $F_m x \ge \frac{1}{2} v$ for all $x \in U$, all $m \ge N$. By Proposition 3.4.12, part (vi)(b), therefore $d(F_m x, Fx) \le K_d \|F_m x - Fx\|$ for all $x \in U$, all $m \ge N$. Furthermore, $d(Fx, Fy) \le K_d \|Fx - Fy\|$ for all $x, y \in U$. To apply Corollary 4.2.4 (ii), we shall show that $(\tilde{x}_n)$ is bounded in (U, d) and F is a generalized contraction on (U, d). For the former observe that $\tilde{x}_{m+k} = F_m \tilde{x}_m$ and $F_m x \ge \frac{1}{2} v$ for all $x \in U$, all $m \ge N$. Therefore $d(\tilde{x}_{m+k}, \tilde{x}_{N+k}) \le K_d \|\tilde{x}_{m+k} - \tilde{x}_{N+k}\| \le 2K_d$ for $m \ge N$.

This shows boundedness of $(\tilde{x}_n)$ in (U, d). To see that F is a generalized contraction we proceed as in part (i) in the proof of Theorem 7.2.2. Since $S = T^k$ is ascending on U, it follows for $x, y \in U, 0 < \lambda, \mu \le 1$ and $\lambda x \le y, \mu y \le x$ that $\phi(\lambda)Sx \le Sy, \phi(\mu)Sy \le Sx$. Consider for $0 < \alpha \le \beta$ given the compact set $I = \{(\lambda, \mu) \in [0,1]^2 \mid e^{-\beta} \le \lambda\mu \le e^{-\alpha}\}$ and the function $l(\lambda, \mu) = \frac{\log(\phi(\lambda)\phi(\mu))}{\log(\lambda\mu)}$ well-defined on I. Since $0 < \lambda\mu < 1$ for $(\lambda, \mu) \in I$ and ϕ is a strictly increasing selfmapping of $]0,1[$ it follows that $\phi(\lambda)\phi(\mu) > \lambda\mu$ and, hence, $l(\lambda, \mu) < 1$. Since ϕ is continuous, l is continuous, too, and $\sup l(I) = \sigma < 1$. Thus, we have

$$d(Fx, Fy) = d(\tilde{S}x, \tilde{S}y) = d(Sx, Sy) \le -\log(\phi(\lambda)\phi(\mu)) \le \sigma(-\log(\lambda\mu))$$

and arrive for $\alpha \le d(x, y) \le \beta$ at $d(Fx, Fy) \le \sigma d(x, y)$, which means F is a generalized contraction on (U, d). Since D is internally closed, the metric space (U, d) is complete by Corollary 3.4.14.

Thus, Corollary 4.2.4 yields $\lim_{n\to\infty} d(\tilde{x}_n, x^*) = 0$, and, hence, $\lim_{n\to\infty} \|\tilde{x}_n - x^*\| = 0$, where x^* is the unique fixed-point of F in U. Furthermore,

$$\|\tilde{T}x^* - x^*\| \le \|\tilde{T}x^* - \tilde{T}\tilde{x}_n\| + \|\tilde{T}\tilde{x}_n - \tilde{T}_n\tilde{x}_n\| + \|\tilde{x}_{n+1} - x^*\|,$$

which implies $\tilde{T}x^* = x^*$. Since x^* is the unique fixed point of F in U, it is the unique fixed point of $\tilde{T}$ in U, too. Equivalently, x^* is the unique solution of $Tx = \lambda x$ in U with $\lambda > 0$. Moreover, if T_n is ray-preserving,

$$\tilde{x}_{n+1} = \tilde{T}_n \circ \cdots \circ \tilde{T}_1 x_1 = \frac{T_n \circ \cdots \circ T_1 x_1}{\|T_n \circ \cdots \circ T_1 x_1\|} = \frac{x_{n+1}}{\|x_{n+1}\|},$$

and we arrive at $\lim_{n\to\infty} \left\| \frac{x_n}{\|x_n\|} - x^* \right\| = 0$. □

Corollary 7.3.2. *For $D = \mathring{K}$ the conclusion $\lim_{n\to\infty} \| \frac{x_n}{\|x_n\|} - x^* \| = 0$ of Theorem 7.3.1, part (ii), holds true if the norm closure $B = \{x \in K \mid \|x\| = 1\}$ of U is assumed to be norm-compact and the requirement on T^k to be ascending on U is replaced by the weaker one (all other assumptions being unchanged)*

$$x, y \in U, \quad 0 < \lambda < 1, \quad \lambda x \lneq y \quad \text{implies} \quad \lambda T^k x < T^k y.$$

Proof. The above follows as in the proof of Theorem 7.3.1, part (ii), by employing part (i) of Corollary 4.2.4 instead of part (ii). Namely, the assumption made implies for $S = T^k$ that $v(Sx, Sy) > v(x, y)$ for $x, y \in U, x \ne y$. Therefore, with $F = \tilde{S}$

$$d(Fx, Fy) < d(x, y) \quad \text{for} \quad x, y \in U, \ x \ne y.$$

Furthermore, since $\tilde{x}_{m+k} = F_m \tilde{x}_m \ge \frac{1}{2} v$ for $m \ge N$ we have that $\tilde{x}_n \in C = \{x \in B \mid x \ge \frac{1}{2} v\} \subseteq U$ for $n \ge N + k$. C is norm-compact by assumption and from Proposition 3.4.12 (vi) (b) it follows that C is compact in (U, d). Thus $(\tilde{x}_n)$ is relatively compact in (U, d) and by Corollary 4.2.4 (i) we arrive at $\lim_{n\to\infty} d(\tilde{x}_n, x^*) = 0$, x^* being the unique fixed point of F in U. As in the proof of Theorem 7.3.1, part (ii), this implies $\lim_{n\to\infty} \| \frac{x_n}{\|x_n\|} - x^* \| = 0$. □

Similarly to the case of weak ergodicity as a corollary we obtain the following specialization to concave operators.

Corollary 7.3.3 (Concave strong ergodicity). *Let $(V, \|\cdot\|)$ be a Banach space containing a convex cone K which is closed and normal with $\overset{\circ}{K} \neq \emptyset$. Let $(T_n)_n$ be a sequence of selfmappings of $\overset{\circ}{K}$ which converges on $U = \{x \in \overset{\circ}{K} \mid \|x\| = 1\}$ uniformly to a concave selfmapping T of $\overset{\circ}{K}$ which is uniformly continuous on U (for $\|\cdot\|$).*

(i) *If for some $u, v \in \overset{\circ}{K}$ it holds $u \leq Tx \leq v$ for all $x \in U$ then the orbit $(\tilde{x}_n)_n$ of the rescaled system*
$$\tilde{x}_{n+1} = \tilde{T}_n \tilde{x}_n, \quad n \geq 1, \quad \tilde{x}_1 = x_1 \in \overset{\circ}{K}$$
converges in norm to the unique solution x^* of $Tx = \lambda x$ in U with $\lambda > 0$.

(ii) *If for some $u, v \in \overset{\circ}{K}, k \geq 1$ it holds $u \leq T^k x$, $Tx \leq v, a \leq \|Tx\|$ for some $a > 0$ for all $x \in U$ and $T, T_n, n \geq 1$, are ray-preserving then for the orbit $(x_n)_n$ given by $x_{n+1} = T_n x_n, x_1 \in \overset{\circ}{K}$ the sequence $(\frac{x_n}{\|x_n\|})_n$ converges in norm to the unique solution x^* of $Tx = \lambda x$ in U with $\lambda > 0$.*

Proof. (i) For $D = \overset{\circ}{K}$, concavity of T together with $u \leq Tx \leq v$ for $x \in U$ implies, as in the proof for part (i) of Corollary 7.2.3, that T is ascending on U. Furthermore, $0 < \|u\| \leq \|Tx\| \leq \|v\|$ for $x \in U$. From Theorem 7.3.1 (ii) the assertion follows for $k = 1$.

(ii) T^k is concave and $T^k x \leq T^k(v) \in \overset{\circ}{K}$. By step (i) T^k is ascending on U. Again, the assertion follows from Theorem 7.3.1 (ii) □

Specializing further to linear operators we obtain in particular the classical strong ergodicity result for non-negative matrices as discussed in Section 7.1.

Corollary 7.3.4 (Linear strong ergodicity). (i) *Let $(V, \|\cdot\|)$ be a Banach space containing a convex cone K which is closed and normal with $\overset{\circ}{K} \neq \emptyset$. Let $(T_n)_n$ be a sequence of ray-preserving selfmappings of $\overset{\circ}{K}$ which converges on $U = \{x \in \overset{\circ}{K} \mid \|x\| = 1\}$ uniformly to a uniformly continuous and linear selfmapping T of $\overset{\circ}{K}$. If for some $u, v \in \overset{\circ}{K}, k \geq 1$, it holds $u \leq T^k x$, $Tx \leq v$ and $a \leq \|Tx\|$ with $a > 0$ for all $x \in U$ then for the orbit $(x_n)_n$ given by $x_{n+1} = T_n x_n, x_1 \in \overset{\circ}{K}$ the sequence $(\frac{x_n}{\|x_n\|})_n$ converges in norm to the unique solution x^* of $Tx = \lambda x$ in U with $\lambda > 0$.*

(ii) *Let $(P_n)_n$ be a sequence of non-negative dxd-matrices without zero row which converges elementwise to a primitive matrix P then for all $1 \leq i, j \leq d$*
$$\lim_{n\to\infty} \frac{(P_n \dots P_1)_{ij}}{\sum\limits_{h=1}^{d} (P_n \dots P_1)_{hj}} = x^*, \quad Px^* = \|Px^*\|_1 x^*,$$
that is the sum-normed columns of the matrix $P_n \dots P_1$ become equal, for n approaching infinity, to the unique sum-normed right eigenvector of P.

Proof. (i) The assertion follows immediately from Corollary 7.3.3 (ii).

(ii) Since P_n has a no zero row the linear mapping given by $T_n x = P_n x$ is a selfmapping of $\overset{\circ}{K}$ for $K = \mathbb{R}^d_+$, $V = (\mathbb{R}^d, \|\cdot\|_1)$ ($\|\cdot\|_1$ the sum-norm). Since $\bar{U}$ is compact for $\|\cdot\|$, the mappings T_n converge uniformly to $Tx = P \cdot x$. Primitivity of P implies $T^k x > 0$ for $x \in \bar{U}$ and, hence, $T^k x \geq u > 0$ for all $x \in \bar{U}$. Furthermore, T is a uniformly continuous selfmapping of $\mathring{K}$ such that $a \leq \|Tx\|$ with $a > 0$ and $Tx \leq v$ for some $0 < v$ and all $x \in \bar{U}$. From step (i) it follows for $x_{n+1} = P_n \cdot x_n, x_1 \in \mathring{K}$ that $\lim_{n\to\infty} \frac{x_n}{\|x_n\|_1} = x^*, Px^* = \|Px^*\|_1 x^*$. This holds also for x_1 the j-th unit vector e_j and from $(x_{n+1})_i = (P_n \dots P_1 e_j)_i$ the assertion follows. □

A further consequence of Theorem 7.3.1 is the following differentiability criterion for strong ergodicity in finite dimensions.

Corollary 7.3.5. *Let $V = \mathbb{R}^d, K = \mathbb{R}^d_+, \|\cdot\|$ any monotone norm on $\mathbb{R}^d$. For a log-convex subset D of $\mathring{K}$ let $(T_n)_n$ be a sequence of selfmappings of D which converges uniformly to a selfmapping T of D, which is uniformly continuous on D. Suppose $0 < u \leq Tx$ for all $x \in D$ and T is differentiable on $\mathring{K}$ such that for the Jacobian J of T^k it holds $|J(x)|x < T^k x$ for all $x \in D$. If the orbit defined by $x_{n+1} = T_n x_n, x_1 \in D$, is contained in a closed and bounded (for the norm) subset of D then it converges in norm to the unique fixed point of T in D.*

Proof. From assumption $|J(x)|x < T^k x, x \in D$, we obtain by Theorem 6.2.1 (ii) $p(T^k x, T^k y) < p(x, y), x \neq y$, that is, T^k is contractive on (D, p). Since bounded sets have compact closure in $(\mathbb{R}^d, \|\cdot\|)$ the assertion follows as in the proof of part (i) of Theorem 7.3.1. □

Remark 7.3.6. For particular cases of Theorem 7.3.1 see [15, 16]. For Corollary 7.3.2 see [16]. For Corollaries 7.3.2 and 7.3.3 in finite dimensions see [15]. For Corollary 7.3.4 in finite dimensions, the classical strong ergodicity, see [44], where the dual result for "forward products" of matrices is proven.

The next section presents an application of strong ergodicity to a non-linear version of a theorem of Poincaré on difference equations. (For examples of strong ergodicity see also Exercises 7, 8, 10.)

7.4 A non-linear version of Poincaré's theorem on non-autonomous difference equations

Consider the following linear difference equation of order n with time-dependent coefficients

$$u(t+n) = p_0(t)u(t) + p_1(t)u(t+1) + \cdots + p_{n-1}(t)u(t+n-1), \quad t \in \mathbb{N}. \tag{7.4.1}$$

Assuming

$$\lim_{t\to\infty} p_i(t) = p_i \quad \text{for all} \quad 0 \leq i \leq n-1 \tag{7.4.2}$$

the question arises on how solutions of equation (7.4.1) are connected to solutions of the corresponding autonomous equation given by

$$u(t+n) = p_0u(t) + p_1u(t+1) + \cdots + p_{n-1}u(t+n-1), \quad t \in \mathbb{N}. \tag{7.4.3}$$

Since the solutions of equation (7.4.3) are given by the roots of the characteristic equation

$$p_0 + p_1\lambda + p_2\lambda^2 + \cdots + p_{n-1}\lambda^{n-1} = \lambda^n, \tag{7.4.4}$$

and the multiplicities of these roots, the question above becomes on how solutions of (7.4.1) are related to the roots of the characteristic equation (7.4.4).

In this respect Poincaré proved in 1885 the following result ([41]; see also [10]):

Poincaré's Theorem. Assume for equation (7.4.1), with coefficients and solutions in $\mathbb{C}$, that beside assumption (7.4.2) it holds for the roots $\lambda_1, \ldots, \lambda_n$ of equation (7.4.4) that

$$|\lambda_i| \neq |\lambda_j| \quad \text{for} \quad i \neq j \tag{7.4.5}$$

($|\cdot|$ the absolute value of a complex number). Then for any solution u of (7.4.1) which is not asymptotically zero there exists an λ_i such that

$$\lim_{t\to\infty} \frac{u(t+1)}{u(t)} = \lambda_i. \tag{7.4.6}$$

A natural question considering Poincaré's theorem is what can be said if the additional assumption (7.4.5) is not being satisfied. Surprisingly, this question has been answered only very recently by M. Pituk [39] with a nice result on **Poincaré's difference systems.** The latter are discrete dynamical systems

$$x(t+1) = \big(A + B(t)\big)x(t), \quad t \in \mathbb{N} \tag{7.4.7}$$

where A and $B(t)$ are complex $n \times n$-matrices and

$$\lim_{t\to\infty} B(t) = 0 \text{ (with respect to some norm } \|\cdot\|.)$$

Obviously, equation (7.4.1) is a special case of a Poincaré difference system where equation (7.4.2) corresponds to equation (7.4.7).

Pituk's Theorem. If x is a solution of (7.4.7) which is not asymptotically zero then there exists an eigenvalue λ of A such that

$$\lim_{t\to\infty} \|x(t)\|^{\frac{1}{t}} = |\lambda|.$$

(For an extension of Pituk's Theorem to linear operators see Exercise 9.)

Both theorems may be interpreted as results on the behavior of solutions of a perturbed system in case the perturbation vanishes in the limit. The difference between the two theorems lies in the two different "growth rates" considered for solutions. The following lemma makes this more explicit.

Lemma 7.4.1. *Let $(x_n)_{n\geq 1}$ be a sequence in the (complex) vector space V with norm $\|\cdot\|$ such that there exists m such that $x_n \neq 0$ for all $n \geq m$.*

For the following types of a growth rate with value $r > 0$ for $(x_n)_{n\geq 1}$ does (a) *imply* (b) *and* (b) *imply* (c):

(a) $\lim\limits_{n\to\infty} \frac{x_n}{r^n}$ *exists in* $V \setminus \{0\}$,

(b) $\lim\limits_{n\to\infty} \frac{\|x_{n+1}\|}{\|x_n\|} = r$,

(c) $\lim\limits_{n\to\infty} \|x_n\|^{\frac{1}{n}} = r$.

Proof. (a) ⇒ (b). $\lim\limits_{n\to\infty} \frac{x_n}{r^n} = \lim\limits_{n\to\infty} \frac{x_{n+1}}{r^{n+1}} \neq 0$ implies

$$\lim_{n\to\infty} \frac{\|x_{n+1}\|}{\|x_n\|} = r \lim_{n\to\infty} \left(\frac{\frac{\|x_{n+1}\|}{r^{n+1}}}{\frac{\|x_n\|}{r^n}} \right) = r.$$

(b) ⇒ (c). Let $a_n = \frac{\|x_{n+1}\|}{\|x_n\| r}$ for $n \geq m$.

By assumption, to $\epsilon > 0$ exists n_1 such that $1 - \epsilon \leq a_n \leq 1 + \epsilon$ for $n \geq n_1$. Since $\frac{\|x_n\|}{r^n} = \frac{\|x_n\|}{\|x_{n-1}\| r} \cdots \frac{\|x_{n_1+1}\|}{\|x_{n_1}\| r} \cdot \frac{\|x_{n_1}\|}{r^{n_1}}$ for $n \geq n_1$, it follows

$$\frac{\|x_n\|^{\frac{1}{n}}}{r} = (a_{n-1} \cdots a_{n_1})^{\frac{1}{n}} \cdot b_n \quad \text{with} \quad b_n = \left(\frac{\|x_{n_1}\|}{r^{n_1}} \right)^{\frac{1}{n}}.$$

Obviously $\lim\limits_{n\to\infty} b_n = 1$. Furthermore $(1 - \epsilon) \leq (a_{n-1} \ldots a_{n_1})^{\frac{1}{n}} \leq (1 + \epsilon)$ for $n \geq n_1 + 1$. Since $1 - \epsilon \leq b_n \leq 1 + \epsilon$ for $n \geq n_2$ with some $n_2 \geq n_1 + 1$ it follows

$$(1 - \epsilon)^2 \leq \frac{\|x_n\|^{\frac{1}{n}}}{r^n} \leq (1 + \epsilon)^2 \quad \text{for } n \geq n_2.$$

This proves $\lim\limits_{n\to\infty} \|x_n\|^{\frac{1}{n}} = r$. □

None of the implications in Lemma 7.4.1 is reversible. This is so even in the particular case of equation (7.4.1) with real and non-negative coefficients, as the following examples show. (For the implicatons in Lemma 7.4.1 see [1].)

Examples 7.4.2. (i) Consider the following particular case of (7.4.1)

$$u(t + 2) = \frac{t + 2}{2t + 1}(u(t) + u(t + 1)), \quad t \in \mathbb{N}.$$

For $p_0(t) = p_1(t) = \frac{t+2}{2t+1}, p_0 = p_1 = \frac{1}{2}$, hence, the characteristic equation is $\frac{1}{2}(1 + \lambda) = \lambda^2$. The roots are $\lambda_1 = 1, \lambda_2 = -\frac{1}{2}$ and have different absolute value whence Poincaré's theorem applies. Especially, for $u(0) = 1, u(1) = 2$ it holds that $\lim\limits_{t\to\infty} \frac{u(t+1}{u(t)} = 1$. Induction shows, however, that $u(t) \geq t + 1$ for all t and $\lim\limits_{t\to\infty} u(t)$ does not converge. Thus, for the sequence $x_t = u(t)$ and $r = 1$, property (b) of Lemma 7.4.1 applies but not property (a).

(ii) Consider the following case of (7.4.1) where the coefficients are even all constant

$$u(t + 4) = \frac{1}{2}(u(t) + u(t + 2)), \quad t \in \mathbb{N}.$$

For $u(0) = 1, u(1) = 2, u(2) = 1, u(3) = 2$ the solution is given by $u(t) = 1$ for t even and $u(t) = 2$ for t odd. Obviously, for the sequence $x_t = u(t)$ and $r = 1$ property (c) of Lemma 7.4.1 applies but not property (b). Actually, this example satisfies the assumption of Pituk's theorem but not those of Poincaré's theorem because ± 1 are roots of the characteristic equation $\lambda^4 = \frac{1}{2}(1 + \lambda^2)$.

From results of the previous section we will obtain a version of Poincaré's theorem for non-linear difference equations and non-linear Poincaré difference systems, respectively. This will be done in the frame work of positive systems within Banach spaces. Under certain assumptions on the unperturbed system it follows that the growth rate as in Poincaré's theorem (type (b)) is **equal for all positive solutions to the dominant eigenvalue.** Contrary to Poincaré's theorem this is true even if some different eigenvalues have equal modulus. Furthermore, the growth rates as in Pituk's theorem (type (c)), too, are equal for all positive solutions to the dominant eigenvalue.

Theorem 7.4.3. *Let $(V, \|\cdot\|)$ be a real Banach space containing a convex cone K which is closed and normal with non-empty interior $\mathring{K}$. Let $T_t, t \geq 0$, and T be positively homogeneous selfmappings of K mapping $\mathring{K}$ into itself. Let*

$$x(t+1) = T_t x(t), \quad t \in \mathbb{N} \tag{7.4.8}$$

be a Poincaré system, that is $\lim_{t\to\infty} T_t x = Tx$ uniformly on $B = \{x \in K \mid \|x\| = 1\}$ and T is continuous on K. In each of the following cases it holds for every solution x of (7.4.8) with $x(0) \in K \setminus \{0\}$ that

$$\lim_{t\to\infty} \frac{x(t)}{\|x(t)\|} = x^* \quad \text{and} \quad \lim_{t\to\infty} \frac{\|x(t+1)\|}{\|x(t)\|} = \lim_{t\to\infty} \|x(t)\|^{\frac{1}{t}} = \lambda^*, \tag{7.4.9}$$

where (x^*, λ^*) is the unique solution of $Tx = \lambda x$ with $x \in B$ and $\lambda > 0$.

Case (i). *T is concave, uniformly continuous on B and there exist $0 < a, 1 \leq k$ and $u, v \in \mathring{K}$ such that*

$$a \leq \|Tx\|, \quad u \leq T^k x, \quad Tx \leq v \quad \text{for all} \quad x \in B.$$

Case (ii). *B is compact (for $\|\cdot\|$) and for some $1 \leq k$*

$$x, y \in B, \quad 0 < \lambda < 1, \quad \lambda x \lneqq y \quad \text{imply that} \quad \lambda T^k x < T^k y.$$

Proof. (1) First we show that in both cases $\lim_{t\to\infty} \frac{x(t)}{\|x(t)\|} = x^*$ for $x(0) \in \mathring{K}$. In case (i) this follows from Corollary 7.3.3 (ii). In case (ii) the assertion will follow from Theorem 7.3.1 (ii) and Corollary 7.3.2. For this we show that, up to concavity, the assumptions of case (i) are satisfied.

Obviously, T is uniformly continuous on B and $C = T^k(B) \subseteq K$ is compact (for $\|\cdot\|$). For $x \in B$, $\frac{1}{2}x \lneqq x$ and by assumption $\frac{1}{2}T^k x < T^k x$. Therefore, $0 < T^k x$ and $C \subseteq \mathring{K}$. For $z \in C$ the set $U(z) = \frac{1}{2}z + \mathring{K}$ is an open neighborhood of z in $(V, \|\cdot\|)$. By compactness there exists a finite covering, $C \subseteq \bigcup_{i=1}^m U(z_i)$. Since $z_i \in \overset{\circ}{K}$ it follows that $\lambda = \min_{1\le i\le m} \lambda(z_1, z_i) > 0$ and, hence, $u = \frac{\lambda}{2} z_1 \in \overset{\circ}{K}$. To $z \in C$ given there exist i and $y \in \overset{\circ}{K}$ such that $z = \frac{1}{2} z_i + y$ and, hence, $z \geq \frac{1}{2} z_i \geq \frac{\lambda}{2} z_1 = u$. This shows $u \leq T^k x$ for all $x \in B$. Furthermore, for $y \in C' = T(B)$ and any $e \in \overset{\circ}{K}$ the set $U(y) = y + e - \overset{\circ}{K}$ is an open neighborhood of y in $(V, \|\cdot\|)$. Again there exists a finite covering $C' \subseteq \bigcup_{j=1}^p U(y_j)$. To $y \in C'$ given there exist i and $w \in \overset{\circ}{K}$ such that $y = y_j + e - w \leq \sum_{i=1}^p y_i + e$. This shows for $v = \sum_{i=1}^p y_i + e \in \overset{\circ}{K}$ that $Tx \leq v$ for all $x \in B$.

Finally, $T0 = 0$ by positive homogeneity and $Tx = 0$ implies $T^k x = T^{k-1} 0 = 0$ and $x = 0$. Therefore, $0 < \|Tx\|$ for all $x \in B$ and by compactness $a \leq \|Tx\|$ for some $0 < a$ and all $x \in B$.

(2) By (1) it holds in both cases, (i) and (ii)

$$0 < a \leq \|Tx\|, \quad u \leq T^k x, \quad Tx \leq v \quad \text{for all} \quad x \in B$$

and not only on $U = \{x \in \overset{\circ}{K} \mid \|x\| = 1\}$. Therefore, according to the proof of Theorem 7.3.1 (ii) and Corollary 7.3.2, we have $\lim_{t\to\infty} \frac{x(t)}{\|x(t)\|} = x^k$ not only for $x(0) \in U$ but for $x(0) \in B$ and, hence, for $x(0) \in K \setminus \{0\}$. It follows

$$\left\| \frac{x(t+1)}{\|x(t)\|} - Tx^* \right\| \leq \left\| T_t \frac{x(t)}{\|x(t)\|} - T \frac{x(t)}{\|x(t)\|} \right\| + \left\| T \frac{x(t)}{\|x(t)\|} - Tx^* \right\|.$$

By assumptions on T_t, T it follows

$$\lim_{t\to\infty} \frac{x(t+1)}{\|x(t)\|} = Tx^* = \lambda^* x^*,$$

and, hence, $\lim_{t\to\infty} \frac{\|x(t+1)\|}{\|x(t)\|} = \lambda^*$. Since $x(t) \in K \setminus \{0\}$ for $x(0) \in K \setminus \{0\}$ from Lemma 7.4.1 it follows that $\lim_{t\to\infty} \|x(t)\|^{\frac{1}{2}} = \lambda^*$. □

Taking up Poincaré's theorem for the linear difference equation (7.4.1) we obtain the following non-linear version within the framework of positivity.

Corollary 7.4.4 (Non-linear Poincaré Theorem). *Let*

$$u(t+n) = f_t(u(t), u(t+1), \ldots, u(t+n-1)), \quad t \in \mathbb{N} \tag{7.4.10}$$

be a non-autonomous difference equation of order n. Assume $f_t \colon \mathbb{R}_+^n \longrightarrow \mathbb{R}_+, f_t(x) > 0$ *for* $x > 0, f_t$ *positively homogeneous. Suppose* $\lim_{t\to\infty} f_t(x) = f(x)$ *uniformly on* $B = \{x \in \mathbb{R}_+^n \mid \|x\| = 1\}$. *(*$\|\cdot\|$ *a monotone norm) where* $f \colon \mathbb{R}_+^n \longrightarrow \mathbb{R}_+$ *with* $f(x) > 0$ *for* $x > 0$ *is a continuous, positively homogeneous mapping which satisfies the following property*

$$0 \leq x \leq y \quad \text{implies} \quad f(x) \leq f(y)$$

and there exist $1 \leq n_1, n_2, \ldots, n_r, n_1 = 1, r \geq 2$ *with* $\gcd\{n - n_1 + 1, \ldots, n - n_r + 1\} = 1$ *such that*

$$0 \leq x \leq y \text{ and } x_{n_i} < y_{n_i} \text{ for some } i \text{ imply } f(x) < f(y).$$

Then for every solution u of (7.4.10) with $(u(0),\dots,u(n-1)) \gneqq 0$ *it holds*

$$\lim_{t\to\infty}\frac{u(t+1)}{u(t)} = \lim_{t\to\infty} u(t)^{\frac{1}{t}} = \lambda^*$$

where λ^* *is the unique positive root of the "characteristic equation"* $f(1,\lambda,\lambda^2,\dots,\lambda^{n-1}) = \lambda^n$.

Proof. Let $T_t x = (x_2,\dots,x_n,f(x))$, $Tx = (x_2,\dots,x_n,f(x))$ for $x \in \mathbb{R}^n_+$. By assumption, T_t and T are positively homogeneous and map the interior of $\mathbb{R}^n_+$ into itself. Furthermore, $\lim_{t\to\infty} T_t x = Tx$ uniformly on B, B is compact and T continuous on B. To apply case (ii) of Theorem 7.4.3 we show that for some $k \geq 1$ T^k is strictly increasing that is, $0 \leq x \lneqq y$ implies $Tx < Ty$. Consider solutions u, v of (7.4.10) with initial conditions $(u(0),\dots,u(n-1)) = x$ and $(v(0),\dots,v(n-1)) = y$. By Lemma 2.5.4 there exists $k \geq 1$ such that $u(t) < v(t)$ for all $k \leq t$. From the definitions of T_t and T it follows by iteration for $t \geq 0$

$$T_t \circ \cdots \circ T_0 x = (u(t+1),\dots,u(t+n)) \quad \text{and} \quad T^t x = (u(t),\dots,u(t+n-1)).$$

Therefore, $T^k x < T^k y$ and all assumptions for case (ii) of Theorem 7.4.3 are satisfied. The eigenequation $Tx = \lambda x$ is equivalent to

$$x_{i+1} = \lambda x_i, \quad 1 \leq i \leq n-1 \quad \text{and} \quad f(x) = x_n$$

which in turn is equivalent to

$$x_{i+1} = \lambda^i x_1, \quad 1 \leq i \leq n-1 \quad \text{and} \quad f(1,\lambda,\lambda^2,\dots,\lambda^{n-1}) = \lambda^n.$$

Therefore, λ^* is the unique positive root of the 'characteristic equation' and $x^* = r(1,\lambda^*,\lambda^{*2},\dots,\lambda^{*(n-1)})$ for some $r > 0$. Furthermore, for $x = (u(0),\dots,u(n-1)) \gneqq 0$ it follows $\lim_{t\to\infty}\frac{x(t)}{\|x(t)\|} = x^*$ and , hence,

$$\lim_{t\to\infty}\frac{u(t+1)}{u(t)} = \lim_{t\to\infty}\frac{x(t)_2}{x(t)_1} = \frac{x_2^*}{x_1^*} = \frac{r\lambda^*}{r} = \lambda^*.$$

Finally, from Lemma 7.4.1 it follows $\lim_{t\to\infty} u(t)^{\frac{1}{t}} = \lambda^*$. □

The corollary applies in particular to minima and maxima of linear equations as the following example shows.

Example 7.4.5. Consider the difference equation $u(t+n) = f_t(u(t),\dots,u(t+n-1))$ with $f_t(x) = \min_{1\leq i\leq m}(a_{i1}(t)x_1 + \cdots + a_{in}(t)x_n)$ for $x \in \mathbb{R}^n_+$ and $a_{ij}(t) \geq 0$. Assume $\lim_{t\to\infty} a_{ij}(t) = a_{ij}$ and let $f(x) = \min_{1\leq i\leq m}(a_{i1}x_1 + \cdots + a_{in}x_n)$. It is easily seen that $\lim_{t\to\infty} f_t(x) = f(x)$ uniformly on B.

Assume the matrix $A = (a_{ij})$ is non-negative with strictly positive columns $j \in J$ where $1 \in J$, $|J| \geq 2$ and $gcd\{n-j+1 \mid j \in J\} = 1$. All assumption of Corollary 7.4.4

are satisfied. (Since $f(x) > 0$ for $x > 0$ we may assume $f_t(x) > 0$ for $x > 0$.) Therefore we obtain for $(u(0), \dots, u(n-1)) \gneqq 0$ that $\lim\limits_{t\to\infty} \frac{u(t+1)}{u(t)} = \lim\limits_{t\to\infty} u(t)^{\frac{1}{t}} = \lambda^*$ where λ^* is the unique positive root of $\min\limits_{1\le i\le m} (a_{i1} + a_{i2}\lambda + \cdots + a_{in}\lambda^{n-1}) = \lambda^n$. A similar result holds in case of $f_t(x) = \max\limits_{1\le i\le m}(a_{i1}(t)x_1 + \cdots + a_{in}(t)x_n)$. For $m = 1$ we are back to the linear case of the Poincaré theorem, this time, however, within a positive framework. Thus, let $m = 2$ and consider for simplicity the numerical example

$$a_{11}(t) = \frac{t}{t+1}, \quad a_{12}(t) = 2 + \frac{1}{t}, \quad a_{21}(t) = 2 + \frac{1}{t^2}, \quad a_{22}(t) = \frac{t+1}{t+2} \quad \text{for} \quad t \ge 1$$

for which $A = \left[\begin{smallmatrix} 1 & 2 \\ 2 & 1 \end{smallmatrix}\right]$. The characteristic equation for λ^* reads $\min\{1 + 2\lambda, 2 + \lambda\} = \lambda^2$. In case of $1 + 2\lambda \le 2 + \lambda$ we have $1 + 2\lambda = \lambda^2$ with roots $\lambda_{1,2} = 1 \pm \sqrt{2}$. None of these roots gives λ^* since $1 + \sqrt{2} > 1$ and $1 - \sqrt{2} < 0$. Therefore, we must have $1 + 2\lambda \ge 2 + \lambda$ and $2 + \lambda = \lambda^2$ with roots $\lambda_{1,2} = \frac{1}{2} \pm \frac{3}{2}$. Since $\frac{1}{2} - \frac{3}{2} < 0$ we conclude with $\lambda^* = 2$ which indeed solves the characteristic equation. Thus, we arrive at $\lim\limits_{t\to\infty} \frac{u(t+1)}{u(t)} = \lim\limits_{t\to\infty} u(t)^{\frac{1}{t}} = 2$, provided $(u(0), u(1)) \gneqq 0$.

Similarly one obtains for $A = \left[\begin{smallmatrix} 1 & 2 \\ 2 & 1 \end{smallmatrix}\right]$ and the case $\max\{1 + 2\lambda, 2 + \lambda\} = \lambda^2$ as unique solution $\lambda^* = 1 + \sqrt{2}$.

As the numerical example indicates to solve the characteristic equation for λ^* in general one faces sets of **inequalities** of real polynomicals in one variable.

A single linear difference as it is considered in Poincaré's theorem is a special case of the above example, taken, however, within the framework of positive systems. Though a rather simple case it allows some interesting observations as shown by the following remarks.

Remarks 7.4.6. Consider the linear difference equation of order n with time-dependent coefficients

$$u(t+n) = p_1(t)u(t) + p_2(t)u(t+1) + \cdots + p_n(t)u(t+n-1),$$

and assume $\lim\limits_{t\to\infty} p_i(t) = p_i$ for all $1 \le i \le n$.

Assume further for $n \ge 2$ all $p_j(t) \ge 0$ and $p_j > 0$ for $j \in J$ where $1 \in J$, $|J| \ge 2$ and $gcd\{n - j + 1 \mid j \in J\} = 1$. Then by Example 7.4.5 we have for $(u(0), \dots, u(n-1)) \gneqq 0$ that $\lim\limits_{t\to\infty} \frac{u(t+1)}{u(t)} = \lim\limits_{t\to\infty} u(t)^{\frac{1}{t}} = \lambda^*$, where λ^* is the unique positive root of the characteristic equation of $p_1 + p_2\lambda + \cdots + p_n\lambda^{n-1} = \lambda^n$.

Thus, within the positive framework the conclusion (7.4.6) in Poincaré's theorem holds in the sharper form that the eigenvalue can be chosen the same for all solutions.

(a) As Examples 7.4.2 (i) show one cannot expect $\lim\limits_{t\to\infty} \frac{u(t)}{\lambda^{*t}}$ to exist in Poincaré's theorem not even when coefficients are positive.

(b) Let $n = 4, p_1 = p_2 = \frac{1}{2}, p_2 = p_4 = 0$. Therefore, $J = \{3, 4\}$ and by the above $\lim\limits_{t\to\infty} \frac{u(t+1)}{u(t)} = \lim\limits_{t\to\infty} u(t)^{\frac{1}{t}} = 1$, where $\lambda^* = 1$ is the unique positive root of $\frac{1}{2} + \frac{1}{2}\lambda = \lambda^4$. Therefore, the conclusion (7.4.6) of Poincaré's theorem does hold, though the assumptions of the theorem are not fulfilled since $\frac{1}{2} + \frac{1}{2}\lambda = \lambda^4$ has two different conjugate roots.

(c) Let $n = 4, p_1 = p_3 = \frac{1}{2}, p_2 = p_4 = 0$. Therefore $J = \{2, 4\}$ and $gcd\{n - j + 1 \mid j \in J\} = 2 \neq 1$, that is the monotonicity assumptions are not satisfied. Indeed, as Examples 7.4.2 (ii) show, $\lim_{t\to\infty} \frac{u(t+1)}{u(t)} = \lambda^*$ does not hold. Since $\frac{1}{2} + \frac{1}{2}\lambda^2 = \lambda^4$ has the roots ± 1 this is in line with Poincaré's theorem. Furthermore, $\lim_{t\to\infty} u(t)^{\frac{1}{t}} = \lambda^*$ does hold for $\lambda^* = 1$ which is in line with Pituk's theorem. This case shows also that even in the restricted domain of positive systems the conclusion of Pituk's theorem may hold whereas the one of Poincaré's theorem does not.

7.5 Price setting in case of technical change

The dynamics of price development in economics has been considered previously in Chapters 1 and 2 (Sections 1.3, 1.4, 2.7). There it has been pointed out that by technical change one arrives at a non-autonomous and non-linear discrete dynamical system given by (see equation 1.4.2)

$$p(t + 1) = k(t)T(t)p(t), \quad t \in \mathbb{N}, \; p(0) \in \mathbb{R}^n_+.$$

Thereby, $p(t)$ is the price vector at t, $k(t) > 0$ a scalar factor and $T(t)\colon \mathbb{R}^n_+ \longrightarrow \mathbb{R}^n_+$ the cost operator given by $T(t)p = c(p, t)$ with a (unit) cost function $c(p, t)$. As a function of prices costs are often concave, for example if there is a choice of techniques. If the technology changes with time then the cost function depends explicitly on time. The latter has been modeled in the previous chapters but for the results obtained we had to assume an autonomous system. With the tools developed in chapter 7 we are now ready to treat technical change and to find out conditions under which the price dynamics is stable in the sense of weak ergodicity. (One can ask, of course, for conditions of strong ergodicity; this, however, is less likely in case of technical change.)

In what follows we shall simplify notation by integrating the scalar factor into the cost function, that is we consider the dynamical system

$$p(t + 1) = T(t)p(t), \quad T(t)p = c(p, t), \quad t \in \mathbb{N}, \quad p(0) \in \mathbb{R}^n_+, \tag{7.5.1}$$

with $T(t)$ a selfmapping of $\mathbb{R}^n_+$. It is quite natural to consider a norm for prices $\|p\| = \sum_{i=1}^n p_i, p \in \mathbb{R}^n_+$. If there is a choice of techniques, minimal costs of producer $1 \le i \le n$ can be specified as (see equation 1.4.1)

$$c_i(p, t) = \inf\{pa + w_i l | (a, l) \in A_i(t)\}. \tag{7.5.2}$$

Thereby, $A_i(t)$ denotes the (non-empty) set of techniques which producer i has at his disposal which consists of pairs (a, l) with $a \in \mathbb{R}^n_+$ being material inputs and $l \ge 0$ the labour input to produce one unit of good i. (pa is the inner product of vectors p and a.)

Furthermore, w_i is the wage paid per hour in the production of good i. The wage can be considered to be independent of prices as a "rigid wage" or to be varying with prices as $w_i = pb^i$ where $b^i \in \mathbb{R}^n_+$ is the "real wage" corresponding to w_i. In the latter case equation (7.5.2) becomes (see also section 1.4)

$$c_i(p,t) = \inf\{p(a + lb^i) | (a,l) \in A_i(t)\}. \tag{7.5.3}$$

(For literature concerning these models as well as for the background in economics see the relevant references given in chapters 1 and 2.) The following result is easily obtained from part (ii) of Corollary 7.2.3.

Proposition 7.5.1. *Consider a price dynamics given by equation* (7.5.1).

(i) *Let the cost function $c(p,t)$ for each t be concave, positively homogeneous and such that $\sup_t c_i(e,t)$ is finite for $1 \le i \le n (e = (1,\dots,1))$. Assume the matrix $D = (d_{ij})$ of minimal expenditures $d_{ij} = \inf_t c_i(e_j,t)$ (e_j the j-th unit vector) is indecomposable with $d_{hh} > 0$ for some h.*
Then there holds weak ergodicity (for $\|\cdot\|$) on $\mathbb{R}^n_+ \setminus \{0\}$ and any two non-zero orbits are asymptotically proportional.

(ii) *The above conclusion holds in particular for a cost function as in equation* (7.5.3) *satisfying the assumptions on matrix D and such that for each i the technology satisfies $A_i(t) \subseteq A_i(t+1)$ for all t or $\sup_t |A_i(t)|$ is finite.*

Proof. (i) For $p = p_1 e_1 + \cdots + p_n e_n$ with $p_i > 0$ for all i and $\|p\| = \sum_{i=1}^n p_i = 1$ concavity of costs implies

$$c_i(p,t) \ge \sum_{j=1}^n p_j c_i(e_j,t) \ge \sum_{j=1}^n p_j d_{ij} = (Dp)_i,$$

that is $c(p,t) \ge Dp$ for all t. By Lemma 2.2.10 the assumptions made imply $D^r > 0$ for some $r \ge 1$. For $K = \mathbb{R}^n_+$ the selfmapping given by $T(t)p = c(p,t)$ is concave and positively homogeneous with $T(t)p \ge Dp$ for all t. Therefore,

$$S(t)p = T(t+r-1) \circ \cdots \circ T(t)p \ge D^r p$$

and

$$(S(t)p)_i \ge (D^r p)_i = \sum_{j=1}^n D^r_{ij} p_j \ge \min_{1 \le j \le n} D^r_{ij} \Big(\sum_{j=1}^n p_j\Big).$$

For the vector u with $u_i = \min_{1 \le j \le n} D^r_{ij}$ one has that $u \in \mathring{K}$ and $u \le S(t)p$ for all t all $p \in K$ with $\|p\| = 1$. Furthermore, $(T(t)p)_i = c_i(p,t) \le c_i(e,t)$ and, hence, for some scalar $k > 0$ it holds $T(t)p \le ku$ for $p \in K$ with $\|p\| = 1$. In particular, for $p = \frac{u}{\|u\|}$, one has that $T(t)u \le k\|u\|u$ for all t and, hence, $u \le S(t)p \le \tilde{k}u$ with some scalar $\tilde{k} > 0$ for all t, all $p \in K$, $\|p\| = 1$. Since $S(t)$ maps $K \setminus \{0\}$ into $\mathring{K}$ the conclusion in (i) follows from part (ii) of Corollary 7.2.3.

(ii) A cost function according to equation (7.5.3) obviously is concave and positively homogeneous. If $A_i(t) \subseteq A_i(t+1)$ then $c_i(p, t+1) \le c_i(p,t)$ and it follows that $\sup_t c_i(e,t)$ is finite. The latter holds, too, in case $\sup_t |A_i(t)|$ is finite. So, part (ii) follows from part (i). □

The following result considers the case of "rigid wages" and is easily obtained from Corollary 7.2.5.

Proposition 7.5.2. *Consider a price dynamics given by equation (7.5.1) with a cost function $c(p,t) = c^m(p,t) + c^w(t)$ consisting of two parts, material costs c^m and wage costs c^w (both non-negative). Suppose $c^m(p,t) > 0$ is for $p > 0$ increasing, differentiable and positively homogeneous of degree $0 \le r \le 1$, that is $c^m(\lambda p, t) = \lambda^r c^m(p,t)$ for $0 \le \lambda$. Suppose further that $\sup_t c_i^m(e,t)$ is finite for all i ($e = (1,\dots,1)$).*

(i) *If $r < 1$ then there holds weak ergodicity (for $\|\cdot\|$) and any two orbits are asymptotically equal (on strictly positive prices).*
(ii) *If $r = 1$ and $\inf_t c_i^w(t) > 0$ for all i then any two bounded (for $\|\cdot\|$) orbits are asymptotically equal and path stability (for $\|\cdot\|$) applies.*

Proof. The price dynamics is driven by $T(t)p = c^m(p,t) + c^w(t)$. For $K = \mathbb{R}^n_+$ differentiation on $\overset{\circ}{K}$ yields

$$\sum_{j=1}^n p_j \left|\frac{\partial (T(t)p)_i}{\partial p_j}\right| = \sum_{j=1}^n p_j \frac{\partial c_i^m(p,t)}{\partial p_j}$$

by taking into account that $c^m(p,t)$ is increasing in p.

By Euler's Theorem on homogeneous functions

$$\sum_{j=1}^n p_j \frac{\partial c_i^m(p,t)}{\partial p_j} = r c_i^m(p,t),$$

and, hence,

$$\sum_{j=1}^n \frac{p_j}{(T(t)p)_i}\left|\frac{\partial (T(t)p)_i}{\partial p_j}\right| = r\,\frac{c_i^m(p,t)}{c_i^m(p,t) + c_i^w(t)}. \qquad (*)$$

(i) If $r < 1$ then for $c = r$ and $D = \overset{\circ}{K}$ the assumptions in case (a) of Corollary 7.2.5 are satisfied and the conclusion in part (i) does follow.

(ii) Let $r = 1$ and, by assumption $c^w(t) \ge u > 0$ for all t. From the assumptions on $c^m(p,t)$ it follows for $p > 0$

$$c^m(p,t) = \|p\| c^m\left(\frac{p}{\|p\|}, t\right) \le \|p\| c^m(e,t) \le \|p\| v \quad \text{for all} \quad t$$

where $v_i = \sup_t c_i^m(e,t)$. Since $u \in \overset{\circ}{K}$ there exists $s > 0$ with $v \le su$ and, hence, $c^m(p,t) \le \|p\| s c^w(t)$ for all $p > 0$, all t. Thus, we arrive at

$$\frac{c_i^m(p,t)}{c_i^m(p,t) + c_i^w(t)} = \frac{1}{1 + \frac{c_i^w(t)}{c_i^m(p,t)}} \le \frac{1}{1 + \frac{1}{\|p\| s}} < 1.$$

This inequality together with $(*)$ shows that the assumptions in case (b) of Corollary 7.2.5 are satisfied for $D = \mathring{K}$ and the conclusion in part (ii) does follow. □

The results on price dynamics obtained we illustrate by some examples.

Examples 7.5.3. (i) Consider the following simple numerical example of two producers, each producing one single good by just one technique, this time, however, the choice of technique depends on time. To make calculations simple assume for producers 1 and 2, respectively, the following techniques:

$$A_1(t) = \{(a(t); l(t))\} = \begin{cases} ((0,1);1), & t \text{ even} \\ ((\frac{1}{2},\frac{1}{2});1), & t \text{ odd} \end{cases}$$

$$A_2(t) = \{a(t); l(t))\} = \begin{cases} ((1,0);2), & t \text{ even} \\ ((2,0);1), & t \text{ odd}. \end{cases}$$

As real wages assume $b^1 = b^2 = (1,1)$.

Regarding the cost functions we obtain for $c_i(e_j,t) = a_j + lb^i_j$ with $(a,l) \in A_i(t)$ that

$$c_1(e_1,t) = \begin{cases} 1, & t \text{ even} \\ \frac{3}{2} & t \text{ odd}, \end{cases} \qquad c_1(e_2,t) = \begin{cases} 2, & t \text{ even} \\ \frac{3}{2}, & t \text{ odd}, \end{cases}$$

$$c_2(e_1,t) = \begin{cases} 3, & t \text{ even} \\ 3, & t \text{ odd}, \end{cases} \qquad c_2(e_2,t) = \begin{cases} 2, & t \text{ even} \\ 1, & t \text{ odd}. \end{cases}$$

For the matrix D we obtain

$$d_{11} = \min\left\{1, \frac{3}{2}\right\} = 1, \quad d_{12} = \min\{3,3\} = 3$$

$$d_{21} = \min\left\{2, \frac{3}{2}\right\} = \frac{3}{2}, \quad d_{22} = \min\{2,1\} = 1.$$

Thus, D is a (strictly) positive matrix and the assumptions of part (ii) in Proposition 7.5.1 are met. Therefore, weak ergodicity as well as asymptotic proportionality do hold for price orbits. We illustrate this result by examining directly the price formation as follows.

Since $A_1(t)$ and $A_2(t)$ consist of one technique only, in this example $c(p,t)$ is linear in p and we have that $T(t)p = c(p,t) = p_1 c(e_1,t) + p_2 c(e_2,t)$. Using the numbers we obtain

$$T(t)p = \begin{cases} p_1 \begin{bmatrix} 1 \\ 3 \end{bmatrix} + p_2 \begin{bmatrix} 2 \\ 2 \end{bmatrix}, & t \text{ even} \\[2ex] p_1 \begin{bmatrix} \frac{3}{2} \\ 3 \end{bmatrix} + p_2 \begin{bmatrix} \frac{3}{2} \\ 1 \end{bmatrix}, & t \text{ odd} \end{cases}$$

or

$$T(t)p = \begin{cases} Ap, & t \text{ even} \\ Bp, & t \text{ odd} \end{cases} \quad \text{with} \quad A = \begin{bmatrix} 1 & 2 \\ 3 & 2 \end{bmatrix}, \quad B = \begin{bmatrix} \frac{3}{2} & \frac{3}{2} \\ 3 & 1 \end{bmatrix}.$$

From $A, B \geq \left[\begin{smallmatrix} 1 & 1 \\ 1 & 1 \end{smallmatrix}\right]$ it follows that each price path for $p(0) \gneq 0$ tends to infinity. Furthermore, $p(2t) = (BA)^t p(0)$ and $p(2t+1) = A(BA)^t p(0)$ for $t \geq 0$.

Since BA is a (strictly) positive matrix we obtain from the classical Frobenius theorem (Theorem 2.4.1) for $p(0) \gneq 0$ arbitrary that $\lim_{t\to\infty} \frac{p(2t)}{\|p(2t)\|} = x^*$ where (x^*, λ^*) is the unique positive solution of $(BA)x^* = \lambda^* x^* (\|x^*\| = 1)$. From $p(2t+1) = Ap(2t)$ it follows $\frac{p(2t+1)}{\|p(2t+1)\|} = \frac{A\frac{p(2t)}{\|p(2t)\|}}{\|A\frac{p(2t)}{\|p(2t)\|}\|}$ and, hence, $\lim_{t\to\infty} \frac{p(2t+1)}{\|p(2t+1)\|} = \frac{Ax^*}{\|Ax^*\|}$. This shows that $\lim_{t\to\infty} \|\frac{p(t)}{\|p(t)\|} - \frac{q(t)}{\|q(t)\|}\| = 0$ for arbitrary $p(0), q(0) \gneq 0$, that is we have weak ergodicity. By Theorem 2.4.1 again, $\lim_{t\to\infty} \frac{p(2t)}{\lambda^{*t}} = x^*$ and, hence, $\lim_{t\to\infty} \frac{p(2t+1)}{\lambda^{*t}} = Ax^*$. It follows, for $p(0), q(0) \gneq 0$ given to $\epsilon > 0$ there exists $s(\epsilon)$ such that for $t \geq s(\epsilon)$

$$\lambda^{*t}(1-\epsilon)x^* \leq p(2t), \quad q(2t) \leq \lambda^{*t}(1+\epsilon)x^*$$

and, hence

$$\frac{1-\epsilon}{1+\epsilon} q(t) \leq p(t) \leq \frac{1+\epsilon}{1-\epsilon} q(t)$$

for $t \geq 2s(\epsilon) + 1$.

These inequalities imply that orbits with $p(0), q(0) \gneq 0$ are asymptotically proportional. Moreover, the inequalities show asymptotical equality. Possibly, does strong ergodicity even hold in this example? If this would be the case we should have that $\frac{Ax^*}{\|Ax^*\|} = x^*$ and, hence, $\lambda^* x^* = (BA)x^* = \|Ax^*\| Bx^*$.

This would mean A and B have a common positive eigenvector. Any positive eigenvector of A must be $x^* = r\left[\begin{smallmatrix}2\\3\end{smallmatrix}\right]$ with $r > 0$ which, however, is not an eigenvector of B. Thus, weak ergodicity does hold for this example but not strong ergodicity.

(ii) The second example addresses another kind of technology, so called Cobb–Douglas technology (see Remark 2.7.2 for the autonomous case). Consider a cost function $c(p,t) = c^m(p,t) + c^w(t)$ where

$$c_i^m(p,t) = k_i(t) \prod_{j=1}^n p_j^{a_{ij}(t)} \quad \text{and} \quad c_i^w(t) = w_i(t) l_i(t) \geq 0$$

with $0 \leq k_i(t), \sup_t k_i(t)$ finite for all i and $0 \leq a_{ij}(t), \sum_{j=1}^n a_{ij}(t) = r \leq 1$ for all i. For $K = \mathbb{R}_+^n$, $c^m(p,t)$ is increasing, differentiable and positively homogeneous of degree r in p on $\overset{\circ}{K}$.

For $r < 1$ by Proposition 7.5.2 (i) it holds weak ergodicity and asymptotical equality on $\overset{\circ}{K}$.

For $r = 1$ suppose $\inf_t w_i(t) l_i(t) > 0$ for all i. Then by Proposition 7.5.2 (ii) any two bounded orbits are asymptotically equal and path stability applies. The assumption of boundedness cannot be simply omitted as the following calculation shows.

Let $k_i(t) = 1$ and $w_i(t) l_i(t) = f(t)$ for all i and all t. The assumptions made in part (ii) of Proosition 7.5.2 are satisfied. Since $(T(t)p)_i = \prod_{j=1}^n p_j^{a_{ij}(t)} + f(t)$ it follows for $\lambda > 0$ and $e = (1, \dots, 1)$ that

$$T(t)(\lambda e) = (\lambda + f(t))e \text{ and} T(s)T(t)e = (1 + f(t) + f(s))e.$$

It follows by induction

$$p(t+1) = \Big(1 + \sum_{i=1}^{t} f(i)\Big)e \quad \text{and} \quad q(t+1) = \Big(\lambda + \sum_{i=1}^{t} f(i)\Big)e.$$

Consider two price orbits starting in $p(0), q(0) > 0$. Since $\inf_t f(t) > 0$ both orbits are unbounded. For $\lambda \neq 1$, however, $\|p(t+1) - q(t+1)\| = |\lambda - 1|\|e\|$ and path stability (for $\|\cdot\|$) does not hold. Similarly, the two orbits are not asymptotically equal. (For strong ergodicity in this model of price setting see Exercise 10.)

7.6 Populations under bounded and periodic enforcement

Consider the non-linear population dynamics in one dimension given by the **Beverton–Holt model** ĥbox to 100pt

$$x(t+1) = \frac{\mu K x(t)}{K + (\mu - 1)x(t)}, \quad x(0) \geq 0, \tag{7.6.1}$$

where $\mu > 1$ is the so called inherent growth rate and $K > 0$ the so called carrying capacity. For the reproduction function $f(x) = \frac{\mu K x}{K + \mu - 1)x}$ one has $x|\frac{f'(x)}{f(x)}| = \frac{K}{K+(\mu-1)x} < 1$ for $x > 0$. Equivalently, f is a cave function, that is, $\frac{f(x)}{x}$ is strictly decreasing but $xf(x)$ is strictly increasing. This means, there is population pressure which, however, is modest since the population decreases not too fast.(For cave functions see also Section 5.3 and the related Exercise 7).

We shall call for any reproduction function f the magnitude $c(x) = x|\frac{f'(x)}{f(x)}|$ the **population pressure for f at state x.** Since $c(x) < 1$ for the Beverton–Holt model the reproduction function has a unique non-zero fixed point $x^* = K$ and $\lim_{t\to\infty} x(t) = x^*$ for each $x(0) > 0$. (This is easily verified and does follow also from Exercise 7 (d) to Section 5.3).

The interaction of a population with its environment and changes in the environment, like seasonal fluctuations, can enforce essential parameters of the population like μ and K to change. To take this enforcement into account, we treat parameters time dependent which for the example considered yields the **non-autonomous Beverton–Holt model**

$$x(t+1) = \frac{\mu(t)K(t)x(t)}{K(t) + (\mu(t) - 1)x(t)}, \quad x(0) \geq 0, \tag{7.6.2}$$

with $\mu(t) > 1, K(t) > 0$ for all $t = 0, 1, \ldots$. An interesting question then is under what conditions the population shows a stability behaviour in the sense of path stability or weak ergodicity. Also interesting is the question of strong ergodicity in the less likely case the time dependent parameters tend to certain values. Particularly relevant is the question whether periodicity of the parameters due to seasonal fluctuations lead to an asymptotically periodic behavior of the population. The latter question has been taken up in a sequence of papers by Cushing/Henson and Elaydi/Sacker, respectively ([7, 8, 11, 12]; see also [25, 31], and see [46] for more general for mathematical population models). These questions are, of course, of interest also for many other kinds of populations and their reproduction functions. It will turn out that for the questions raised the population pressure plays a fundamental role.

In the following we shall consider the general situation of several populations which depend on each other in their development within a changing environment. The above non-autonomous Beverton–Holt model as well as others constitute the special case of just one population. Suppose there are n populations with $x_i(t)$ the level of population i at time t which develop according to

$$x_i(t+1) = f_i(t, x_1(t), \dots, x_n(t)), \quad x(0) = (x_1(0), \dots, x_n(0)) > 0 \tag{7.6.3}$$

for $i = 1, \dots, n$ and $t = 0, 1, \dots,$ where f_i is the **extended reproduction function** of population i which takes into account also the levels of populations other than i. Suppose further for each i and t differentiability of $f_i(t, \cdot)$ in the interior of $\mathbb{R}^n_+$ and let D be a subset of the interior such that $x(0) \in D$ implies $x(t) \in D$ for all t. We define

$$c_{ij}(t,x) = \frac{x_j}{f_i(t,x)} \left| \frac{\partial f_i(t,x)}{\partial x_j} \right| \tag{7.6.4}$$

to be the **population pressure of population j on population i** and

$$c_i(t,x) = \sum_{j=1}^{n} c_{ij}(t,x) \tag{7.6.5}$$

the **total population pressure on population i** in state x at time t. (For an interpretation similar to the one dimensional case see Figure 6.1 and Exercise 5 to Chapter 6.)

The following theorem provides answers to the three questions mentioned considering path stability, periodicity, and strong ergodicity.

Theorem 7.6.1. *For the non-autonomous population system* (7.6.3) *with D log-convex the following properties hold true.*

(i) *If for each population i the total population pressure $c_i(t,x)$ is bounded for $t \geq 0, x \in D$ by $c_i < 1$ then all orbits $(x(t))$ starting in D are asymptotically equal. If at least one orbit has compact closure in D (for the norm) then path stability holds (for the norm).*
For the weaker assumption $c_i(t,x) \leq c_i(x) < 1, t \geq 0, x \in D$, asymptotic equality holds for bounded orbits and path stability holds for orbits with compact closure in D.

(ii) *Suppose D is internally closed and the population system (7.6.3) is periodic with period k, that is $f_i(t+k, \cdot) = f_i(t, \cdot)$ for $t \geq 0$ and $k \geq 1$ minimal. If $c_i(t,x) \leq c_i < 1$ for $t \geq 0, x \in D$ then each orbit $(x(t))$ starting in D converges to a unique k-cycle $(y^0, y^1, \dots, y^{k-1})$ in D given by $y_i^{j+1} = f_i(j, y^j)$ for $0 \leq j \leq k-1, 1 \leq i \leq n$. For the weaker assumption $c_i(t,x) \leq c_i(x) < 1, t \geq 0, x \in D$, the convergence to the cycle holds for each orbit with compact closure in D.*

(iii) *Suppose for each i there exists a differentiable function f_i on $\overset{\circ}{K}$ with bounded derivative on D such that $f_i(t,\cdot)$ converges in t uniformly on D to $f_i(\cdot)$ and $f_i(\cdot)$ has a strictly positive lower bound on D. If for each i the total pressure for $f_i(\cdot)$ satisfies $c_i(x) < 1$ for each $x \in D$ then each orbit with compact closure in D converges (in norm) to the unique population equilibrium $x^* \in D$ defined by $f_i(x^*) = x_i^*$ for $1 \le i \le n$.*

Proof. (i) Applying Corollary 7.2.5, case (a), to selfmappings T_t of int $\mathbb{R}^n_+$ defined by $(T_tx)_i = f_i(t,x)$ and $S_t = T_t$ one obtains for $x \in D$

$$A(x) = \sup_t \max_i \sum_{j=1}^{n} c_{ij}(t,x) = \sup_t \max_i c_i(t,x) \le \max_i c_i < 1.$$

This proves the first part of property (i). The second part follows from case (b) of Corollary 7.2.5.

(ii) By the mean value theorem it follows (as in the proof of Corollary 7.2.5 for cases (a) and (b), respectively)

$$p(T_tx, T_ty) \le cp(x,y) \quad \text{for} \quad x,y \in D \quad \text{and} \quad c = \max_i c_i < 1$$

and, for the weaker assumption $c_i(x) < 1$,

$$p(T_tx, T_ty) < p(x,y) \quad \text{for} \quad x,y \in D, \ x \neq y.$$

For $Tx = T_{k-1} \circ \ldots \circ T_0x$ for $x \in D$ it follows that T is a contraction in the first case and a contractive mapping in the second case. Since int $\mathbb{R}^n_+$ is internally complete and, by assumption, D internally closed the metric space (D,p) is complete. In the first case it follows from Banach's contraction principle that T has a fixed point $x^* \in D$. For the second case consider an orbit $(x(t))$ with compact closure (for the norm) in D. There exists a subsequence $(x(t_k))$ converging in norm to $x^* \in D$ and, since norm topology and part topology coincide on $\overset{\circ}{K}$, this convergence holds for p, too. Thus, $\omega(x(0)) \neq \emptyset$ in (D,p) and from Lemma 4.1.2 (b) it follows that T has a fixed point $x^* \in D$.

Let $y^{t+1} = T_t \circ \ldots \circ T_0x^*$ the orbit starting in x^* and $y^0 = x^*$. By periodicity, for $t = nk + i, 0 \le i < k-1$

$$y^{t+1} = T_{nk+i} \circ \cdots \circ T_{nk}T^nx^* = T_i \circ \cdots \circ T_0x^*.$$

Since the orbit (y^t) is finite, from (i) it follows for any orbit $(x(t))$ and any orbit with compact closure in D, respectively, that $\lim_{t\to\infty} \|x(t) - y^t\| = 0$.

Thus, for the cycle $C = (y^0, y^1, \ldots, y^{k-1})$ we have that $\lim_{t\to\infty} \inf_{c\in C} \|x(t) - c\| = 0$ which proves (ii).

(iii) For $(T_tx)_i = f_i(t,x)$ and $(Tx)_i = f_i(x)$ it follows that (T_t) converges uniformly to T on D. Since f_i has bounded derivative on D one has that T is uniformly continuous on D. Furthermore, $Tx \ge u$ for all $x \in D$ and some $u > 0$. The total population pressure of f_i in $x \in D$ is $\sum_{j=1}^n \frac{x_j}{f_i(x)} |\frac{\partial f_i(x)}{\partial x_j}| < 1$ and, hence, $\sum_{j=1}^n |\frac{\partial (Tx)_i}{\partial x_j}| x_j < (Tx)_i$. From Corollary 7.3.5

follows for $k = 1$ the convergence of orbits to the unique fixed point x^* of T in D which proves property (iii). □

Theorem 7.6.1 we illustrate by two examples in one dimension.

Examples 7.6.2. (i) (Non-autonomous Beverton–Holt model). The extended reproduction function according to equation (7.6.2) is $f(t,x) = \frac{\mu(t)K(t)x}{K(t)+(\mu(t)-1)x}$ for $x > 0$ with $\mu(t) > 1, K(t) > 0$ for $t = 0, 1, \ldots$. For the (total) population pressure on the population we obtain

$$c(t,x) = \frac{x}{f(t,x)}\left|\frac{\partial f(t,x)}{\partial x}\right| = \frac{K(t)}{K(t) + (\mu(t) - 1)x}.$$

Suppose

$$\inf_t K(t) > 0 \quad \text{and} \quad \inf_t \frac{\mu(t) - 1}{K(t)} > 0. \tag{7.6.6}$$

Choose $0 < a \leq \inf_t K(t)$ and put $D = [a, \infty[$.

For $x \geq a$ one has that

$$\mu(t)K(t)x - K(t)a \geq (\mu(t) - 1)K(t)x \geq (\mu(t) - 1)ax,$$

and, hence,

$$f(t,x) = \frac{\mu(t)K(t)x}{K(t) + (\mu(t) - 1)x} \geq a,$$

that is, $f(t,\cdot)$ maps D into itself. Furthermore, from assumptions (7.6.6) it follows that $c(t,x) \leq c < 1$ for all $t \geq 0$, all $x \in D$. Since D is log-convex it follows from property (i) of Theorem 7.6.1 that all orbits starting in D are asymptotically equal. For $x(0) > 0$, $y(0) > 0$ given choose $a \leq \min\{x(0), y(0)\}, 0 < a \leq \inf_t K(t)$ to obtain that all orbits $(x(t))$ with $x(0) > 0$ are asymptotically equal. This holds also in case all orbits are unbounded as for example $\mu(t) = t + 2, K(t) = t + 1$ which satisfies assumptions (7.6.6).

Considering property (ii) suppose $\mu(t + k, \cdot) = \mu(t, \cdot), K(t + k, \cdot) = K(t, \cdot)$ for all t, some $k \geq 1$. From Theorem 7.6.1 (ii) it follows that all orbits $(x(t))$ with $x(0) \geq a$ converges to the same k-cycle. As above this follows also for any $x(0) > 0$.

Considering property (iii) suppose $\lim_{t\to\infty} \mu(t) = \mu > 1$ and $\lim_{t\to\infty} K(t) = K > 0$. For $\alpha(t) = \frac{\mu(t)-1}{K(t)}, \alpha = \frac{\mu-1}{K}$ and $f(x) = \lim_{t\to\infty} f(t,x) = \frac{\mu x}{1+\alpha x}$ it follows that

$$|f(x) - f(t,x)| \leq \frac{|\mu - \mu(t)|x}{(1 + \alpha x)(1 + \alpha(t)x)} + \frac{|\mu\alpha(t) - \mu(t)\alpha|x^2}{(1 + \alpha x)(1 + \alpha(t)x)},$$

which yields uniform convergence of $f(t,\cdot)$ to $f(\cdot)$ on $D = [a, \infty[, 0 < a \leq \inf_t K(t)$. Furthermore, $\inf_{x\in D} f(x) > 0$ and $|f'(x)| = \frac{\mu}{(1+\alpha x)^2} \leq \mu$ on $\mathring{K}$.

The population pressure of f at $x \in D$ is given by $\frac{1}{1+\alpha x} < 1$ for $x > 0$. Thus, Theorem 7.6.1 (iii) implies that each orbit with compact closure in D converges to $x^* = K$ the unique fixed point of f in $\mathring{K}$. This holds, obviously, for every orbit with compact closure in the positive reals.

For the Beverton–Holt model orbits can be unbounded also if the assumptions (7.6.6) are met. If, however, in addition to assumptions (7.6.6) one has that $\sup_t \mu(t) < \infty$ then all orbits are bounded. For, $x(t+1) = \frac{\mu(t)K(t)}{\frac{K(t)}{x(t)}+(\mu(t)-1)} \leq \mu(t)\frac{K(t)}{(\mu(t)-1)}$ and $\sup_t \frac{K(t)}{\mu(t)-1} < \infty$ by assumption. Therefore, the compact closure of an orbit is contained in $D = [a, \infty[$ for some $a > 0$ and path stability holds for the norm. In case of periodicity, every orbit then converges to a common cycle also in the case of the weak assumption on population pressure. Furthermore, strong ergodicity holds under the augmented assumptions for each orbit.

To illustrate the case of periodicity further, consider the simple case of two seasons only, say spring and autumn with equal growth rates $\mu_0 = \mu_1 = \mu > 1$ but different positive capacities $K_0 \neq K_1$. The assumptions (7.6.6) as well as $\sup_t \mu(t) < \infty$ are satisfied, it follows that each orbit converges to a common cycle. This cycle is given by $\{x^*, f_0(x^*)\}$ where x^* is the unique fixed point of $f_1 \circ f_0$. One obtains $x^* = \frac{(\mu+1)K_0K_1}{K_1+K_0\mu}$ which depends not only on the two capacities but also on the common growth rate. The fixed points of f_0 and f_1 are K_0 and K_1, respectively. Since $K_0 \neq K_1$ we cannot have $f_0(x^*) = x^*$ which means that the cycle is a 2-cycle. This means in particular, we do have path stability but not strong ergodicity.

(ii) **(Non-autonomous Hassell–May Model).** The extended reproduction function is given by

$$f(t,x) = \frac{\lambda(t)x}{(1+a(t)y)^{b(t)}} \quad \text{for} \quad x > 0$$

with $\lambda(t), a(t), b(t) > 0$ for $t = 0, 1, \ldots$. For the (total) population pressure on the population one obtains

$$c(t,x) = \frac{x}{f(t,x)}\left|\frac{\partial f(t,x)}{\partial x}\right| = \frac{|1+a(t)(1-b(t))x|}{1+a(t)x}.$$

Suppose $\alpha = \inf_t a(t) > 0, \beta = \inf_t b(t) > 0,$

$$\gamma = \sup_t a(t) < \infty \text{ and } \sup_t b(t) \leq 2 \tag{7.6.7}$$

Then, for $b(t) \leq 1$, or $b(t) \geq 1$ and $a(t)(1-b(t))x \leq 1$

$$c(t,x) = 1 - \frac{a(t)b(t)x}{1+a(t)x} \leq 1 - \frac{\alpha\beta x}{1+\alpha x} < 1 \quad \text{for} \quad x > 0$$

and, for $b(t) \geq 1$ and $a(t)(b(t)-1) \geq 1$

$$c(t,x) = \frac{a(t)(b(t)-1)x-1}{1+a(t)x} \leq \frac{a(t)x-1}{1+a(t)x} \leq \frac{\gamma x-1}{1+\gamma x} < 1 \quad \text{for} \quad x > 0.$$

Defining $c(x) = \max\{1-\frac{\alpha\beta x}{1+\alpha x}, \frac{\gamma x-1}{1+\gamma x}\}$ we arrive at $c(t,x) \leq c(x) < 1$ for $x > 0$. Since $D = \operatorname{int} \mathbb{R}_+$ is log-convex, from Theorem 7.6.1, part (i), if follows that bounded orbits are asymptotically equal and path stability (for the norm) holds for orbits with compact closure in D. Considering periodicity, part (ii) of Theorem 7.6.1 implies that all orbits with compact closure in D converge to a common cycle. As for strong ergodicity, property (iii) does not supply meaningful conditions on the parameters beside the case $b(\cdot) = 1$ which comes back to the Beverton–Holt model.

For further examples see Exercise 11. For an example of a coupled system of two Beverton–Holt populations see Exercise 12.

Exercises

1. (a) Show for $V = \mathbb{R}^n$ and $K = \mathbb{R}^n_+$ that none of the three asymptotic properties of linkedness, proportionality and equality is equivalent to another one.
   (b) Find two sequences $(x_k), (y_k)$ with $\|x_k\| = \|y_k\| = 1$ for all k and $\lim_{k\to\infty} \|x_k - y_k\| = 0$ for which none of the above asymptotic properties does hold ($\|\cdot\|$ some norm on $\mathbb{R}^n$).
2. Let $\mathcal{C}([0,1])$ be the space of all real valued continuous functions on the unit interval and
   $$(T_n f)(u) = \int_{[0,1]} k(u,v)(f(v))^{l_n} dv \quad \text{for} \quad u \in [0,1].$$
   Thereby, $k\colon [0,1] \times [0,1] \longrightarrow \mathbb{R}_+$ is a continuous strict positive kernel, $f \in \mathcal{C}_+([0,1])$ and $l_n \in [0,1], n \geq 1$, with $l = \sup_n l_n < 1$. Consider the positive dynamical system given by
   $$f_{n+1} = T_n f_n, f_1 \in \mathcal{C}_+([0,1]).$$
   (a) Prove that any two paths $(f_n), (g_n)$ are asymptotically equal.
   (b) Find conditions on the kernel such that path stability holds for the sup-norm.
3. [36] Let a non-linear and non-autonomous Leslie model given by
   $$T_t \begin{bmatrix} x_1 \\ \vdots \\ x_n \end{bmatrix} = \begin{bmatrix} \sum_{i=1}^{n} b_i(t)(\sqrt{x_i} + 1) \\ s_1(x_1 + 1) \\ \vdots \\ s_{n-1}(x_{n-1} + 1) \end{bmatrix}$$
   with birth rates $b_i(t) \geq \underline{b} > 0$ and survival rates $s_i \geq \underline{s} > 0$.
   (a) Show that there exists a bounded orbit.
   (b) Prove path stability (for any norm).
   (c) Illustrate path stability for $n = 3$ by computer simulation for some chosen example of birth and survival functions.
4. Let a Leslie model given by
   $$T_t \begin{bmatrix} x_1 \\ x_2 \end{bmatrix} = \begin{bmatrix} x_1 + (1 - \frac{1}{t})\sqrt{x_2} \\ \frac{1}{2} x_1 \end{bmatrix} \quad \text{and let} \quad S_t = T_{t+1} \circ T_t.$$

(a) Prove that case (b) of Corollary 7.2.5 does apply.
(b) Show that case (a) of this corollary is not applicable.
(c) Show that unbounded orbits do exist.

5. Consider a “reversed” continued fraction given by $x(t+1) = f_t(x(t)), t = 0, 1, \ldots,$ $x(t) \in \mathbb{R}_+$ where

$$f_t(x) = \begin{cases} \frac{1}{1+x}, & t \text{ even} \\ \frac{2}{2+x}, & t \text{ odd.} \end{cases}$$

(a) Use Corollary 7.2.5 to prove path stability, that is $\lim_{t\to\infty} |x(t) - y(t)| = 0$ for any $x(0), y(0) \in \mathbb{R}_+$.
(b) Determine the limit set of $(x(t))$ for $x(0) \in \mathbb{R}_+$.

6. (a) Consider the non-autonomous affine dynamical system $x(t+1) = f_t(x(t)), t = 0, 1, \ldots, x(t) \in \mathbb{R}_+$ where $f_t(x) = a_t x + b_t$ with $a_t, b_t \in \mathbb{R}_+$ and $\sup_t a_t < 1$. Prove path stability (for the absolute value on $\mathbb{R}$).
(b) Let $\omega_s(x; (a_t), (b_t))$ denote the limit set for a path with $x(0) = x$. Show that the “Cantor dust”, that is the union of all $\omega_s(x; (a_t), (b_t))$ over all possible sequences with $(a_t, b_t) = (\frac{1}{3}, \frac{2}{3})$ or $(a_t, b_t) = (\frac{1}{3}, 0)$ is independent of x.
(c) Consider the system $x(t+1) = f_t(x(t)), t = 0, 1, \ldots, x(t) > 0$ where $f_t(x) = a_t\sqrt{x} + b_t$ with $a_t > 0, b_t \geq 0$. Show that all orbits are asymptotically equal and find sequences $(a_t), (b_t)$ for which path stability does not hold.
(d) Let for the system in (c) $\omega_s(x; (a_t), (b_t))$ be the limit set for a path with $x(0) = x > 0$. Show that the “non-linear Cantor dust” defined as in (b) (for the system in (c)) is independent of x.

7. (a) Prove strong ergodicity for the non-linear and non-autonomous Leslie model given by

$$T_t \begin{bmatrix} x_1 \\ \vdots \\ x_n \end{bmatrix} = \begin{bmatrix} \sum_{i=1}^{n} b_i(t) x_i^\alpha \\ s_1(t) x_1^\alpha \\ \vdots \\ s_{n-1}(t) x_{n-1}^\alpha \end{bmatrix},$$

where $0 < \alpha < 1$ and $b_i = \lim_{t\to\infty} b_i(t) > 0, s_i = \lim_{t\to\infty} s_i(t) > 0$.
(b) Compute the equilibrium solution (λ^*, x^*) in terms of $b_i, s_i (\|x^*\| = 1$ for the l_1-norm).

8. (a) Let $A_t\colon \mathbb{R}_+^n \longrightarrow \mathbb{R}_+^n, t = 0, 1, \ldots,$ be a sequence of cone mappings which converges, uniformly on bounded sets, to a continuous mapping A and let (a_t) be a sequence in $\mathbb{R}_+^n$, which converges to some $a > 0$. If there exists $b \in \mathbb{R}_+^n$ such that $0 < A(b) < b$, then each orbit given by $x(t+1) = A_t(x(t)) + a_t$ with $x(0) > 0$ which is bounded, converges to the unique solution x^* of $A(x) + a = x$.

(b) Find a sequence of cone mappings $A_t\colon \mathbb{R}^n_+ \longrightarrow \mathbb{R}^n_+$ which converges, uniformly on bounded sets, to a continuous mapping A, and find a sequence (a_t) which converges to some $a > 0$ such that no orbit converges to a solution of $A(x) + a = x$.

9. Prove the following extension of Pituk's Theorem to linear operators in a positive setting. Let $(V, \|\cdot\|)$ be a Banach space, K a closed and normal convex cone with non-empty interior $\mathring{K}$. Consider the Poincaré difference system

$$x(t+1) = (T + S(t))x(t), \quad x(0) \in K$$

where T and $S(t)$ are bounded linear operators on V which leave K invariant. Suppose $T(B) \subseteq [u, v] \subseteq \mathring{K}(B = \{x \in K \mid \|x\| = 1\}, [u, v] = \{x \in K \mid u \leq x \leq v\})$ and $\lim_{t\to\infty} \|S(t)\| = 0$ for the pertubation $S(t)$ of T.
Then

$$\lim_{t\to\infty} \|x(t)\|^{\frac{1}{t}} = \lambda^* \quad \text{for all} \quad x(0) \neq 0,$$

where $\lambda^* > 0$ is the unique positive eigenvalue of T (with eigenvector in K).

10. Consider a dynamical system of price setting given by

$$p(t+1) = c^m(p(t), t) + c^w(t), t = 0, 1, \ldots, p(t) \in \mathbb{R}^n_+$$

with material costs $c^m(p,t)_i = k(t)\prod_{j=1}^n p_j^{a_{ij}(t)}$ and labor costs $c_i^w(t)$. Assume $0 \leq a_{ij}(t), \sum_{j=1}^n a_{ji}(t) = r \in]0,1[, c_i^w(t) \geq 0$.
Assume further, $\lim_{t\to\infty} a_{ij}(t) = a_{ij}$, $\lim_{t\to\infty} k(t) = k$ and $\lim_{t\to\infty} c_i^w(t) = c_i^w > 0$.

(a) Prove strong ergodicity, that is $\lim_{t\to\infty} p(t) = p^*$ for each bounded orbit with $p(0) > 0$.

(b) Which price vectors p^* can be asymptotically reached by controlling asymptotically the labor costs?

11. [31] Prove for the following non-autonomous population models $f(t,\cdot)\colon D \longrightarrow D, D = \{x \in \mathbb{R} \mid x > 0\}$, path stability as well as the existence of a globally stable cycle under the conditions specified.

(a) Riccati model:

$$f(t,x) = \frac{a(t) + b(t)x}{c(t) + d(t)x}.$$

Path stability if $0 < \inf_t \frac{a(t)d(t)}{b(t)c(t)} \leq \sup_t \frac{a(t)d(t)}{b(t)c(t)} < \infty$.
Existence of a stable k-cycle if f is k-periodic and the coefficients are strictly positive.

(b) **Power Riccati model:**

$$f(t,x) = \frac{a(t) + b(t)x^{r(t)}}{c(t) + d(t)x^{s(t)}}$$

with non-negative coefficients, $c(t) + d(t) > 0$.

Path stability if $\sup_t r(t), \sup_t s(t) < 1$.

Existence of a stable k-cycle if f is k-periodic and $r(t), s(t) < 1$ for all t.

(c) **Maynard Smith model:**

$$f(t,x) = \frac{\lambda(t)x}{1 + (\lambda(t) - 1)x^{b(t)}}.$$

If $\inf_t \lambda(t) > 1$ then path stability holds and $x^* = 1$ is a globally stable fixed point.

12. Consider the following system of two coupled Beverton–Holt populations

$$f_1(t,x) = \sqrt{x_2}\frac{\mu K(t)x_1}{K(t) + (\mu - 1)x_1}$$

$$f_2(t,x) = \sqrt{x_1}\frac{\nu L(t)x_2}{L(t) + (\nu - 1)x_2}, \quad x_1 > 0,\ x_2 > 0.$$

Assume there are two seasons, $K(t) = K_0, L(t) = L_0$ for t even and $K(t) = K_1, L(t) = L_1$ for t odd.

Suppose, $K = \min\{K_0, K_1\} \geq 1, L = \min\{L_0, L_1\} \geq 1$ and $\mu \geq 2K + 1, \nu \geq 2L + 1$.

(a) Prove for $f(t,x) = (f_1(t,x), f_2(t,x))$ that $f(t,\cdot)$ maps for each t the set $D = \{x = (x_1, x_2) \in \mathbb{R}^2 \mid 1 \leq x_1, 1 \leq x_2\}$ into itself.

(b) Prove that each orbit given by $x(t+1) = f(t, x(t)), x(0) \in D$ is bounded.

(c) Show path stability for the system on D.

(d) Show that all paths $(x(t)), x(0) \in D$ converge to a unique 2-cycle $\{x^*, f(0,x^*)\}$ and compute x^*.

(e) Explore by computer simulations the behavior of $x(t))$ for $x_1(0) < 1, x_2(0) < 1$.

Bibliography

[1] R. Abu–Saris, S. Elaydi and S. Jang. Poincaré types solutions of systems of difference equations. *J. Math. Anal. Appl.*, 275: 69–83, 2002.

[2] Y.-Z. Chen. Inhomogeneous iterates of contraction mappings and non-linear ergodic theorems. *Nonlinear Analysis*, 39: 1–10, 2000.

[3] Y.-Z. Chen. Path stability and non-linear weak ergodic theorems. *Trans. Amer. Math. Soc.*, 352: 5279–5292, 2000.

[4] A. J. Coale. How the age distribution of a population is determined. *Cold Spring Harbor Symposia on Quantitative Biology*, 22:83–89, 1957.

[5] J. E. Cohen. Ergodic theorems in demography. *Bull. Amer. Math. Soc.*, 1:275–295, 1979.

[6] J. E. Cohen. Contractive inhomogeneous products of non-negative matrices. *Math. Proc. Camb. Phil. Soc.*, 86:351–364, 1979.

[7] J. M. Cushing. A strong ergodic theorem for some non-linear matrix models for structural population growth. *Natur. Resource Modeling*, 3:331–357, 1989.

[8] J. M. Cushing and S. M. Henson. A periodically forced Beverton–Holt equation. *J. Diff. Equ. and Appl.*, 8:1119–1120, 2002.

[9] W. Doeblin. Le cas discontinu des probabilités en chain. *Publ. Fac. Sci. Univ. Masaryk*, 236, 1937.

[10] S. Elaydi. *An Introduction to Difference Equations. Second Edition*. Springer-Verlag, New York, 1999.

[11] S. Elaydi and R. J. Sacker. Non-autonomous Beverton–Holt equations and the Cushing–Henson conjectures. *J. Diff. Equ. Appl.*, 11:337–347, 2005.

[12] S. Elaydi and R. J. Sacker. Global stability of periodic orbits of non-autonomous difference equations in population biology and the Cushing–Henson conjecture. In B. Aulbach et al., editors, *Proceedings of the Eighth International Conference on Difference Equations and Applications*, Chapman & Hall/CRC 2005, 113–126.

[13] G. Farkas. On asymptotics of solutions of Poincaré difference systems. *J. Diff. Equ. Appl.*, 7:183–191, 2001.

[14] S. Friedland. Convergence of products of matrices in projective space. *Lin. Alg. Appl.*, 413:247–263.

[15] T. Fujimoto and U. Krause. Strong ergodicity for strictly increasing non-linear operators. *Lin. Alg. Appl.*, 71:101–112, 1985.

[16] T. Fujimoto and U. Krause. Asymptotic properties for inhomogeneous iterations of non-linear operators. *SIAM J. Math. Anal.*, 19:841–853, 1988.

[17] T. Fujimoto and U. Krause. Stable inhomogeneous iterations of non-linear positive operators on Banach spaces. *SIAM J. Math. Anal.*, 25:1195–1202, 1994.

[18] M. Golubitsky, E.B. Keeler and M. Rothschild. Convergence of the age structure: applications of the projective metric. *Theoret. Pop. Biology*, 7:84–93, 1975.

[19] J. Hajnal. The ergodic properties of non-homogeneous finite Markov chains. *Proc. Camb. Phil. Soc.*, 52:67–77, 1956.

[20] J. Hajnal. Weak ergodicity in non-homogeneous Markov chains. *Proc. Camb. Phil. Soc.*, 54:233–246, 1958.

[21] D. J. Hartfiel. *Nonhomogeneous Matrix Products.* World Scientific Publ. New Jersey etc., 2002.

[22] H. Inaba. Weak ergodicity of population evolution processes. *Math. Biosciences*, 96:195–219, 1989.

[23] H. Inaba. *Functional Analytic Approach to Age-structured Population Dynamics.* Ph.D., Rijksuniversiteit van Leiden, 180 p., 1989.

[24] V. I. Istratescu. *Fixed Point Theory. An Introduction.* D. Reidel Publ., Dordrecht, 1981.

[25] D. Jillson. Insect populations respond to fluctuating environments. *Nature*, 288:699–700, 1980.

[26] A. N. Kolmogoroff. Über die analytischen Methoden in der Wahrscheinlichkeitsrechnung. *Math. Ann.*, 104:415–458, 1931.

[27] U. Krause. Path stability of prices in a non-linear Leontief model. *Ann. Oper. Res.*, 37:141–148, 1992.

[28] U. Krause. Stability trichotomy, path stability, and relative stability for positive non-linear difference equations of higher order. *J. Diff. Equ. Appl.*, 1:323–346, 1995.

[29] U. Krause. Positive non-linear systems in economics. In T. Maruyama and W. Takahashi, editors. *Nonlinear and Convex Analysis in Ecnomic Theory.* Springer-Verlag, Berlin etc., 1995, 181–195.

[30] U. Krause. A theorem of Poincaré type for non-autonomous non-linear difference equations. In S. Elaydi et al., editors. *Advances in Difference Equations, Proceedings of the Second International Conference on Difference Equations.* Gordon and Breach Publ., Amsterdam, 1997, 107–117.

[31] U. Krause. Stability of non-autonomous population models with bounded and periodic enforcement. J. Diff. Equ. Appl., 15:649–658, 2009.

[32] N. Kruse. *Semizyklen und Kontraktivität nichtlinearer positiver Differenzengleichungen und Anwendungen in der Populationsdynamik.* Ph.D. thesis, Universität Bremen, dissertation.de,

Berlin, 1999.

[33] A. Lopez. *Problems in Stable Population Theory.* Office of Population Research. Princeton, 1961.

[34] P. Martens. *Asymptotische Eigenschaften nichtstationärer Operatorfolgen im nichtlinearen Fall.* Ph.D., Universität Erlangen-Nürnberg, 62 p., 1989.

[35] T. Nesemann. A non-linear extension of the Coale–Lopez theorem. *Positivity*, 3:135–148, 1999.

[36] T. Nesemann. *Stability Behavior of Positive Nonlinear Systems with Applications to Economics.* Ph.D., Universität Bremen, 165 p., Wissenschaftlicher Verlag Berlin, Berlin 1999.

[37] R. D. Nussbaum. Some non-linear weak ergodic theorems. *SIAM J. Math. Anal.*, 21:436–460.

[38] M. Pituk. Asymptotic behavior of a Poincaré recurrence system. *J. Approx. Theory*, 91:226–243, 1997.

[39] M. Pituk. More on Poincaré's and Perron's theorems for difference equations. *J. Diff. Equ. Appl.*, 8:201–216, 2002.

[40] C. Pötzsche. *Geometric Theory of Discrete Nonautonomous Dynamical Systems.* Springer-Verlag, Berlin etc., 2010.

[41] H. Poincaré. Sur les equations linéaires aux différentielles ordinaire et aux différences finies. *Amer. J. Math.*, 7:203–258, 1885.

[42] E. Seneta. On the historical development of the theory of finite inhomogeneous Markov chains. *Proc. Camb. Phil. Soc.*, 74:507–513, 1973.

[43] E. Seneta. Coefficients of ergodicity: structure and applications. *Adv. Appl. Prob.*, 11:576–590, 1979.

[44] E. Seneta. *Non-negative Matrices and Markov Chains.* Springer, Berlin, Revised Printing, 2006.

[45] H. R. Thieme. Asymptotic proportionality (weak ergodicity) and conditional asymptotic equality of solutions to time-heterogeneous sublinear difference and differential equations. *J. Differential Equ.*, 73:237–268, 1988.

[46] H. R. Thieme. *Mathematics in Population Biology.* Princeton University Press, Princeton and Oxford, 2003.

[47] W. F. Trench. Asymptotic behavior of solutions of Poincaré difference equations. *Proc. Amer. Math. Soc.*, 119:431–438, 1993.

8 Dynamics of interaction: opinions, mean maps, multi-agent coordination, and swarms

In previous chapters we met already applications of the theory of positive dynamical systems to questions of interaction within groups of actors. For example, the price setting of economic actors based on the prices set by other actors or cooperative systems modeled by differential equations. In this chapter we investigate systematically the dynamics of interaction as it has been addressed in recent years within quite diverse fields. An essential feature thereby is that interaction takes place as the formation of means or of averages or of convex combinations. For example, in the most simple case of interaction given by a single matrix A, averaging means that A is nonnegative with each row summing up to one. A fundamental result for such a (row)-stochastic matrix states that the powers of A converge to a matrix B with equal rows if and only if some power of A is scrambling. Thereby, a stochastic matrix is **scrambling** if any two rows have a strictly positive entry in a common column. A special case of this result is the famous Basic Limit Theorem for regular Markov chains. In that case a row of B corresponds to the equilibrium distribution of the Markov chain. For linear interaction Markov chains figure as the most prominent example. In Section 8.1, beside the scrambling property, other and weaker structures will be investigated as, for example, coherent matrices and Sarymsakov matrices.

Section 8.2 exhibits models of how a group of individuals, called agents, can reach a **consensus** by themselves, both for linear and for nonlinear interaction. In the linear case consensus will be asymptotically reached precisely if some power of the underlying matrix is scrambling. In the more realistic nonlinear case interaction depends on the state in that an agent takes only opinions into account which are not too distinct from his own. This model of **opinion dynamics under bounded confidence** has during the last years found a lot of attention across the disciplines, ranging from physics over electrical engineering and biology to economics and sociology (see the references given).

A general form of nonlinear interaction is treated in Section 8.3 as a **mean map** T (or compromise map) which sends a collection $x = (x^1, \dots, x^n)$ of points into another collection whithin the convex hull of $x^1, \dots, x^n$. The interaction is only local in the sense that $T_i x$ may depend only on a subset of "neighbors" of i in state x. It will be shown that the iterates of T converge to consensus, a collection of points being equal, if T satisfies a shrinking property. A particular example is a **Gauss soup** where each component T_i is given by a weighted arithmetic or geometric mean (or other means, too).

Time-dependent interaction in nonautonomous positive systems is the topic of Section 8.4. Different from Chapter 7, however, the infinitely many matrices $A(t)$ are assumed to be stochastic and asymptotic behavior is with respect to consensus. It is shown that the latter will be approached if the strength of interaction does not vanish "too fast" and the structure of interaction becomes not "too loose" in the course of time. For a tool often used, a **theorem of Wolfowitz** on infinite products of stochastic matrices, an extended version will be presented and proved.

The results obtained in Section 8.4 are used in Section 8.5 for interactions called broadly multiagent coordination. A fundamental condition on local interaction to get consensus is the **principle of the third agent** (printh) which, roughly, says that the neighbors of of any two agents have one agent in common. In case there is no (global) consensus it is still possible that consensus holds locally within subgroups. This is true in particular for reciprocal interaction. It is here where the results on opinion dynamics under bounded confidence as appearing in Section 8.2, as well as extensions, will be proven.

In Section 8.6 previous results will be used to investigate swarm dynamics. The motivating question in behind is how a group of birds is able to coordinate themselves flying in a **swarm** together. The latter means that the birds by local interaction approach asymptotically the same velocity and their relative distances do converge. The recently much discussed Cucker–Smale model of bird flocking in discrete time is treated. Another model which requires less conditions on local interaction is developed. Actually, swarm dynamics is not confined to birds, fishes or other animals but appears also in jams of people, distributed computing in networks or selforganizing groups of robots.

8.1 Scrambling matrices

The notion of a scrambling matrix has been introduced by J. Hajnal in his analysis of the weak ergodicity in non-homogeneous Markov chains, where he explains the term as follows:

"A scrambling matrix is one in which the probabilities of transition from different initially states are not all in distinct columns, but, as it were, scrambled." ([39, p. 235]. For scrambling matrices and the history behind this concept see [40, 91, 93].)

Definition 8.1.1. A non-negative $n \times n$-matrix $A = (a_{ij})$ is (row-)**stochastic** if all rows sum up to 1. A, not necessarily stochastic, is **scrambling** if for any two rows i and j there exists a column $k = k(i,j)$ such that $a_{ik} > 0$ and $a_{jk} > 0$. Equivalently, AA' is a strictly positive matrix (A' being the transposed matrix of A).

Compared with two other important notions already dealt with in Section 2.4, that of a primitive and indecomposable matrix, respectively, there is no direct relationship. More precisely, a scrambling matrix need neither be indecomposable nor primitive as the example $A = \begin{bmatrix} 1 & 0 \\ \frac{1}{2} & \frac{1}{2} \end{bmatrix}$ shows; conversely, an indecomposable or primitive stochastic matrix need not be scrambling as the examples $A = \begin{bmatrix} 0 & 1 \\ 1 & 0 \end{bmatrix}$ and $A = \begin{bmatrix} \frac{1}{2} & 0 & \frac{1}{2} \\ \frac{1}{3} & \frac{1}{3} & \frac{1}{3} \\ 0 & 1 & 0 \end{bmatrix}$ show.

In the following we shall characterize a stochastic and scrambling matrix by the way it operates on $\mathbb{R}^d$. Let $A = (a_{ij})$ be a stochastic $n \times n$-matrix and let $x = (x^1, \dots, x^n)$ be a collection of points in $\mathbb{R}^d (d \geq 1)$.

Define

$$f_i(x) = \sum_{k=1}^{n} a_{ik} x^k \quad \text{for} \quad 1 \leq i \leq n.$$

Obviously,

$$\operatorname{conv}\{f_1(x), \dots, f_n(x)\} \subseteq \operatorname{conv}\{x^1, \dots, x^n\}$$

where $\operatorname{conv} M = \{\sum_{m \in M} \alpha_m m \mid 0 \leq \alpha_m, \sum_{m \in M} \alpha_m = 1\}$ denotes the **convex hull** of a subset M of $\mathbb{R}^d$.

Let $\|\cdot\|$ denote an arbitrary but fixed norm on $\mathbb{R}^d$ and denote by $\Delta M = \sup\{\|m - m'\| \mid m, m' \in M\}$ the **diameter** of a subset M of $\mathbb{R}^d$. Notice that

$$\Delta \operatorname{conv} M = \Delta M \quad \text{for} \quad M \subseteq \mathbb{R}^d.$$

Obviously, $\Delta \operatorname{conv} M \geq \Delta M$. To see $\Delta \operatorname{conv} M \leq \Delta M$ observe that

$$\sum_{m \in M} \alpha_m m - \sum_{m' \in M} \beta_{m'} m' = \sum_{m, m' \in M} \alpha_m \beta_{m'} (m - m').$$

Using the Hilbert metric H on $\operatorname{int} \mathbb{R}^n_+$ we shall characterize a scrambling matrix also by the way it operates multiplicatively, that is we consider the selfmapping g of $\operatorname{int} \mathbb{R}^n_+$ given by

$$g_i(x) = \prod_{j=1}^{n} x_j^{a_{ij}}, \quad 1 \leq i \leq n, \quad x = (x_1, \dots, x_n) \in \operatorname{int} \mathbb{R}^n_+.$$

Theorem 8.1.2. *Let $A = (a_{ij})$ be a stochastic $n \times n$-matrix.*

A (i) *The following equality does hold*

$$\frac{1}{2} \max_{1 \leq i,j \leq n} \sum_{k=1}^{n} |a_{ik} - a_{jk}| = 1 - \min_{1 \leq i,j \leq n} \sum_{k=1}^{n} \min\{a_{ik}, a_{jk}\}. \tag{8.1.1}$$

(ii) *Let $c(A) \in [0, 1]$ be the quantity defined by equation (8.1.1) and let $f_i(x) = \sum_{k=1}^{n} a_{ik} x^k$. Then $c(A)$ is the smallest constant c such that*

$$\Delta \operatorname{conv}\{f_1(x), \dots, f_n(x)\} \leq c \cdot \Delta \operatorname{conv}\{x^1, \dots, x^n\} \tag{8.1.2}$$

for all collections $x = (x^1, \dots, x^n)$ with $x^i \in \mathbb{R}^d$.

(iii) *Let g be the selfmapping of $\operatorname{int} \mathbb{R}^n_+$ defined by $g_i(x) = \prod_{j=1}^{n} x_j^{a_{ij}}$ for $1 \leq i \leq n$. Then $c(A)$ is the smallest constant c such that*

$$H(g(x), g(y)) \leq c \cdot H(x, y) \tag{8.1.3}$$

for all collections $x = (x_1, \dots, x_n), y = (y_1, \dots, y_n)$ in $\operatorname{int} \mathbb{R}^n_+$.

B *The following statements are equivalent.*
(a) *A is scrambling.*
(b) *A has contraction property* (8.1.2) *for some* $c < 1$.
(c) *A has contraction property* (8.1.3) *for some* $c < 1$.
(d) *A has the following shrinking property*

$$\operatorname{conv}\{f_1(x), \ldots, f_n(x)\} \subsetneq \operatorname{conv}\{x^1, \ldots, x^n\} \tag{8.1.4}$$

for all collections $x = (x^1, \ldots, x^n)$ *with* $x^i \in \mathbb{R}^d$ *and not all* x^i *being equal.*

Proof. First we address part **A.**
(i) For $a, b \in \mathbb{R}$ one has

$$|a - b| + 2\min\{a, b\} = a + b.$$

Since $A = (a_{ij})$ is stochastic this yields for $1 \le i, j \le n$.

$$\sum_{k=1}^{n} (|a_{ik} - a_{jk}| + 2\min\{a_{ik}, a_{jk}\}) = 2$$

which proves equation (8.1.1).

(ii) First we show inequality (8.1.2) for $c = c(A)$. Let $\lambda_{hk} = a_{hk} - \min\{a_{ik}, a_{jk}\}$ for $h = i, j$. We have that $\lambda_{hk} \ge 0$ and $\sum_k \lambda_{ik} = \sum_k \lambda_{jk} = r_{ij}$ with $r_{ij} = 1 - \sum_k \min\{a_{ik}, a_{jk}\}$. With $\alpha_{hk} = \frac{\lambda_{hk}}{r_{ij}}$ for $r_{ij} > 0$ it holds $\sum_k \alpha_{ik} = \sum_k \alpha_{jk} = 1$. Therefore,

$$\begin{aligned}\|f_i(x) - f_j(x)\| &= \|\sum_k (a_{ik} - a_{jk})x^k\| = \|\sum_k (\lambda_{ik} - \lambda_{jk})x^k\| \\ &= r_{ij}\|\sum_k \alpha_{ik}x^k - \sum_k \alpha_{jk}x^k\|\end{aligned}$$

which implies, for $r_{ij} > 0$,

$$\|f_i(x) - f_j(x)\| \le r_{ij}\Delta\operatorname{conv}\{x^1, \ldots, x^n\} \quad \text{for all} \quad i, j.$$

For $r_{ij} = 0$ we must have that $a_{ik} = a_{jk}$ for all k and the above inequality holds trivially. Using $\Delta\operatorname{conv}M = \Delta M$ for $M \subseteq \mathbb{R}^d$ we arrive at

$$\Delta\operatorname{conv}\{f_1(x), \ldots, f_n(x)\} \le \max_{1 \le i,j \le n} r_{ij} \cdot \Delta\operatorname{conv}\{x^1, \ldots, x^n\}$$

which proves (8.1.2) with $c = c(A)$.

Conversely, suppose inequality (8.1.2) holds for some c and all x. For i, j fixed choose $\bar{x}$ defined by $\bar{x}^k = \frac{1}{2}e$ if $a_{ik} \ge a_{jk}$ and $\bar{x}^k = -\frac{1}{2}e$ if $a_{ik} < a_{jk}$ for some $e \in \mathbb{R}^d \setminus \{0\}$. It follows

$$\|f_i(\bar{x}) - f_j(\bar{x})\| \le \Delta\operatorname{conv}\{f_1(\bar{x}), \ldots, f_n(\bar{x})\} \le c\Delta\{\bar{x}^1, \ldots, \bar{x}^k\} \le c\|e\|.$$

Now, $f_i(\bar{x}) - f_j(\bar{x}) = \sum_k (a_{ik} - a_{jk})\bar{x}^k = \frac{1}{2}\sum_k |a_{ik} - a_{jk}|e$ and, hence

$$\frac{1}{2}\sum_k |a_{ik} - a_{jk}|\,\|e\| \leq c\|e\|.$$

Since i, j were arbitrary chosen, it follows from inequality (8.1.1) that

$$c(A) = \frac{1}{2}\max_{1\leq i,j\leq n}\sum_{k=1}^{n} |a_{ik} - a_{jk}| \leq c.$$

This proves (ii).

(iii) From Definition 2.1.8 of the Hilbert metric on $\operatorname{int}\mathbb{R}^n_+$ we have

$$\begin{aligned} H(x,y) &= -\log(\min_i \tfrac{x_i}{y_i}\cdot\min_i \tfrac{y_i}{x_i}) \\ &= -\log\min_j \frac{x_j}{y_j} + \log\max_i \frac{x_i}{y_i} = \max_{1\leq i,j\leq n}\left(\log\frac{x_i}{y_i} - \log\frac{x_j}{y_j}\right). \end{aligned}$$

From the definition of g we obtain $\log\frac{g_i(x)}{g_i(y)} = \sum_k a_{ik}\log\frac{x_k}{y_k}$ and, hence,

$$H(g(x), g(y)) = \max_{1\leq i,j\leq n}\sum_k (a_{ik} - a_{jk})\log\tfrac{x_k}{y_k}.$$

Let $z \in \mathbb{R}^n$ be defined by $z_k = \log\frac{x_k}{y_k}$. For $f_i(z) = \sum_k a_{ik}z_k$ and x, y given the inequality

$$H(g(x), g(y)) \leq cH(x,y) \tag{$*$}$$

is equivalent to

$$\max_{1\leq i,j\leq n}|f_i(z) - f_j(z)| \leq c\max_{1\leq i,j,\leq n}|z_i - z_j|. \tag{$**$}$$

Since $\Delta\operatorname{conv}M = \Delta M$, from part (ii), for $d = 1$ we see that $(*)$ holds for $c = c(A)$. Furthermore, if $(*)$ holds for all $x, y \in \operatorname{int}\mathbb{R}^n_+$ then $(**)$ holds for all $z \in \mathbb{R}^n$ (choose, for example, $x_k = e^{z_k}, y_k = 1$). By part (ii) again we get $c(A) \leq c$.

Consider now part B of Theorem 8.1.2. By equation (8.1.1) we have that A is scrambling if and only if $c(A) < 1$. Therefore, part A (ii) yields the equivalence of (a) and (b). Similarly, part A (iii) yields the equivalence of (a) and (c). Since $\Delta\operatorname{conv}\{x^1, \ldots, x^n\} = 0$ if and only if the x^i are all equal we have that (b) implies (d).

To complete the proof of part B we show that (d) implies (a). Suppose A is not scrambling. Then there exist i and j such that $\min\{a_{ik}, a_{jk}\} = 0$ for all k. Let $I = \{1 \leq k \leq n \mid a_{ik} = 0\}$. Since A is stochastic there exist h and l such that $a_{ih} > 0$ and $a_{jl} > 0$ and, hence, $a_{il} = 0$. Therefore, $\emptyset \neq I \subsetneq \{1, \ldots, n\}$. Let a, b two different points in $\mathbb{R}^d$ and define $x^k = a$ for $k \notin I$ and $x^k = b$ for $k \in I$. It follows

$$f_i(x) = \sum_{k\notin I} a_{ik}x^k = a \quad\text{and}\quad f_j(x) = \sum_{k\in I} a_{jk}x^k = b.$$

Thus, $\operatorname{conv}\{x^1, \ldots, x^n\} = \operatorname{conv}\{a, b\} \subseteq \operatorname{conv}\{f_1(x), \ldots, f_n(x)\}$. Since the x^k are not all equal the shrinking property does not hold. Therefore, (d) implies (a). This completes the proof of Theorem 8.1.2. □

The above theorem characterizes a scrambling matrix by certain contraction and shrinking properties, respectively. Notice that whereas the contraction property (8.1.3) means a proper contraction property in the sense of metric spaces, this is not the case with property (8.1.2) which states only that the diameter of certain sets is contracting.

From Theorem 8.1.2 we obtain some further useful properties for scrambling matrices.

Corollary 8.1.3. *Let A und B stochastic $n \times n$-matrices.*

(i) *If A is scrambling then AB and BA are scrambling. Furthermore, any finite product of stochastic $n \times n$-matrices is scrambling whenever at least one of these matrices is scrambling.*

(ii) *A is scrambling if and only if for any stochastic B and any $x \in \mathbb{R}^n$*

$$BAx = x \text{ implies } x = (r, \dots, r)' \quad \text{for some} \quad r \in \mathbb{R}. \tag{8.1.5}$$

Proof. (i) Let for a $n \times n$-matrix $M = (m_{ij})$

$$f_M(x)_i = \sum_{j=1}^{n} m_{ij} x_j, \, x = (x_1, \dots, x_n)' \in \mathbb{R}^n.$$

For

$$\begin{aligned} f_{AB}(x) &= f_A(y) \quad \text{with} \quad y = f_B(x) \\ f_{BA}(x) &= f_B(z) \quad \text{with} \quad z = f_A(x) \end{aligned}$$

it follows from Theorem 8.1.2 part A

$$\begin{aligned} \Delta\text{conv}\{f_{AB}(x)_1, \dots, f_{AB}(x)_n\} &= \Delta\text{conv}\{f_A(y)_1, \dots, f_A(y)_n\} \\ &\le c(A)\Delta\text{conv}\{y_1, \dots, y_n\} \le c(A)\Delta\text{conv}\{x_1, \dots, x_n\} \end{aligned}$$

and

$$\begin{aligned} \Delta\text{conv}\{f_{BA}(x)_1, \dots, f_{BA}(x)_n\} &= \Delta\text{conv}\{f_B(z)_1, \dots, f_B(z)_n\} \\ &\le \Delta\text{conv}\{z_1, \dots, z_n\} \le c(A)\Delta\text{conv}\{x_1, \dots, x_n\}. \end{aligned}$$

For A scrambling we have $c(A) < 1$ and, hence, by theorem 8.1.2 part B it follows that AB and BA are scrambling. Furthermore, a finite product of stochastic $n \times n$-matrices where, say, A is scrambling is of the form AB or BA or BAC. The above implies that the finite product is scrambling.

(ii) First, we show for a scrambling matrix A that $Ax = x$ implies $x = (r, \dots, r)'$. If $Ax = x$ then for $f(y) = Ay$ it follows $\text{conv}\{f_1(x), \dots, f_n(x)\} = \text{conv}\{x_1, \dots, x_n\}$ and, by part B of Theorem 8.1.2, $x = (r, \dots, r)'$. For A scrambling and B stochastic by part (i) BA is scrambling and, hence, BA has property (8.1.5). Conversely, assume property (8.1.5) and suppose $\text{conv}\{x_1, \dots, x_n\} \subseteq \text{conv}\{f_1(x), \dots, f_n(x)\}$. Then there exist $b_{ij} \ge 0$ with $\sum_{j=1}^n b_{ij} = 1$ for all i such that for each $1 \le i \le n$

$$x_i = \sum_{j=1}^{n} b_{ij} f_j(x) = \sum_{j=1}^{n} b_{ij} \Big(\sum_{k=1}^{n} a_{jk} x_k \Big) = \sum_{k=1}^{n} \Big(\sum_{j=1}^{n} b_{ij} a_{jk} \Big) x_k,$$

that is, $x = BAx$. B being stochastic from property (8.1.5) we must have $x = (r, \dots, r)'$. Thus property (8.1.4) in part B of Theorem 8.1.2 is satisfied and A must be scrambling. □

The next theorem presents a most important feature of scrambling matrices. This theorem generalizes also our earlier result from linear Perron–Frobenius theory, Theorem 2.4.1 (iii) (c), which states, for the case of a stochastic matrix, primitivity of A implies the existence of $\lim_{k\to\infty} A^k$ with all rows being equal. This is a fundamental property of Markov chains (A' being stochastic), also referred to as ergodic theorem for primitive Markov chains in [92] or basic limit theorem for regular Markov chains in [76]. See also the infinite generalization of this theorem obtained previously in Corollary 5.2.8.

Theorem 8.1.4. *For a stochastic matrix A it exists $\lim_{k\to\infty} A^k = B$ with all rows of B being equal if and only if a power of A is scrambling. In that case for the row b determining B the transpose b' is the unique normalized eigenvector for the eigenvalue 1 of the transposed matrix A'.*

Proof. (i) Let a power A^p be scrambling. Let $f(x) = Ax, x \in \mathbb{R}^n$, and $C(k) = \operatorname{conv}\{f^k(x)_1, \dots, f^k(x)_n\}$ for x fixed, $k \ge 0$, f^k the k-th iterate of f. Obviously, $C(k+1) \subseteq C(k)$ for all k and $C = \bigcap_{k\ge 0} C(k) \ne \emptyset$ by compactness of $C(k)$. From part A of Theorem 8.1.2 we have that

$$\Delta C(k+p) \le c(A^p)\Delta C(k) \quad \text{with} \quad c(A^p) < 1.$$

Therefore, $\lim_{k\to\infty} \Delta C(k) = 0$ and for $c = c(x) \in C$ it follows $\lim_{k\to\infty} f^k(x)_i = c$ for all $1 \le i \le n$ because of $|c - f^k(x)_i| \le \Delta C(k)$ for all k. In particular, for x the j-th unit vector e_j we obtain $\lim_{k\to\infty} f^k(e_j)_i = c(e_j)$ for all $1 \le i, j \le n$.

If B denotes the matrix with each row equal to $b = (c(e_1), \dots, c(e_n))$ it follows that $\lim_{k\to\infty} (A^k)_{ij} = B_{ij}$ for all $1 \le i, j \le n$. Furthermore, since $BA = \lim_{k\to\infty} A^{k+1} = B$ it follows that $\sum_{k=1}^n b_k a_{kj} = b_j$, that is, $A'b' = b'$ and $\sum_{j=1}^n b_j = 1$. If $A'x = x$ with $\sum_{k=1}^n x_k = 1$ then $x = \lim_{k\to\infty} A'^k = B'x$, that is $x_j = b_j \sum_{k=1}^n x_k = b_j$. Thus, b' is the unique normalized eigenvector for the eigenvalue 1 of A'.

(ii) Assume $\lim_{k\to\infty} A^k = B$ where B has all rows equal. If A^p is not scrambling for all $p \ge 1$ then for each p there exist $i(p)$ and $j(p)$ such that for the entries $a^{(p)}_{i(p),k}$ and $a^{(p)}_{j(p),k}$ of A^p we must have that $\min\{a^{(p)}_{i(p),k}, a^{(p)}_{j(p),k}\} = 0$. Since $i(p), j(p) \in \{1, \dots, n\}$ there exists a sequence $(p_r)_r$ such that $i(p_r) = i$ and $j(p_r) = j$ for all $r \in \mathbb{N}$. From $\lim_{k\to\infty} A^k = B$ it follows that $\lim_{r\to\infty} A^{p_r} = B$ and, hence, $\lim_{r\to\infty} a^{(p_r)}_{ik} = b_{ik}$ and $\lim_{r\to\infty} a^{(p_r)}_{jk} = b_{jk}$ for all k.

Thus, $\min\{b_{ik}, b_{jk}\} = \lim_{r\to\infty} \min\{a^{(p_r)}_{ik}, a^{(p_r)}_{jk}\} = 0$.

Since all rows of B are equal we must have that $b_{ik} = b_{jk} = 0$ for all k. This, however, is a contradiction since each row of B sums up to 1. Therefore, $\lim_{k\to\infty} A^k = B$ where B has equal rows implies that A^p is scrambling for some $p \ge 1$. □

Since for a primitive matrix some power is strictly positive and, hence, scrambling the above theorem implies the mentioned fundamental theorem on Markov chains, where b' is also called the equilibrium or stationary distribution. As already seen, a scrambling matrix need not be primitive and, hence, Theorem 8.1.4 sharpens the fundamental theorem on Markov chains.

In the following we analyze the relationship of scrambling matrices to related kinds of matrices such as Markov matrices and Sarymsakov matrices of which the former notion is stronger than that of a scrambling matrix and the latter ia a weaker one. For that reason we introduce a little calculus for the positivity structure of matrices. Define for a stochastic $n \times n$-matrix $A = (a_{ij})$ and a non-empty subset M of $\{1, \dots, n\}$

$$s(M) = \{j \in \{1, \dots, n\} \mid a_{ij} > 0 \quad \text{for some} \quad i \in M\}. \tag{8.1.6}$$

Since A is stochastic, s maps the set of non-empty subsets of $\{1, \dots, n\}$ into itself and we can define **iterates of** s for $M \neq \emptyset$ by

$$s^0(M) = M, s^{k+1}(M) = s(s^k(M)) \quad \text{for} \quad k \geq 0.$$

For $M = \{i\}$ we abbreviate $s^k(\{i\})$ by $s^k(i)$. In the following we list some elementary properties of the mapping $s(\cdot)$ which will be used later on.

Properties of $s(\cdot)$ 8.1.5. *Let M, M' be non-empty subsets of $\{1, \dots, n\}$.*
(i) $M \subseteq M'$ *implies* $s(M) \subseteq s(M')$
(ii) $s(M \cup M') = s(M) \cup s(M')$
(iii) $s(M \cap M') \subseteq s(M) \cap s(M')$
(iv) $M \cap M' \neq \emptyset$ *implies* $s(M) \cap s(M') \neq \emptyset$
(v) *For $p \geq 1, k \in s^p(M)$ if and only if there exists a* **chain of length p from M to k,** *that is there exist $k_1, \dots, k_p$ in $\{1, \dots, n\}, k_1 \in M$ with*

$$a_{k_1,k_2} > 0, a_{k_2,k_3} > 0, \dots, a_{k_p,k} > 0.$$

Equivalently, $k \in s^p(M)$ if and only if $a_{ik}^{(p)} > 0$ for some $i \in M(A^p = (a_{ij}^{(p)}))$.

Proof. Properties (i) to (iv) are obvious. For (v), $k \in s^p(M)$ is equivalent to $a_{k_p,k} > 0$ for some $k_p \in s^{p-1}(M)$. In turn, $k_p \in s^{p-1}(M)$ is equivalent to $a_{k_{p-1},k_p} > 0$ for some $k_{p-1} \in s^{p-2}(M)$. By iteration there exist $k_{p-2}, \dots, k_2, k_1$ such that $k_2 \in s(M), a_{k_1,k_2} > 0$ for some $k_1 \in M$. Furthermore, $a_{ik}^{(p)} > 0$ for some $i \in M$ is equivalent to the existence of a sequence $i = k_1, \dots, k_p$ such that $a_{k_1,k_2} > 0, \dots, a_{k_p,k} > 0$. □

Definition 8.1.6. Let A be a stochastic matrix. A is a **Markov matrix** if A has a (strictly) positive column.

A is a **Sarymsakov matrix** if for any two non-empty subsets M and M' of $\{1, \dots, n\}$ with $s(M) \cap S(M') = \emptyset$ it holds that

$$|M \cup M'| < |s(M) \cup s(M')|$$

($|M|$ the number of elements in a finite set M). Call $\emptyset \neq M \subseteq \{1,\dots,n\}$ **saturated** if $s(M) \subseteq M$. A is called **coherent** if any two saturated subsets have a non-empty intersection.

Proposition 8.1.7. *Consider the following properties for a stochastic $n \times n$-matrix A:*
(1) *Markov matrix;*
(2) *scrambling matrix;*
(3) *Sarymsakov matrix;*
(4) *A^{n-1} is scrambling;*
(5) *coherent matrix.*
(i) *The following implications do hold:*

$$(1) \Rightarrow (2) \Rightarrow (3) \Rightarrow (4) \Rightarrow (5).$$

Furthermore, if A is scrambling then A^{n-1} is a Markov matrix.
(ii) *None of the implications in (i) can be reversed in general.*
(iii) *Some power of A is scrambling if for any $i,j \in \{1,\dots,n\}$ there exist $k = k(i,j) \in \{1,\dots,n\}$ and $p_i, p_j \geq 0$ such that $a_{ik}^{(p_i)} > 0, a_{jk}^{(p_j)} > 0$ and $a_{kk} > 0$.*
If A has a positive diagonal then properties (3), (4), (5) are equivalent.

Proof. (i) Obviously, a Markov matrix has to be scrambling. Let A be scrambling and M, M' two non-empty subsets of $\{1,\dots,n\}$. There exist $i \in M, j \in M'$ and, since A is scrambling, it holds $s(i) \cap s(j) \neq \emptyset$. Properties 8.1.5 (i) implies $s(i) \cap s(j) \subseteq s(M) \cap s(M')$ and, hence, $s(M) \cap s(M') \neq \emptyset$. Thus, A is a Sarymsakov matrix. Next, let A be a Sarymsakov matrix. Let M, M' non-emtpy subsets of $\{1,\dots,n\}$ and assume $s^{n-1}(M) \cap s^{n-1}(M') = \emptyset$. The definition of a Sarymsakov matrix yields

$$|s^{n-2}(M) \cup s^{n-2}(M')| + 1 \leq |s^{n-1}(M) \cup s^{n-1}(M')|.$$

By property (iv) of $s(\cdot)$ this step can be iterated and we arrive finally at

$$|s^0(M) \cup s^0(M')| + n - 1 \leq |s^{n-1}(M) \cup s^{n-1}(M')|.$$

Since $s^0(M) \cup s^0(M') = M \cup M', M \cap M' = \emptyset$ and $M, M' \neq \emptyset$ we must have that $2 \leq |s^0(M) \cup s^0(M')|$ and, hence,

$$n + 1 \leq |s^{n-1}(M) \cup s^{n-1}(M')|,$$

which, however, is impossible. This shows that

$$s^{n-1}(M) \cap s^{n-1}(M') \neq \emptyset.$$

Especially, for $M = \{i\}, M' = \{j\}$ we obtain $k \in s^{n-1}(i)$ and $k \in s^{n-1}(j)$. By property (v) of s this means $a_{ik}^{(n-1)} > 0$ and $a_{jk}^{(n-1)} > 0$, that is, A^{n-1} is scrambling.

Finally, let A^p be scrambling for some $p \geq 1$. By property (v) of s we have for any i,j that $s^p(i) \cap s^p(j) \neq \emptyset$. Let M, M' saturated sets with, say, $i \in M, j \in M'$. M being saturated it follows $s(i) \subseteq M$and, by iteration, $s^p(i) \subseteq M$. Similarly, $s^p(j) \subseteq M'$ and we obtain $\emptyset \neq s^p(i) \cap s^p(j) \subseteq M \cap M'$. This shows that A is coherent and the first part of (i) is proven. For the second part of (i) we must show that for A scrambling the power A^{n-1} is a Markov matrix. By induction over k we show that for any collection $M_1, \dots, M_k$, $k \geq 2$ of non-empty subsets of $\{1, \dots, n\}$ satisfying $M_i \cap M_{i+1} \neq \emptyset$ for $1 \leq i \leq k-1$ we must have

$$s^{k-2}(M_1) \cap s^{k-2}(M_2) \cap \dots \cap s^{k-2}(M_k) \neq \emptyset. \qquad (*)$$

For $k = 2$ assertion $(*)$ amounts to $M_1 \cap M_2 \neq \emptyset$ which is true by assumption. Suppose $(*)$ holds for some $k \geq 2$ and any collection $M_1, \dots, M_k$ satisfying $M_i \cap M_{i+1} \neq \emptyset$. Then there exist

$$i \in s^{k-2}(M_1) \cap \dots \cap s^{k-2}(M_k) \quad \text{and} \quad j \in s^{k-2}(M_2) \cap \dots \cap s^{k-2}(M_k) \cap s^{k-2}(M_{k+1})$$

where M_{k+1} satisfies $M_k \cap M_{k+1} \neq \emptyset$. For A scrambling we have that $s(i) \cap s(j) \neq \emptyset$. Using property (i) of s we get

$$s(i) \cap s(j) \subseteq s^{k-1}(M_1), s^{k-1}(M_2), \dots, s^{k-1}(M_k)$$

as well as

$$s(i) \cap s(j) \subseteq s^{k-1}(M_2), s^{k-1}(M_3), \dots, s^{k-1}(M_k), s^{k-1}(M_{k+1}),$$

which implies

$$\emptyset \neq s(i) \cap s(j) \subseteq s^{k-1}(M_1) \cap \dots \cap s^{k-1}(M_k) \cap s^{k-1}(M_{k+1}).$$

This proves assertion $(*)$.

Especially, for $M_i = s(i)$, $1 \leq i \leq n$, $k = n$ we have that $M_i \cap M_{i+1} \neq \emptyset$ for $1 \leq i \leq n-1$ because A is scrambling. Thus, for this case, $(*)$ yields.

$$s^{n-1}(1) \cap s^{n-1}(2) \cap \dots \cap s^{n-1}(n) \neq \emptyset.$$

That is, there exists $k \in s^{n-1}(i)$ for all $1 \leq i \leq n$, which, by property (v) of s, means that $a_{ik}^{(n-1)} > 0$ for all i. Thus, the k-th column of A^{n-1} is positive which proves part (i) of Proposition 8.1.7.

(ii) For counter-examples proving (ii) see Examples 8.1.8 below.

(iii) By assumption $k \in s^{p_i}(i), k \in s^{p_j}(j), k \in s(k)$. For $p = \max_{1 \leq h \leq n} p_h$ it follows that $k \in s^{p-p_i}(k)$ and, hence, $k \in s^{p-p_i}(s^{p_i}(i)) = s^p(i)$ and, similarly, $k \in s^p(j)$. Thus, $s^p(i) \cap s^p(j) \neq \emptyset$ for the p above and all i, j. Therefore, A^p is scrambling.

Suppose, finally, A has a positive diagonal. Then for any non-empty subset M of $\{1, \dots, n\}$ it holds $M \subseteq s(M)$. Therefore, M is saturated if and only if $s(M) = M$. Let M, M' two non-empty subsets of $\{1, \dots, n\}$ such that $s(M) \cap s(M') = \emptyset$. By property (iv) of s we must have that $M \cap M' = \emptyset$. If A is a coherent matrix then M and M' cannot be both saturated and, hence, $M \cup M' \subsetneq s(M) \cup s(M')$.

Therefore, $|M \cup M'| < |s(M) \cup s(M')|$ and A is a Sarymsakov matrix. Together with (i) it follows that (3), (4), (5) must be equivalent. □

The notions introduced we illustrate by some simple examples which also provide counterexamples to the reversal of the implications as mentioned in part (ii) of Proposition 8.1.7.

Examples 8.1.8. (a)

$$A = \begin{bmatrix} \frac{1}{2} & \frac{1}{2} & 0 \\ 0 & \frac{1}{2} & \frac{1}{2} \\ \frac{1}{2} & 0 & \frac{1}{2} \end{bmatrix}$$

is scrambling with a positive diagonal. A is not a Markov matrix, and, hence in Proposition 8.1.7 property (2) does not imply (1).

(b)

$$A = \begin{bmatrix} 1 & 0 & 0 \\ \frac{1}{2} & \frac{1}{2} & 0 \\ 0 & \frac{1}{2} & \frac{1}{2} \end{bmatrix}$$

is not scrambling and has a positive diagonal. Since

$$A^2 = \begin{bmatrix} 1 & 0 & 0 \\ \frac{3}{4} & \frac{1}{4} & 0 \\ \frac{1}{4} & \frac{1}{2} & \frac{1}{4} \end{bmatrix},$$

A^2 is a Markov matrix and, hence, A^2 is scrambling. Part (iii) of Proposition 8.1.7 shows that A is a Sarymsakov matrix. Thus property (3) does not imply (2). Examples (a) and (b) also show that even in case of a positive diagonal neither (1) nor (2) are equivalent to one of (3), (4), (5).

(c) For $A = \begin{bmatrix} 1&0&0 \\ 1&0&0 \\ 0&1&0 \end{bmatrix}$ the power $A^2 = \begin{bmatrix} 1&0&0 \\ 1&0&0 \\ 1&0&0 \end{bmatrix}$ is a Markov matrix and, hence, a scrambling matrix. A is not a Sarymsakov matrix. For $M = \{3\}, M' = \{1, 2\}$ one has that $s(M) = \{2\}, s(M') = \{1\}$ and, therefore, $s(M) \cap s(M') = \emptyset$ but

$$|s(M) \cup s(M')| = |\{1, 2\}| < |M \cup M'|.$$

This shows, property (4) does not imply (3).

(d) For $A = \begin{bmatrix} 0&1 \\ 1&0 \end{bmatrix}$ no power is a scrambling matrix. A is a coherent matrix since $M = \{1, 2\}$ is the only saturated set. Thus, property (5) does not imply (4).

(e) Examples (b) and (c) show that A^2 may be a Markov matrix for A not scrambling. Thus, the additional implication in part (i) of Proposition 8.1.7 cannot be reversed and, as example (b) shows, this is true even in case of a positive diagonal.

(f) Considering part (iii) of Proposition 8.1.7, for the example

$$A = \begin{bmatrix} 1 & 0 & 0 \\ 1 & 0 & 0 \\ 0 & \frac{1}{2} & \frac{1}{2} \end{bmatrix},$$

it holds $a_{11} > 0, a_{21} > 0$ and $a_{31}^{(2)} > 0$. Therefore, with $k(i,j) = 1$ for all i, j, a power of A must be scrambling; indeed, A^2 is scrambling. A is, however, not a Sarymsakov matrix. For this let $M = \{1, 2\}, M' = \{3\}$, in which case $s(M) = \{1\}, s(M') = \{2, 3\}$ and

$$|s(M) \cup s(M')| = |\{1, 2, 3\}| = |M \cup M'|.$$

Thus, although (4) implies (3) if there is a positive diagonal, this need not be the case if just one entry in the diagonal is not positive.

(g) Part (iii) of the proposition shows that positive entries in the diagonal of A play its role. It is, however, by no means necessary for a scrambling matrix to have a positive entry in the diagonal as the example

$$A = \begin{bmatrix} 0 & \frac{1}{2} & \frac{1}{2} \\ \frac{1}{2} & 0 & \frac{1}{2} \\ \frac{1}{2} & \frac{1}{2} & 0 \end{bmatrix}$$

shows.

Remarks 8.1.9. (1) The concept of a Sarymsakov matrix has been defined in [91] and [40] for arbitrary nonnegative matrices, too.

(2) Scrambling matrices can be investigated also in terms of eigenvalues. In [34] the following result is proven: For a stochastic matrix A it holds $\lim_{k\to\infty} A^k = B$ with all rows of B being equal if and only if

(a) A has beside 1 no other eigenvalue of absolute value 1,

(b) 1 is a simple root of the characteristic equation of A.

By Exercise 2 below this result does follow from Theorem 8.1.4. The formula in Exercise 2 (a) provides a calculation by the entries of A for the second eigenvalue which plays an important role for the rapid mixing of Markov chains. (See for the latter [5, Chapter 10].)

(3) As mentioned already, the contraction property (8.1.2) is different from a contraction with respect to the metric defined by a norm, that is $\|f(x) - f(y)\| \le c\|x - y\|$ for some $c < 1$, all collections x, y. For example, the matrix A in Examples 8.1.8(a) is scrambling and, hence, contraction property (8.1.2) holds for any norm but for each norm $\|Ae - A0\| = \|e - 0\|$, where $e = (1, \ldots, 1)'$.

The analysis of scrambling matrices carried out in this section we want to extend later on (see Section 8.3) to certain nonlinear mappings. For this reason we did not consider an analysis in terms of eigenvalues but put emphasis on the contraction and shrinking properties of scrambling matrices. The latter turn out to be fruitful to handle "scrambling processes" tending to a "consensus" as in the cases of opinion dynamics and swarm dynamics to be considered in the following sections.

8.2 Consensus formation and opinion dynamics under bounded confidence

Consider a group of experts who have to assess a certain magnitude like the world's wheat production in the year 2030. Each of the experts has his own expertise but is open to some extent to revise it in case of opposing expertises by his colleagues. Knowing the revisions may lead to further revisions and the question occurs whether this iterative process of opinion making will tend to a consensus among the experts concerning the value of the magnitude under consideration.

Methods to deal with a problem like this were developed around 1960 as Opinion Pool [97] and Delphi-method [25] and later on as a simple Matrix Model ([26, 69]; see [36] for a survey). Within a time-dependent matrix model an agreement algorithm was developed around 1980 – this time for the communication among electronic processors [8, 98, 99]. Denote by $x_i(t) \in \mathbb{R}$ the assessment made by expert $i \in \{1,\dots,n\}$ at time $t \in \{0,1,2,\dots\}$ of the magnitude under consideration. Let $x_i(t+1)$ be the revised opinion of expert i by taking the assessments $x_j(t)$ of the previous period into account with certain weights a_{ij}. If $A = (a_{ij})$ denotes the matrix of weights then the **matrix model** is

$$x(t+1) = Ax(t) \text{ for } t = 0, 1, \dots, \tag{8.2.1}$$

where the matrix A is stochastic since weights are nonnegative and add up to 1; $x(\cdot)$ is the column-vector with components $x_i(\cdot)$.

The main question is whether the experts will reach a consensus among themselves, that is

$$\lim_{t\to\infty} x_i(t) = c \quad \text{for } \mathbf{all} \quad i \in \{1,\dots,n\}, \tag{8.2.2}$$

where the **consensus** c depends on initial opinions $x(0)$.

From Theorems 8.1.2 and 8.1.4 of the previous section we obtain the following answer.

Theorem 8.2.1. *The following statements are equivalent.*

(i) *A consensus will be reached for each* $x(0) \in \mathbb{R}^n$.

(ii) *Some power of A is scrambling.*

(iii) *The map given by $f(x) = Ax$ has an iterate f^p such that for some $c < 1$*

$$\left(\max_{1\le i\le n} f^p(x)_i - \min_{1\le i\le n} f^p(x)_i\right) \le c\left(\max_{1\le i\le n} x_i - \min_{1\le i\le n} x_i\right)$$

for all $x = (x_1,\dots,x_n)' \in \mathbb{R}^n$.

(iv) *There exists an iterate f^q such that*

$$\left[\min_{1\le i\le n} f^q(x)_i, \max_{1\le i\le n} f^q(x)_i\right] \subsetneqq \left[\min_{1\le i\le n} x_i, \max_{1\le i\le n} x_i\right]$$

for all $x = (x_1,\dots,x_n)' \in R^n$*, except for a consensus where the* x_i *are all equal.*

Proof. Since $x(t) = A^t x(0)$, reaching a consensus means $\lim_{t\to\infty}(A^t x(0))_i = c(x(0))$ for all i. This holds for all $x(0) \in \mathbb{R}^n$ if and only if $\lim_{t\to\infty}(A^t)_{ij} = c(e_j)$ for all $1 \leq i,j \leq n$ where e_j is the j-th unit vector. The matrix $B = (b_{ij})$ with $b_{ij} = c(e_j)$ is a stochastic matrix with equal rows. From Theorem 8.1.4 it follows that a consensus is reached for all $x(0)$ if and only if a power A^p is scrambling. This proves the equivalence of (i) and (ii). The equivalence of (ii), (iii) and (iv) follows from part B of Theorem 8.1.2. For this notice that for $d = 1$ and $y = (y_1, \ldots, y_n)' \in \mathbb{R}^n$ one has that

$$\operatorname{conv}\{y_1, \ldots, y_n\} \text{ is the interval } \left[\min_{1\leq i\leq n} y_i, \max_{1\leq i\leq n} y_i\right]$$

and

$$\Delta\operatorname{conv}\{y_1, \ldots, y_n\} = \max_{1\leq i\leq n} y_i - \min_{1\leq i\leq n} y_i. \qquad \square$$

Using Proposition 8.1.7 we obtain the following sufficient condition for a consensus.

Corollary 8.2.2. *A consensus will be reached for each $x(0) \in \mathbb{R}^n$ if for any two experts i and j there exists a third one $k = k(i,j)$ and $p_i, p_j \geq 0$ such that $a_{ik}^{(p_i)} > 0$, $a_{jk}^{(p_j)} > 0$ and $a_{kk} > 0$.*

Proof. By Proposition 8.1.7 (iii) a power of A must be scrambling which by Theorem 8.2.1 yields the conclusion. $\square$

Whereas Theorem 8.1.2 provides equivalent conditions for reaching a consensus, in the literature, e.g., in [26] and [69], mainly sufficient conditions are given. For the sufficient condition supplied in Corollary 8.2.2 the assumption $a_{ik}^{(p_i)} > 0$ means that there exists a chain of length p_i from i to k, that is there exist $k_1, \ldots, k_{p_i}$ in $\{1, \ldots, n\}$ with $k_1 = i$ such that $a_{k_1k_2} > 0, \ldots, a_{k_{pi},k} > 0$ (see part (v) of Properties of $s(\cdot)$ 8.1.5). Using a formulation from [69] we can rephrase the sufficient condition in Corollary 8.2.2 by saying that for any two experts there exist **chains of respect** to a third one who **respects himself.**

An interesting result in [69, Theorem 7.2] states that consensus will be reached if there exists a chain of respect from every expert to a particular expert k who respects himself. This result follows immediately from Corollary 8.2.2. The latter, however, admits an expert k with self-respect to depend on i and j. Without the assumption on self-respect Corollary 8.2.2 may fail as Examples 8.1.8 (d) exhibits. Indeed, as Proposition 8.1.7 (iii) shows positive entries on the diagonal of A and, hence, experts with self-respect play an important role. On the other hand, as Examples 8.1.8 (g) shows, a matrix A with zero diagonal may be scrambling and, by Theorem 8.2.1, consensus will be reached although none of the experts possesses self-respect.

Now we want to introduce another model which is nonlinear and more realistic in that experts do not trust necessarily all the other experts – depending on the assessment made by other experts. In the following we outline the model of bounded confidence, which was introduced in 2000 ([54] and developed further in 2002 [43]) and has found much attention since then. (In [43] also the history and background of the model are considered and references to the literature are given. See also Remarks 8.2.6 below.) We will present some major results but without giving a proof since in the next chapters we develop a general framework which yields these results as special cases. (See Theorem 8.5.7 and the consequences drawn from it.)

Using the language of multi-agent systems we consider n agents $i = 1, \dots, n$ where $x_i(t) \in \mathbb{R}_+$ for $t = 0, 1, \dots$ denotes the **opinion** of agent i at time t. In making up his opinion in the next period, agent i takes into account from the previous period the opinions of those agents he is confident in. More precisely, depending on an **opinion profile** $x = (x_1, \dots, x_n) \in \mathbb{R}_+^n$ the **confidence set** of agent i is given by

$$I(i, x) = \{1 \le j \le n \mid |x_i - x_j| \le \epsilon\}$$

where $\epsilon > 0$ is a certain **confidence level** assumed to be equal for all agents. The **dynamics of opinion formation under bounded confidence** is given by

$$x_i(t+1) = |I(i, x(t))|^{-1} \sum_{j \in I(i,x(t))} x_j(t) \tag{8.2.3}$$

for $i = 1, \dots, n, t = 0, 1, \dots$ and given initial opinions $x(0) \in \mathbb{R}_+^n$.

It is easy to solve this system for $n = 2$ but for the general case no analytical solution is available. Although there are still many questions open concerning the behavior of solutions of model (8.2.3) many results have been obtained. Thereby, the following concept plays a major role. A **chain of confidence** of agent i to agent j from period s to period $t > s$ is a sequence of agents $(i_0, i_1, \dots, i_{t-s})$ such that $i_0 = i, i_{t-s} = j$ and $i_r \in I(i_{r-1}, x(t-r))$ for all $1 \le r \le t-s$.

Theorem 8.2.3. *If for any two agents i and j there exists a third agent $k = k(i,j)$ such that a chain of confidence goes from i to k and from j to k, from s to $s+h$ for some fixed $h \ge 1$ and all s, then consensus will be reached in finite time, that is for some $T \in \mathbb{N}$*

$$x_i(t) = c \quad \text{for all} \quad 1 \le i \le n \quad \text{and} \quad t \ge T.$$

where the consensus c depends on initial conditions.

An immediate consequence of this theorem is the following result.

Corollary 8.2.4. *Consensus will be reached in finite time if for all i, j and $t \ge T$*

$$I(i, x(t)) \cap I(j, x(t)) \neq \emptyset.$$

Without making any assumptions on model (8.2.3) it can be shown that always in finite time a **fragmentation of opinions** will be reached.

Theorem 8.2.5. *For the model of bounded confidence an opinion fragmentation is reached in finite time, that is there is a disjunctive decomposition of $\{1, \dots, n\}$ into* *non-empty subsets A_j, $1 \le j \le k$ such that for some T*

$$x_i(t) = c_j \quad \textit{for all} \quad i \in A_j \quad \textit{and} \quad t \ge T$$

where the partial consensus c_j of agents in group A_j depends on initial conditions.

For proofs of the results above see [28, 43, 44, 54, 71, 72].

As mentioned already there are still unsolved problems concerning the model (8.2.3) of bounded confidence. One such problem considers the distance between two opinion clusters according to Theorem 8.2.5 which, backed by computer simulations is conjectured to be roughly equal to 2ϵ (see [9]).

A major question is, how, in case of reaching a consensus, this consensus depends on intial condition $x(0)$. For the linear model discussed above this question is easily answered. From Theorem 8.1.4 it follows

$$c(x) = \sum_{j=1}^{n} b_j x_j,$$

where $b' = (b_1, \dots, b_n)'$ is the unique normalized eigenvector of A'.

Opinion dynamics is since some years quite an expanding field in its own. We give some further references with emphasis on the model of bounded confidence as described above.

Remarks 8.2.6. Various models of opinion dynamics are presented in [47], with applications to sociology in [51, Chapter 8]. Various aspects and applications of the model of bounded confidence have been considered and explored further in [9, 10, 28, 31–33, 44, 45, 47, 55–57, 61, 62, 71–75, 78, 87].

A model of bounded confidence which is driven randomly by pairwise interaction of agents is developed in [27, 102]. Computer simulations have shown similarities in the dynamics of this model and the one presented here (see, for example, [72, 74]).

Among recent contributions concerning the model (8.2.3) we mention [79] which addresses the dynamics in case the agents have different confidence levels; see also [74, 96] where the "phase transition" between consensus and fragmentation is investigated by computer simulations and [21] where, within a broader framework, an algorithmic approach is pursued to get for the dynamics bounds on the convergence rates.

8.3 Mean processes, mean structures and the iteration of mean maps

The action of a stochastic matrix A on a vector x can be viewed as formation of arithmetic means $(Ax)_i = \sum_{j=1}^{n} a_{ij}x_j$ with weights a_{ij}. It has the nice property that

$$\min_{1\leq j\leq n} x_j \leq (Ax)_i \leq \max_{1\leq j\leq n} x_j \quad \text{for all} \quad 1 \leq i \leq n \tag{8.3.1}$$

which is shared also by nonlinear means like the geometric mean given by $(Gx)_i = \prod_{1\leq j\leq n} x_j^{a_{ij}}$ or the power mean given by

$$(Px)_i = \left(\sum_{j=1}^{n} a_{ij}x^r\right)^{\frac{1}{r}}$$

for $r \neq 0$. (For more on those concrete means see below.) Actually, inequalities (8.3.1) are precisely what one would expect of means – a value between the extremes. Mappings satisfying these inequalities are called **abstract means** in [12].

Since for $x \in \mathbb{R}^n$ the closed interval $[\min_{1\leq j\leq n} x_j, \max_{1\leq j\leq n} x_j]$ equals the convex hull of $x_1, \ldots, x_n$ one might extend inequalities (8.3.1) to higher dimensions as $(Ax)_i \in \text{conv}\{x_1, \ldots, x_n\}$. This has been considered already in Section 8.1 for stochastic matrices A. In this section the analysis will be extended to nonlinear mappings called mean maps. Actually, the extension to nonlinear mappings in higher dimensions (that is $x_i \in \mathbb{R}^d$) is needed to treat opinion dynamics (for example, the model (8.2.3) in Section 8.2) and other models of interaction dynamics.

Definition 8.3.1. Let S be a non-empty convex subset of $\mathbb{R}^d (d \geq 1)$ and S^n the cartesian product of n copies of S. A sequence $(x(t)), t \in \{0, 1, \ldots\}$ with $x(t) \in S^n (n \geq 1)$ is called a **mean process on S^n** if $x^i(t + 1) \in \text{conv}\{x^1(t), \ldots x^n(t)\}$ for all $1 \leq i \leq n$, all $t \geq 0$. Or, in short,

$$\text{conv}\{x(t + 1)\} \subseteq \text{conv}\{x(t)\} \quad \text{for all} \quad t \geq 0 \tag{8.3.2}$$

where $\text{conv}\{x\} = \text{conv}\{x^1, \ldots, x^n\}$ for $x = (x^1, \ldots, x^n) \in S^n$.

We speak of a **mean structure M on S^n** if for each $x \in S^n$ a mean process $(x(t))$ with $x(0) = x$ is specified. A selfmaping T of S^n is a **mean map on S^n** if its iterates defined by $x(t) = T^t x, x \in S^n$ is a mean structure on S^n.

Obviously, the composition of mean maps is a mean map, too. By this simple principle a huge variety of mean maps can be generated from a given set of concrete means. If $T(t), t \in \{0, 1, \ldots\}$ is an infinite sequence of mean maps then by $x(t + 1) = T(t)x(t)$ a mean process is defined which is not given by a mean map. A mean process, in particular a mean map, can be described by stochastic matrices as follows. By equation (8.3.2) there exist a stochastic matrix $A(t)$ such that $x(t + 1) = A(t)x(t)$, and vice versa. Thus, the asymptotic analysis of a mean process amounts to that of an infinite product of stochastic matrices. This point of view will be taken up in the next section.

The main aim of this section is to find for a mean map T conditions which yield the following property, which occurs already in Section 8.1 and as consensus in Section 8.2:

$$(P) \lim_{t\to\infty} T^t x = \bar{c}(x)$$

for all x, where $\bar{c}(x) = (c(x), \ldots, c(x)), c(x) \geq 0$.

The following lemma will be crucial to obtain such conditions.

Lemma 8.3.2. *Let $(x(t))$ be a mean process on S^n and ω the limit set of $(x(t))$.*

(i) *For $y \in \omega$ it holds $\operatorname{conv}\{y\} = \bigcap_{t=0}^{\infty} \operatorname{conv}\{x(t)\} \neq \emptyset$.*

(ii) *For $C = \bigcap_{t=0}^{\infty} \operatorname{conv}\{x(t)\}$ and all $1 \leq i \leq n$*

$$\lim_{t\to\infty} \inf_{c\in C} \|x^i(t) - c\| = 0$$

($\|\cdot\|$ be any norm on $\mathbb{R}^d$.)

Proof. Let $C(t) = \operatorname{conv}\{x(t)\}$ for $t \geq 0$. From the definition of a mean process we have $C(t+1) \subseteq C(t)$ for all $t \geq 0$. Since all the $C(t)$ are non-empty and compact it follows that $C = \bigcap_{t\geq 0} C(t)$ is non-empty and compact, too.

(i) Since $x(t)$ is contained in the compact set $C(0)^n$ the limit set ω is non-empty. Let $y \in \omega, y^i = \lim_{s\to\infty} x^i(t_s)$ for $1 \leq i \leq n$. Obviously, $x^i(t_s) \in C(t_s) \subseteq C(t)$ for $t \leq t_s$ and, hence, $y^i \in C(t)$ for all i, all t. Therefore, $y^i \in C$ and $\operatorname{conv}\{y\} \subseteq C$. For the converse let $x \in C$ and $\delta > 0$ be given. To $y \in \omega$ exist a sequence (t_s) and s_0 such that $\|x^i(t_s) - y\| \leq \delta$ for all $s \geq s_0$, all i. From $x \in C \subseteq C(t_{s_0})$ we have $x = \sum_{i=1}^n \alpha_i x^i(t_{s_0})$ with $\alpha_i = \alpha_i(x, t_{s_0}) \geq 0$ and $\sum_{i=1}^n \alpha_i = 1$. Thus,

$$\|x - \sum_{i=1}^n \alpha_i y^i\| = \|\sum_{i=1}^n \alpha_i (x^i(t_{s_0}) - y^i)\| \leq \sum_{i=1}^n \alpha_i \,\delta = \delta\,.$$

Since $\delta > 0$ is arbitrary and $\operatorname{conv}\{y\}$ is closed it follows $x \in \operatorname{conv}\{y\}$ which proves (i).

(ii) Let $y \in \omega$ and for $\delta > 0$ given $\|x^i(t_s) - y^i\| \leq \delta$ for $s \geq s_0$, all i. For $t \geq t_{s_0}$ it holds $x^i(t) \in C(t) \subseteq C(t_{s_0})$ and for a convex combination $x^i(t) = \sum_{j=1}^n \alpha_j(i,t) x^j(t_{s_0})$. Let $c(i,t) = \sum_{j=1}^{n=1} \alpha_j(i,t) y^j$.

By part (i) $C = \operatorname{conv}\{y\}$ and $c(i,t) \in C$ for all $i, t \geq t_{s_0}$. This yields

$$\|x^i(t) - c(i,t)\| = \|\sum_{j=1}^n \alpha_j(i,t)(x^j(t_{s_0}) - y^j)\| \leq \sum_{j=1}^n \alpha_j(i,t)\delta = \delta\,.$$

Thus,

$$\inf_{c\in C} \|x^i(t) - c\| \leq \|x^i(t) - c(i,t)\| \leq \delta \text{ all } i, \text{all } t \geq t_{s_0}\,.$$

This shows part (ii). □

In Section 8.1, Theorem 8.1.2 part B, a scrambling matrix has been characterized by a shrinking property with respect to convex hulls. Inspired by this we define a shrinking property for the nonlinear case as follows.

Definition 8.3.3. A mean process $(x(t))$ on S^n is said to be **shrinking for *t*** if $\operatorname{conv}\{x(t)\} \subsetneq \operatorname{conv}\{x(0)\}$.

The process is **shrinking** if it it is shrinking for some t.

A mean structure M on S^n is **shrinking at *x* for *t*** if the mean process specified for $x = x(0)$ is shrinking for t.

The mean structure is **shrinking at *x*** if it is shrinking at x for some t.

A mean map T is said to have a property as above if it holds for the mean structure given by T.

Shrinking is not possible at $x \in \mathbf{diag}\boldsymbol{S^n} = \{\bar{c} = (c, \dots, c) | c \in S\}$.

If $\dot{\boldsymbol{M}}$ denotes for $M \subset S^n$ the set $\dot{M} = M \setminus \operatorname{diag}S^n$ then shrinking is possible only at points $x \in \dot{\boldsymbol{S}}^{\boldsymbol{n}}$.

Using the language above our earlier result on stochastic matrices can be rephrased by saying that A is scrambling if and only if the linear mapping induced by A is shrinking at each point of $\dot{S}^n (S = \mathbb{R}^d)$. Our main result on stochastic matrices, Theorem 8.1.4, can be rephrased by saying that $\lim_{t\to\infty} A^t x = \bar{c}(x) = (c(x), \dots, c(x))$ for each x if and only if the map induced by A is shrinking at each $x \in \dot{S}^n$ globally for some common t_0.

The following theorem presents an extension of this result to mean maps in general and to mean structures, too.

Theorem 8.3.4. *Let S be a non-empty convex subset of $\mathbb{R}^d$.*

(i) *For a mean process $(x(t))$ with $x(0) = x \in S^n$ it holds $\lim_{t\to\infty} x(t) = \bar{c}(x)$ if and only if $\omega(x) \cap \operatorname{diag}S^n \neq \emptyset$. In this case the process must be shrinking for $x \in \dot{S}^n$.*

(ii) *For a mean structure M on S^n $\lim_{t\to\infty} x(t) = \bar{c}(x)$ for all $x \in S^n$ does hold if and only if for each $x \in \dot{S}^n$ the structure M is shrinking at x and $\omega(x)$ is invariant, that is $y \in \omega(x)$ implies $y(t) \in \omega(x)$ for all t.*

(iii) *For a continuous mean map T on S^n $\lim_{t\to\infty} T^t x = \bar{c}(x)$ for all $x \in S^n$ does hold if and only if T is shrinking at each $x \in \dot{S}^n$.*

Proof. (i) Obviously, if $\lim_{t\to\infty} x(t) = \bar{c}(x)$ then $\omega(x) = \{\bar{c}(x)\} \subseteq \operatorname{diag}S^n$. Conversely, if $y \in \omega(x) \cap \operatorname{diag}S^n$ then $\operatorname{conv}\{y\}$ is a singleton. From Lemma 8.3.2 it follows that $C = \{c\}$ and $\lim_{t\to\infty} \|x^i(t) - c\| = 0$ for all i, that is $\lim_{t\to\infty} x(t) = \bar{c}$. It remains to show that $\lim_{t\to\infty} x(t) = \bar{c}(x)$ implies $(x(t))$ is shrinking for $x \in \dot{S}^n$. Suppose $\operatorname{conv}\{x(t)\} = \operatorname{conv}\{x\}$ for all t. Then there exist $0 \le \alpha_{ij}(t), \sum_{j=1}^n \alpha_{ij}(t) = 1$ such that

$$x^i = \sum_{j=1}^{n} \alpha_{ij}(t) x^j(t) \quad \text{for all } i, \quad \text{all } t.$$

Since $t \mapsto (\alpha_{ij}(t)_{1\le i,j\le n}$ is a bounded sequence in $\mathbb{R}^{n\times n}$ there exists a sequence $(t_k)_k$ such that $\lim_{k\to\infty} \alpha_{ij}(t_k) = \alpha_{ij}$ for all $1 \le i, j \le n$. From $\lim_{t\to\infty} x(t) = \bar{c}(x)$ and, hence, $\lim_{k\to\infty} x(t_k) = \bar{c}(x)$

we obtain for all i

$$x^i = \lim_{k\to\infty}\sum_{j=1}^{n}\alpha_{ij}(t_k)x^j(t_k) = \sum_{j=1}^{n}\alpha_{ij}c(x) = c(x).$$

Therefore, $x \in \mathrm{diag}S^n$ which shows that $(x(t))$ is shrinking for $x \in \dot{S}^n$.

(ii) Let $\lim_{t\to\infty} x(t) = \bar{c}(x)$ for all $x \in S^n$. By (i) the structure M is shrinking at each $x \in \dot{S}^n$. Since, by assumption, $\omega(x) = \{\bar{c}(x)\}$ and $(x(t))$ is a mean process $\omega(x)$ is invariant for each $x \in S^n$. Conversely, suppose M is shrinking and $\omega(x)$ is invariant for each $x \in \dot{S}^n$. If $y \in \omega(x)$ then by Lemma 8.3.2 $\mathrm{conv}\{y\} = \mathrm{conv}\{y(t)\}$ for all t. Therefore, $y \notin \dot{S}^n$ and, hence, $y \in \mathrm{diag}S^n$ and we have $\omega(x) \subseteq \mathrm{diag}S^n$. Part (i) implies $\lim_{t\to\infty} x(t) = \bar{c}(x)$ for all $x \in S^n$.

(iii) For a continuous mean map on S it holds $T(\omega(x)) \subseteq \omega(x)$ for all $x \in S^n$. For the mean structure on S^n given by $x(t) = T^t x, x \in S^n$, part (ii) implies part (iii). □

Later on applications of this theorem will be given to soups made of various well-known means and to opinion dynamics (in Section 8.5). Before doing so some examples and counter-examples will shed some light on the assumptions used in the theorem.

Examples and counter-examples 8.3.5. (1) Parts (i) and (ii) apply also to non-continuous maps as the following simple example will show. For $S = \mathbb{R}, n = 2$, let T be the selfmapping of S^2 given by

$$T(x_1,x_2) = \begin{cases} (x_1,x_1) & \text{if } x_1 < x_2 \\ (x_2,x_2) & \text{otherwise,} \end{cases}$$

for $x = (x_1,x_2) \in S^2$.

Obviously, T is a mean map which is not continuous on S^2. If $x_1 < x_2$ then $Tx = (x_1,x_1)$ and $T^t x = (x_2,x_2)$ for $t \geq 2$. If $x_1 \geq x_2$ then $T^t x = (x_2,x_2)$ for all $t \geq 1$. Therefore, $\omega(x) = (x_2,x_2)$ for $x \in S^2$. Part (i) of the theorem yields $\lim_{t\to\infty} T^t x = \bar{c}(x)$ for all $x \in S^2$, which is obvious in this simple example. The example shows also that the assumption on $\omega(x)$ made in part (ii), though implied by continuity, is weaker than continuity. Furthermore, it is easy to check that T is shrinking at each $x \in \dot{S}^2$ (for $t = 1$). This shows that the equivalence stated in part (iii) may hold without continuity.

(2) In contrast to example (1) the following case of a non-continuous mean map shows that the equivalence stated in part (iii) of the theorem can fail if T is not continuous. For $S = \mathbb{R}, n = 3$, let T be the selfmapping of S^3 given by

$$T(x_1,x_2,x_3) = \begin{cases} (x_1,x_2,\frac{1}{2}x_3 + \frac{1}{2}\min\{x_1,x_2\}) \text{ if } x_3 < \min\{x_1,x_2\} \\ (x_1,x_1,x_1) \text{ otherwise.} \end{cases}$$

Obviously, T is a non-continuous mean map. T is shrinking at each $x \in \dot{S}^3$ for $t = 1$ as the following calculation shows. Let $\min z = \min\{z_1,\dots,z_n\}$ for short and similarly for $\max z$.

In case of $x_3 < \min\{x_1, x_2\}$ we have

$$\min Tx = \frac{1}{2}x_3 + \frac{1}{2}\min\{x_1, x_2\}, \max Tx = \max\{x_1, x_2\}$$

which implies

$$\operatorname{conv}\{Tx\} = \left[\frac{1}{2}x_3 + \frac{1}{2}\min\{x_1, x_2\}, \max\{x_1, x_2\}\right] \subsetneq [x_3, \max\{x_1, x_2\}] = \operatorname{conv}\{x\}.$$

Thus, T is shrinking at x for $t = 1$.

In case of $x_3 \geq \min\{x_1, x_2\}$ we have

$$\min Tx = \max Tx = \{x_1\},$$

and, hence,

$$\operatorname{conv}\{Tx\} = [x_1, x_1] \subsetneq [\min x, \max x] = \operatorname{conv}\{x\}$$

holds trivially for $x \in \dot{S}^3$. This demonstrates T is shrinking at each $x \in \dot{S}^3$ for $t = 1$.

It is easily checked that in case of $x^3 < \min\{x_1, x_2\}$

$$T^t x = (x_1, x_2, f(t)),$$

with $f(t) = \frac{1}{2^t}x_3 + (1 - \frac{1}{2^t})\min\{x_1, x_2\})$. Therefore, in that case, $\lim_{t\to\infty} T^t x = (x_1, x_2, \min\{x_1, x_2\})$. This shows $\lim_{t\to\infty} T^t x \notin \operatorname{diag} S^3$ for $x_3 < \min\{x_1, x_2\}, x_1 \neq x_2$. Therefore, the equivalence stated in part (iii) fails.

Furthermore, considering part (ii), the assumption of invariance is not satisfied. Let $x \in S^3$ with $x_3 < \min\{x_1, x_2\}, x_1 \neq x_2$. By the above $\omega(x) = \{(x_1, x_2, \min\{x_1, x_2\})\}$ but $T^t(x_1, x_2, \min\{x_1, x_2\}) = (x_1, x_1, x_1)$ for all $t \geq 1$ and $(x_1, x_1, x_1) \notin \omega(x)$ since $x_1 \neq x_2$.

(3) As a simple example of a nonlinear mean map consider for $S = \mathbb{R}_+, n = 2$, the selfmapping of S^2 defined by $T(x_1, x_2) = (\alpha x_1 + (1-\alpha)x_2, x_1^\beta x_2^{(1-\beta)})$ for $0 \leq \alpha, \beta \leq 1$. For a more general setting see the Gauss soup later on. The particular case of $\alpha = \beta = \frac{1}{2}$ is that of the famous **arithmetic-geometric mean.** Since $\min x \leq T_i x \leq \max x$ for $i = 1, 2$ we have $\operatorname{conv}\{Tx\} \subseteq \operatorname{conv}\{x\}$ and T is a mean map. Whether T is shrinking or not depends on parameters α and β. First, consider the case $0 < \alpha, \beta < 1$. If $x_1 \neq x_2$ then $\alpha x_1 + (1 - \alpha)x_2 < \max x$ as well as $x_1^\beta x_2^{1-\beta} < \max x$. Therefore, T is shrinking at each $x \in \dot{S}^2$ for $t = 1$. Since T is continuous from part (iii) of Theorem 8.3.4 it follows that $\lim_{t\to\infty} T^t x = \bar{c}(x)$ for each $x \in S^2$.

Notice that T is a concave selfmapping of $K = \mathbb{R}_+^2$ but none of the Concave Perron Theorems (2.1.11, 2.2.11) does apply because neither $Tx > 0$ for all $x \gneqq 0$ nor T is weakly idecomposable.

Consider next the cases where $\alpha, \beta \in \{0, 1\}$. Then $Tx = (x_1, x_2)$ or $Tx = (x_2, x_1)$ and T is not shrinking at any $x \in \dot{S}^2$. The remaining cases to discuss are $0 < \alpha < 1$ and $\beta \in \{0, 1\}$ or $0 < \beta < 1$ and $\alpha \in \{0, 1\}$. In the first case T shrinks at each $x \in \dot{S}^2$. In the second case, if $\alpha = 0$, T shrinks at each $x \in \dot{S}^2$, except for $x = (0, r)$ with $0 < r$. If $\alpha = 1$,

T shrinks at $x \in \dot{S}^2$ with the exception of $x = (r, 0)$ with $r > 0$. In this example, T may shrink at some points but not on others (in $\dot{S}^2$).

To find the value of $\bar{c}(x)$, provided it exists, is not as easy. In the classical case $\alpha = \beta = \frac{1}{2}$ it is well-known that $c(x)$ is given by a complete elliptic integral of the first kind

$$c(x) = \frac{\pi}{2}\left[\int_0^{\frac{\pi}{2}} \frac{d\phi}{\sqrt{x_1(0)^2\cos^2\phi + x_2(0)^2\sin^2\phi}}\right]^{-1},$$

which shows in no way any similarity with the means initially given. For general $0 < \alpha, \beta < 1$ I do not know of any such formula.

(4) Another simple example of a nonlinear mean map is for $S = \mathbb{R}_+, n = 3$, given by

$$T(x_1, x_2, x_3) = (\min\{x_1, x_2\}, x_3, x_1).$$

T is a concave selfmapping of $\mathbb{R}^3_+$ which is positively homogeneous and seems therefore to be a candidate for concave Perron–Frobenius Theory as considered in Chapter 2. It satisfies, however, not the conditions of the First or Second Concave Perron Theorem (Theorems 2.1.11 and 2.2.11, respectively) because neither $Tx > 0$ for $x \gneqq 0$ nor $T_h e_h > 0$ for some $1 \le h \le 3$. Nevertheless, from part (iii) of Theorem 8.3.4 it follows $\lim_{t\to\infty} T^t x = \bar{c}(x)$ for all $x \in \mathbb{R}^3_+$ and $x^* = \bar{c}(x)$ is a fixed point of T. To see this we show that T is shrinking at each $x \in \dot{S}^3$ for $t = 4$. From the definition of T we obtain

$$T^2 x = (\min\{x_1, x_2, x_3\}, x_1, \min\{x_1, x_2\}).$$

Thus, for $y = T^2 x$

$$\begin{aligned} T^4 x = T^2 y &= (\min\{y_1, y_2, y_3\}, y_1, \min\{y_1, y_2\}) \\ &= (\min\{x_1, x_2, x_3\}, \min\{x_1, x_2, x_3\}, \min\{x_1, x_2, x_3\}). \end{aligned}$$

Therefore, $\operatorname{conv}\{T^4 x\} = \{\min\{x_1, x_2, x_3\}\}$ which equals $\operatorname{conv}\{x\}$ if and only if $x \in \operatorname{diag}\mathbb{R}^3_+$. This shows that T is shrinking at each $x \in \dot{S}^3$ for $t = 4$. Actually, this does not hold for $t \le 3$ since T is not shrinking at $x = (1, 0, 0)$ for $t = 2$ and not shrinking at $x = (1, 1, 0)$ for $t = 3$.

(5) Examples 3 and 4 show in particular that a mean map can, for some t, shrink at all points or shrink at some points but not at others. We examine in more detail the linear case that is $Tx = Ax$, where $x \in \mathbb{R}^n, A \in \mathbb{R}^{n\times n}_+$ stochastic. Let x be any point, $x \notin \operatorname{diag}\mathbb{R}^n$. $\operatorname{conv}\{Ax\} = \operatorname{conv}\{x\}$ is equivalent to

$$\min_i \sum_j a_{ij} x_j \le \min x \quad \text{and} \quad \max_i \sum_j a_{ij} x_j \ge \max x$$

which is equivalent to

$$\min_i \sum_j a_{ij}(x_j - \min x) \le 0 \quad \text{and} \quad \max_i \sum_j a_{ij}(x_j - \max x) \ge 0.$$

Let $I_0(x) = \{j \mid \min x < x_j\}$ and $I_1(x) = \{j \mid x_j < \max x\}$.

Since $x \notin \operatorname{diag}\mathbb{R}^n$ these two sets are non-empty. Therefore, $\operatorname{conv}\{Ax\} = \operatorname{conv}\{x\}$ holds if and only if there exist i_0, i_1 such that $a_{i_0j} = 0$ for $j \in I_0(x)$ and $a_{i_1j} = 0$ for $j \in I_1(x)$. This shows that in general a stochastic matrix will, for $t = 1$, be shrinking at certain points and not shrinking at others. From Theorem 8.1.2 part **B** we know already that a stochastic matrix is, for some t, shrinking at all $x \notin \operatorname{diag}\mathbb{R}^n$ if and only if A^t is scrambling. What are the matrices on the opposite side, that is those stochastic matrices which are not shrinking at any $x \notin \operatorname{diag}\mathbb{R}^n$? Suppose $\operatorname{conv}\{Ax\} = \operatorname{conv}\{x\}$ for all $x \notin \operatorname{diag}\mathbb{R}^n$. For $x = e^k$, the k-th unit vector in $\mathbb{R}^n$, we obtain by the above $I_0(x) = \{k\}$ and $I_1(x) = \{1, \ldots, n\} \setminus \{k\}$ and there exist i_0, i_1 such that $a_{i_0k} = 0, a_{i_1j} = 0$ if $j \neq k$. Since A is stochastic we must have $a_{i_1k} = 1$. Defining $\sigma(k) = i_1$ we have $a_{\sigma(k)k} = 1, a_{\sigma(k)j} = 0$ if $j \neq k$. This defines a selfmapping σ of $\{1, \ldots, n\}$ for which $k \neq l$ implies $\sigma(k) \neq \sigma(l)$ because of $a_{\sigma(k)l} = 0$ and $a_{\sigma(l)l} = 1$. In other words σ is a permutation of $\{1, \ldots, n\}$ for which $a_{\sigma(k)j} = \delta_{kj}$ for all $1 \le k, j \le n$, δ being the Kronecker symbol. Therefore, A must be a permutation matrix, where for a permutation τ a **permutation matrix $A(\tau)$** is defined by $a_{ij}(\tau) = \delta_{\tau(i)j}$. In the above $A = A(\tau)$ for $\tau = \sigma^{-1}$.

Conversely, if $A = A(\tau)$ then $(Ax)_i = \sum_{j=1}^n a_{ij}(\tau)x_j = x_{\tau(i)}$ and $\operatorname{conv}\{Ax\} = \operatorname{conv}\{x\}$ for all $x \in \mathbb{R}^n$. We conclude that a stochastic matrix is, for $t = 1$, not shrinking at any $x \notin \operatorname{diag}\mathbb{R}^n$ if and only if A is a permutation matrix. It follows that A is not shrinking at any $x \notin \operatorname{diag}\mathbb{R}^n$ if and only if A is a permutation matrix.

To draw conclusions from Theorem 8.3.4 we analyze in the following the crucial assumption of a shrinking behaviour. For this we introduce the concepts of neighbor and neighborhood. Let $(x(t))$ be a mean process on S^n where S is a non-empty convex subset of $\mathbb{R}^d$. The defining property $\operatorname{conv}\{x(t+1)\} \subseteq \operatorname{conv}\{x(t)\}$ implies the existence of a stochastic matrix $A(t)$ such that $x(t+1) = A(t)x(t)$ for all $t \ge 0$. Note that these matrices are not uniquely determined. With respect to a fixed sequence of matrices $A(t)$ we call for $i \in \{1, \ldots, n\}$ and $t \ge 0$

$$N(i,t) = \{j \in \{1, \ldots, n\} \mid a_{ij}(t) > 0\},$$

a set of **neighbors of *i* at *t***, and

$$U(i,t) = \{x^j(t) \mid j \in N(i,t)\},$$

a **neighborhood of *i* at *t***.

To the collection of all the sets $N(i,t)$ and $U(i,t)$ we will refer also as a **neighboring system of $x(t)$**. Since $A(t)$ is stochastic the sets $N(i,t)$ and $U(i,t)$ are non-empty. For a sequence $\tau = (t_1, \ldots, t_r)$ with $t_i \ge 0$ and the matrix $B(\tau) = A(t_1) \cdots A(t_r)$ we define more general

$$N(i,\tau) = \{j \mid b_{ij}(\tau) > 0\}$$

as **neighbors of *i* via τ** and

$$U(i,\tau) = \{x^j(t_r) \mid j \in N(i,\tau)\}$$

as **neighborhood of *i* via τ**.

The following lemma describes shrinking by neighborhoods and presents criteria for shrinking.

Lemma 8.3.6. *Let $(x(t))$ be a mean process with $x(0) = x$.*

(i) *The process is for none $t \geq 0$ shrinking at $x \in \dot{S}^n$ if and only if for each extreme point x^i of $\operatorname{conv}\{x\}$ and each $\tau = (t+p,\dots,t+1,t)$ there exists $k = k(i,\tau) \in \{1,\dots,n\}$ such that the neighborhood $U(k,\tau)$ consists only of x^i.*

(ii) *Consider for a neighboring system and a sequence $\tau_0 = (t_0 + p_0,\dots,t_0+1,t_0)$ the following properties*
   (a) *$N(i,\tau_0) \cap N(j,\tau_0) \neq \emptyset$ for all $1 \leq i,j \leq n$.*
   (b) *$U(i,\tau_0) \cap U(j,\tau_0) \neq \emptyset$ for all $1 \leq i,j \leq n$.*
   (c) *For all $1 \leq i,j \leq n$ with $U(i,\tau_0) \neq U(j,\tau_0)$ it holds $|U(i,\tau_0)| \geq 2$ or $|U(j,\tau_0)| \geq 2$.*

   Then (a) implies (b) and (b) implies (c) and the process $(x(t))$ is shrinking at $x \in \dot{S}^n$ for some t if one of the above properties does hold.

Proof. (i) By definition $(x(t))$ is not shrinking at x for t if and only if $\operatorname{conv}\{x(t)\} = \operatorname{conv}\{x\}$. If this is the case, then any extreme point x^i of $\operatorname{conv}\{x\}$ must be equal to some $x^k(t)$. If $(x(t))$ is for none t shrinking at x, then to τ and extreme point x^i given, there exists $k = k(i,\tau)$ such that $x^i = x^k(t+p+1)$. Since

$$x(t+p+1) = A(t+p)\cdots A(t+1)A(t)x(t) = B(\tau)x(t),$$

it follows

$$x^i = x^k(t+p+1) = \sum_{j=1}^{n} b_{kj}(\tau)x^j(t).$$

Because of $\operatorname{conv}\{x\} = \operatorname{conv}\{x(t)\}$ the point x^i is an extreme point of $\operatorname{conv}\{x(t)\}$ and, hence, $x^j(t) = x^i$ if $b_{kj}(\tau) > 0$. This proves $U(k,\tau) \subseteq \{x^i\}$ and $U(k,\tau) = \{x^i\}$ since $U(k,\tau) \neq \emptyset$.

For the converse choose $p = 0$ and $\tau = (t)$ and for an extreme point x^i of $\operatorname{conv}\{x\}$ a $k = k(i,\tau)$ such that $U(k,\tau) = \{x^i\}$. It follows $x^i = x^j(t)$ for some $j \in N(k,\tau)$ and $x^i \in \operatorname{conv}\{x(t)\}$ because of $\tau = (t)$. This holds for all extreme points of $\operatorname{conv}\{x\}$ and, hence, $\operatorname{conv}\{x\} \subseteq \operatorname{conv}\{x(t)\}$, that is $\operatorname{conv}\{x(t)\} = \operatorname{conv}\{x\}$, t being arbitrary.

(ii) (a) implies (b) by definition of $U(i,\tau_0)$. (b) implies (c) since for $A = U(i,\tau_0)$, $B = U(j,\tau_0)$ such that $A \cap B \neq \emptyset$ and $A \cap B \subsetneq A$ it follows that

$$|A| = |A \cap B| + |A \setminus (A \cap B)| \geq 2.$$

Furthermore, suppose the process is for none $t \geq 0$ shrinking at $x \in \dot{S}^n$. Then $\operatorname{conv}\{x\}$ must have at least two different extreme points, say x^{i_1} and x^{i_2}. Part (i) yields $U(k_1,\tau_0) = \{x^{i_1}\}$ and $U(k_2,\tau_0) = \{x^{i_2}\}$ with $1 \leq k_1,k_2 \leq n$ which contradicts property (c). This proves part (ii). □

With the help of Lemma 8.3.6 from Theorem 8.3.4 we obtain immediately the following result.

Theorem 8.3.7. *Let T be a mean map on S^n for a non-empty and convex subset S of $\mathbb{R}^d$. It holds $\lim_{t\to\infty} T^t x = \bar{c}(x)$ for all $x \in S^n$ if T is continuous and for each $x \in \dot{S}^n$ the mean process given by $x(t) = T^t x$ has a neighboring system with one of the properties* (a), (b), (c) *of Lemma 8.3.6.*

From this theorem we obtain the following result which demonstrates the role of scrambling matrices also for general mean maps.

Corollary 8.3.8. *Let $Tx = A(x)x, x \in S^n$, S a non-empty convex subset of $\mathbb{R}^d$ and $A(x)$ a stochastic matrix the entries of which depend continuously on x. It holds*

$$\lim_{t\to\infty} T^t x = \bar{c}(x) \quad \textit{for all} \quad x \in S^n$$

in each of the following cases.

(a) *For each $x \in \dot{S}^n$ exist $t = t(x)$ and $p = p(x)$ such that on the orbit of x given by $x(t) = T^t x$ the product $A(x(t+p)) \cdots A(x(t))$ is scrambling.*
(b) *For each $x \in \dot{S}^n$ at least one of the matrices $A(x(t))$ has a scrambling power and all these matrices are of the same type , that is $a_{ij}(x(t)) = 0$ if and only if $a_{ij}(x(0)) = 0$ where $1 \leq i, j \leq n, t \geq 0$.*
(c) *All matrices $A(x)$ are of the same type and $A(x_0)$ has a scrambling power for some $x_0 \in S^n$.*

Proof. By assumption T is a continuous mean map on S^n to which we will apply Theorem 8.3.7.

(a) Fix $x(0) = x \in \dot{S}^n$. A neighboring system for the mean process given by $x(t) = T^t x$ is defined by taking $A(x(t))$ as $A(t)$. For $\tau = (t + p, \ldots, t)$ then $B(\tau) = A(x(t + p)) \cdots A(x(t))$. By definition of $N(i, \tau)$ the condition $N(i, \tau) \cap N(j, \tau) \neq \emptyset$ for $1 \leq i, j \leq n$ means that $B(\tau)$ is scrambling. Thus, Theorem 8.3.7 implies the conclusion for (a).

(b) This follows from the above. Let $A(x(t_0))$ be a matrix, the p-th power of which is scrambling. Since all matrices $A(x(t))$ are of the same type the product $A(x(t + p)) \cdots A(x(t))$ is scrambling for each t.

(c) is a special case of (b). □

The following consequence of Corollary 8.3.8 provides a criterion in terms of the Jacobian for differentiable mean maps.

Corollary 8.3.9. *Let T be a mean map on $\operatorname{int} \mathbb{R}^n_+$ which is positively homogeneous and continuously differentiable. If there exist a $p \geq 1$ such that for all x the Jacobian $J(x)$ of T is stochastic at x and products taking for any p points are scrambling, then $\lim_{t\to\infty} T^t x = \bar{c}(x)$ for all $x \in \operatorname{int} \mathbb{R}^n_+$.*

Proof. Since T is positively homogeneous by Euler's Theorem $T_i x = \sum_{j=1}^{n} \frac{\partial T_i}{\partial x_j}(x)x_j$ for all i, that is $Tx = J(x)x$. The assertion follows from Corollary 8.3.8 (a). □

Particular examples of mean maps are given by the well-known means as arithmetic mean, geometric mean, harmonic mean, power mean, Lehmer mean, and many others. All these means we admit in a weighted version.

Definition 8.3.10. Let a_k, $1 \le k \le n$ be any weights that is $0 \le a_k$ and $\sum_{k=1}^{n} a_k = 1$.

A mapping $f\colon \operatorname{int} \mathbb{R}_+^n \to \mathbb{R}_+$ is a weighted mean called

- **arithmetic mean** if

$$f(x) = \sum_{k=1}^{n} a_k x_k, \ x = (x_1, \dots, x_n)' \in \mathbb{R}^n;$$

- **geometric mean** if

$$f(x) = \prod_{k=1}^{n} x_k^{a_k};$$

- **harmonic mean** if

$$f(x) = \Big(\sum_{k=1}^{n} a_k x_k^{-1}\Big)^{-1};$$

- **power mean** if for some $r \in \mathbb{R} \setminus \{0\}$

$$f(x) = \Big(\sum_{k=1}^{n} a_k x_k^{r}\Big)^{\frac{1}{r}}$$

(also called Hölder mean);

- **Lehmer mean** if for some $r \in \mathbb{R}$

$$f(x) = \Big(\sum_{k=1}^{n} a_k x_k^{r+1}\Big)\Big(\sum_{k=1}^{n} a_k x_k^{r}\Big)^{-1}$$

(also called contraharmonic mean).

With the exception of the Lehmer mean all these means can be looked at as cases of a power mean.

This is obvious for the arithmetic mean, taking $r = 1$. Similarly, the harmonic mean is a power mean for $r = -1$. It is not difficult to see that

$$\lim_{r\to 0}\Big(\sum a_k x_k^r\Big)^{\frac{1}{r}} = \prod_{k=1}^{n} x_k^{a_k},$$

and, hence, the geometric mean can be viewed as a power mean for $r = 0$. In the following we consider selfmappings of $\operatorname{int} \mathbb{R}_+^n$ where each component mapping is given by one of the means above. Since for all these means we have that

$$\min_{1\le k\le n} x_k \le f(x) \le \max_{1\le k\le n} x_k, \quad x \in \mathbb{R}_+^n$$

the mixture T of these means is a mean map as defined earlier. Since such a mean map stirs up various means, call it a soup. A particular case is a Gauss soup, where only the arithmetic mean and geometric mean are involved. A special case, the famous arithmetic-geometric mean we discussed already as axample (3) in Examples and counter-examples 8.3.5. More precisely, we have the following definition.

Definition 8.3.11. Let $A \in \mathbb{R}_+^{n\times n}$ be a stochastic matrix. A selfmapping T of $\operatorname{int}\mathbb{R}_+^n$ is a **soup based on A** if for each $1 \leq i \leq n$ the component mapping T_i is one of the means in Definition 8.3.10 with the i-th row of A as weights. T is a **Gauss soup** if T_i is for each i either an arithmetic or a geometric mean.

On soups we have the following fundamental result.

Theorem 8.3.12. *Let T be a selfmapping of* $\operatorname{int}\mathbb{R}_+^n$.

(i) *If T is a soup based on a scrambling matrix then $\lim\limits_{t\to\infty} T^t x = \bar{c}(x)$ for all $x \in \operatorname{int}\mathbb{R}_+^n$.*

(ii) *If T is a soup based on a matrix which has a scrambling power, then $\lim\limits_{t\to\infty} T^t x = \bar{c}(x)$ for all $x \in \operatorname{int}\mathbb{R}_+^n$ provided the soup does not contain a Lehmer mean as component as it is the case for a Gauss soup.*

Proof. As already remarked T is a mean map and T is obviously continuous on $\operatorname{int}\mathbb{R}_+^n$. To apply Theorem 8.3.4 (iii) we show that T is shrinking, for $t = 1$, at each $x \in \operatorname{int}\mathbb{R}_+^n$.

(i) Let $\min y = \min\limits_{1\leq i\leq n} y_i$, $\max y = \max\limits_{1\leq i\leq n} y_i$ for $y \in \operatorname{int}\mathbb{R}_+^n$. By definition T is shrinking for $t = 1$ at $x \in \mathbb{R}_+^n$ if and only if $[\min Tx, \max Tx] = \operatorname{conv}\{Tx\} \subsetneq \operatorname{conv}\{x\} = [\min x, \max x]$ that is, either $\min Tx \neq \min x$ or $\max Tx \neq \max x$. To prove this we show that $T_i x = \min x$ together with $T_j x = \max x$ for any $1 \leq i \neq j \leq n$ implies $\min x = \max x$. This will follow from the properties of each of the means considered

$$\begin{aligned} f(x) &= \min x, a_k > 0 \quad \text{for some } k \text{ implies } x_k = \min x \quad \text{and} \\ f(x) &= \max x, a_l > 0 \quad \text{for some } l \text{ implies } x_l = \max x. \end{aligned} \tag{$*$}$$

Namely, assume properties $(*)$ to hold for each mean f. Since by assumption the matrix A on which T is based is scrambling, for $i \neq j$ given there exists a k such that $a_{ik} > 0$ and $a_{jk} > 0$ which together with $T_i x = \min x$ and $T_j x = \max x$ yields $x_k = \min x$ and $x_k = \max x$, that is $\min x = \max x$.

It remains to show properties $(*)$ for each mean f as in Definition 8.3.10.

Consider first a power mean

$$f(x) = \left(\sum_{k=1}^{n} a_k x_k^r\right)^{\frac{1}{r}}, \quad r \neq 0.$$

If $f(x) = \min x$ then $\sum_k a_k x_k^r = (\min x)^r$ and $\sum_k a_k (x_k^r - (\min x)^r) = 0$. If $a_k > 0$ then $x_k^r = (\min x)^r$ and $x_k = \min x$. The same reasoning applies in case of $f(x) = \max x$. This covers for $r =$ and $r = -1$ the cases of an arithmetic mean and harmonic mean, respectively. Consider next a geometric mean, $f(x) = \prod_{k=1}^n x_k^{a_k}$. If $f(x) = \min x$ then

$\prod_k x_k^{a_k} = \prod_k (\min x)^{a_k}$. If $a_k > 0$ then because of $x \in \operatorname{int} \mathbb{R}_+^n$ we must have that $x_k^{a_k} = (\min x)^{a_k}$ that is $x_k = \min x$. Similar for $f(x) = \max x$.

As the last case consider a Lehmer mean

$$f(x) = \Big(\sum_{k=1}^{n} a_k x_k^{r+1}\Big)\Big(\sum_{k=1}^{n} a_k x_k^r\Big)^{-1}.$$

If $f(x) = \min x$, then $\sum_k a_k x_k^{r+1} = \sum_k a_k x_k^r \min x$ and $\sum_k a_k x_k^r (x_k - \min x) = 0$. If $a_k > 0$ then because of $x \in \operatorname{int} \mathbb{R}_+^n$ we must have $x_k = \min x$. Similar for $f(x) = \max x$. Thus, properties (∗) hold for each mean considered which proves part (i).

(ii) Let T be a soup based on A and containing no Lehmer mean as a component. We show that for all $t \geq 0$ T^t is a soup based on A^t. Let T_i be a component given by a power mean,

$$T_i x = \Big(\sum_{k=1}^{n} a_{ik} x_k^r\Big)^{\frac{1}{r}}$$

for all $x \in \operatorname{int} \mathbb{R}_+^n$. Suppose for some t we have that

$$T_i^t x = \Big(\sum_k a_{ik}^{(t)} x_k^r\Big)^{\frac{1}{r}}, \quad a_{ik}^{(t)}$$

being the entries of A^t. Then

$$\begin{aligned} T_i^{t+1} &= T_i^t(Tx) = \Big(\sum_k a_{ik}^{(t)} (T_k x)^r\Big)^{\frac{1}{r}} = \Big[\sum_k a_{ik}^{(t)} \Big(\sum_h a_{kh} x_h^r\Big)\Big]^{\frac{1}{r}} \\ &= \Big[\sum_h \Big(\sum_k (a_{ik}^{(t)} a_{kh})\Big) x_h^r\Big]^{\frac{1}{r}} = \Big(\sum_h a_{ih}^{(t+1)} x_h^r\Big)^{\frac{1}{r}}. \end{aligned}$$

By induction this proves that T_i^t is a power mean with weights given by the i-th row of A^t. This covers for $r = 1$ and $r = -1$ arithmetic and harmonic mean, respectively. The remaining case to be considered is a geometric mean, $T_i x = \prod_{k=1}^n x_k^{a_{ik}}$. Similarly as the above we obtain

$$\begin{aligned} T_i^{t+1} x &= T_i^t(Tx) = \prod_k (T_k x)^{a_{ik}^{(t)}} = \prod_k \Big(\prod_h x_h^{a_{kh}}\Big)^{a_{ik}^{(t)}} \\ &= \prod_h \prod_k x_h^{a_{ik}^{(t)} a_{kh}} = \prod_h x_h^{(\sum_k a_{ik}^{(t)} a_{kh})}. \end{aligned}$$

By induction T_i^t is a geometric mean with weights given by the i-th row of A^t. Thus, by induction for each t the iterate T^t is a soup based on A^t. If A^p is scrambling for some p from part (i) we obtain that $\lim_{t\to\infty} T^{pt} x = \bar{c}(x)$, for all $x \in \operatorname{int} \mathbb{R}_+^n$. Since T is continuous we get $\lim_{t\to\infty} T^{pt+s} x = T^s \bar{c}(x) = \bar{c}(x)$ which proves the desired result. □

The conclusion in part (i) of Theorem 8.3.12 still holds if we stir up a soup even further. To see this we use the following simple lemma which extends what we know by Corollary 8.1.3 about scrambling matrices to shrinking mean maps.

Lemma 8.3.13. *Let R, T be mean maps on S^n.*

(i) *The compositions $R \circ T$ and $T \circ R$ are mean maps on S^n and are both shrinking, for $t = 1$, at each $x \in \dot{S}^n$ if this is the case for T. Furthermore, any finite composition of mean maps on S^n is shrinking, for $t = 1$, at each $x \in \dot{S}^n$ whenever this is true for one of these mean maps.*

(ii) *T is shrinking, for $t = 1$, any $x \in \dot{S}^n$ if and only if for each mean map R and each x it holds that $(R \circ T)x = x$ implies $x = \bar{c}$ for some $c \in S$.*

Proof. (i) Obviously, $R \circ T$ and $T \circ R$ are mean maps. If T is shrinking, for $t = 1$, at x then $\operatorname{conv}\{(R \circ T)x\} \subseteq \operatorname{conv}\{Tx\} \subsetneqq \operatorname{conv}\{x\}$. If T is shrinking, for $t = 1$, at Rx then $\operatorname{conv}\{(T \circ R)(x)\} \subsetneqq \operatorname{conv}\{Rx\} \subseteq \operatorname{conv}\{x\}$. If $x \in \dot{S}^n$ but $Rx \notin \dot{S}^n$, that is $Rx = \bar{c}$ for some $c \in S$, then $(T \circ R)x = T\bar{c} = \bar{c}$ and, hence, $\operatorname{conv}\{(T \circ R)x\} = \{c\} \subsetneqq \{x\}$. Thus, $T \circ R$ is shrinking on $\dot{S}^n$. This proves the first statement in (i). The second statement follows immediately.

(ii) If T is shrinking, for $t = 1$, at all $x \in \dot{S}^n$ and R a mean map such that $(R \circ T)x = x$, then $\operatorname{conv}\{x\} = \operatorname{conv}\{(R \circ T)x\} \subseteq \operatorname{conv}\{Tx\}$ and, hence $x \notin \dot{S}^n$ that is $x = \bar{c}$ for some $c \in S$. Conversely, suppose $\operatorname{conv}\{Tx\} = \{x\}$ for $x \in S^n$. Then $x^i \in \operatorname{conv}\{Tx\}$ for all i and there exists an $n \times n$-stochastic matrix B such that for $Ry = By$ it follows $x = (R \circ T)x$. R being a mean map from the assumption it follows $x = \bar{c}$ for some $c \in S$. This shows that T is shrinking, for $t = 1$, at any $x \in \dot{S}^n$. □

Using this lemma Theorem 8.3.4 yields the following result.

Corollary 8.3.14. *For a soup T on S^n and any stir up P, that is P is a composition of continuous mean maps containing T, it holds* $\lim \lim_{t\to\infty} P^t x = \bar{c}(x)$ *for all $x \in S^n$ if the soup is not too viscous, that is for each mean map R and each $x \in \dot{S}^n$ it holds $(R \circ T)x \neq x$. The latter condition is equivalent for T to be shrinking, for $t = 1$, at each $x \in \dot{S}^n$ and is satisfied if T is based on a scramling matrix.*

Proof. If T is shrinking, for $t = 1$, on $\dot{S}^n$ then by Lemma 8.3.13 (i) this holds for the composition P, too. From Theorem 8.3.4 (iii) it follows that $\lim_{t\to\infty} P^t x = \bar{c}(x)$ on S^n. By Lemma 8.3.13 (ii) T is shrinking, for $t = 1$, on $\dot{S}^n$ if and only if for each mean map and each $x \in \dot{S}^n$ $(R \circ T)x \neq x$. If T is based on a scrambling matrix then according to the proof of Theorem 8.3.12 T is, for $t = 1$, shrinking on $\dot{S}^n$. □

Examples 8.3.15. (1) An obvious consequence of Corollary 8.3.14 is that $\lim_{t\to\infty} P^t x = \bar{c}(x)$ holds on S^n also for a weighted soup P, that is $P_i x = \sum_{j=1}^n b_{ij} T_j x$ where $B = (b_{ij})$ is a stochastic matrix and T based on a scrambling matrix. For T consisting of Lehmer means only and A primitive this result can be found in [84, Proposition 3.4].

(2) Let T be a soup based on a scrambling matrix A. Consider a selfmapping R of $\operatorname{int} \mathbb{R}^n_+$ with $R_i x$ equals $\min\{x_j \mid j \in I(i)\}$ or $\max\{x_j \mid j \in I(i)\}$ where $I(i)$ is a non-empty subset of $\{1, \dots, n\}$ for each i. By Lemma 8.3.13 each zigzag mapping P is shrinking, for $t = 1$, on $\dot{S}^n$, where P is defined to be a composition of T and finitely many mappings R

of the type as above. Corollary 8.3.14 yields that $\lim_{t\to\infty} P^t x = \bar{c}(x)$ on S^n. Particular cases are $P_i x$ equals

$$\min\left\{\sum_{j=1}^{n} a_{kj} \mid k \in I(i)\right\} \quad \text{or} \quad \max\left\{\sum_{j=1}^{n} a_{kj} \mid k \in I(i)\right\} \quad \text{and} \quad Q_i x = \sum_{j=1}^{n} a_{ij} m(j)$$

where $I(i)$ is a non-empty subset of $\{1,\dots,n\}$ and $m(j) = \min\{x_k \mid k \in I(j)\}$ or $m(j) = \max\{x_k \mid k \in I(j)\}$. (See also the treatment of zigzag mappings in Corollary 5.3.6 (iii) and Examples 5.4.2 (iii), where without assuming stochastic matrices just the existence of (absolute) stable fixed points has been shown.)

(3) The following example which stems from population biology we take from [85, equations 3.6.4] (see also [70, p. 160]). Let T be the following selfmapping of int $\mathbb{R}^n_+$

$$Tx = \begin{bmatrix} a_1x_1 & + & b_1\vartheta(x_1,x_2) & + & c_1\vartheta(x_1,x_4) & + & d_1\vartheta(x_2,x_3) \\ a_2x_2 & + & b_2\vartheta(x_1,x_2) & + & c_2\vartheta(x_1,x_4) & + & d_2\vartheta(x_2,x_3) \\ a_3x_3 & + & b_3\vartheta(x_3,x_4) & + & c_3\vartheta(x_1,x_4) & + & d_3\vartheta(x_2,x_3) \\ a_4x_3 & + & b_4\vartheta(x_3,x_4) & + & c_4\vartheta(x_1,x_4) & + & d_4\vartheta(x_2,x_3) \end{bmatrix}, \qquad (*)$$

where $a_j, b_j, c_j, d_j \geq 0$ for $1 \leq j \leq 4$ and ϑ is one half of the harmonic mean, $\vartheta(s,t) = \frac{1}{2}h(s,t) = \frac{st}{s+t}$ for $s,t > 0$. By detailed analysis in [85] complicated conditions are given which determine exactly when T has an eigenvector in int $\mathbb{R}^4_+$ (see [70]; actually the analysis in [85] allows also a_j to be negative). We shall prove that under certain assumptions on the coefficients it will follow that $\lim_{t\to\infty} T^t x = \bar{c}(x)$ on int $\mathbb{R}^n_+$. To apply Theorem 8.3.4 (iii) we first assume that $a_j + \frac{1}{2}(b_j + c_j + d_j) = 1$ for all $1 \leq j \leq 4$, which assures that T is a continuous mean map on S^n for $S = \text{int}\,\mathbb{R}_+$. It remains to show that T is shrinking on $\dot{S}^n$ For this we proceed as in the proof for Theorem 8.3.12 (i) by showing that $T_i x = \min x$ together with $T_j x = \max x$ for $1 \leq i \neq j \leq 4$ implies $\min x = \max x$. The i-th component of T has by $(*)$ the form

$$T_i x = a_i x_i + \frac{1}{2} b_i h(\cdot,\cdot) + \frac{1}{2} c_i h(\cdot,\cdot) + \frac{1}{2} d_i h(\cdot,\cdot). \qquad (**)$$

Thus, $T_i x$, though not one of the means as in Definition 8.3.10 it is a combination of two of them, of arithmetic and harmonic means. As with the latter we proceed with $(**)$ and obtain for $T_i x = \min x$, depending on which of the coefficients is positive, that either $x_i = \min x$ or $h(s,t) = \min x$, in which case we must have that $s = t = \min x$. Similar for $T_i x = \max x$. Instead of carrying out the details we illustrate the method by a numerical example. (See also Exercise 8.) Suppose the matrix $M = [a\ b\ c\ d]$ is given by

$$M = \begin{bmatrix} 0 & 1 & 0 & 1 \\ 0 & 1 & 1 & 0 \\ \frac{1}{2} & 0 & 1 & 0 \\ 0 & 0 & 1 & 1 \end{bmatrix} \quad \text{and} \quad Tx = \begin{bmatrix} \frac{1}{2}h(x_1,x_2) & + & \frac{1}{2}h(x_2,x_3) \\ \frac{1}{2}h(x_1,x_2) & + & \frac{1}{2}h(x_1,x_4) \\ \frac{1}{2}x_3 & + & \frac{1}{2}h(x_1,x_4) \\ \frac{1}{2}h(x_1,x_4) & + & \frac{1}{2}h(x_2,x_3) \end{bmatrix}.$$

The method described above for (∗∗) yields in this case if $T_i x = \min x$ or $T_i x = \max x$ for $i = 1: x_1 = x_2 = x_3$, for $i = 2: x_1 = x_2 = x_4$, for $i = 3: x_1 = x_3 = x_4$ and for $i = 4: x_1 = x_2 = x_3 = x_4$.

This shows that $T_i x = \min x$ and $T_j x = \max x$ must give $\min x = \max x$. Thus T is shrinking, though the matrix M is not scrambling. This proves the conclusion wanted. It should be pointed out, however, that the assumptions made are stronger than in [85]; by the assumption $a_j + \frac{1}{2}(b_j + c_j + d_j) = 1$ for $1 \le j \le 4$ the mere question for an eigenvector in $\operatorname{int} \mathbb{R}^n_+$ becomes trivial.

In concluding this section we relate results obtained to those in the literature.

Remarks 8.3.16. (1) In [12] a continuous mapping $f: \operatorname{int} \mathbb{R}^n_+ \to \mathbb{R}$ is called an **abstract mean** if $\min x \le f(x) \le \max x$ (actual, the definition is given for $n = 2$). An **abstract mean** is called **strict,** if $f(x) = \min x$ or $f(x) = \max x$ holds precisely if all components of x are equal. In case of $\lim\limits_{t\to\infty} T^t x = \bar{c}(x)$, T a mean map, the common limit $\bar{c}(x)$ is called the **compound** of the component mappings T_i (in case of $n = 2$). It is shown in [12, Theorem 8.2], for $n = 2$ that a compound exists if one of the component mappings is strict. This is, even for $n = 2$, just a special case of a shrinking map. (Actually, a further assumption is made to establish monotonicity of the iterates.) In general strictness is neither necessary nor sufficient for $\lim\limits_{t\to\infty} T^t x = \bar{c}(x)$ to hold on $\operatorname{int} \mathbb{R}^n_+$ (see Exercise 4).

(2) Considering Definition 8.3.3, a mean map which is shrinking, for $t = 1$, at each $x \in \dot{S}^n$ is called a **compromise map** in [59]. There part (ii) of Theorem 8.3.4 is shown for such maps by using arguments similar to Lemma 8.3.2. See also [60], where only an iterate of the mean map is required to be a compromise map. Maps similar to mean maps or compromise maps are considered in [81]. For the results obtained there a condition of strict convexity is used which we do not assume. (See Examples and counter-examples 8.3.5 (4) and Exercise 12.)

(3) Corollary 8.3.8 (c) is proven in [84, Proposition 3.3] under the stronger assumption that $A(x_0)$ is primitive.

In [84, 85] as well as [70] the iteration of nonlinear means is analyzed where each component of a selfmapping f of $\mathbb{R}^n_+$ is a positive linear combination of power means on $\operatorname{int} \mathbb{R}^n_+$, including the limit case of a geometric mean. Various conditions are given which guarantee $\lim\limits_{t\to\infty} f^t(x) = \lambda(x)u$ on $\operatorname{int} \mathbb{R}^n_+$ with $\lambda(x) > 0$, $u \in \operatorname{int} \mathbb{R}^n_+$ an eigenvector of f. In [84, Proposition 3.4] such a result is proven if the components of f are convex combinations of Lehmer means (see Examples 8.3.15 (1)). Also the model from biology given in Examples 8.3.15 (3) is analyzed in detail in [85] for the existence of eigenvectors in $\operatorname{int} \mathbb{R}^n_+$.

(4) The area of means and their iterations is as old as it is fascinating and the literature on it is quite widespread. The book [12] is a beautiful account which presents many examples and hints at the literature. The article [2] is very illuminating and treats also means in infinite dimensions. A classic is [15] which treats a lot of famous exam-

ples in a systematic manner. All these references address also the history of the subject and give further references.

(5) For Gauss soups see [44, 59, 60]. Stated in the language used here it is proven by different reasoning, in [29] and [37] that for a soup with geometric mean and power mean as components it holds $\lim_{t\to\infty} T^t x = \bar{c}(x)$ provided the soup is based on a stochastic matrix which has all its rows equal and strictly positive. This result is a special case of Theorem 8.3.12, of part (i) as well as of part (ii).

(6) As already remarked, to determine for a mean map T the limit $\bar{c}(x)$, provided it exists, can be very difficult and is known only in a few cases. It is easy for T given by a scrambling stochastic matrix. For, by Theorem 8.1.4 $\lim_{k\to\infty} A^k = B$ where all rows of B are equal to vector b the transpose of which is the unique normalized eigenvector of the transpose A'. From this it follows $c(x) = \sum_{j=1}^n b_j x_j$ for all x. In the most simple case of a Gauss soup, $n = 2$ and based on

$$A = \begin{bmatrix} \frac{1}{2} & \frac{1}{2} \\ \frac{1}{2} & \frac{1}{2} \end{bmatrix},$$

$c(x)$ is given by a complete elliptic integral of the first kind (see Examples and counterexamples 8.3.5 (3); for a proof see [12]). Following Gauss, later on Borchardt [11] investigated the case $n = 4$ and the following selfmapping f of int $\mathbb{R}^n_+$ given by composing arithmetic and geometric mean in the following manner:

$$f_1(x) = \frac{1}{4}(x_1 + x_2 + x_3 + x_4), \qquad f_2(x) = \frac{1}{2}(\sqrt{x_1 x_2} + \sqrt{x_3 x_4}),$$

$$f_3(x) = \frac{1}{2}(\sqrt{x_1 x_2} + \sqrt{x_2 x_4}), \qquad f_4(x) = \frac{1}{2}(\sqrt{x_1 x_4} + \sqrt{x_2 x_3}).$$

This, of course, is a mean map and it is easily confirmed that f is, for $t = 1$, shrinking at each $x \in \operatorname{int} \mathbb{R}^4_+$. (See a generalization in Exercise 9.) Therefore, by Theorem 8.3.4 it holds $\lim_{t\to\infty} T^t x = \bar{c}(x)$ for all x. Borchardt proved that $c(x)$ is given by an integral over a Kummer's quartic surface [2, 11].

(7) Considering the determination of $c(x)$ there is a close relationship to invariants for T (first integrals), that is mappings H: int $\mathbb{R}^n_+ \to$ int $\mathbb{R}^n_+$ such that $H(Tx) = H(x)$ for all $x \in \operatorname{int} \mathbb{R}^n_+$. If H is a continuous invariant for T and $\lim_{t\to\infty} T^t x = \bar{c}(x)$ then $H(x) = H(\bar{c}(x))$. Therefore, knowing an invariant H one possibly could calculate $c(x)$ from it. Conversely, in case of $\lim_{t\to\infty} T^t x = \bar{c}(x)$, $H(x) = \bar{c}(x)$ yields an invariant. Actually, the latter is the only continuous mean map H which is invariant. For, $H(Tx) = H(x)$ for all x implies $H(x) = \lim_{t\to\infty} H(T^t x) = H(\bar{c}(x)) = \bar{c}(x)$. (This is called the " Invariance Principle" in [12].) For examples for the use of integrals see Exercises 12, 13.

8.4 Infinite products of stochastic matrices: path stability, convergence and a generalized theorem of Wolfowitz

A mean process $(x(t))$ on S^n, that is $\text{conv}\{x(t+1)\} \subseteq \text{conv}\{x(t)\}$ for all $t \geq 0$ by definition, can be equivalently described by $x(t+1) = A(t)x(t)$ with stochastic matrices $A(t)$ for $t \geq 0$. (See Definition 8.3.1 and the remarks thereafter.) Therefore, $x(t+1) = A(t)A(t-1)\ldots A(1)A(0)x(0)$ and the question whether $\lim_{t\to\infty} x(t) = \bar{c}(x(0)) \in \text{diag}S^n$ becomes one of the matrix products $A(t)\cdots A(0)$ tending to a matrix with all rows equal. To find conditions for the latter is the main aim of the present section. For the special case $A(t) = A$ for all t we know already from Theorem 8.1.4 as a necessary and sufficient condition that a power of A has to be scrambling. One might expect, therefore, as condition for $A(t)\cdots A(0)$ to tend to a matrix with equal rows that products of the matrices $A(s)$ should be scrambling. Whereas such a condition is necessary it is by no means sufficient. What is needed, moreover, is a structure of being scrambling which is not weakened too fast for t tending to infinity. This phenomenon is illustrated by the following simple examples which also foreshadows the general relationships as set out in Theorem 8.4.2 and later results.

Examples 8.4.1. We consider two examples of sequences of 2×2-matrices $A(t)$, stochastic and acting on $\mathbb{R}$.

Example A. Let

$$A(t) = \begin{bmatrix} 1 & 0 \\ \frac{1}{t} & 1-\frac{1}{t} \end{bmatrix}$$

for $t \geq 2$. One confirms easily by induction that

$$P(t) = A(t)A(t-1)\ldots A(2) = \begin{bmatrix} 1 & 0 \\ 1-\frac{1}{t} & \frac{1}{t} \end{bmatrix} \quad \text{for} \quad t \geq 2.$$

Therefore, $\lim_{t\to\infty} P(t) = \left[\begin{smallmatrix} 1 & 0 \\ 1 & 0 \end{smallmatrix}\right]$ – a matrix with all rows equal.

Example B. Let

$$A(t) = \begin{bmatrix} 1 & 0 \\ \frac{1}{t^2} & \left(1-\frac{1}{t^2}\right) \end{bmatrix}$$

for $t \geq 2$. One confirms easily by induction that

$$P(t) = A(t)A(t-1)\ldots A(2) = \begin{bmatrix} 1 & 0 \\ 1-\alpha(t) & \alpha(t) \end{bmatrix} \quad \text{with} \quad \alpha(t) = \prod_{i=2}^{t}\left(1-\frac{1}{i^2}\right).$$

Also by induction follows $\alpha(t) = \frac{1}{2}(1+\frac{1}{t})$ and, hence,

$$\lim_{t\to\infty} P(t) = \begin{bmatrix} 1 & 0 \\ \frac{1}{2} & \frac{1}{2} \end{bmatrix},$$

a matrix with not all rows equal.

In both examples the matrices $A(t)$ are all scrambling as well as the limits $\lim_{t\to\infty} P(t)$ which, however, are quite different considering equality of rows. In the following we analyze this phenomenon under various aspects. Roughly speaking, the point of difference is that the convergence in example B is "too fast".

Let $c(\cdot)$ be the coefficient for matrices introduced in Theorem 8.1.2 A. For example A we have $c(P(t)) = \frac{1}{t}$ which converges to 0, whereas for example B we have $c(P(t)) = \alpha(t)$ which does not converge to 0. This is related to $\sum_{t=2}^{\infty} \frac{1}{t} = \infty$ versus $\sum_{t=2}^{\infty} \frac{1}{t^2} < \infty$. (See also Example 8.4.10 and Theorem 8.4.5 for the general case.)

The difference between examples A and B can also be recognized by checking for path stability. Consider the mean process $(x(t))$ defined by $x(t+1) = A(t)x(t)$ and, hence $x(t+1) = P(t)x(2)$. For example A we obtain $x(t+1) = (x_1, (1-\frac{1}{t})x_1 + \frac{1}{t}x_2)$ where $x = x(2)$. Therefore, $\|x_1(t+1) - x_2(t+1)\| = \frac{1}{t}\|x_1 - x_2\|$ converges to 0 for $t \to \infty$. That is, path stability holds with respect to any norm on $D = \mathbb{R}$ in the sense of Definition 7.1.3. For example B, on the other hand, we obtain $x(t+1) = (x_1, (1 - \alpha(t))x_1 + \alpha(t)x_2)$ and, hence, $\|x_1(t+1) - x_2(t+1)\| = \alpha(t)\|x_1 - x_2\|$ which shows because of $\lim_{t\to\infty} \alpha(t) = \frac{1}{2}$ that path stability does not hold.

Path stability can also be considered for internal metrics in the sense of Chapter 3. For this we need a convex cone K in the interior of which the mean process can be embedded. Let $X = \operatorname{conv}\{x_1, x_2\} \subseteq \mathbb{R}$ which we embed into int $\mathbb{R}^2_+$ as follows. Define $\tilde{x} = \{(a + x, r) \mid x \in X\}$, with a such that $a + x > 0$ for all $x \in X$ and $r = \sup\{a + x \mid x \in X\} > 0$. For $y = (y_1, y_2) \in \mathbb{R}^2$ by $\|y\| = \max\{|y_1|, |y_2|\}$ a norm monotone on $\mathbb{R}^2_+$ is defined. Obviously, for $\tilde{x} = (a + x, r), x \in X$ it holds $\|\tilde{x}\| = \max\{|a + x|, r\} = r$. To compute an internal metric we consider the order function $\lambda(\cdot,\cdot)$ on $\mathbb{R}^2_+$. If $u, v \in X$ then

$$\lambda(\tilde{v}, \tilde{u}) = \min_{1\le i\le 2} \frac{\tilde{u}_i}{\tilde{v}_i} = \min\left\{\frac{a+u}{a+v}, \frac{r}{r}\right\}.$$

This gives

$$\min\{\lambda(\tilde{v}, \tilde{u}), \lambda(\tilde{u}, \tilde{v})\} = \min\left\{\frac{a+u}{a+v}, \frac{a+v}{a+u}, 1\right\} = \frac{a + \min\{u, v\}}{a + \max\{u, v\}}.$$

In case of example A we obtain for $u = x_1(t+1), v = x_2(t+1)$

$$\min\{\lambda(\tilde{v}, \tilde{u}), \lambda(\tilde{u}, \tilde{v})\} = \frac{a + \min\{x_1, (1-\frac{1}{t})x_1 + \frac{1}{t}x_2\}}{a + \max\{x_1, (1-\frac{1}{t})x_1 + \frac{1}{t}x_2\}},$$

which converges to 1 and, hence, for the part metric p $\lim_{t\to\infty} p(\tilde{x}_1(t), \tilde{x}_2(t)) = 0$. That is, path stability for p on int $\mathbb{R}^2_+$. In contrast, a similar calculation for example B yields for $u = x_1(t+1), v = x_2(t+1)$

$$\min\{\lambda(\tilde{v}, \tilde{u}), \lambda(\tilde{u}, \tilde{v})\} = \frac{a + \min\{x_1, (1-\alpha(t))x_1 + \alpha(t)x_2\}}{a + \max\{x_1, (1-\alpha(t))x_1 + \alpha(t)x_2\}},$$

which converges because of $\lim_{t\to\infty} \alpha(t) = \frac{1}{2}$ to a limit different from 1 for $x_1 \neq x_2$. Therefore, $\lim_{t\to\infty} p(\tilde{x}_1(t), \tilde{x}_2(t)) \neq 0$ for $x_1 \neq x_2$ and path stability for p does not hold.

For the Hilbert metric we obtain for $u, v \in X$

$$\lambda(\tilde{v}, \tilde{u}) \cdot \lambda(\tilde{u}, \tilde{v}) = \min\left\{\frac{a+u}{a+v}, 1\right\} \cdot \min\left\{\frac{a+v}{a+u}, 1\right\}$$
$$= \frac{a + \min\{u, v\}}{a + \max\{u, v\}}.$$

Thus, Hilbert metric and part metric are equal in this case and the same distinction as for the part metric applies.

Finally, we may check the two examples for assymptotic equality in the sense of Definition 7.1.1.

In case of example A we have, for $t \geq 2$, $\gamma(t)\tilde{x}_1 \leq \tilde{x}_2(t+1) \leq \tilde{\gamma}(t)\tilde{x}_1$ for $\gamma(t) = 1 - \frac{1}{t}\frac{|x_1 - x_2|}{a+x_1}$, $\tilde{\gamma}(t) = 1 + \frac{1}{t}\frac{|x_1-x_2|}{a+x_1}$ which proves that the sequences $(\tilde{x}_1(t))$ and $(\tilde{x}_2(t))$ are assyptotically equal. This is not so for example B, since, for $t \geq 2$, the existence of $\gamma(t)$ and $\tilde{\gamma}(t)$ as above would require $\frac{\gamma(t)}{\tilde{\gamma}(t)} \leq \alpha < 1$ for t big enough and, hence, $(\tilde{x}_1(t))$ and $(\tilde{x}_2(t))$ are not even asymptotically linked.

The next theorem provides necessary and sufficient conditions for a mean process to converge to a point on the diagonal diagS^n. The theorem shows in particular that this convergence is equivalent to path stability, for a norm or for any of the internal metrics taken with respect to an appropriate convex cone. The case of stochastic matrices differs here greatly from the case of general nonnegative matrices. For the latter weak ergodicity or path stability and strong ergodicity do not coincide. Compare in particular the results in Chapter 7 (Corollaries 7.2.4 and 7.3.4) to part (iv) of the following theorem.

Theorem 8.4.2. *Let $(A(t))$ be a sequence of stochastic matrices, $M(t, s) = A(t + s) \cdots A(s)$ for $s, t \geq 0$ and $(x(t))$ the mean process defined by $x(t+1) = A(t)x(t)$ for $t \geq 0$, $x(0) = x \in S^n$ where S is a non-empty convex subset of $\mathbb{R}^d$.*

(i) *For $s \geq 0$, $x \in S^n$ fixed, $\lim_{t\to\infty} x(t+s) \in \operatorname{diag} S^n$ holds if and only if*
$$\lim_{t\to\infty} \|x^i(t+s) - x^j(t+s)\| = 0 \quad \text{for all} \quad 1 \leq i, j \leq n$$
($\|\cdot\|$ any norm on $\mathbb{R}^d$, $x(t) = (x^1(t), \ldots, x^n(t))$).

(ii) *For $s \geq 0$ fixed, $\lim_{t\to\infty} M(t, s)x \in \operatorname{diag} S^n$ holds for all $x \in \mathbb{R}^d$ if and only if $\lim_{t\to\infty} c(M(t, s)) = 0$.*

(iii) *For $s \geq 0$, $x \in S^n$ fixed suppose there exists a lineless closed convex cone $\mathcal{K}$ in $\mathbb{R}^d$ which contains* conv$\{x\}$ *in its interior and which admits a norm $\|\cdot\|$ on $\mathbb{R}^d$ monotone for $\mathcal{K}$. Then $\lim_{t\to\infty} x(t+s) \in \operatorname{diag} S^n$ holds if and only if one of the following equivalent properties applies.*
   (a) *For any $i, j \in \{1, \ldots, n\}$ the sequences $(x^i(t+s))$ and $(x^j(t+s))$ are asymptotically equal (for $\mathcal{K}$).*
   (b) *For any $i, j \in \{1, \ldots, n\}$ and any internal metric m on $\mathcal{K}$ it holds $\lim_{t\to\infty} m(x^i(t+s), x^j(t+s)) = 0$. In case of the Hilbert metric assume the norm constant on* conv$\{x\}$.

(iv) *For $s \geq 0$ fixed, $\lim_{t\to\infty} M(t,s)x = \bar{c}(x) > 0$ holds for all $x \in \mathbb{R}^n_+ \setminus \{0\}$ if and only if one of the following equivalent properties applies.*
   (a) $\lim_{t\to\infty} \frac{M(t,s)_{ij}}{\sum_{k=1}^{n} M(t,s)_{ik}} = v_j^{(s)} > 0$ *for all* $1 \leq i,j \leq n$,
   (b) $\lim_{t\to\infty} \frac{M(t,s)_{ik}}{M(t,s)_{jk}} = 1$ *for all* $1 \leq i,j,k \leq n$.
   (The entries of $M(t,s)$ being finally positive.)

Proof. To simplify assume without loss $s = 0$.

(i) Obviously, $\lim_{t\to\infty} x(t) \in \operatorname{diag} S^n$ implies $\lim_{t\to\infty} \|x^i(t) - x^j(t)\| = 0$. For the converse let $C(t) = \operatorname{conv}\{x(t)\}$ and $C = \bigcap_{t\geq 0} C(t)$. Since $\Delta C(t) = \max_{1\leq i,j\leq n} \|x^i(t) - x^j(t)\|$ it follows $\lim_{t\to\infty} \Delta C(t) = 0$ and, hence, $\Delta C = 0$. That is $C = \{c\}$ and from Lemma 8.3.2 (ii) we obtain $\lim_{t\to\infty} x^i(t) = c$ for all i.

(ii) For $x(t+1) = M(t,0)x$ we obtain $\Delta\operatorname{conv}\{x(t+1)\} \leq c(M(t,0))\Delta\operatorname{conv}\{x\}$ from Theorem 8.1.2 A (ii). Therefore, $\lim_{t\to\infty} c(M(t,0)) = 0$ implies $\lim_{t\to\infty} \Delta\operatorname{conv}\{x(t+1)\} = 0$ and by part (i) it follows $\lim_{t\to\infty} M(t,0)x \in \operatorname{diag} S^n$. Conversely, if the latter holds for each x then $A = \lim_{t\to\infty} M(t,0)$ exists and all rows of A are equal. This shows $\lim_{t\to\infty} c(M(t,0)) = c(A) = 0$.

(iii) Since $\operatorname{conv}\{x\}$ is compact in $\operatorname{int} K$, from Proposition 3.4.12 (vi) if follows that all internal metrics for $\mathcal{K}$ are on $\operatorname{conv}\{x\}$ equivalent to the metric given by $\|\cdot\|$. Therefore, (b) holds for any internal metric if and only if $\lim_{t\to\infty} \|x^i(t) - x^j(t)\| = 0$ for all i,j. By part (i) the latter is equivalent to $\lim_{t\to\infty} x(t) \in \operatorname{diag} S^n$. By Lemma 7.1.2 (i) property (a) is equivalent to property (b) for the part metric.

(iv) Let $e(k) \in \mathbb{R}^n_+$ be the k-th unit vector. Obviously, $\lim_{t\to\infty} \sum_{k=1}^{n} M(t,0)_{ik} x_k = c(x)$ for all $x \in \mathbb{R}^n_+$ is equivalent to $\lim_{t\to\infty} M(t,0)_{ik} = c_k$ for all $1 \leq k \leq n$ where $c_k = c(e(k))$. Therefore, $\lim_{t\to\infty} M(t,0)x = \bar{c}(x) > 0$ for all $x \in \mathbb{R}^n_+ \setminus \{0\}$ is equivalent to $\lim_{t\to\infty} M(t,0)_{ik} = c_k > 0$ for all i, all k. Since $M(t,0)$ is stochastic, the latter is equivalent to $\lim_{t\to\infty} \frac{M(t,0)_{ij}}{\sum_k M(t,0)_{ik}} = c_j$ for all i,j. This proves the equivalence of property (a) and $\lim_{t\to\infty} M(t,0)x = \bar{c}(x) > 0$ for all $x \in \mathbb{R}^n_+ \setminus \{0\}$. Obviously, property (a) implies property (b), with strictly positive entries of $M(t,0)$ for t big enough. Finally, suppose property (b) and let $\gamma(t) = \min_{i,j,k} \frac{M(t,0)_{ik}}{M(t,0)_{jk}}$ for t big enough. It follows

$$1 - c(M(t,0)) = \min_{i,j} \sum_k \min\{M(t,0)_{ik}, M(t,0)_{jk}\} \geq \min_j \sum_k M(t,0)_{jk} \cdot \min\{1, \gamma(t)\}$$

and, hence, $1 - c(M(t,0)) \geq \min\{1,\gamma(t)\}$. Now, $\lim_{t\to\infty} \gamma(t) = 1$ implies $\lim_{t\to\infty} c(M(t,0)) = 0$.

Part (ii) yields $\lim_{t\to\infty} M(t,0)x = \bar{c}(x)$ and due to the positive entries of $M(t,0)$ we must have $\bar{c}(x) > 0$. This proves part (i) to (iv) of the theorem. □

By part (ii) of Theorem 8.4.2 we have that $\lim_{t\to\infty} M(t,0)x \in \operatorname{diag} S^n$ for all x is equivalent to $\lim_{t\to\infty} c(M(t,s)) = 0$ for all $s \geq 0$. Our next purpose is to find criteria for the latter to hold which then, later on, can be used when dealing with opinion dynamics as well as with swarms. Obviously, for $c(A(t)\cdots A(s))$ to converge to 0 the matrix $A(t)\cdots A(s)$ must be scrambling for t big enough. This, however, is not sufficient as shown already in Examples 8.4.1. Necessary and sufficient conditions will be presented in Theorem 8.4.5. One condition is geometrical and employs the distance $\rho_A(x)$ between conv$\{x\}$ and its subset conv$\{Ax\}$ in connection with the notion of a simple set, a finite set most simple next to that of the diagonal.

Definition 8.4.3. A non-empty set $X \subseteq \dot{S}^n$ is **simple** if for some $e \in \mathbb{R}^d$, $\|e\| = 1$

$$X = \{x = (x^1, \ldots, x^n)\} \mid x^i = \pm e \quad \text{for each} \quad 1 \leq i \leq n\}.$$

Obviously, a simple set must be finite. In Theorem 8.1.2 A we obtained a characterization of $c(A)$ as a contraction factor with respect to the diameters of all sets conv$\{x\}$. The following lemma describes $c(A)$ by the action of A on just one arbitrary chosen simple set.

Lemma 8.4.4. *For any stochastic matrix A and any simple set X it holds*

$$c(A) = 1 - \frac{1}{2} \min_{x \in X} \rho_A(x), \tag{8.4.1}$$

where

$$\rho_A(x) = \sup_{y \in \text{conv}\{x\}} \inf_{z \in \text{conv}\{Ax\}} \|y - z\| \tag{8.4.2}$$

for $x \in S^n$, $\|\cdot\|$ *a norm on* $\mathbb{R}^d$.

Proof. Let A and X be given. For $x \in X$ the sets conv$\{x\}$ and conv$\{Ax\}$ are intervals in $\mathbb{R}e$ and, therefore, $\rho_A(x) = \Delta\text{conv}\{x\} - \Delta\text{conv}\{Ax\}$ (ΔM the diameter of a set M). By definition $x \in \dot{S}^n$ and, hence, $\Delta\text{conv}\{x\} = 2$. Furthermore,

$$\begin{aligned}\Delta\text{conv}\{Ax\} &= \max_{i,j} \|\sum_k a_{ik}x^k - \sum_k a_{jk}x^k\| \\ &\leq \max_{i,j} \sum_k |a_{ik} - a_{jk}| \cdot \max_k \|x^k\|\end{aligned}$$

which yields $\Delta\text{conv}\{Ax\} \leq 2c(A)$ using Theorem 8.1.2 A. Thus, we obtain

$$\rho_A(x) = 2 - \Delta\text{conv}\{Ax\} \geq 2 - 2c(A)$$

and

$$\min_{x \in A} \rho_A(x) \geq 2(1 - c(A)).$$

For the reverse inequality consider indices i and j with different corresponding rows of A and let $I = \{k \mid a_{ik} \geq a_{jk}\}$. Obviously, $I \subsetneq \{1, \ldots, n\}$. Define $y \in X$ by $y^k = e$ for $k \in I$ and $y^k = -e$ for $k \notin I$. We have that

$$\sum_{k=1}^{n} (a_{ik} - a_{jk})y^k = \Big(\sum_{k \in I} (a_{ik} - a_{jk}) - \sum_{k \notin I} (a_{ik} - a_{jk})\Big)e = \sum_{k=1}^{n} |a_{ik} - a_{jk}|e,$$

which implies

$$\min_{x\in X}\rho_A(x) \le \rho_A(y) = 2 - \Delta\text{conv}\{Ay\} \le 2 - \|\sum_{k=1}^{n}(a_{ik} - a_{jk})y^k\| \le 2 - \sum_{k=1}^{n}|a_{ik} - a_{jk}|.$$

Since the latter inequality holds trivially in case the rows for i and j are equal, we obtain

$$\min_{x\in X}\rho_A(x) \le 2 - \min_{i,j}\sum_{k=1}^{n}|a_{ik} - a_{jk}| = 2(1 - c(A)).$$

This proves equation (8.4.1) □

Theorem 8.4.5. *Let $(A(t))$ be a sequence of stochastic matrices.*

(i) $\lim_{t\to\infty} c(A(t+s)\dots A(s)) = 0$ *holds for each $s \ge 0$ if and only if there exists a set $\mathfrak{J}$ of disjunct intervals $I = [a,b]$, $a \le b$, in $\mathbb{N}$ such that for $B(I) = A(b)A(b-1)\dots A(a+1)A(a)$ it holds*

$$\sum_{I\in\mathfrak{J}}\Big(1 - c(B(I))\Big) = \infty, \tag{8.4.3}$$

or, equivalently,

$$\sum_{I\in\mathfrak{J}}\min_{x\in X}\rho_{B(I)}(x) = \infty \tag{8.4.4}$$

for some simple set X.

(ii) (a) *Condition (8.4.2) is satisfied for each $I = [a,b] \in \mathfrak{J}$ if*

$$B(I) = A(b)\cdots A(a) \text{ is scrambling, and}$$
$$\sum_{I\in\mathfrak{J}}\alpha(I) = \infty \tag{8.4.5}$$

where $\alpha(I) = \alpha(b)\cdots\alpha(a)$ with $\alpha(t) = \min_{i,j}\{a_{ij}(t) \mid a_{ij}(t) > 0\}$.

(b) *Condition (8.4.4) is satisfied if the set of matrices $B(I)$ is* ***equiproper on a simple set X*** *that is, for each $x \in X$ there exists $\delta(x) > 0$ such that*

$$\rho_{B(I)}(x) \ge \delta(x) \quad \text{for all} \quad I \in \mathfrak{J}. \tag{8.4.6}$$

Proof. (i) Suppose $\lim_{t\to\infty} c(A(t+s)\dots A(s)) = 0$ for each $s \ge 0$ and let $0 < \epsilon < 1$ be given. Define inductively a set of intervals $I_k = [a_k, b_k]$ in $\mathbb{N}$ as follows. For $k = 1$ let $a_1 = 0$ and $b_1 = t + a_1$ with a t such that $c(A(t)\dots A(0)) \le \epsilon$. If $I_k = [a_k, b_k]$ for $k \ge 0$ choose $a_{k+1} = b_k + 1$, $b_{k+1} = t + a_{k+1}$ with a t such that $c(A(t + a_{k+1})\dots A(a_{k+1})) \le \epsilon$. Obviously, all the intervals I_k are disjunct. For $B(I_k) = A(b_k)\dots A(a_k)$ one has that $c(B(I_k)) \le \epsilon$ for all $k \ge 1$. Therefore, $\sum_{k=1}^{\infty}(1 - c(B(I_k))) \ge \sum_{k=1}^{\infty}(1-\epsilon) = \infty$ which proves (8.4.2). Conversely, suppose condition (8.4.2) holds for a set $\mathfrak{J}$ of intervals I. Because of $c(AB) \le c(A)c(B)$ for any two stochastic matrices A, B (by Theorem 8.1.2 A(ii)) it holds for $s \ge 0$ that $c(A(t)\dots A(s)) \le \prod_{I\in\mathfrak{J}(t)} c(B(I))$ where $\mathfrak{J}(t) \subseteq \mathfrak{J}$ consits of all $I \subseteq [s,t]$. For any finite subset $\mathfrak{J}'$ of $\mathfrak{J}$ one has

$$\prod_{I\in\mathfrak{J}'}c(B(I))\cdot\sum_{I\in\mathfrak{J}'}((1 - c(B(I))) \le \prod_{I\in\mathfrak{J}'}c(B(I))\cdot\prod_{I\in\mathfrak{J}'}(2 - c(B(I))) \le 1$$

and, hence,

$$c(A(t)\dots A(s)) \le \Big[\sum_{I\in\mathfrak{J}(t)} (1 - c(B(I)))\Big]^{-1}.$$

Since only finitely many $I \in \mathfrak{J}$ can intersect $[0, s]$ from (8.4.2) it follows that $\lim_{t\to\infty} c(A(t)\cdots A(s)) = 0$. This proves the case for condition (8.4.2). The equivalence of conditions (8.4.2) and (8.4.4) follows immediately from Lemma 8.4.4.

(ii) This part will follow from part (i). For (a) let $\bar{M}$ denote the incidence matrix of a matrix $M \in \mathbb{R}_+^{n\times n}$, that is $\bar{m}_{ij} = 1$ if $m_{ij} > 0$ and $\bar{m}_{ij} = 0$ if $m_{ij} = 0$. Obviously, $M \ge \alpha(M)\bar{M}$ where $\alpha(M) = \min_{i,j}\{m_{ij} \mid m_{ij} > 0\}$. Therefore, $B(I) \ge \alpha(B(I))\bar{B}(I)$. If $\alpha(I) = \alpha(b)\dots\alpha(a)$ then $\alpha(B(I)) \ge \alpha(I)$ and, hence, $B(I) \ge \alpha(I)\bar{B}(I)$. For $\beta(I) = \min_{i,j}\sum_{k=1}^n \min\{\bar{b}_{ik}, \bar{b}_{jk}\}$ it follows $1 - c(B(I)) \ge \alpha(I)\beta(I)$ for all $I \in \mathfrak{J}$. Since $B(I)$ is scrambling by assumption the same is true for $\bar{B}(I)$ which, however, need not be stochastic. This shows $\beta(I) > 0$ for all $I \in \mathfrak{J}$. Since for each I the incidence matrix $\bar{B}(I)$ consists of 0 and 1 there are only finitely many different matrices $\bar{B}(I)$ for $I \in \mathfrak{J}$. Thus, because of $\inf_{I\in\mathfrak{J}}\beta(I) = \beta > 0$ we finally get

$$\sum_{I\in\mathfrak{J}}(1 - c(B(I))) \ge \beta\sum_{I\in\mathfrak{J}}\alpha(I) = \infty.$$

As for case (b) condition (8.4.6) implies $\min_{x\in X}\rho_{B(I)}(x) \ge \min_{x\in X}\delta(x) > 0$ since a simple set is finite. Therefore, condition (8.4.4) is satisfied. This proves part (ii) of the theorem. □

Corollary 8.4.6. *Let $(A(t))$ be a sequence of stochastic matrices such that for some $p \in \mathbb{N}$ each product of p consecutive matrices is scrambling. If for some t_0 from $a_{ij}(t) > 0$ it follows that $a_{ij}(t) \ge \frac{1}{t^{\frac{1}{p}}}$ for all $t \ge t_0$ and all $i, j \in \{1, \dots, n\}$ then*

$$\lim_{t\to\infty} c(A(t+s)\dots A(s)) = 0 \quad \textit{for all} \quad s \ge 0.$$

Proof. Let $I = [a, b]$ an interval in $\mathbb{N}$ with $a = kp, b = (k+1)p - 1$. The set $\mathfrak{J}$ of all the I for $k \ge 0$ consists of disjunct intervals. We will apply Theorem 8.4.5 (ii). By assumption $B(I) = A(b)\dots A(a)$ is scrambling. Let $\alpha(t) = \min_{i,j}\{a_{ij}(t) \mid a_{ij}(t) > 0\}$ for $t \ge t_0$ and $\alpha(t) = 0$, otherwise. It follows for $\alpha(I) = \alpha(b)\dots\alpha(a)$

$$\alpha(I) \ge \frac{1}{b^{\frac{1}{p}}}\cdots\frac{1}{a^{\frac{1}{p}}} \ge \left(\frac{1}{b^{\frac{1}{p}}}\right)^p = \frac{1}{b} \quad \text{in case of } a \ge t_0$$

and $\alpha(I) = 0$, otherwise. If $k \ge k_0 = \max\{t_0, p\}$ then

$$\alpha(I) \ge \frac{1}{(k+1)p - 1} \ge \left(\frac{1}{p + t_0}\right)\frac{1}{k} \quad \text{and, hence,}$$

$$\sum_{I\in\mathfrak{J}}\alpha(I) \ge \frac{1}{p + t_0}\sum_{k\ge k_0}\frac{1}{k} = \infty.$$

From Theorem 8.4.5 (ii) the assertion of the corollary follows. □

Examples 8.4.1 do illustrate the above corollary. In case of example A the corollary yields convergence for $p = 1$. In case of example B, however, the corollary does not apply for any $p \geq 1$. Indeed, we know that in this case convergence does not hold.

The crucial condition in Corollary 8.4.6 requiring the existence of a p such that all products of p matrices are scrambling is connected, as will be seen in the following, to the following property.

Definition 8.4.7. A set M of matrices from $\mathbb{R}_+^{n\times n}$ has the **Wolfowitz property** or **W-property** for short if each finite product of matrices from M has a power which is scrambling. This definition extends the fundamental property in Section 8.1 for a stochastic matrix to possess a scrambling power to collections of several matrices.

Lemma 8.4.8. *Let M be a set of stochastic $n \times n$-matrices.*
(i) *There exists $p(M) \in \mathbb{N}$ such that all products of $p(M)$ matrices from M are scrambling if and only if M has the W-property. In that case $p(M)$ can be chosen to be $(2^n - 1)^n + 1$.*
(ii) *If each $A \in M$ is a Sarymsakov matrix then M has the W-property and $p(M)$ according to (i) can be chosen to be $n - 1$.*

Proof. Call $A = (a_{ij}), B = (b_{ij})$ from $\mathbb{R}_+^{n\times n}$ equivalent, $A \sim B$ if for any pair $(i,j) a_{ij} = 0$ is equivalent to $b_{ij} = 0$. Obviously, "$\sim$" is an equivalence relation on $\mathbb{R}_+^{n\times n}$, indeed it coincides with the equivalence relation introduced earlier in Section 3.2 (Definition 3.2.1) for the convex cone $K = \mathbb{R}_+^{n\times n}$; the equivalence classes coincide with the parts of K. Obviously, in case of $A \sim B$ matrix A is scrambling if and only if this is true for B. Furthermore, $A \sim B$ and $C \in \mathbb{R}_+^{n\times n}$ implies $AC \sim BC$. The number q of equivalence classes (parts) of stochastic matrices in $\mathbb{R}_+^{n\times n}$ is $q = (2^n - 1)^n$ which can be seen as follows. Face for a stochastic matrix a particular row and replace positive entries by 1. There are 2^n possibilities for a row consisting of $0's$ and $1's$. Since the matrix is stochastic we cannot have a zero-row which leaves $2^n - 1$ possibilities. For the n rows then we get $(2^n - 1)1^n$ possibilities.

(i) Suppose first, M has the W-property. Let $p = q + 1$ and $P = A_1 \ldots A_p$ with $A_i \in M$. Of the $q + 1$ products $A_1, A_1A_2, \ldots, A_1A_2 \ldots A_p$ at least two must be equivalent, that is there exist $1 \leq i < j \leq p$ such that $P_1 = A_1 \ldots A_i$ and $P_2 = A_1 \ldots A_j$ are equivalent.This means $P_1 \sim P_1P_3$ where $P_3 = A_{i+1} \ldots A_j$. By assumption P_3^k is scrambling for some k. By iteration from $P_1 \sim P_1P_3$ it follows that $P_1 \sim P_1P_3^k$. By Corollary 8.1.3, $P_1P_3^k$ is scrambling and, hence, P_1 is scrambling. From $P = P_1A_{i+1} \ldots A_p$ it follows that P is scrambling, too. This shows that any product of $p(M) = p$ factors from M is scrambling.

Conversely, suppose the latter for some $p \in \mathbb{N}$. Let P be a product of k factors from M. In case of $k \geq p$ we have $P = P_1P_2$ with P_1 scrambling and, hence, P is scrambling, too. In case of $k < p$ there exists $m \in \mathbb{N}$ such that $mk = p + p', p' \geq 0$. Therefore, $P^m = P_1P_2$ where P_1 and P_2 are products of p and p' factors from M, respectively. By assumption P_1 is scrambling and, hence, P^m is scrambling, too. This shows that M has the W-property.

(ii) We show that the product of $n-1$ Sarymsakov matrices of order n is scrambling. For this we proceed as in Proposition 8.1.7 to obtain (4) from (3). Let $A(1),\dots,A(n-1)$ be S-matrices and let for $\emptyset \neq M \subseteq \{1,\dots,n\}$
$s_k(M) = \{j \in \{1,\dots,n\} \mid a_{ij}(k) > 0 \text{ for some } i \in M\}$.

Similarly, for $A(1)\dots,A(k)$ let

$$S_k(M) = \{j \in \{1,\dots,n\} \mid (A(1)\dots A(k))_{ij} > 0 \text{ for some } i \in M\}.$$

It is easily seen that $S_k(M) = s_k(S_{k-1}(M))$ for $1 \leq k \leq n-1$ where $S_0(M) = M$. We show that $S_{n-1}(M) \cap S_{n-1}(M') \neq \emptyset$ for any two non-empty disjoint subsets M and M' of $\{1,\dots,n\}$. Suppose on the contrary $S_{n-1}(M) \cap S_{n-1}(M') = \emptyset$ for some M, M'. Since $A(n-1)$ is an S-matrix we have that

$$|S_{n-2}(M) \cup S_{n-2}(M')| < |s_{n-1}(S_{n-2}(M)) \cup s_{n-1}(S_{n-2}(M'))|$$

and $S_{n-2}(M) \cap S_{n-2}(M') = \emptyset$ by Properties 8.1.5 (iv). Thus,

$$|S_{n-2}(M) \cup S_{n-2}(M')| + 1 \leq |S_{n-1}(M) \cup S_{n-1}(M')|,$$

and by iteration

$$|S_0(M) \cup S_0(M')| + n - 1 \leq |S_{n-1}(M) \cup S_{n-1}(M')|.$$

The latter inequality implies

$$n + 1 = 2 + n - 1 \leq |M \cup M'| + n - 1 \leq |S_{n-1}(M) \cup S_{n-1}(M')| \leq n$$

which is impossible. This proves $S_{n-1}(M) \cap S_{n-1}(M') \neq \emptyset$.Chosing $M = \{i\}, M' = \{j\}$ for $i \neq j$ this shows there exists $k \in S_{n-1}(\{i\})$ and $k \in S_{n-1}(\{j\})$. Therefore, for $A = A(1)\dots A(n-1)$ we have $A_{ik} > 0$ and $A_{jk} > 0$ for some k which proves that A is scrambling. □

Using this lemma from Theorem 8.4.5 and Corollary 8.4.6, we obtain the following **generalized Wolfowitz Theorem** of which the original Wolfowitz Theorem proven in [103] is the special case of a finite set $\{A(t)\}$.

Theorem 8.4.9. *Let $(A(t))$ be a sequence of stochastic $n \times n$-matrices.*
(i) *If $\{A(t)\}$ has the Wolfowitz property with $p(M)$ as in Lemma 8.4.8 then $\lim_{t\to\infty} c(A(t+s)\dots A(s)) = 0$ for all $s \geq 0$ holds provided the following condition is satisfied. There exists a set $\mathfrak{J}$ of disjoint intervals $I = [a,b]$ in $\mathbb{N}$ with $b - a \geq p(M) - 1$ such that $\sum_{I\in\mathfrak{J}} \alpha(I) = \infty$ where $\alpha(I) = \sum_{t\in I} \alpha(t)$, $\alpha(t) = \min_{i,j}\{a_{ij}(t) \mid a_{ij}(t) > 0\}$.*
(ii) *The condition in (i) is especially satisfied in the following cases*
  (a) *$\alpha(t) \geq \frac{1}{t^{\frac{1}{p}}}$ for $p = p(M), t \geq t_0 \in \mathbb{N}$,*
  (b) *$\alpha(t) \geq \alpha > 0$ for $t \geq t_0 \in \mathbb{N}$,*
  (c) *$\{A(t)\}$ is finite.*
(iii) *If $A(t)$ is a Sarymsakov matrix for each $t \geq t_0 \in \mathbb{N}$ then $\lim_{t\to\infty} c(A(t+s)\dots A(s)) = 0$ holds for all $s \geq 0$ if the condition in (i) (or in (ii) (a)) is satisfied for $p(M) = n - 1$.*

Proof. Each matrix $B(I) = A(b)\dots A(a)$ consists of at least $p(M)$ factors and is scrambling by Lemma 8.4.8. Therefore, parts (i) and (iii) follow from Theorem 8.4.5 together with Lemma 8.4.8. Corollary 8.4.6 yields case (a) in (ii) which trivially implies cases (b) and (c). □

The following examples illustrate this theorem.

Examples 8.4.10. (a) Whereas Wolfowitz' theorem [103, p. 733] assumes a finite set of matrices, Theorem 8.4.9 allows also for infinite sets. Consider the following example which slightly extends Examples 8.4.1. For $r > 0$ given let

$$A(t) = \begin{bmatrix} 1 & 0 \\ \frac{1}{t^r} & \left(1 - \frac{1}{t^r}\right) \end{bmatrix} \quad \text{for} \quad t \geq 2.$$

Since all $A(t)$ are scrambling (and Sarymsakov matrices, too) the infinite set $\{A(t)\}$ has the Wolfowitz property with $p(M) = 1$. For $t \geq t_0 = 2^{\frac{1}{r}}$ we have that $\alpha(t) = \frac{1}{t^r}$. Therefore, from case (a) of Theorem 8.4.9 (ii) $\lim\limits_{t\to\infty} c(A(t)\dots A(2)) = 0$ follows for $r \leq 1$. To see that this bound is sharp consider $r > 1$. By induction it follows

$$A(t)\cdots A(2) = \begin{bmatrix} 1 & 0 \\ 1 - q(t) & q(t) \end{bmatrix} \quad \text{with} \quad q(t) = \prod_{i=2}^{t} \left(1 - \frac{1}{i^r}\right).$$

Since the sequence $(q(t))$ is decreasing on $\mathbb{R}_+$ the limit q exists. It is not difficult to see that $q > 0$ (cf. [53, p. 96/97]). Therefore,

$$\lim_{t\to\infty} A(t)\cdots A(2) = \begin{bmatrix} 1 & 0 \\ 1 - q & q \end{bmatrix} \quad \text{and}$$

$$\lim_{t\to\infty} c(A(t)\cdots A(2)) = c(\lim_{t\to\infty} A(t)\cdots A(2)) = 1 - (1 - q)) = q > 0.$$

For case (b) of Theorem 8.4.9 (ii) see also [93, Theorem 4.19]. As the above example shows $\lim\limits_{t\to\infty} c(A(t)\dots A(2)) = 0$ may hold also in case $\alpha(t)$ is not positively bounded from below.

(b) For the case of on single matrix A the Wolfowitz property means that a power of A is scrambling. By Theorem 8.1.4 we have already seen that the latter is equivalent to $\lim\limits_{k\to\infty} A^k$ being a matrix with equal rows which in turn is equivalent to $\lim\limits_{t\to\infty} c(A^t) = 0$. For the next simple case of two stochastic matrices A and B one might expect, therefore, that for a sequence $(A(t))$ consisting of A and B only $\lim\limits_{t\to\infty} c(A(t)\dots A(0)) = 0$ holds if both A and B have a scrambling power. This, however, is not true as the following example demonstrates. Let

$$A = \begin{bmatrix} 1 & 0 & 0 \\ 1 & 0 & 0 \\ 0 & 1 & 0 \end{bmatrix} \quad \text{and} \quad B = \begin{bmatrix} 0 & 1 & 0 \\ 0 & 0 & 1 \\ 0 & 0 & 1 \end{bmatrix}.$$

Neither A nor B is scrambling but

$$A^2 = \begin{bmatrix} 1 & 0 & 0 \\ 1 & 0 & 0 \\ 1 & 0 & 0 \end{bmatrix} \quad \text{and} \quad B^2 = \begin{bmatrix} 0 & 0 & 1 \\ 0 & 0 & 1 \\ 0 & 0 & 1 \end{bmatrix}$$

both are scrambling. Furthermore,

$$AB = \begin{bmatrix} 0 & 1 & 0 \\ 0 & 1 & 0 \\ 0 & 0 & 1 \end{bmatrix}, \quad (AB)^2 = \begin{bmatrix} 0 & 1 & 0 \\ 0 & 1 & 0 \\ 0 & 0 & 1 \end{bmatrix} = AB \quad \text{and} \quad BAB = \begin{bmatrix} 0 & 1 & 0 \\ 0 & 0 & 1 \\ 0 & 0 & 1 \end{bmatrix} = B.$$

This shows that none of these matrices is scrambling. Define $A(t) = A$ if t is odd and $A(t) = B$ if t is even. It follows that $c(A(t) \cdots A(0)) = 1$ for all $t \geq 0$. The finite set $\{A, B\}$ does not possess the Wolfowitz property though A^2 and B^2 are scrambling. (See also [39, p. 235] and [103, p. 734].)

According to part (ii) of Lemma 8.4.8 at least one of the two matrices cannot be a Sarymsakov matrix; actually, neither A nor B is a Sarymsakov matrix (see also Examples 8.1.8 (c)).

To conclude this section we connect the results obtained to the literature.

Remarks 8.4.11. (1) Theorem 8.4.2 shows that for a mean process the convergence to a point on the diagonal is equivalent to path stability. In part (i) with respect to a norm, in part (iii) with respect to internal metrics. Part (iv) shows that for stochastic matrices weak and strong ergodicity coincide. More precisely, this equivalence holds for row-stochastic matrices $A(t)$ and backward products of those, that is $A(t) \ldots A(s)$ for $t > s$ (see also [93, Theorem 4.17]). Such an equivalence does not hold for nonnegative matrices in general as shown in part (i) of Examples 7.5.3. For nonnegative matrices one might change from a backward product to a forward product, that is $A(s) \ldots A(t)$ for $t > s$ by taking the transpose of matrices. In case of row stochastic matrices this means, however, to switch to column-stochastic matrices. (For the two kinds of products as well as weak and strong ergodicity see [93] and the earlier Corollaries 7.2.4(ii) and 7.3.4 (ii).)

(2) The first result in Theorem 8.4.5(i) concerning condition (8.4.2) goes back to [39, Theorem 3]. See also [93, Theorem 4.18] with references to the work of J. Hajnal and W. Doeblin. Theorem 8.4.5 (ii) (a) generalizes [57, Theorem 2]. For a weaker version see [93, Theorem 4.19]. In [59, p. 3] the term **proper compromise mapping** is used for a mean map which is shrinking at each point not on the diagonal. For a mean map given by a stochastic matrix this means that the matrix must be scrambling. In [75, p. 94] a family F of mean maps is called **equiproper** if for each $x \in \dot{S}^n$ the distance between $\mathrm{conv}\{x\}$ and $\mathrm{conv}\{f(x)\}$ is bounded from below by $\delta(x) > 0$ for all $f \in F$. There it is proved [75, Theorem 1] that for F equiproper and equicontinuous $f_t \circ f_{t-1} \circ \ldots \circ f_1(x)$ converges to a point on the diagonal, where $f_t \in F, x \in S^n$. As part (ii)(b) of Theorem 8.4.5 shows, in case of mean maps given by stochastic matrices the condition of

equicontinuity can be omittted and the condition of being equiproper is required only for products $B(I)$ on just one simple, hence, finite subset.

(3) The proof that the product of $(n-1)$ Sarymsakov matrices of order n is scrambling (in the proof of Lemma 8.4.8 (ii)) follows [40, Theorem 4.8]. The original theorem of Wolfowitz [103, Theorem p. 733] states, in our language, that for a finite set $\{A(t)\}$ having the Wolfowitz property it follows that $\lim_{t\to\infty} c(A(t)\dots A(s)) = 0$. The proof of Lemma 8.4.8 (i) follows in part [103, Lemma 4].

In [103] a stochastic matrix A is called a **SIA matrix** if $\lim_{k\to\infty} A^k$ is a matrix with equal rows; other terms used sometimes are that of an ergodic or regular matrix. By Theorem 8.1.4, these notions are equivalent to A having a power which is scrambling. Wolfowitz' Theorem has many applications, see, for example [52, 90], which, however, are restricted by the assumption of finitely many matrices. Considering mean processes one faces in a natural way infinitely many matrices which requires an extension of Wolfowitz' Theorem. The most simple form is given perhaps by part (ii) (b) of Theorem 8.4.9 which says that for a sequence $(A(t))$ of infinitely many matrices with the Wolfowitz property and $\min^+ A(t) \geq \alpha > 0$ for all t one has that $\lim_{t\to\infty} c(A(t+s)\dots A(s) = 0$ for all $s \geq 0$ $(*)$.

A different extension of Wolfowitz' Theorem to infinitely many matrices is obtained in [46, Proposition 1] under the assumptions that all matrices are type-symmetric with positive diagonal and a connected graph associated to the sequence. It is remarked that such an extension can be useful in dealing with quite general nonlinear systems. Using our terminology this remark states that a solution of a nonlinear discrete system converges to consensus if it is a mean process satisfying, for example, the above extension $(*)$ of Wolfowitz' Theorem. Actually, in Section 8.3 mean processes have been used to handle nonlinear systems as the arithmetic-geometric mean amd, more general, Gauss soups.

8.5 Multi-agent coordination and opinion dynamics

In this section we come back to the question of consensus formation in opinion dynamics as treated already in Section 8.2. Actually, we shall generalize the framework and consider interaction and coordination of agents which, beside humans in a social setting comprises swarms of birds, electronic networks of sensors or groups of robots seeking for a rendezvous. Often the term "multi-agent coordination" is used to cover these and other quite diverse areas. (See for example [8, 16, 21, 47, 51, 64, 79, 81, 98, 99].)

Let $N = \{1,\dots,n\}$ be a finite set of agents with states in a multidimensional state space S which is assumed to be a non-empty convex subset of $\mathbb{R}^d$. Denote by $x^i(t)$ the state of agent $i \in N$ at time $t \in \mathbb{N} = \{0,1,2,\dots\}$. By the column vector $x(t) = (x^1(t),\dots,x^n(t))$, $t \in \mathbb{N}$ and $x(0) \in S^n$ a dynamical system in discrete time is defined on

S^n which we assume to be a **mean process** in the sense of Section 8.3 that is

$$x^i(t+1) \in \operatorname{conv}\{x^1(t), \ldots, x^n(t)\} \quad \text{for} \quad i \in N,\ t \in \mathbb{N} \tag{8.5.1}$$

Our focus will be on the **convergence to consensus**, that is

$$\lim_{t\to\infty} x^i(t) = c(x(0)) \quad \text{for all} \quad i \in N \tag{8.5.2}$$

or, equivalently, $\lim_{t\to\infty} x(t) \in \operatorname{diag} S^n$, where $\operatorname{diag} S^n$ is the diagonal of S^n defined by the points $\bar{c} = (c, \ldots, c)$ for $c \in S$.

As examples we considered in Section 8.2 opinion formation according to the matrix model $x(t+1) = A(t)$ and opinion formation under bounded confidence given by

$$x_i(t+1) = |I(i, x(t))|^{-1} \sum_{j \in I(i,x(t))} x_j(t), \quad i \in N,\ t \in \mathbb{N}$$

where $I(i,x) = \{j \in N \mid |x_i - x_j| \le \epsilon\}$ with $\epsilon > 0$. In both examples the state space is one-dimensional, $S = \mathbb{R}$ and $S = \mathbb{R}_+$, respectively, and the dynamics is given by a mean process with initial opinion profile $x(0)$.

The results obtained in the previous section we shall apply to multi-coordination in general. A special case will be opinion formation in a generalized setting.

Considering the interaction of agents we introduce the following concept of the strength or intensity of interaction based on the extraction of scrambling matrices with entries 0 and 1 only.

Definition 8.5.1. Let $\lambda(\cdot,\cdot)$ be the order function on the convex cone $\mathbb{R}_+^{n\times n}$ of nonnegative matrices and $\mathfrak{M}$ the subset of $\mathbb{R}_+^{n\times n}$ consisting of all scrambling matrices with entries 0 and 1. The scrambling **strength of A** $\in \mathbb{R}_+^{n\times n}$ is

$$\mu(A) = \max\{\lambda(M,A) \mid M \in \mathfrak{M}\} = \max\{\mu \in \mathbb{R}_+ \mid \mu M \le A, M \in \mathfrak{M}\}.$$

The following properties are useful when dealing with the strength of $A(t)$, which measures the strenght of interaction in multi-agent coordination.

Lemma 8.5.2. *Let $A = (a_{ij}) \in \mathbb{R}_+^{n\times n}$.*

(i)

$$\mu(A) = \min_{i,j\in N} \max_{k\in N} \min\{a_{ik}, a_{jk}\}. \tag{8.5.3}$$

(ii)

$$\mu(A) \le 1 - c(A) \le n\mu(A) \quad \textit{for } A \textit{ stochastic}. \tag{8.5.4}$$

(iii) *For $B \in \mathbb{R}_+^{n\times n}, r \in \mathbb{R}_+$*

$$\begin{aligned} &\text{(a)} \quad A \le B \textit{ implies } \mu(A) \le \mu(B)\\ &\text{(b)} \quad \mu(rA) = r\mu(A)\\ &\text{(c)} \quad \mu(A)\mu(B) \le \mu(AB). \end{aligned} \tag{8.5.5}$$

(iv) *Define*

$$\mu^+(A) = \min_{i,j\in N} \max_{k\in N}\{\min\{a_{ik}, a_{jk}\} \mid \min\{a_{ik}, a_{ij}\} > 0\} \tag{8.5.6}$$

(where $\mu^+(A) = 0$ *in case of* $\min\{a_{ik}, a_{jk}\} = 0$ *for all* i, j, k*).*
Then $\mu(A) \le \mu^+(A)$ *and equality holds if A is scrambling; if A is not scrambling* $\mu(A) < \mu^+(A)$ *is possible.*

Proof. (i) Let $\nu = \min_{i,j\in N} \max_{k\in N} \min\{a_{ik}, a_{jk}\}$. Suppose $\mu M \le A$ for $\mu \in \mathbb{R}_+, M \in \mathfrak{M}$. Since M is scrambling, for $i, j \in N$ there exists $k \in N$ such that $m_{ik} = m_{jk} = 1$. Therefore, $\mu \le \min\{a_{ik}, a_{jk}\}$ and, hence, $\mu \le \nu$ which proves $\mu(A) \le \nu$. For the converse define a matrix M with $m_{ij} = 1$ if $a_{ij} \ge \nu$ and $m_{ij} = 0$, otherwise. Obviously, $\nu M \le A$. M is scrambling since for i, j given, there exists k such that $\nu \le \min\{a_{ik}, a_{jk}\}$ and, hence, $m_{ik} = m_{jk} = 1$. Therefore, $\mu(A) \ge \nu$ which proves $\mu(A) = \nu$.

(ii) Using the definition of $c(A) = 1 - \min_{i,j\in N} \sum_{k\in N} \min\{a_{ik}, a_{jk}\}$ (see equation (8.1.1) in Theorem 8.1.2) one has from (i) that $\mu(A) \le \min_{i,j} \sum_k \min\{a_{ik}, a_{jk}\} = 1 - c(A)$.

Furthermore,

$$1 - c(A) = \min_{i,j} \sum_k \min\{a_{ik}, a_{jk}\} \le \min_{i,j}\{n \max_k \min\{a_{ik}, a_{jk}\}\}$$

and, by (i), $1 - c(A) \le n\mu(A)$.

(iii) The first two properties follow immediately from the definition of $\mu(A)$. For property (c) let $A, B \in \mathbb{R}_+^{n\times n}$ and $\mu M \le A, \nu P \le B$ with $\mu, \nu \in \mathbb{R}_+, M, P \in \mathfrak{M}$. It follows $\mu\nu MP \le AB$ where $\mu\nu \in \mathbb{R}_+$ and MP is a scrambling matrix. Define $C \in \mathbb{R}_+^{n\times n}$ with $c_{ij} = 1$ if $(MP)_{ij} > 0$ and $c_{ij} = 0$, otherwise. Since MP is scrambling C is scrambling, too. If $(MP)_{ij} > 0$ then $(MP)_{ij} \ge 1 = c_{ij}$ and if $(MP)_{ij} = 0$ then $c_{ij} = 0$ and $(MP)_{ij} \ge c_{ij}$. Therefore, $C \le MP$ and $\mu\nu C \le \mu\nu MP \le AB$. Therefore, $\mu(AB) \ge \mu\nu$ which proves $\mu(A)\mu(B) \le \mu(AB)$.

(iv) From the definition of $\mu^+(A)$ and (i) it follows $\mu(A) \le \mu^+(A)$ and $\mu(A) = \mu^+(A)$ if A is scrambling. For $\mu(A) < \mu^+(A)$ consider for example the stochastic matrix

$$A = \begin{bmatrix} 1 & 0 & 0 \\ \frac{1}{2} & \frac{1}{2} & 0 \\ 0 & \frac{1}{2} & \frac{1}{2} \end{bmatrix}.$$

A is not scrambling and $\mu(A) = 0$ whereas $\mu^+(A) = \frac{1}{2}$. □

From the previous section we obtain the following result on the convergence to a consensus for multi-agent coordination.

Theorem 8.5.3. *Consider multi-agent coordination among n agents given by a mean process* $(x(t))$ *on* S^n *and let* $x(t + 1) = A(t)x(t)$, $x(0) \in S^n$ *and* $A(t)$ *a stochastic matrix for* $t \in \mathbb{N}$*. It holds convergence to a consensus if there exists a set* $\mathfrak{I}$ *of disjoint intervals*

$I = [a, b] \subseteq \mathbb{N}$, $a \leq b$ such that for $B(I) = A(b) \dots A(a)$

$$\sum_{I \in \mathfrak{I}} \mu(B(I)) = \infty, \tag{8.5.7}$$

or, equivalently, for an infinite family $\mathfrak{I}^+$ in $\mathfrak{I}$ $B(I)$ is scrambling for all $I \in \mathfrak{I}^+$ and $\sum_{I \in \mathfrak{I}^+} \mu^+(B(I)) = \infty$.

Proof. The result will follow from Theorem 8.4.5 (i). Condition (8.4.4) of that Theorem is by Lemma 8.5.2 (ii) equivalent to $\sum_{I \in \mathfrak{I}} \mu(B(I)) = \infty$. Together with part (ii) of Theorem 8.4.2 this shows that convergence to consensus is guaranteed by condition (8.5.7).

Let $\mathfrak{I}^+ = \{I \in \mathfrak{I} \mid \mu(B(I)) > 0\}$. From Lemma 8.5.2 it follows that Condition (8.5.7) implies $B(I)$ to be scrambling for $I \in \mathfrak{I}^+$ and $\sum_{I \in \mathfrak{I}^+} \mu^+(B(I)) = \infty$. Obviously, the latter two properties for some $\mathfrak{I}^+$ in $\mathfrak{I}$ implies condition (8.5.7). □

As Theorem 8.5.3 exhibits the condition (8.5.7) is crucial for convergence to consensus. The condition means that in some sense the intensity of interaction should not be too weak and we will analyze the condition further in that direction. Before doing so, however, we shall in the following remarks mention and analyze some measures different from $\mu(\cdot)$ which were important in the history of inhomogeneous Markov chains. The connection to the latter is seldom reflected in the recent literature on multi-agent coordination.

Remarks 8.5.4. (1) For the asymptotic analysis of inhomogeneous Markov chains so called **coefficients of ergodicity** $\tau(\cdot)$ play an important role. Thereby $\tau(A)$ for a stochastic matrix A is just a number in $[0, 1]$ which depends continuously on A. It is proper if $\tau(A) = 0$ if and only if all rows of A are equal (see [93, Definition 4.6]).

Relevant examples are for $A = (a_{ij}) \in \mathbb{R}_+^{n \times n}$

$$\lambda(A) = \max_{j \in N} \min_{i \in N} a_{ij} \quad \text{and} \quad \delta(A) = \sum_{j \in N} \min_{i \in N} a_{ij}. \tag{8.5.8}$$

Also $\mu(A)$ just introduced and $c(A)$, considered earlier, are coefficients of ergodicity. From the definitions one obtains easily the following relationsships

$$\begin{aligned} &\lambda(A) \leq \delta(A) \leq 1 - c(A) \\ \text{and} \quad &\lambda(A) \leq \mu(A), \delta(A) \leq n\lambda(A). \end{aligned} \tag{8.5.9}$$

One does not have necessarily $\delta(A) \leq \mu(A)$ or $\mu(A) \leq r\delta(A)$ with $r \geq 0$ as can be seen from the examples

$$A = \begin{bmatrix} \frac{1}{2} & \frac{1}{2} \\ \frac{1}{2} & \frac{1}{2} \end{bmatrix} \quad \text{and} \quad A = \begin{bmatrix} 0 & \frac{1}{2} & \frac{1}{2} \\ \frac{1}{2} & 0 & \frac{1}{2} \\ \frac{1}{2} & \frac{1}{2} & 0 \end{bmatrix},$$

respectively. Since for the second example $\lambda(A) = 0$ and $\delta(A) = 0$ and $\mu(I) = 0$ for the identity matrix, the coefficients of ergodicity $\mu(\cdot), \lambda(\cdot), \delta(\cdot)$ are not proper. One verifies easily that $c(\cdot)$ is proper, another proper coefficient would be $1 - \delta(\cdot)$.

(2) According to E. Seneta ([93, p. 145], [91]) the coefficient $\lambda(A)$ is related to the work of A.A. Markov (and S.N. Bernstein), the coefficient $1-\delta(A)$ to the work of W. Doeblin and the coefficient $c(A)$ to that of R. L. Dobrushin. (In [91, p. 137] $\delta(A)$ is denoted by $\alpha(A)$, see also [93, p. 137]). Weak ergodicity (in the sense of Kolmogorov, see Section 7) holds for a sequence of stochastic matrices (P_k) if $\sum_{k=1}^{\infty}\lambda(P_k)=\infty$, a result which is known also as **Markov's Theorem.** (This result is with respect to forward products but applies also to backward products, cf. [91, pp. 153].)

The little known work of W. Doeblin is addressed in [91] and related to the work of J. Hajnal. According to E. Seneta, W. Doeblin arrived at the conclusion, **Doeblin's assertion,** that weak ergodicity holds for (P_k) if and only if there exists a strictly increasing sequence $(i_j), j=1,2,\ldots$ in $\mathbb{N}$ such that $\sum_{j=1}^{\infty}\delta(T(I_j))=\infty$ where $I_j=[i_j+1, i_{j+1}]$ and $T(I)=P_b\ldots P_a$ for $I=[a,b], a\le b$.

(3) The results of 2. can be obtained from Theorem 8.4.5 (i) by using the inequalities (8.5.9). (For simplicity we consider backward products.) For any collection $\mathfrak{I}$ of disjunct intervals $I=[a,b]$ in $\mathbb{N}$ and $B(I)=P_b\ldots P_a$ we have that $\sum_{I\in\mathfrak{I}}\lambda(B(I))=\infty$ is equivalent to $\sum_{I\in\mathfrak{I}}\delta(B(I))=\infty$.

Because of $\lambda(A)\le 1-c(A)$ from $\sum_{I\in\mathfrak{I}}\lambda(B(I))=\infty$ it follows that $\sum_{I\in\mathfrak{I}}(1-c(B(I)))=\infty$. The converse, however, is not true as can be seen from

$$P_k=\begin{bmatrix}0 & \frac{1}{2} & \frac{1}{2}\\ \frac{1}{2} & 0 & \frac{1}{2}\\ \frac{1}{2} & \frac{1}{2} & 0\end{bmatrix}\quad\text{for all } k.$$

To see this, let $\mathfrak{I}$ be the collection of intervals $I=[a,a]$ for $a\in\mathbb{N}$. Obviously, $c(P_k)=\frac{1}{2}$ and $\sum_{I\in\mathfrak{I}}(1-c(B(I)))=\infty$. On the other hand, $\lambda(B(I))=\lambda(P_k)=0$ and $\sum_{I\in\mathfrak{I}}\lambda(B(I))=0$. By selecting a different collection $\tilde{\mathfrak{I}}$, however, from $\sum_{I\in\mathfrak{I}}(1-c(B(I)))=\infty$ it follows $\sum_{I\in\tilde{\mathfrak{I}}}\lambda(B(I))=\infty$. For, from $\sum_{I\in\mathfrak{I}}(1-c(B(I)))=\infty$ we obtain, using Theorems 8.4.5 and 8.4.2, that $\lim_k P_{k+l}\ldots P_l=Q_l$ for all l and Q_l a stochastic matrix with equal rows. Therefore, $\lim_k \delta(P_{k+l}\ldots P_l)=\delta(Q_l)=1$ for all l. As in the proof of Theorem 8.4.5 (i) it follows the existence of a collection $\tilde{\mathfrak{I}}=\{I_k\}$ such that $\sum_{I\in\tilde{\mathfrak{I}}}\delta(B(I))=\infty$ and, hence, $\sum_{I\in\tilde{\mathfrak{I}}}\lambda(B(I))=\infty$. We arrive at the conclusion that $\lim_k P_{k+l}\ldots P_l$ is for each l a stochastic matrix with equal rows if and only if for some collection $\mathfrak{I}$ one has $\sum_{I\in\mathfrak{I}}\lambda(B(I))=\infty$ or, equivalently, $\sum_{I\in\mathfrak{I}}\delta(B(I))=\infty$. Notice, however, concerning Markov's Theorem as well as Doeblin's assertion, $\sum_{k=1}^{\infty}\lambda(P_k)=\infty$, or $\sum_{k=1}^{\infty}\delta(P_k)=\infty$, is a sufficient but not a necessary condition. As the above reasoning shows, to obtain a sufficient and necessary condition the sum has to be taken over $\lambda(B(I))$, or $\delta(B(I))$, that is by lumping the P_k together.

(4) Considering the ergodic coefficients $\lambda(A), \delta(A), \mu(A)$ it is a particular feature of the scrambling strength $\mu(A)$ that it dominates the other two, that is $\lambda(A), \delta(A) \leq r\mu(A)$ for some $r > 0$ and all stochastic matrices A. Neither $\lambda(A)$ nor $\delta(A)$ dominates the other two. As a consequence whenever for a collection $\mathfrak{J}$ the criterion $\sum_{I\in\mathfrak{J}} p(B(I)) = \infty$ is for $p = \lambda$ or $p = \delta$ conclusive for convergence to consensus the same is true for $p = \mu$. The converse, however, does not hold in general, that is it may happen that $\sum_{I\in\mathfrak{J}} p(B(I)) = \infty$ for $p = \mu$ but neither for $p = \lambda$ nor $p = \delta$. An example is given by

$$P_k = P = \begin{bmatrix} 0 & \frac{1}{2} & \frac{1}{2} \\ \frac{1}{2} & 0 & \frac{1}{2} \\ \frac{1}{2} & \frac{1}{2} & 0 \end{bmatrix}$$

for all k and $\mathfrak{J}$ the collection of intervals $I = [a, a]$, $a \in \mathbb{N}$. Of course, there must exist a collection $\tilde{\mathfrak{J}}$ such that $\sum_{I\in\tilde{\mathfrak{J}}} \lambda(B(I)) = \infty$ and $\sum_{I\in\tilde{\mathfrak{J}}} \delta(B(I)) = \infty$. In this simple case one has $\lambda(P^2) = \frac{1}{4}$ and $\delta(P^2) = \frac{3}{4}$ and, hence, a possible choice for $\tilde{\mathfrak{J}}$ is the set of all intervals $[2m, 2m+1]$ for $m \in \mathbb{N}$.

(5) Another often used coefficient of ergodicity is $\min^+ A = \min_{i,j}\{a_{ij} \mid a_{ij} > 0\}$. Obviously, for A scrambling $\min^+ A \leq \mu^+(A)$ and by Theorem 8.5.3 convergence to consensus holds if for some collection $\mathfrak{J}$ one has that $(B(I))$ is scrambling and $\sum_{I\in\mathfrak{J}} \min^+ B(I) = \infty$ (see also part (ii) of Theorem 8.4.5). It may happen, however, that all $B(I)$ are scrambling but $\sum_{I\in\mathfrak{J}} \min^+ B(I) < \infty$ and convergence to consensus does hold, nevertheless. For example, consider the following slight variation of Examples 8.4.1 part B. Let

$$A(t) = \begin{bmatrix} 1 & 0 \\ 1 - \frac{1}{t^2} & \frac{1}{t^2} \end{bmatrix}$$

for $t \geq 2$ and $\mathfrak{J}$ the collection of $I = \{m\}$ for $m \geq 2$. Obviously, $B(I) = A(m)$ is scrambling, $\min^+ B(I) = \frac{1}{m^2}$ and $\sum_i I \in \mathfrak{J} \min^+ B(I) < \infty$. Considering scrambling strength, however, we have $\mu^+(B(I)) = \mu(B(I)) = 1 - \frac{1}{m^2}$ and $\sum_{I\in\mathfrak{J}} \mu^+(B(I)) = \sum_{m=2}^{\infty}(1 - \frac{1}{m^2}) = \infty$. Therefore, convergence to consensus holds by Theorem 8.5.3.

As mentioned already the crucial condition (8.5.7) in Theorem 8.5.3 for reaching a consensus may be interpreted as the intensity of interaction becoming not too weak. For one part, the requirement of scrambling matrices $B(I)$ may be seen as letting the structure of interaction of agents becoming not too loose in the course of time. And as the other part the condition $\sum_{I\in\mathfrak{J}} \mu^+(B(I)) = \infty$ requires the intensity or strength of interaction vanishing not too fast. In some models, for example in the one of bounded confidence, the latter requirement is fulfilled in that the intensity is positively bounded from below, in particular $\min^+ A(t) \geq \alpha > 0$ for all t. (See below for further examples of this kind.) In those cases the condition for reaching a consensus reduces to assume all matrices $B(I)$ for $I \in \mathfrak{J}$ to be scrambling. It is this requirement which we shall examine in more detail in the following. The multi-agent coordination we will describe by local interaction in terms of neighbors as introduced already in Section 8.3.

Let $(x(t))$ be a mean process on S^n given by $x(t+1) = A(t)x(t)$ with $x(0) \in S^n$ and stochastic matrices $A(t)$ for $t \in \mathbb{N}$. For $i \in N$ the **set of neighbors of *i* at *t*** is $N(i,t) = \{j \in N \mid a_{ij}(t) > 0\}$. For a sequence $\tau = (t_1, \ldots, t_r)$ in $\mathbb{N}$ and $B(\tau) = A(t_1) \ldots A(t_r)$ the **set of neighbors of *i* via τ** is $N(i,\tau) = \{j \in N \mid b_{ij}(\tau) > 0\}$. We say there exists a **chain of neighbors of agent *i* to agent *j* via $\tau = (t_1, \ldots, t_r)$** if $j \in N(i,\tau)$, or explicitly, if there exists a sequence of agents $(i_0, i_1, \ldots i_r)$ with $i_0 = i, i_r = j$ such that

$$a_{i_0 i_1}(t_1) > 0, a_{i_1 i_2}(t_2) > 0, \ldots, a_{i_{r-1}, i_r}(t_r) > 0.$$

In this language, for $I = [a,b]$, $\tau = (b, b-1, \ldots, a+1, a)$ the matrix product $B(I) = B(\tau) = A(b) \ldots A(a)$ is scrambling if and only if for any two agents $i, j \in N$ there exists a third one k such that chains of neighbors via τ exist from i to k and from j to k. (See also the chains of confidence in case of opinion dynamics under bounded confidence.) This concept of the principle of the third agent we generalize as follows.

Definition 8.5.5. A sequence $(A(t))$ of stochastic matrices satisfies the **principle of the third agent** or **printh** for short **on a sequence τ** in $\mathbb{N}$ if for any two $i, j \in N$ there exist finite subsequences $\tau(i,j)$ and $\sigma(i,j)$ of τ such that

$$N(i, \tau(i,j)) \cap N(j, \sigma(i,j)) \neq \emptyset.$$

In case of $\tau = (b, b-1, \ldots, a+1, a)$ for $a \leq b$ we will instead of printh on sequence τ also speak of printh on $[a,b]$.

Obviously, if $A(t_1) \ldots A(t_r)$ is scrambling then printh holds on $\tau = (t_1, \ldots, t_r)$, for this just take $\tau(i,j) = \sigma(i,j) = \tau$. The converse, however, is not true in general, as can be seen already in simple cases as the following one. Taking up Examples 8.4.10, case (b), let $A = \begin{bmatrix} 1&0&0 \\ 1&0&0 \\ 0&1&0 \end{bmatrix}$, $B = \begin{bmatrix} 0&1&0 \\ 0&0&1 \\ 0&0&1 \end{bmatrix}$ and define a sequence $(A(t))$ by $A(t) = A$ if t is odd and $A(t) = B$ if t is even. This sequence satisfies printh on $\tau = (0,1,2)$ by chosing $\tau(i,j) = \sigma(i,j) = (0,2)$ for any $1 \leq i,j \leq 3$ because $B^2 = \begin{bmatrix} 0&0&1 \\ 0&0&1 \\ 0&0&1 \end{bmatrix}$ is scrambling. The product $A(0)A(1)A(2) = BAB = \begin{bmatrix} 0&1&0 \\ 0&0&1 \\ 0&0&1 \end{bmatrix}$, however, is not scrambling.

In a similar manner one obtains that the sequence $(A(t))$ just defined satisfies printh on all sequences $(t, t+1, t+2)$ for $t \in \mathbb{N}$. Though this seems to be a structure of interaction not too loose, we know from case (b) of Examples 8.4.10 that convergence to consensus does not hold. The following lemma shows that the principle of the third agent implies, other than in the example above, the scrambling property in case of matrices with a positive diagonal.

Lemma 8.5.6. (i) *Let $A(1), A(2), \ldots, A(p)$ stochastic $n \times n$-matrices with positive diagonal and let $\tau = (k_1, k_2, \ldots, k_q)$ a subsequence of $(1, 2, \ldots, p)$. Then for any $i, j \in N = \{1, \ldots, n\}$*

$$[A(k_1)A(k_2) \ldots A(k_q)]_{ij} > 0 \quad \textit{implies} \quad [A(1)A(2) \ldots A(p)]_{ij} > 0.$$

(ii) *Let $(A(t))$ a sequence of stochastic $n \times n$-matrices with positive diagonal and let $\tau = (t_1, \ldots, t_r)$ a sequence in $\mathbb{N}$.*

If printh holds on τ then the matrix $A(t_1)\dots A(t_r)$ is scrambling.

Proof. (i) Suppose $[A(k_1)\dots A(k_q)]_{ij} > 0$ for $i,j \in N$. Then there exists a sequence $(i_1,\dots,i_{q-1})$ such that $a_{ii_1}(k_1) > 0, a_{i_1,i_2}(k_2) > 0,\dots,a_{i_{q-1},j}(k_q) > 0$. We augment this sequence by terms $a_{rr}(t) > 0$ to obtain a sequence over $1,2,\dots,p$. Consider $1 \le l \le p$. If $l < k_1$ choose $a_{ii}(1) > 0, a_{ii}(2) > 0,\dots,a_{ii}(k_1-1)$. If $k_1 < l < k_2$ then augment by $a_{i_1i_1}(k_1+1), a_{i_1i_1}(k_1+2),\dots a_{i_1i_1}(k_2-1)$. Similar, for $k_s < l < k_{s+1}$, with $2 \le s \le q-1$. If $k_q < l$ then augment by $a_{jj}(k_q+1) > 0, a_{jj}(k_q+2) > 0,\dots,a_{jj}(p) > 0$. This gives a sequence $(j_1,\dots,j_{p-1})$ such that $a_{ij_1}(1) > 0, a_{j_1j_2}(2) > 0,\dots,a_{j_{p-1},j}(p) > 0$ and, hence, $[A(1)A(2)\dots A(p)]_{ij} > 0$.

(ii) If printh holds on τ then for i,j given there exist subsequences $\tau(i,j), \sigma(i,j)$ of τ and $k \in N$ such that $k \in N(i,\tau(i,j))$ and $k \in N(j,\sigma(i,j))$. If $\tau(i,j) = (k_1,\dots,k_q)$ then $[A(k_1)\dots A(k_q)]_{ik} > 0$ and part (i) implies $[A(t_1)A(t_2)\dots A(t_r)]_{ik} > 0$. Similarly, from $k \in N(j,\sigma(i,j))$ it follows that $[A(t_1)A(t_2)\dots A(t_r)]_{jk} > 0$. Thus, $A(t_1)A(t_2)\dots A(t_r)$ is scrambling. □

With the help of this lemma from Theorem 8.5.3 we obtain the following result.

Theorem 8.5.7. *Consider multi-agent coordination among n agents given by a mean process $(x(t))$ on S^n and let $x(t+1) = A(t)x(t), x(0) \in S^n$ and $A(t)$ a stochastic matrix for $t \in \mathbb{N}$. There holds convergence to consensus provided the following assumptions are met:*

(a) *$A(t)$ has a positive diagonal for $t \ge \underline{t}$ for some $\underline{t} \in \mathbb{N}$.*
(b) *There exists a sequence $t_1 < t_2 < t_3 < \cdots$ in $\mathbb{N}$ such that the principle of the third agent holds on $[t_k+1, t_{k+1}]$ for $k \ge \underline{k}$ for some $\underline{k} \in \mathbb{N}$.*
(c)
$$\sum_{k \ge \underline{k}} \beta_k^{(t_{k+1}-t_k)} = \infty, \tag{8.5.10}$$

where β_k is the smallest positive entry in all the matrices $A(t)$ for $t \in [t_k+1, t_{k+1}]$

Proof. Let $I_k = [t_k+1, t_{k+1}]$ for $k \ge \underline{k}$ such that $t_k \ge \underline{k}$. Obviously, the collection $\mathfrak{I}$ of the I_k consists of disjoint intervals in $\mathbb{N}$. From Lemma 8.5.6 (ii) we obtain that $B(I_k) = A(t_{k+1})\dots A(t_k+1)$ is scrambling for $I_k \in \mathfrak{I}$. Furthermore,

$$\mu^+(B(I_k)) \ge \min_{i,j}\{b_{ij}(I_k) \mid b_{ij}(I_k) > 0\} \ge \beta_k^{|I_k|},$$

and, since $|I_k| = t_{k+1} - t_k$, we obtain by assumption $\sum_{I_k \in \mathfrak{I}} \mu^+(B(I_k)) = \infty$. From Theorem 8.5.3 convergence to consensus does follow. □

In what follows we draw several interesting consequences from Theorem 8.5.7. The first one presents an extension of Theorem 8.2.3 on opinion dynamics under bounded confidence.

Corollary 8.5.8. *Let for a convex subset S of $\mathbb{R}^d$ a mean process on S^n given by*

$$x^i(t+1) = |I(i,x(t))|^{-1} \sum_{j\in I(i,x(t))} x^j(t) \tag{8.5.11}$$

for $i \in N = \{1,\dots,n\}, t \in \mathbb{N}, x(0) \in S^n$ and $I(i,x) \subseteq N$ for $x \in S^n$ such that $i \in I(i,x)$.

(i) *Convergence to consensus holds provided for any $i,j \in N$ there exists $k = k(i,j) \in N$ such that chains of confidence go from i to k and from j to k from s to $s + h(s)$ with $1 \le h(s) \le h \in \mathbb{N}$ for all $s \in \mathbb{N}$.*

(ii) *Assume in addition there exists $\delta > 0$ such that*

$$\|x^i - x^j\| \le \delta \quad \textit{implies} \quad j \in I(i,x) \tag{8.5.12}$$

for each $x \in S^n, i,j \in N$.

Then consensus will be reached in finite time, that is for some $T \in \mathbb{N}$ $x^i(t) = c$ for all $i \in N$ and $t \ge T$.

Proof. (i) The process defined by (8.5.11) we write as $x(t+1) = A(t)x(t)$ with $a_{ij}(t) = |I(i,x(t))|^{-1}$ for $j \in I(i,x(t))$ and $a_{ij}(t) = 0$, otherwise. Since $i \in I(i,x)$ each $A(t)$ has a positive diagonal. Let $t_l = (l-1)h$ for $l \ge 1$. A chain of confidence of agent i to agent j from period s to $t > s$ means a chain of neighbors of i to j via $\tau = (t-1,\dots,s)$, that is $j \in N(i,\tau)$. By assumption $k \in N(i,\tau) \cap N(j,\tau)$ for $\tau = (h(s)+s-1,\dots,s)$. For $s = t_l + 1$ the sequence τ is contained in $[t_l+1, t_{l+1}]$ and therefore printh holds on $[t_l+1, t_{l+1}]$ for each l. To apply Theorem 8.5.7 it remains to show that $\sum_{l\in\mathbb{N}} \beta_l^{(t_{l+1}-t_l)} = \infty$. This follows, since $t_{l+1} - t_l = h$ and $\beta_l \ge \frac{1}{n}$ for all l by the definitions of t_l and $a_{ij}(t)$, respectively. Thus, convergence to consensus follows from Theorem 8.5.7.

(ii) By part (i) for $\delta > 0$ according to (ii) there exists $T \in \mathbb{N}$ such that $\|x^i(t) - x^j(t)\| \le \delta$ for all $i,j \in N, t \ge T$. By assumption (8.5.12) $j \in I(i,x(t))$ for all $i,j \in N, t \ge T$ and, hence, $I(i,x(t)) = N$ for all $i \in N, t \ge T$. Therefore

$$x^i(t+1) = \frac{1}{n}\sum_{j=1}^{n} x^j(t) \quad \text{for all} \quad i \in N, \ t \ge T.$$

With $c = \frac{1}{n}\sum_{j=1}^{n} x^j(T)$ it follows by induction over s that $x^i(T+s) = c$ for all $i \in N, s \ge 0$. ☐

This corollary contains Theorem 8.2.3 for the special case $d = 1, I(i,x) = \{j \in N \mid \|x^i - x^j\| \le \epsilon\}$ and $h(s)$ constant for all s. In this case $I(i,x)$ is symmetric in that $j \in I(i,x)$ is equivalent to $i \in I(j,x)$. This symmetry is not required in Corollary 8.5.8 which therefore allows for **heterogeneous levels of confidence** that is $I(i,x) = \{j \in N \mid \|x^i - x^j\| \le \epsilon_i\}$ with different $\epsilon_i > 0$. In this more general case condition (8.5.12) is satisfied, too, by taking $\delta = \min_{i\in N} \epsilon_i$. (For a detailed analysis of this case, called "the heterogeneous HK model", see [79].)

In Corollary 8.5.8 the condition (8.5.10) of Theorem 8.5.7 is fulfilled in a simple matter in that $t_{k+1} - t_k$ is bounded from above by a constant and β_k is bounded from below by a positive constant. The same applies to the following consequences of Theorem 8.5.7 which were obtained in the literature by different proofs and which we collect in the following corollary. (Different from the other parts, part (i), however, allows for β_k tending to 0.)

Corollary 8.5.9. *For a sequence of stochastic matrices $A(t), t \in \mathbb{N}$, with positive diagonal for $t \geq \underline{t}, \underline{t} \in \mathbb{N}$, the mean process on S^n given by $x(t+1) = A(t)x(t), x(0) \in S^n$ converges to consensus in each of the following cases.*

(i) *There exists a sequence (r_k) in $\mathbb{N}$ with $1 \leq r_{k+1} - r_k \leq r$ for all k and some $p \in \mathbb{N}$ such that*
  - *$A(r_{k+p}) \dots A(r_{k+1})$ is scrambling for $k \geq \underline{k} \in \mathbb{N}$,* *and*
  - *$\sum_{k \geq \underline{k}} \beta_k^{pr} = \infty$ where β_k is the smallest positive entry in all the matrices $A(t)$ for $kpr + 1 \leq t \leq (k+1)pr$.*

(ii) *There exists $B \in \mathbb{N}$ and for each $t \in \mathbb{N}$ an agent $m(t)$ such that*
  - *for each i a chain of neighbors leads from i to $m(t)$ via a sequence in $[t, t+B]$,* *and*
  - *$\min^+ A(t) \geq \alpha > 0$ for all $t \in \mathbb{N}$.*

(iii) *Following an arbitrary time there is a chain of neighbors from any agent to any other via a sequence in $\mathbb{N}$ and*
  - *$\min^+ A(t) \geq \alpha > 0$ for all $t \in \mathbb{N}$,* *and*
  - *there exists $B \in \mathbb{N}$ such that for any two agents i, j with $a_{ij}(s) > 0$ for infinitely many $s \in \mathbb{N}$ and for any $t \in \mathbb{N}$ there exists $v = v(i,j,t)$, $0 \leq v \leq B$ with $a_{ij}(t+v) > 0$ (condition of "bounded intercommunication intervals").*

Proof. (i) Let $t_k = kpr$. Since $r_{i+1} - r_i \leq r$ there exist points $r_{q+p}, \dots, r_{q+1}$ in $[t_k, t_{k+1}]$. By assumption $A(r_{q+p}) \dots A(r_{q+1})$ is scrambling for $q \geq \underline{k}$ and printh holds on $[t_k + 1, t_{k+1}]$ for $k \geq \underline{k}$. Since $t_{k+1} - t_k = pr$ and $\sum_{k \geq \underline{k}} \beta_k^{pr} = \infty$ it follows $\sum_{k \geq \underline{k}} \beta_k^{t_{k+1} - t_k} = \infty$ and convergence to consensus follows from Theorem 8.5.7.

(ii) Let $t_k = k(B+1)$. The assumptions imply in particular that printh holds on $[t, t+B]$ for each $t \in \mathbb{N}$. Especially, for $t = t_k + 1$, printh holds on$[t_k + 1, t_k + 1 + B] = [t_k + 1, t_{k+1}]$. Since $t_{k+1} - t_k = B + 1$ and $\beta_k \geq \alpha$ for all k it follows $\sum_{k \in \mathbb{N}} \beta_k^{t_{k+1} - t_k} = \infty$. Thus, the assertion follows from Theorem 8.5.7.

(iii) Let E be the set of all pairs (i,j) for which $a_{ij}(s) > 0$ holds for infinitely many $s \in \mathbb{N}$. There exists $t' \in \mathbb{N}$ such that $a_{ij}(t) = 0$ for $i, j \notin E, t \geq t'$. By assumption for any two agents i and j there is a chain of neighbors $a_{ii_1}(s_1) > 0, \dots, a_{i_r j}(s_{r+1}) > 0$ for $s_i \geq t'$. Furthermore, by the condition of bounded intercommunication intervals, for each t exists $s_i' \in [t, t+B]$ such that $a_{ii_1}(s_1') > 0, \dots, a_{i_r j}(s_{r+1}') > 0$. Therefore, printh holds on $[t, t+B]$ for each t. In particular, for $t_k = k(B+1)$ printh holds on $[t_k + 1, t_{k+1}]$ for all k. Because of $t_{k+1} - t_k = B + 1$ and $\beta_k \geq \alpha$ for all t it follows $\sum_{k \in \mathbb{N}} \beta_k^{t_{k+1} - t_k} = \infty$ and the assertion follows from Theorem 8.5.7. □

In contrast to the assumptions made for the last two corollaries, Theorem 8.5.7 does allow also β_k to tend to 0 and $t_{k+1} - t_k$ to tend to infinity. Actually, what condition (8.5.10) in Theorem 8.5.7 requires is some interplay of the intensity of interaction, modeled by β_k, and the structure of interaction, modeled by chains of neighbors on $[t_k + 1, t_{k+1}]$. Condition (8.5.10) can be satisfied for an intensity decreasing to 0 as long as $t_{k+1} - t_k$ increases not too much, as for example in case of $t_{k+1} - t_k \leq c$. (As in case A of Examples 8.4.1 with $t_{k+1} - t_k = 1$.) Considering the structure of interaction, $t_{k+1} - t_k$ may tend to infinity for β_k decreasing not too fast, in particular for $\beta_k \geq \alpha > 0$ for all k. Moreover, condition (8.5.10) allows β_k tending to 0 – if this is not too fast – and, $t_{k+1} - t_k$ tending to infinity – if this is slow enough (see Example 8.5.14 below and Exercise 15). Roughly speaking, condition (8.5.10) holds if the intensity of interaction does not decrease too fast and the structure of interaction does not become too loose.

In what follows we derive two more results from Theorem 8.5.7 which allow the intensity of interaction to decrease to zero. Thereby, we make assumptions on the saturated subsets in the sense of Definition 8.1.6 with respect to the matrices $A(t)$. For the first result we assume that each matrix $A(t)$ is coherent, that is any two saturated subsets for $A(t)$ have a non-empty intersection.

Corollary 8.5.10. *For a sequence of stochastic matrices $A(t), t \in \mathbb{N}$, which are coherent and possess a positive diagonal for $t \geq \underline{t}, \underline{t} \in \mathbb{N}$ the mean process on S^n given by $x(t+1) = A(t)x(t), x(0) \in S^n$ does converge to consensus, provided for some $k_0 \in \mathbb{N}$*

$$\sum_{k \geq k_0} \beta_k^{n-1} = \infty,$$

where β_k is the smallest positive entry in all the matrices $A(t)$ for $k(n-1) \leq t \leq (k+1)(n-1)$.

Proof. Let $t_k = k(n-1), k \in \mathbb{N}$. By Proposition 8.1.7 (iii) for $t \geq \underline{t}$ each matrix $A(t)$ is a Sarymsakov matrix. Lemma 8.4.8 yields that any product of $n-1$ matrices $A(t), t \geq \underline{t}$, is scrambling. Therefore, $A(t_{k+1}) \dots A(t_k + 1)$ is scrambling for $k \geq \underline{k}$ with $\underline{k}(n-1) \geq \underline{t}$ and printh holds on $[t_k + 1, t_{k+1}]$ for $k \geq \underline{k}$. By the assumptions made on the β_k convergence to consensus follows from Theorem 8.5.7. □

The earlier Examples 8.4.1 illustrate this result for $n = 2$.

In both cases A and B all matrices $A(t), t \geq 2$, are coherent with positive diagonal. In case A one has $\beta_k = \frac{1}{(k+1)(n-1)}$ and, hence, the condition on the β_k in the corollary is satisfied. In case B one has $\beta_k = [\frac{1}{(k+1)(n-1)}]^2$ and the condition on the β_k is not satisfied – in accordance with the fact that convergence to consensus does not hold.

For the second result we employ a new condition introduced by J.M. Hendrickx and J. N. Tsitsiklis in [49].

Definition 8.5.11. A stochastic matrix $n \times n$-matrix A is **cut-balanced** if for any subset $\emptyset \neq M \subsetneq N = \{1, \dots, n\}$ there exist $i \in M, j \notin M$ with $a_{ij} > 0$ if any only if there exist $i' \in M, j' \notin M$ with $a_{j'i'} > 0$.

This condition means that a group M of agents being influenced by the other ones does also influence the other ones. Obviously, a type-symmetric matrix A, that is $a_{ij} > 0$ is equivalent to $a_{ji} > 0$, and, in particular, a symmetric matrix is cut-balanced. More general, A is cut-balanced if there exists a strictly positive vector w such that $wA = wA'$ (A' being the transpose of A), which includes double stochastic matrices (see [49, Proposition 1]).

We need the following lemma which in part (i) describes cut-balance in terms of saturated sets and which in part (ii) yields that for cut-balanced matrices the $\min^+$ of arbitrary long products can be bounded from below by finitely many products.

Lemma 8.5.12. (i) *A stochastic matrix A is cut-balanced if and only if A and its transpose A' have the same saturated sets, or, equivalently, for each saturated set of A, the complement (if $\neq \emptyset$) is saturated, too.*

(ii) *Let $A(1), \dots, A(p)$ stochastic $n \times n$ matrices with positive diagonal which are cut-balanced. Then there exist $1 \le k_i \le p, 1 \le i \le q$, such that*

$$min^+(A(p)\dots A(1)) \ge min^+A(k_q)\dots min^+A(k_1), \tag{8.5.13}$$

with $q \le n^2 - n + 1$.

Proof. (i) By Definition 8.1.6 a set $\emptyset \neq M \subsetneq N$ is saturated for A if $i \in M$ and $a_{ij} > 0$ implies $j \in M$. Therefore, M is not saturated precisely if $a_{ij} > 0$ for some $i \in M$ and $j \notin M$. It follows that the cut-balance condition is equivalent to the condition that M is not saturated for A if and only if M is not saturated for A'. Furthermore, M is saturated for A precisely if $N \setminus M$ is saturated for A'. This shows part (i).

(ii) Let for a stochastic matrix $P(A) = \{(i,j) \in N \times N \mid a_{ij} > 0\}$. For $A(k,1) = A(k)\dots A(1)$ Lemma 8.5.6 implies $P(A(k,1)) \subseteq P(A(k+1,1))$. Let

$$q(k) = |\{1 \le j \le k | P(A(j,1)) \subsetneq P(A(j+1,1))\}| \quad \text{for} \quad 1 \le k \le p-1.$$

Since $A(1,1)$ has at most $n^2 - n$ zeros and for each j in the definition of $q(k)$ at least one zero of $A(j,1)$ turns into a positive entry of $A(j+1,1)$ we must have $q(k) \le n^2 - n$ for each k. Now we show by induction over p the inequality (8.5.13) holds for $q = q(p) + 1$. The assertion is trivial for $p = 1$. For the step from p to $p + 1$ we distinguish two cases.

First case: Assume $P(A(p,1)) \subsetneq P(A(p+1,1))$.

Obviously, $\min^+(A(p+1,1)) \ge \min^+ A(p+1) \cdot \min^+ A(p,1)$ and $q(p+1) = q(p) + 1$. By induction hypothesis we have $1 \le k_i \le p, 1 \le i \le q = q(p+1)$ such that $\min^+ A(p,1) \ge \min^+ A(k_q)\dots\min^+ A(k_1)$. Putting $k_{q+1} = p + 1$ we obtain inequality (8.5.13) for $p + 1$ with $q + 1 = q(p) + 1 + 1 = q(p+1) + 1 \le n^2 - n + 1$.

Second case: Assume $P(A(p,1)) = P(A(p+1),1))$. Let $A = A(p+1), B = A(p,1)$. Then $P(AB) = P(B)$. We shall show that

$$\min^+(AB) \ge \min^+B, \tag{$*$}$$

which by induction hypothesis proves inequality (8.5.13) for $p + 1$. Up to now we have not yet used the cut-balanced condition which will be done now to prove $(*)$.

Suppose $(AB)_{ik} > 0$ for some $(i,k) \in N \times N$. From $P(AB) = P(B)$ we get $b_{ik} > 0$. If $M = \{j \in N \mid b_{jk} = 0\}$ then

$$(AB)_{ik} = \sum_{j \notin M} a_{ij} b_{jk} \geq \min{}^{+} B \sum_{j \notin M} a_{ij}.$$

We show that $\sum_{j \notin M} a_{ij} = 1$ for $i \notin M$ which will prove $(*)$. Because of $b_{ik} > 0$ we have $M \neq N$ and, without loss, $M \neq \emptyset$. M is saturated for A because from $a_{jh} > 0$ for $j \in M$ we have $b_{jk} = 0$ and, hence, $(AB)_{jk} = 0$ which implies $b_{hk} = 0$, that is $h \in M$. For A cut-balanced from part (i) we obtain that M is saturated for A', too. Therefore, $j \in M$and $a_{ij} > 0$ imply $i \in M$, that is $a_{ij} = 0$ for $i \notin M, j \in M$.Thus, $\sum_{j \in M} a_{ij} = 0$ for $i \notin M$, and, hence, $\sum_{j \notin M} a_{ij} = 1$ for $i \notin M$. This finishes the proof of the lemma. □

With the help of this lemma, from Theorem 8.5.7 we obtain the following corollary.

Corollary 8.5.13. *For a sequence of stochastic matrices $A(t), t \in \mathbb{N}$, which are cut-balanced and possess a positive diagonal for $t \geq \underline{t}, \underline{t} \in \mathbb{N}$ the mean process on S^n given by $x(t+1) = A(t)x(t), x(0) \in S^n$ does converge to consensus, provided the following conditions are met,*

- *there exists a sequence $t_1 < t_2 < \cdots$ in$\mathbb{N}$ such that the principle of the third agent holds on $[t_k + 1, t_{k+1}]$ for $k \geq k_0 \in \mathbb{N}$,*
- *it holds*

$$\sum_{k \geq k_0} \beta_k^{n(n-1)+1} = \infty,$$

where β_k is the smallest positive entry in all the matrices $A(t)$ for $t_k + 1 \leq t \leq t_{k+1}$.

Proof. Similarly as in the proof of Theorem 8.5.7 the assertion follows from Theorem 8.5.3. Let $I_k = [t_k + 1, t_{k+1}]$ and $B(I_k) = A(t_{k+1}) \ldots A(t_k + 1)$ for $k \geq k_0$ and $t_k \geq \underline{t}$. $B(I_k)$ is scrambling by Lemma 8.5.6 (ii) and from Lemma 8.5.12 (ii) we have that

$$\min{}^{+} B(I_k) \geq \min{}^{+} A(k_q) \ldots \min{}^{+} A(k_1),$$

with $t_k + 1 \leq k_i \leq t_{k+1}, 1 \leq i \leq q \leq n(n-1) + 1$. Therefore, $\min^{+} B(I_k) \geq \beta_k^{n(n-1)+1}$ and, hence, $\sum_{k \geq \underline{k}} \min^{+} B(I_k) = \infty$ with $\underline{k} \geq k_0, t_{\underline{k}} \geq \underline{t}$. Since for any stochastic matrix $\mu^{+}(A) \geq \min^{+} A$ all assumptions of Theorem 8.5.3 are satisfied. □

The printh assumptions in the above corollary cannot simply be omitted as the case, where all $A(t)$ are equal to the identity matrix, shows for which all assumptions of Corollary 8.5.13 with the exception of the printh assumption are satisfied. The conditions of coherence and cut-balance assumed in Corollaries 8.5.10 and 8.5.13 are in some sense opposite to each other. Whereas coherence requires any two saturated sets to have a non-empty intersection, cut-balance requires saturated sets to have a saturated complement. Of course, both assumptions may hold together which, however, happens precisely if the respective matrix has N as the only saturated set.

The following example illustrates Corollary 8.5.13 and presents the case already mentioned of convergence to consensus though $\min^{+} A(t)$ decreases to 0 and $t_{k+1} - t_k$ increases to ∞.

Example 8.5.14. Let $t_1 < t_2 < t_3 < \cdots$ an arbitrary sequence in $\mathbb{N}$ and define for $n = 3$

$$A(t) = \begin{bmatrix} \alpha(t) & 1-\alpha(t) & 0 \\ 0 & \alpha(t) & 1-\alpha(t) \\ 1-\alpha(t) & 0 & \alpha(t) \end{bmatrix} \quad \text{for } t = t_k$$

and

$$A(t) = \begin{bmatrix} \alpha(t) & 1-\alpha(t) & 0 \\ \alpha(t) & 1-\alpha(t) & 0 \\ 0 & 0 & 1 \end{bmatrix} \quad \text{for } t \neq t_k$$

with $0 < \alpha(t) < 1$ for all $t \in \mathbb{N}$.

$A(t)$ has a positive diagonal for $t \in \mathbb{N}$. For $t = t_k$ the matrix $A(t)$ is cut-balanced because it is double stochastic. ($A(t)$ is neither symmetric nor type-symmetric.) For $t \neq t_k$ the matrix $A(t)$ is cut-balanced, too, though not necessarily double stochastic, since the saturated sets are (beside $N = \{1, 2, 3\}$) the sets $\{1, 2\}$ and $\{3\}$ which are complements of each other. Since $A(t)$ is for $t = t_k$ scrambling, printh holds on $[t_k + 1, t_{k+1}]$. Printh does, however, not hold on proper sub-intervals.

Suppose now $t_k = k^2$ and $\alpha(t) = t^{-\frac{1}{14}}$ for $t \geq 1$. Obviously, $t_{k+1} - t_k = 2k + 1$ increases to ∞ for increasing k. For β_k we have that $\beta_k \geq (t_{k+1})^{-\frac{1}{14}} = (k+1)^{-\frac{1}{7}}$ and since $n(n-1) + 1 = 7$ for $n = 3$ we obtain $\sum_k \beta_k^{n(n-1)+1} \geq \sum_k \frac{1}{k+1} = \infty$.

All assumptions of Corollary 8.5.12 being satisfied convergence to consensus does hold. Thus, in this case consensus is approached though intensity $\min^+ A(t)$ of interaction tends to 0 and the structure of interaction weakens steadily since $t_{k+1} - t_k$ tends to infinity.

In the results so far we addressed the question of convergence to consensus. Now we turn to the more general question if by relaxing conditions convergence can still be obtained, in particular with a consensus on certain subgroups. Actually, considering opinion dynamics under bounded confidence this is what Theorem 8.2.5 states in this case without any further assumptions. For the more general model of multi-agent coordination we need some assumptions and additional concepts. The concepts of saturating sets and coherence introduced in Section 8.1 for a single matrix we extend to sequences of matrices as follows.

Definition 8.5.15. For a sequence $(A(t))$ of stochastic $n \times n$-matrices a non-empty subset M of $N = \{1, \ldots, n\}$ is called **saturated** if for any $i \in M$ and $j \in N$ such that for each $t \in \mathbb{N}$ exists $t' \geq t$ with $a_{ij}(t') > 0$ it follows that $j \in M$. The sequence $(A(t))$ is **coherent** if any two saturated sets for $(A(t))$ have a non-empty intersection.

The following lemma helps in dealing with these concepts.

Lemma 8.5.16. *Let $(A(t))$ be sequence of stochastic matrices.*

(i) *There exists $t^* \in \mathbb{N}$ such that M is saturated for the sequence $(A(t))$ if and only if M is saturated for each $A(t), t \geq t^*$, or equivalently $a_{ij}(t) = 0$ for all $i \in M, j \notin M, t \geq t^*$.*

(ii) *Assume all* $A(t)$ *possess a positive diagonal for* $t \geq \underline{t}, \underline{t} \in \mathbb{N}$. *Given* $i \in N$ *there exists a smallest for* $(A(t))$ *saturated set* $M(i)$ *containing* i *and*

$$M(i) = \bigcap_{t \geq 0} \bigcup_{t \prec \tau} N(i, \tau), \tag{8.5.14}$$

where for $\tau = (t_1, \ldots, t_r)$ $t \prec \tau$ *means that* $t \leq t_j$, $1 \leq j \leq r$.

Proof. (i) Let M be saturated for $(A(t))$. If $i \in M, j \notin M$ then by definition there exists $t_*(i,j)$ such that $a_{ij}(t) = 0$ for $t \geq t_*(i,j)$. For $t_*(M) = \max_{i,j} t_*(i,j)$ and $t^* = \max\{t_*(M) \mid M$ saturated for $(A(t))\}$ it follows $a_{ij}(t) = 0$ for $i \in M, j \notin M, t \geq t^*$. Conversely, let the latter hold for some $t^* \in \mathbb{N}$. Assume $i \in M, j \in N$ such that for each t exists $t' \geq t$ with $a_{ij}(t') > 0$. By assumption for $t = t^*$ we cannot have $j \notin M$, that is $j \in M$. Furthermore, for a matrix A a set $M \neq \emptyset$ is saturated precisely if $a_{ij} = 0$ for $i \in M, j \notin M$.

(ii) Let M saturated for $(A(t))$ with $i \in M$. By definition for t given exists $t' \geq t$ such that $N(i, t') \subseteq M$. By iteration for t exists $t \prec \tau$ such that $N(i, \tau) \subseteq M$. Therefore, for $M(i)$ given by equation (8.5.14) it follows $M(i) \subseteq M$. It remains to show that $M(i)$ is saturated for $(A(t))$ and contains i. Since $A(t)$ has a positive diagonal we must have $i \in M(i)$. Furthermore, let $j \in M(i)$ and $k \in N$ such that for each t exists $t' \geq t$ with $a_{jk}(t') > 0$. For t given $j \in N(i, \tau)$ for some $t \prec \tau$. Since $a_{jk}(t') > 0$ for some $t' \geq t$ it follows for the sequence $\tau' = (\tau, t')$ that $k \in N(i, \tau')$ with $t \prec \tau'$. This shows that $k \in M(i)$ and, hence, $M(i)$ is saturated for $(A(t))$. □

By using this lemma we prove the following result.

Theorem 8.5.17. *Let* $(A(t))$ *be a sequence of stochastic matrices which are cut-balanced, possess a positive diagonal for* $t \geq \underline{t}, \underline{t} \in \mathbb{N}$ *with* $\min^+ A(t) \geq \alpha > 0$ *for all* $t \geq \underline{t}$. *Then for the mean process on* S^n *given by* $x(t+1) = A(t)x(t), x(0) \in S^n$ *there exists a decomposition*

$$N = M_1 \,\dot{\cup} \cdots \dot{\cup}\, M_r \tag{8.5.15}$$

of N *into disjoint non-empty subsets* M_i*such that*

$$\lim_{t \to \infty} x^j(t) = c_i(x(0)) \quad \text{for all} \quad j \in M_i, \ 1 \leq i \leq r \tag{8.5.16}$$

Proof. (i) In a first step we derive the decomposition (8.5.15). Let $\emptyset \neq M \subsetneq N$ be saturated for $(A(t))$. Since all $A(t)$ are cut-balanced for $t \geq \underline{t}$ it follows from Lemma 8.5.12 (i) and Lemma 8.5.16 that $N \setminus M$ is saturated for $(A(t)$ and $t \geq \max\{\underline{t}, t^*\}$. From this we obtain for $i, j \in N$ and $M(i), M(j)$ as in Lemma 8.5.16 (ii)

$$i \in M(j) \quad \text{implies} \quad j \in M(i) \tag{$*$}$$

for $i \in M(j)$ and $j \notin M(i)$ imply $M(j) \subseteq N \setminus M(i)$ since $N \setminus M(i)$ is saturated for $(A(t))$ which, however, contradicts $i \in M(j)$.

Now, suppose $M(i) \cap M(j) \neq \emptyset$ for $i, j \in N$. For $k \in M(i), k \in M(j)$ from $(*)$ we obtain $i \in M(k), j \in M(k)$ and, hence, $M(i) = M(k) = M(j)$. Let $M_1, \ldots, M_r$ the pairwise different sets among the sets $M(i), i \in N$. From what we have just shown, we must have $M_p \cap M_q = \emptyset$ for $p \neq q$ and since N is the union of the $M(i)$ we arrive at the decomposition $N = M_1 \dot{\cup} \ldots \dot{\cup} M_r$ as wanted.

(ii) Let for $i \in N$ fixed $M = M(i), \bar{t} = \max\{\underline{t}, t^*\}$ and $A_M(t)$ the matrix with entries $a_{ij}(t)$ for $i,j \in M$. Since M is saturated by Lemma 8.5.16 (i) we have that $a_{ij}(t) = 0$ for $i \in M, j \notin M, t \geq \bar{t}$. Therefore, $\sum_{j \in M} a_{ij}(t) = 1$ for $i \in M, t \geq \bar{t}$ and the matrices $A_M(t)$ are stochastic for $t \geq \bar{t}$. Furthermore, for $i \in M, t \geq \bar{t}$

$$x^i(t+1) = \sum_{j=1}^{n} a_{ij}(t)x^j(t) = \sum_{j \in M} a_{ij}(t)x^j(t).$$

Thus, the sequence $(A_M(t))$ of stochastic matrices defines a mean process on $S^m, m = |M|$, by

$$y(t+1) = A_M(t)y(t) \quad \text{with} \quad y(t) = (y_i(t)) \in S^m \tag{8.5.17}$$

for $t \geq \bar{t}, y(\bar{t}) \in S^m$.

We shall apply to this mean process Corollary 8.5.13. Concerning the assumptions $A_M(t)$ with $t \geq \bar{t}$ obviously has a positive diagonal. Also, $A_M(t)$ is cut-balanced for $t \geq \bar{t}$ which can be seen as follows. Let $\emptyset \neq U \subsetneq M$ be saturated for matrix $A_M(t), t \geq \bar{t}$ that is $a_{ij}(t) = 0$ for $i \in U, j \in M \setminus U$. Since $A(t)$ is cut-balanced for $t \geq \bar{t}$ and $a_{ij}(t) = 0$ for $i \in M, j \in N \setminus M$ we obtain $a_{ij}(t) = 0$ for $i \in U, j \in N \setminus U$ and, hence, $a_{ij}(t) = 0$ for $i \in N \setminus U, j \in U$. Thus $A_M(t)$ is cut-balanced for $t \geq \bar{t}$.

(iii) Next we show that for the mean process given by (8.5.17) printh holds on $[t_k + 1, t_{k+1}]$ for some sequence (t_k) in $\mathbb{N}$. From Lemma 8.5.16 (ii) we have that, $i \in N$ fixed, $M = M(i) = \bigcap_{t \geq 0} \bigcup_{t \prec \tau} N(i, \tau)$, where neighborhoods $N(i,\tau)$ are defined for $(A(t))$. Therefore, for $j \in M$ and $t \geq 0$ given there exists $t \prec \tau(t,j)$ such that $j \in N(i, \tau(t,j))$. Since N is finite there exists $m(t) \geq t$ such that $\tau(t,j) \subseteq [t, m(t)]$ for all $j \in N, t \geq 0$.

Let $t_{k+1} = m(t_k + 1) + 1$ for $k \geq 1, t_1 = 0$. Obviously, $t_k < t_{k+1}, [t_k + 1, m(t_k + 1)] \subseteq [t_k + 1, t_{k+1}]$. This shows that for $(A_M(t)), t \geq \bar{t}$, printh holds on $[t_k + 1, t_{k+1}]$.

(iv) Since by assumption $\min^+ A(t) \geq \alpha > 0$ for all $t \geq \bar{t}$ also the last assumption of Corollary 8.5.13 is satisfied for the mean process given by (8.5.16). Thus we obtain $\lim_{t\to\infty} x^j(t) = \lim_{t\to\infty} y^j(t) = c$ for all $j \in M$. If applied to each $M = M(i) = M_k$ we obtain assertion (8.5.16) which proves the theorem. □

The following lemma gives a description of the cut-balanced property which enables an interpretation of Theorem 8.5.17 and of its consequences.

Lemma 8.5.18. *Let A be a stochastic $n \times n$-matrix and $s(M) = \{j \in N \mid a_{ij} > 0$ for some $i \in M\}$ for $\emptyset \neq M \subseteq N$. A is cut-balanced if and only if for any $i,j \in N$ and iterations s^k of s*

$$i \in s^k(j) \quad \textit{for some} \quad k \in \mathbb{N} \quad \textit{implies} \quad j \in s^l(i) \quad \textit{for some} \quad l \in N \tag{8.5.18}$$

Proof. The set $M(j) = \bigcup_{k \geq 0} s^k(j)$ is the smallest saturated set (for A) which contains i (cf. Definition 8.1.6).

(1) Suppose A is cut-balanced and let $i \in s^k(j)$. If $j \in N \setminus M(i)$ then $M(j) \subseteq N \setminus M(i)$ by Lemma8.5.12 (i). This is a contradiction to $i \in M(j)$, and hence, $j \in M(i)$, that is $j \in s^l(i)$ for some l.

(2) Suppose condition 8.5.18 and let $M \subsetneq N$ be saturated. If $i \in s(N \setminus M)$ then $i \in s(j)$ for some $j \in N \setminus M$ and by (8.5.18) $j \in s^l(i)$ for some l. Since $j \in N \setminus M$ we must have $i \in N \setminus M$. This shows $s(N \setminus M) \subseteq N \setminus M$ and $N \setminus M$ is saturated. Lemma 8.5.12 (i) yields that A is cut-balanced. □

The above descritpion of cut-balanced suggests the following definition.

Definition 8.5.19. A mean process $x(t+1) = A(t)x(t)$ on S^n is **reciprocal at $t \in \mathbb{N}$** if the stochastic matrices $A(\cdot)$ satisfy the following conditions

- $i \in N(i, s)$ for all $i \in N$, all s,

  and
- $i \in N(j, \tau)$ for some τ implies $j \in N(i, \sigma)$ for some σ,

where τ, σ are finite sequences in N consisting of t only.

In the special case of $\tau = \sigma = (t)$, that is $i \in N(j, t)$ implies $j \in N(i, t)$ the process is said to be **mutual at t**.

From Theorem 8.5.17 and Lemma 8.5.18 we obtain the following result.

Theorem 8.5.20. *Let $x(t+1) = A(t)x(t), x(0) \in S^n$ be a mean process which is reciprocal at each $t \geq \underline{t} \in \mathbb{N}$ and such that* $\min^+ A(t) \geq \alpha > 0$ *for $t \geq \underline{t}$.*

(i) *The formulas* (8.5.15) *and* (8.5.16) *of Theorem 8.5.17 do hold.*

(ii) *Convergence to consensus (on N) does hold for each $x(0) \in S^n$ where $S = \mathbb{R}^d$ if and only if for all $i, j \in N, t \in \mathbb{N}$ a chain of neighbors from i to j exists via $\tau = (t_1, \ldots, t_r)$ with $t_i \geq t$ for $1 \leq i \leq r$.*

Proof. (i) For a sequence $\tau = (t, \ldots, t)$ of length k one has $i \in N(j, \tau)$ if and only if $i \in s^k(j)$ where $s = s(t)$ is defined with respect to $A(t)$. For a reciprocal process therefore $A(t)$ has a positive diagonal and $i \in s^k(j)$ implies $j \in s^l(i)$. By Lemma 8.5.18, $A(t)$ is cut-balanced for $t \geq \underline{t}$. Thus (i) follows from Theorem 8.5.17.

(ii) For $M(i) = \bigcap_{t \geq 0} \bigcup_{t < \tau} N(i, \tau)$ the condition in (ii) means that $j \in M(i)$ for all $i, j \in N$. Therefore, this condition is equivalent to $M(i) = M(j)$ for all $i, j \in N$ and, by part (i), equivalent for $N = M(i)$ for all $i \in N$. Since the decomposition (8.5.14) does depend only on $(A(t))$ and not on $x(0)$, the condition in (ii) implies convergence to consensus for each $x(0) \in S^n$ with $S = \mathbb{R}^d$. Conversely, suppose the latter. Let $M = M(i)$ for some $i \in N$ and choose $x(0) \in S^n$ with $x(0)^j = 1$ for $j \in M$ and $x(0)^j = 0$ for $j \notin M$. Let $y(t+1) = A_M(t)y(t)$ be the mean process defined on M as in the proof for Theorem 8.5.17, part (2). Since $y(0) = 1_M$ on M it follows that on $My(t)$ and, hence, $x(t)$, converges to a consensus of value 1. Since $M \subsetneq N$ part (i) implies a $M(j) \subseteq N \setminus M$. Since $x(0)^j = 0$ for $j \notin M$ we must have that $x(t)$ converges to 0 on $M(j)$. This is a contradiction and we must have, therefore, $N = M(i)$ for all $i \in N$. By the above this implies the condition given in (ii). □

Since in case of opinion dynamics under bounded confidence $N(i,t) = \{j \in N \mid \|x^i(t) - x^j(t)\| \le \epsilon\}$ the corresponding process is mutual. Theorem 8.5.20 is applicable to this case and yields, for example, Theorem 8.2.5 presented earlier without proof. The following result covers even more general cases of bounded confidence.

Theorem 8.5.21. *Let for a convex subset S of* $\mathbb{R}^d$ *a mean process on* S^n *given by*

$$x^i(t+1) = |I(i,x(t))|^{-1} \sum_{j \in I(i,x(t))} x^j(t) \tag{8.5.19}$$

for $i \in N, t \in \mathbb{N}, x(0) \in S^n$ *and* $I(i,x) \subseteq N$ *for* $x \in S^n$.

(i) *If this process is reciprocal then there exists a decomposition* $N = M_1 \dot\cup \dots \dot\cup M_r$ *into disjoint non-empty subsets* M_i *such that*

$$\lim_{t\to\infty} x^j(t) = c_i(x(0)) \quad \text{for all} \quad j \in M_i, \ \ 1 \le i \le r. \tag{8.5.20}$$

Assume in addition there exists $\delta > 0$ *such that*

$$\|x^h - x^j\| \le \delta \quad \text{implies} \quad j \in I(h,x) \tag{8.5.21}$$

for each $x \in S^n$ *and all* $h, j \in M_i, 1 \le i \le r$. *Then consensus on each* M_i *will be reached in finite time, that is for some* $T \in \mathbb{N}$

$$x^j(t) = c_i \quad \text{for all} \quad j \in M_i, \ \ 1 \le i \le r, \ \ t \ge T.$$

(ii) *Assume* $i \in I(i,x(t))$ *for all* $i \in N, t \in \mathbb{N}$. *Consider the following properties of process* (8.5.19) *for some* $\underline{t} \in \mathbb{N}$

(a) *for each* $t \ge \underline{t}$ *exists a numbering* $N = \{i_1, \dots, i_n\}$ *(dependent on t) such that*

$$i_1 \in I(i_2, x(t)), i_2 \in I(i_3, x(t)), \dots, i_{n-1} \in I(i_n, x(t)),$$

(b) *for each* $t \ge \underline{t}$ *and* $i, j \in N$ *given there exists a chain from i to j or from j to i where the latter means there exist* $i_1, \dots i_p$ *(dependent on* t, i, j*) such that*

$$i \in I(i_1, x(t)), \dots, i_p \in I(j, x(t)),$$

(c) *for each* $t \ge \underline{t}$ *and* $i, j \in N$ *given exist* $k \in N$ *and chains from i to k and from j to k as in* (b) *(dependent on* t, i, j*).*

Then property (a) *implies* (b), (b) *implies* (c) *and property* (c) *implies for process* (8.5.19) *and* $x(0)$ *given convergence to consensus (on N). Properties* (a), (b), (c) *are all equivalent to convergence to consensus, provided the condition* (8.5.21) *holds for each* $x \in S^n$ *and all* $h, j \in N$. *In this case convergence holds in finite time.*

Proof. Let $x(t+1) = A(t)x(t)$ with $a_{ij}(t) = |I(i,x(t))|^{-1}$ for $j \in I(i,x(t))$ and $a_{ij}(t) = 0$ otherwise.

(i) It holds $\min^+ A(t) \ge \frac{1}{n}$ for all t and for neighborhoods $N(i,t) = I(i,x(t))$ the decomposition with statement (8.5.20) follows from Theorem 8.5.20 (i). The statement on finite time follows from condition (8.5.21) as in Corollary 8.5.8 (ii).

(ii) First, assume property (a) and let $i, j \in N$. Given a numbering $N = \{i_1, \dots, i_n\}$ we must have $i = i_r, j = i_q$ and, without loss, $r < q$. Then $i_{r+1}, \dots, i_q$ yield a chain as required in (b). Property (b) implies, obviously, property (c). To see that property (c) implies convergence to consensus we show that property (c) forces each matrix $A(t), t \geq \underline{t}$, to be coherent. For this let M, M' be non-empty subsets of N which are saturated for $A(t)$ with $t \geq \underline{t}$ fixed. If $i \in M, j \in M'$ then by property (c) there exist $k \in N$ and chains from i to k and from j to k. Since M, M' are saturated we obtain $k \in M \cap M'$. Thus, $A(t)$ is coherent for each $t \geq \underline{t}$. Because of $i \in I(i, x(t)), t \in \mathbb{N}$, $A(t)$ has a positive diagonal. From Corollary 8.5.10 convergence to consensus follows.

Finally, suppose (8.5.21) holds for each $x \in S^n$ and $h, j \in N$. If convergence to consensus holds then for $\delta > 0$ given $\|x^h(t) - x^j(t)\| \leq \delta$ for all $h, j \in N$ and $t \geq T$ for some $T \in \mathbb{N}$. Condition (8.5.21) implies $j \in I(h, x(t))$ for all $h, j \in N, t \geq T$ which yields (a) with $\underline{t} = T$. Convergence holds in finite time as for Corollary 8.5.8 (ii). □

Theorem 8.5.21 admits also for heterogeneous confidence levels ϵ_i as well as for other asymmetric confidence intervals $I(i, x)$. This as well as some other aspects we illustrate by the following examples.

Examples 8.5.22. (1) As mentioned already, Theorem 8.5.21 yields in particular Theorem 8.2.5 which was presented for motivation but without a proof. Theorem 8.5.21 admits, moreover, a generalization to different confidence levels ϵ_i (and to multidimensional opinions as well). Let in Theorem 8.5.21 $I(i, x) = \{j \in N \mid \|x^i - x^j\| \leq \epsilon_i\}$ for $\epsilon_i > 0, 1 \leq i \leq n$, given. Obviously, $i \in I(i, x)$ and condition (8.5.21) is met for $\delta = \min_{i \in N} \epsilon_i$. By part (i) a result as in Theorem 8.2.5 does hold (including convergence in finite time) provided the process is reciprocal with neighborhoods $N(i, t) = I(i, x(t))$. (For a particular case see (2) below.)

Of course, in Theorem 8.2.5 the process is automatically mutual and no additional assumption is needed. Without requiring the process to be reciprocal, part (ii) of Theorem 8.5.21 provides a characterization of convergence for heterogeneous confidence levels. In the particular case of one dimension and $\epsilon_i = \epsilon$ property (a) of part (ii) is well-known. It amounts to an **ϵ-profile** or **ϵ-chain**, that is for $x = x(t)$ exists a numbering $N = \{i_1, \dots, i_n\}$ such that $x_{i_1} \leq x_{i_2} \leq \dots \leq x_{i_n}$ and $x_{i_{k+1}} - x_{i_k} \leq \epsilon$ for $1 \leq k \leq n - 1$. Therefore, in this special case the equivalence of property (a) and convergence to consensus means that the latter holds if and only if for t big enough $x(t)$ is an ϵ-profile (see [28, 43, 54]). Actually, if $x(t)$ is an ϵ-profile for all $t \geq \underline{t}$ it must be an ϵ-profile for all $t \in \mathbb{N}$. In particular, for convergence to consensus $x(0)$ has to be an ϵ-profile. In case of $2 \leq n \leq 4$ this condition is sufficient, too. For $n \geq 5$ this is no longer true (see [54] and Exercise 14).

(1) Consider the mean process (8.5.19) in Theorem 8.5.21 for $S = \mathbb{R}$ and $I(i, x) = \{j \in N \mid |x_i - x_j| \leq \epsilon_i\}, i \in N$. Assume for the confidence levels that $\epsilon_1 > \epsilon_2 > \dots > \epsilon_{n-1}$ and $\epsilon_1 + \epsilon_2 + \dots + e_{n-1} \leq \epsilon_n$. Let $x \in \mathbb{R}^n$ be given by $x_i = \sum_{j=1}^{i-1} \epsilon_j$ for $2 \leq i \leq n$ and $x_1 = 0$. One finds that $I(i, x) = \{i, i + 1\}$ for $1 \leq i \leq n - 1$ and $I(n, x) = \{1, \dots, n\}$. Therefore, if

$x(t) = x$ the transition matrix to $x(t+1)$ is given by

$$A = \begin{bmatrix} \frac{1}{2} & \frac{1}{2} & 0 & \cdots & \cdots & 0 \\ 0 & \frac{1}{2} & \frac{1}{2} & 0 & \cdots & 0 \\ \vdots & & & & & \vdots \\ 0 & 0 & \cdots & \cdots & \frac{1}{2} & \frac{1}{2} \\ \frac{1}{n} & \frac{1}{n} & \cdots & \cdots & \frac{1}{n} & \frac{1}{n} \end{bmatrix}.$$

This shows that the process is reciprocal at t (but not mutual).

(2) Beside its application to bounded confidence, Theorem 8.5.20 covers also opinion dynamics for the simple matrix model (8.2.1). For $A(t) = A$ all $t \in \mathbb{N}$ the corresponding mean process is reciprocal precisely if A has a positive diagonal and $(A^p)_{ij} > 0$ for some $p = p(i,j)$ implies $(A^q)_{ji} > 0$ for some $q = q(i,j)$. By part (i) of Theorem 8.5.20 for any such matrix N has a decomposition into subsets on which convergence to consensus holds. This applies especially to a type-symmetric or double stochastic matrix. By part (ii), if the process given by A is reciprocal, then convergence to consesus holds for all $x(0) \in \mathbb{R}^d$ precisely if a power of A is irreducible. Since in this case A has a positive diagonal, the latter is equivalent to A being primitive. From an earlier result (Theorem 8.1.4) we know for any stochastic matrix that powers converge to a consensus if and only if the matrix has a scrambling power. Indeed, one easily verifies directly that a matrix is primitive with positive diagonal if and only if it has a scrambling power and the corresponding mean process is reciprocal. For this equivalence the latter condition cannot be omitted. Of course, a primitive matrix must have a scrambling power, the converse, however, is not true, not even in case of a positive diagonal as the example

$$A = \begin{bmatrix} 1 & 0 & 0 \\ \frac{3}{4} & \frac{1}{4} & 0 \\ \frac{1}{4} & \frac{1}{2} & \frac{1}{4} \end{bmatrix}$$

shows. Indeed, the mean process for this matrix is not reciprocal since $A_{21} > 0$ but $(A^q)_{12} = 0$ for all $q \in \mathbb{N}$. This again shows that assuming a stochastic matrix to be primitive is stronger than requiring a power to be scrambling.

(3) Theorem 8.5.21 allows beside neighborhoods given by a norm, as discussed in example (1), also more general neighborhoods. Consider on the state space $S \subseteq \mathbb{R}^d$ instead of a metric given by a norm some **valuation** $v\colon S \times S \to V$ and let $I(i,x) = \{j \in N \mid v(x^i, x^j) \in V_0\}$, where V_0 is a non-empty subset of the abstract valuation set V serving as a measure of confidence. Such an abstract framework can be attacked by Theorem 8.5.21. To illustrate this we treat the particular case where $V = \mathbb{R}^m$ and $v(x,y) = f(x) - f(y)$ with $f\colon S \to \mathbb{R}^m, \emptyset \neq V_0 \subseteq \mathbb{R}^m$. Thus $I(i,x) = \{j \in N \mid f(x^i) - f(x^j) \in V_0\}$ for $i \in N, x \in S^n$. A natural candidate for V_0 to consider is a closed convex cone K in $\mathbb{R}^m$. Since $0 \in K$ we have that $i \in I(i,x)$. Suppose for the ordering induced by K that for

each $t \geq \underline{t}$ a numbering exists such that $f(x^{i_1}) \leq f(x^{i_2}) \leq \cdots \leq f(x^{i_n})$ for $x = x(t)$. Part (ii) of Theorem 8.5.21 then yields convergence to consensus . For $V_0 = K$ the process is reciprocal precisely if for each monotone chain $f(x^i) \leq f(x^{i_1}) \leq \cdots \leq f(x^j)$ with $x = x(t)$ there exists a monotone chain from j to i, too.

Another choice of V_0 would be a subset V_0 of R^m containing 0 and symmetric with respect to 0, that is $V_0 \subseteq -V_0$. Such a set needs not to be convex in which case it cannot be given by a norm. Nevertheless, the process is mutual and Theorem 8.5.21 is applicable.

Still another choice of V_0 is an interval $[a, b]$ for $a, b \in \mathbb{R}^m$ that is $V_0 = \{y \in \mathbb{R}^m \mid a_i \leq y_i \leq b_i \text{ for } 1 \leq i \leq m\}$. Suppose f is continuous and $a_i < 0 < b_i$ for $1 \leq i \leq m$. Then $i \in I(i, x)$ and there exists $\delta > 0$ such that for $x \in S^n$ from $\|x^i - x^j\| \leq \delta$ it follows $f(x^i) - f(x^j) \in V_0$. Thus, in this setting part (i) of Theorem 8.5.21 yields a decomposition of N into subsets on which convergence to consensus does hold, provided the process is reciprocal for V_0. By Theorem 8.5.21 convergence to consensus on N does hold for such a V_0 precisely if for each $x = x(t), t \geq \underline{t}$, a numbering $N = \{i_1, \ldots, i_n\}$ exists such that

$$a \leq f(x^{i_{k+1}}) - f(x^{i_k}) \leq b \quad \text{for} \quad 1 \leq k \leq n-1. \tag{$*$}$$

Very special but still interesting cases of $(*)$ are the following ones. Consider first $f\colon S \to \mathbb{R}$ and the cone $V_0 = \{r \in \mathbb{R} \mid r \leq 0\}$. One might think of f as attaching a reward to state x or as giving the number of all agents being in state x. Then $I(i, x) = \{j \in N \mid f(x^i) \leq f(x^j)\}$ and for the mean process given by equation (8.5.19) an agent takes all agents with a higher reward than his own into account or all agents having more followers than he has. Since $f(x^i(t))$ is a real number there exists for each t a numbering as required by property (a) in Theorem 8.5.21. Thus, the agents' opinions will converge to a consensus.

As the second special case consider $S = \mathbb{R}$, f the identity on $\mathbb{R}$ and $V_0 = \{y \in \mathbb{R} \mid -\epsilon_l \leq y \leq \epsilon_r\}$ with $0 < \epsilon_l, \epsilon_r$. V_0 is an asymmetric confidence interval when the confidence level ϵ_l to the left differs from the confidence level ϵ_r to the right. Being a special case of $(*)$ convergence to consensus holds precisely if for each $t \geq \underline{t}$ a numbering $N = \{i_1, \ldots, i_n\}$ exists such that $-\epsilon_l \leq x^{i_{k+1}}(t) - x^{i_k}(t) \leq \epsilon_r$. Opinion dynamics under bounded confidence in this asymmetric case has been extensively explored by computer simulations in [43].

To conclude, we add further remarks connecting the results of this sectin to those obtained in the literature.

Remarks 8.5.23. (1) As remarked already, it is difficult in general to compute the consensus $\tilde{c}(x(0))$ in dependence of initial conditions $x(0) \in S^n$. One has, however, the following sensitivity property for $x(0), y(0) \in S^n$ and $\tilde{c}(x(0)), \tilde{c}(y(0))$ provided the latter exist

$$\|\tilde{c}(x(0)) - \tilde{c}(y(0))\| \leq \max_{1 \leq i,j \leq n} \|x^i(0) - y^j(0)\|.$$

This follows immediately from

$$\tilde{c}(x(0)) \in \operatorname{conv}\{x^1(0), \dots, x^n(0)\}, \tilde{c}(y(0)) \in \operatorname{conv}\{y^1(0), \dots, y^n(0)\}$$

$$\text{and, hence, } \tilde{c}(x(0)) - \tilde{c}(y(0)) \in \operatorname{conv}\{x^i(0) - y^j(0) \mid 1 \le i, j \le n\}.$$

(2) To link the results on convergence to consensus of this section to those obtained in the literature, we collect some criteria which arise immediately from our results. (Only multi-agent coordination in discrete time will be addressed.)

For a sequence $(A(t))$ of stochastic matrices let a mean process given on S^n for a convex subset S of $\mathbb{R}^d$ by $x(t+1) = A(t)x(t), x(0) \in S^n$. In the literature mainly the one dimensional case is treated, that is $S \subseteq \mathbb{R}$. As it is often assumed in the literature suppose $\min^+ A(t) \ge \alpha > 0$ for all $t \in \mathbb{N}$. Then each of the following criteria does assure convergence of $x(t)$ to consensus.

(a) There exists a collection $\mathfrak{J}$ of disjoint intervals $I = [a, b] \subseteq \mathbb{N}$ such that $A(b) \dots A(a)$ is scrambling and $\sum_{I \in \mathfrak{J}} \alpha^{|I|} = \infty$.
(b) $\{A(t)\}$ has the Wolfowitz property.

Another often made assumption is that all $A(t)$ have a positive diagonal. Assuming this together with $\min^+ A(t) \ge \alpha > 0$ the following criteria yield convergence to consensus.

(c) The principle of the third agent (printh) holds on an infinite collection $\mathfrak{J}$ of disjoint intervals $I \subseteq \mathbb{N}$ with $|I| \le p$ for some $p \in \mathbb{N}$.
(d) All $A(t)$ are cut-balanced and printh holds on an infinite collection of disjoint intervals in $\mathbb{N}$.
(e) The process is reciprocal and connectivity holds in the sense that a chain of neighbors exists for any two agents across infinitely many intervals in $\mathbb{N}$.

Criterion (a) follows from Theorem 8.5.3, (b) follows from Theorem 8.4.9, (c) follows from Theorem 8.5.7, (d) follows from Corollary 8.5.13 and (e) follows from Theorem 8.5.20.

(3) In [72, Theorem 3.2.37] various conditions are considered which yield convergence to consensus if $\min^+ A(t) \ge \alpha > 0$. Criterion (a) settles one of the cases where a collection of intervals I_s in $\mathbb{N}$ is assumed with $|I_s| \le T \log(\log s)$ for some $T \in \mathbb{N}$ and all $s \in \mathbb{N}, s \ge 2$. Actually, criterion (a) allows to weaken this condition to

$$|I_s| \le T(\log s + \log_2 s + \dots + \log_{p+1} s),$$

where $\log_k s$ is the iterated logarithm given by $\log_{k+1} s = \log(\log_k s)$ and $\log_0 s = s$ $(s \ge s_0 \in \mathbb{N}.)$ This follows easily from the divergence of the so called **Abelian series** [53, p. 63]

$$\sum_{s \ge 2} (s \log s \log_2 s \log_3 s \dots \log_p s)^{-1}$$

by putting $T = -(\log \alpha)^{-1}$. The cases 3 and 4 in [72, Theorem 3.2.37] do follow from criteria (b) and (c), respectively.

(4) A major impact in the area of multi-agent coordination had the article [100] in which a simple but intricate model was explored by means of computer simulations. This article, however, does not supply an analytical explanation of the quite surprising phenomenon of consensus. Such an explanation was undertaken in [52] for the **Vicsek's model** as the model in [100] has been named afterwards. The main result in [52, Theorem 2] can be obtained from criterion (c). In [52] it is assumed in addition that $A(t)$ is type-symmetric and there are only finitely many $A(t)$. The latter assumption is made in order to apply the (original) Theorem of Wolfowitz. Actually [52] analyses a simplified version of Vicsek's model (see [46, 47, 81]) which is equivalent to the model of bounded confidence as in [54]. Therefore, the main result in [52] can be obtained from Corollary 8.5.8, too. The result in [52] has been extended in [88] by not requiring type-symmetric and allowing for more general matrices $A(t)$. It is assumed, however, that the entries of $A(t)$ are taken from a fixed finite set. To apply Wolfowitz' Theorem it is, moreover, assumed there are only finitely many $A(t)$. The main result in [88, Theorem 3.10] for discrete time can be obtained also from criterion (c). Another extension of [52] can be found in [46] where, to cover a proper infinite sequence $(A(t))$, an infinite version of Wolfowitz's theorem is developed. For the latter in [46, Proposition 1] the matrices $A(t)$ are assumed to be type-symmetric which makes criteria (d) and (e) applicable (see also Remarks 8.4.11 (3)). Various extensions of Viscek's model can be found in [13].

(5) Much earlier to the Vicsek model and its analysis later on is the pioneering work of J.N. Tsitsiklis [98, 99]. Actually, the results obtained there are more general than the ones obtained in [52]. (Cf. [6, 8, 47]). Furthermore, the "agreement algorithm" obtained allows also for delays. Corollary 8.5.9 (iii) is proven invoking an assumption called "bounded intercommunication intervals". Corollary 8.5.9 (ii) can be found in [47, Corollary 9.1]. There it follows from [47, Theorem 9.2] which itself can be obtained from criterion (a) in (2). Whereas parts (ii) and (iii) of Corollary 8.5.9 require $\min^+ A(t) \geq \alpha > 0$, part (i) admits also for $\min^+ A(t)$ approaching 0 [57].

The notion of a cut-balanced matrix is introduced in [49]. Theorem 8.5.17 is proved in [49, Theorem 2] for the onedimensional case $S = R$. For type-symmetric matrices, a particular case of cut-balanced matrices, Theorem 8.5.17 was proven in [71, Theorem 2], [72, Theorem 3.2.39], and [46, Theorem 1].

Concerning the crucial Lemma 8.5.12 (ii) an optimal lower bound of $n-1$ in case of type-symmetric matrices has been obtained in [21, Theorem 2.5]. [46] discusses also the question of necessary conditions in case of type-symmetric matrices with positive diagonal. (See part (ii) of Theorem 8.5.20 for such conditions.) Theorem 8.5.21 provides sufficient and necessary conditions in case of bounded confidence which cover in particular the well-known characterization of convergence to consensus in one dimension by ϵ-chains or the lacking of a "split" [28, 54, 72]). See similarly [95, Theorem 1] and the neat concept of condensation introduced therein. See also [104].

A major question in opinion dynamics under bounded confidence is what happens if confidence levels are different. Theorem 8.5.21 admits for those but does not give a conclusive answer. The question has been extensively investigated by computer simulations in [79] leading to the conjecture that every trajectory converges to a limiting opinion vector [79, Conjecture 2.1 and 2.2].

(6) The field of multi-agent coordination and opinion dynamics has been rapidly developed during the last years. Surveys can be found in [16, 47, 73, 86]. Topics not considered in this section are, among others, the cases of differentiable system [86], systems with delays [8], infinitely many agents [10], random variables [22], convergence rates and speed of algorithms [13, 21], non-convex domains [1, 90].

8.6 Swarm dynamics

In this last section we shall use results obtained to analyse the dynamics of swarms of birds and other animals. Doing so we need first to examine for the convergence to consensus in multi-agent coordination the rate of convergence, what we have not done yet. In case the intensity of interaction is bounded from below by a positive constant one expects convergence to be exponential. This is not true in general where convergence can be rather slow. More precisely we prove the following theorem which supplements Theorem 8.5.3 with respect to the rate of convergence.

Theorem 8.6.1. *Let $(A(t))$ be a sequence of $n \times n$-stochastic matrices and let $\mathfrak{I}$ a sequence of disjoint intervals $I_k \subseteq \mathbb{N}, k \geq 1$, with $t_1 < t_2 < \cdots$ for $t_k = \max\{t \in \mathbb{N} \mid t \in I_k\}$.*

(i) *Let for $I \in \mathfrak{I}, I = [a, b], B(I) = A(b) \ldots A(a)$ and $\rho_i = c(B(I_i)) \ldots c(B(I_1))$. Then*

$$\sum_{s=0}^{t_k-1} c(A(s) \ldots A(0)) \leq \sum_{i=0}^{k-1} (t_{i+1} - t_i)\rho_i, \tag{8.6.1}$$

where $t_0 = 0, \rho_0 = 1$.

(ii) *Let $x(t+1) = A(t)x(t), t \in \mathbb{N}, x(0) \in S^n$ where S is a non-empty convex subset of $\mathbb{R}^d$. If $\sum_{k=1}^{\infty} \mu(B(I_k)) = \infty$ then $\lim\limits_{t\to\infty} x^i(t) = c$ for all $i \in N$ with $c = c(x(0)) \in S$ and*

$$\max_{i\in N} \|x^i(t) - c\| \leq \rho_k \max_{i,j\in N} \|x^i(0) - x^j(0)\| \tag{8.6.2}$$

for $t \geq t_k, k \geq 1$.

(iii) *If for some $\underline{k} \in \mathbb{N}, \alpha > 1$*

$$\mu(B(I_k)) \geq \frac{\alpha}{k} \quad \text{for} \quad k \geq \underline{k}, \tag{8.6.3}$$

then $\sum_{k=1}^{\infty} \rho_k < \infty$.

Especially, if $\mu(B(I_k)) \geq \beta > 0$ for $k \geq \underline{k}$ then the convergence to consensus is exponential in the sense that

$$\max_{i\in N} \|x^i(t) - c\| \leq (1-\beta)^{k-\underline{k}} \max_{i,j\in N} \|x^i(0) - x^j(0)\| \tag{8.6.4}$$

for $t \geq t_k, k \geq \underline{k}$.

Proof. (i) Let $a(s) = c(A(s)\dots A(0)), s \in \mathbb{N}$. If $s \geq t_i$ then $a(s) \leq c(B(I_i)) \cdot c(B(I_{i-1}))\dots c(B(I_1)) = \rho_i$ because of $c(AB) \leq c(A)c(B)$ and $c(A) \in [0,1]$. Therefore, in case of $t \geq t_k$

$$\begin{aligned}\sum_{s=0}^{t_k-1} a(s) &= \sum_{s=0}^{t_1-1} a(s) + \sum_{s=t_1}^{t_2-1} a(s) + \dots + \sum_{s=t_{k-1}}^{t_k-1} a(s) \\ &\leq t_1 + (t_2 - t_1)\rho_1 + \dots + (t_k - t_{k-1})\rho_{k-1}.\end{aligned}$$

This proves inequality (8.6.1).

(ii) Theorem 8.5.3 yields $\lim_{t\to\infty} x^i(t) = c$ for all $i \in N$. Furthermore, since $x(t+1) = A(t)\dots A(0)x(0)$ from Theorem 8.1.2 A (ii) it follows that $\Delta x(t+1) \leq c(A(t)\dots A(0))\Delta x(0)$. (Thereby, for $y \in S^n$ one has $\Delta y = \Delta\{y^1,\dots,y^n\} = \Delta\mathrm{conv}\{y\} = \max_{i,j\in N}\|y^i - y^j\|$.) From $x^i(t+1+s) \in \mathrm{conv}\{x(t+1)\}$ for $s \geq 0$ it follows that $c \in \mathrm{conv}\{x(t+1)\}$ for all $t \in \mathbb{N}$. Thus

$$\max_{i\in N}\|x^i(t+1) - c\| \leq \Delta\mathrm{conv}\{x(t+1)\} \leq c(A(t)\dots A(0))\Delta x(0).$$

Since by step (i) $a(t) \leq \rho_k$ for $t \geq t_k$ this proves inequality (8.6.2).

(iii) By definition of ρ_i and Lemma 8.5.2 (ii) we have

$$\rho_{k+1} = c(B(I_{k+1}))\rho_k \leq (1 - \mu(B(I_{k+1}))\rho_k,$$

and, hence, by assumption $\rho_{k+1} \leq (1 - \frac{\alpha}{k+1})\rho_k$ for $k \geq \underline{k}$. Since $\alpha > 1$ from Raabe's Test [53, p. 136] we obtain $\sum_{k=1}^{\infty} \rho_k < \infty$. Especially, if $\mu(B(I_k)) \geq \beta > 0$ for $k \geq \underline{k}$ we have $\rho_{k+1} \leq (1-\beta)\rho_k$ for $k \geq \underline{k}$. Therefore,

$$\rho_{k+1} \leq (1-\beta)^{k+1-\underline{k}}\rho_{\underline{k}} \leq (1-\beta)^{k+1-\underline{k}} \quad \text{for} \quad k \geq \underline{k}.$$

From step (ii) we obtain for $t \geq t_k, k \geq \underline{k}$

$$\max_{i\in N}\|x^i(t) - c\| \leq (1-\beta)^{k-\underline{k}}\max_{i,j\in N}\|x^i(0) - x^j(0)\|,$$

that is, inequality (8.6.4). □

Considering now swarms of birds or other self-organized groups of animals, the main assumption we shall make on the movement of the animals is that they "match velocity with nearby neighbors". Many observations and experiments with real animals suggest such a behavior. (See the instructive review [68] for experiments, biological roots and principles concerning the organized flight of birds.) Further, in computer simulations such an assumption is usual since the pioneering work of C. Reynolds [89] on "boids" (artificial birds). By the assumption of velocity matching, the averaging of velocities, the framework developed in the previous section provides useful tools to investigate swarms. Of course, swarms thereby function as an abstract concept, like the concept of an ideal gas, and should not be confounded with swarms outside in nature. Nevertheless, it is the aim of such a model to point out essential features of real swarms.

Consider a number n of birds in $\mathbb{R}^3$, or without much more effort, in a convex region S in some $\mathbb{R}^d$. Assuming discrete time $t \in \mathbb{N}$ let $x^i(t)$ in S denote the **position of bird *i* at time *t*.** The **velocity $v^i(t)$ of bird *i* at time *t*** is given by $v^i(t) = x^i(t+1) - x^i(t)$. The velocity matching or, according to [89], the "alignment" which "steers towards the average heading of local flockmates" we model by a convex combination of velocities, that is $v^i(t+1) \in \operatorname{conv}\{v^1(t), \dots, v^n(t)\}$ for each bird $i \in N$, each $t \in \mathbb{N}$. Thus the velocity vectors $v(t)$ form a mean process in the sense of Definition 8.3.1. Equivalently, there exists a sequence of stochastic matrices $A(t)$ such that $v(t+1) = A(t)v(t)$ for the vectors $v(t)$ of velocities $v^i(t)$. Thus we arrive at the following **swarm model** for $x(t), v(t) \in S^n$

$$\begin{aligned} x(t+1) - x(t) &= v(t) \\ v(t+1) &= A(t)v(t), \end{aligned} \tag{8.6.5}$$

with $t \in \mathbb{N}$ and initial conditions $x(0), v(0) \in S^d$.

What makes the birds a swarm is that all birds tend asymptotically to the same velocity, in other words, a consensus in terms of velocities. Considering the positions one would not speak of a swarm if the relative position $x^i(t) - x^j(t)$ of any two birds would tend to infinity. Stronger than this boundedness of relative positions we require for a swarm relative positions to converge asymptotically. It is this property for which we will need Theorem 8.6.1 on the speed of convergence because the position $x^i(t)$ is an accumulation of all the velocities $v^i(s)$ from 0 up to t. For the swarm model (8.6.5) "local flockmates" or "nearby neighbors" play a role which refers to structure and intensity of interaction. The following main result of the present section states that the birds will form a swarm if the interaction among birds is coherent often enough and the intensity of interaction within certain intervals does not approach zero too fast. It is remarkable that to form a swarm birds need not follow always definite rules, except the one of matching velocities. In other words, the birds can form a swarm even in case of interruptions or pertubations. Later on we will interprete the result below and show how coherence corresponds to certain flight regimes as line formations, especially V or J formations. Also, the result below generalizes Corollary 8.5.10 in that coherence is required only for certain periods.

Theorem 8.6.2. *For the swarm model* (8.6.5) *let $(A(t))$ be a sequence of stochastic matrices with positive diagonal and such that for a sequence $r_0 = 0 \le r_1 < r_2 < \dots$ of time periods $A(r_k)$ is coherent for $k = 1, 2, \dots$.*

Let for $k = 1, 2, \dots$ $I_k = [r_{(k-1)(n-1)+1}, r_{k(n-1)}]$ an interval in $\mathbb{N}$, $p_k = |I_k|$ and $m_k = \min\{\min^+ A(t) | t \in I_k\}$.

Then for arbitrary initial conditions $x(0), v(0) \in S^d$ the following properties do hold.

(i) *If $\sum_{k\ge1} m_k^{p_k} = \infty$ then $\lim\limits_{t\to\infty} v^i(t) = v^*(v(0))$ for all $i \in N$.*

(ii) (a) *For any* $i, j \in N$ *and* $t \in \mathbb{N}$,

$$\|x^i(t) - x^j(t)\| \le \|x^i(0) - x^j(0)\| + t\max_{h,l\in N}\|v^h(0) - v^l(0)\|. \tag{8.6.6}$$

(ii) (b) *If* $r = \sup_k (r_{k+1} - r_k) < \infty$ *and for some* $\underline{k}$ *and* $\alpha > 1$ *it holds* $m_k^{p_k} \ge \frac{\alpha}{k}$ *for* $k \ge \underline{k}$ *then*

$$\lim_{t\to\infty} (x^i(t) - x^j(t)) = x^*(i, j, x(0), v(0)) \quad \text{for} \quad i, j \in N. \tag{8.6.7}$$

Proof. (i) To apply Theorem 8.5.3 we show that $B(I) = A(b)\ldots A(a)$ is scrambling for $I = [a, b]$ and, for k fixed, $a = r_{(k-1)(n-1)+1}, b = r_{k(n-1)}$. The interval I in $\mathbb{N}$ contains the $(n-1)$ values $t_i = r_{(k-1)(n-1)+i}$ for $1 \le i \le n-1$. By assumption the matrices $A(t_i), 1 \le i \le n-1$ are coherent. Since these matrices have a positive diagonal they are Sarymsakov matrices by Proposition 8.1.7 (iii). By Lemma 8.4.8 (ii) the product $A(t_{n-1})\ldots A(t_1)$ is scrambling. The sequence $\sigma = (t_i)$ for $1 \le i \le n-1$ is contained in $\tau = [a, b]$ and we have for $i, j \in N$ and $\tau(i, j) = \sigma(i, j) = \sigma$ that

$$N(i, \tau(i, j)) \cap N(j, \sigma(i, j)) \ne \emptyset \quad \text{where} \quad N(h, \sigma) = \{l | [A(t_{n-1})\ldots A(t_1)]_{hl} > 0\}.$$

Thus, the principle of the third agent holds on τ and $A(b)\ldots A(a)$ is scrambling by Lemma 8.5.6 (ii). Since $B(I)$ is scrambling we have that

$$\mu^+(B(I)) \ge \min{}^+(B(I)) \ge \min{}^+A(b)\ldots\min{}^+A(a),$$

and, hence, $\mu^+(B(I_k)) \ge m_k^{|I_k|} = m_k^{p_k}$. Theorem 8.5.3 yields property (i).

(ii) (a) For any $t \ge 1$ we have from model (8.6.5)

$$x^i(t) - x^j(t) = x^i(0) - x^j(0) + \sum_{s=0}^{t-1} (v^i(s) - v^j(s)).$$

Also from the model

$$v(t+1) = A(t)\ldots A(0)v(0) \text{ and, by Theorem 8.1.2 (ii),}$$
$$\Delta v(t+1) \le c(A(t)\ldots A(0))\Delta v(0) \le \Delta v(0) = \max_{h,l\in N}\|v^h(0) - v^l(0)\|.$$

This shows (ii) (a).

(ii) (b) This property follows from Theorem 8.6.1 applied to $v(t+1) = A(t)v(t)$. By assumption we have for $1 \le i < j$

$$r_j - r_i = r_j - r_{j-1} + \cdots + r_{i+1} - r_i \le (j-i)r.$$

By definition of I_k we have for $t_k = \max\{t \in \mathbb{N} | t \in I_k\} = r_{k(n-1)}$ and, hence, $t_{i+1} - t_i \le (n-1)r$. Part (i) of Theorem 8.6.1 yields

$$\sum_{s=0}^{t_k-1} c(A(s)\ldots A(0)) \le (n-1)r\sum_{i=0}^{k-1} \rho_i.$$

From $v(s+1) = A(s)\dots A(0)v(0)$ we have for $i,j \in N$

$$\|v^i(s+1) - v^j(s+1)\| \le \Delta v(s+1) \le c(A(s)\dots A(0))\Delta v(0).$$

Putting together, with $w(s) = v^i(s) - v^j(s)$ for i,j fixed

$$\sum_{s=0}^{t_k-1} \|w(s+1)\| \le \left[\sum_{s=0}^{t_k-1} c(A(s)\dots A(0))\right]\Delta v(0) \le (n-1)r\Delta v(0)\sum_{i=0}^{k-1}\rho_i.$$

From part (iii) of Theorem 8.6.1 we obtain by assumption that $\sum_{i=0}^{\infty}\rho_i < \infty$ and because of $t_1 < t_2 < \dots$ we must have that $\sum_{s=0}^{\infty}\|w(s)\| < \infty$. Therefore, $\lim_{t\to\infty}\sum_{s=0}^{t} w(s)$ exists and the conclusion in (ii) (b) follows from $x^i(t) - x^j(t) = x^i(0) - x^j(0) + \sum_{s=0}^{t-1} w(s)$. □

For the particular case where all matrices $A(t)$ are scrambling the proof of Theorem 8.6.2 yields the following simpler variant of Theorem 8.6.2.

Corollary 8.6.3. *For the swarm model* (8.6.5) *let* $(A(t))$ *be a sequence of scrambling matrices.*

(i) *If* $\sum_{t\in\mathbb{N}} \min^+ A(t) = \infty$ *then* $\lim_{t\to\infty} v^i(t) = v^*(v(0))$ *for all* $i \in N$.

(ii) (a) *For any* $i,j \in N$ *and* $t \in \mathbb{N}$,

$$\|x^i(t) - x^j(t)\| \le \|x^i(0) - x^j(0)\| + t\max_{h,l\in N}\|v^h(0) - v^l(0)\|.$$

(ii) (b) *If for some* $\underline{t}$ *and* $\alpha > 1$ *it holds* $\min^+ A(t) \ge \frac{\alpha}{t}$ *for* $t \ge \underline{t}$ *then*

$$\lim_{t\to\infty}(x^i(t) - x^j(t)) = x^*(i,j,x(0),v(0)) \quad \text{for} \quad i,j \in N.$$

Proof. Since $A(t)$ is scrambling we can apply directly Theorem 8.5.3 with $I_t = \{t\}$. Since $\mu^+(B(I_t)) \ge \min^+ A(t)$ Theorem 8.5.3 yields (i). Parts (ii)(a) and (ii)(b) follow as in Theorem 8.6.2. □

The following examples show that swarm formation is possible also if the intensity of interaction goes to zero and they demonstrate also the importance of the assumption $\alpha > 1$ for the convergence of the relative positions.

Examples 8.6.4. (a) Consider a swarm model as in Theorem 8.6.2 and assume $r = \sup_k (r_{k+1} - r_k) < \infty$ and $\min^+ A(t) \ge \alpha\left[\frac{(n-1)r}{t}\right]^p$ for $t \ge \underline{t}$ with $p = \frac{1}{(n-2)r+1}$, $\alpha > 0$. If $t \in I_k$ then $\frac{1}{t} \ge \frac{1}{r_{k(n-1)}}$. From $r_j - r_i \le (j-i)r$ for $1 \le i < j$ we obtain $r_j \le jr$. Therefore, $\frac{1}{t} \ge \frac{1}{k(n-1)r}$ for $t \in I_k$ and $m_k \ge \alpha\left[\frac{(n-1)r}{k(n-1)r}\right]^p = \frac{\alpha}{k^p}$. Furthermore,

$$p_k = |I_k| = r_{k(n-1)} - r_{(k-1)(n-1)+1} + 1 \le (n-2)r + 1 = \frac{1}{p}.$$

It follows that $m_k^{p_k} \ge \alpha^{p_k} \cdot \frac{1}{k} \ge \frac{\alpha}{k}$. From Theorem 8.6.2 it follows $\lim_{t\to\infty} v^i(t) = v^*$ for $\alpha > 0$ and $\lim_{t\to\infty}(x^i(t) - x^j(t)) = x^*(i,j)$ for $\alpha > 1$. Obviously, $\min^+ A(t)$ goes to zero for t tending to infinity.

(b) Consider the following special case of the example above. Let $r_k = k, k \geq 1$. Then $r = 1, I_k = [(k-1)(n-1)+1, k(n-1)]$ and $p_k = n-1$. The assumption in (a) becomes $\min^+ A(t) \geq \alpha(\frac{n-1}{t})^{\frac{1}{n-1}}$. In this case all matrices $A(t)$ are coherent and $\min^+ A(t)$ goes to zero as $(\frac{1}{t})^{n-1}$ for $\alpha > 0$. Again, swarm formation takes place. Actually, in case of $r_k = k$ part (i) of Theorem 8.6.2 gives back Corollary 8.5.10.

(c) Consider the following variation of Examples 8.4.1 A: $A(t) = \left[\begin{smallmatrix} 1 & 0 \\ \frac{\alpha}{t} & (1-\frac{\alpha}{t}) \end{smallmatrix}\right]$ for $\alpha > 0, t > \alpha$. These matrices are of the kind considered above and we could apply Theorem 8.6.2. Since these matrices are scrambling it is easier to apply Corollary 8.6.3. From part (i) we obtain for the swarm model (8.6.5) that $\lim_{t\to\infty} v^i(t) = v^*$ for $\alpha > 0$. Actually, from Examples 8.4.1 A we know this convergence already for $\alpha = 1$. In this case we calculate directly $v_1(t+1) - v_2(t+1) = \frac{1}{t}(v_1(2) - v_2(2))$ for $t \geq 2$. Therefore, for $t \geq 2$

$$x_1(t+1) - x_2(t+1) = x_1(2) - x_2(2) + \Big(\sum_{s=2}^{t} \frac{1}{s-1}\Big)(v_1(2) - v_2(2)).$$

This is in accordance with part (ii) (a), however, the relative position of the two birds does not converge for $v_1(2) \neq v_2(2)$. According to part (ii) (b) of Corollary 8.6.3 the relative position does converge for $\alpha > 1$. Let us check this directly for $\alpha = 2$. Similar as in Examples 8.4.1 A we obtain by induction $A(t)\dots A(3) = \left[\begin{smallmatrix} 1 & 0 \\ 1-a(t) & a(t) \end{smallmatrix}\right]$ where $a(t) = (1-\frac{2}{t})\dots(1-\frac{2}{3})$ for $t \geq 3$. One finds

$$\sum_{s=3}^{t}(v_1(s) - v_2(s)) = \Big(\sum_{s=3}^{t} a(s-1)\Big)(v_1(0) - v_2(0)).$$

Since $\sum_{s=3}^{t} a(s-1)$ converges to some w we obtain $\lim_{t\to\infty}(x_1(t) - x_2(t)) = x_1(0) - x_2(0) + w(v_1(0) - v_2(0))$.

(d) Different from example (c) the next example shows that swarm formation can happen also for $\alpha \leq 1$. Let $A(t) = \left[\begin{smallmatrix} \frac{1}{t} & (1-\frac{1}{t}) \\ 1 & 0 \end{smallmatrix}\right]$ for $t \geq 2$. Since $A(t)$ is scrambling and $\min^+ A(t) = \frac{1}{t}$ from Corollary 8.6.3 it follows that velocities converge to a common value. (Notice, $A(t)$ has no positive diagonal which is admitted in Corollary 8.6.3.) For $w(t) = v_1(t) - v_2(t)$ one finds $w(t+1) = -(1-\frac{1}{t})w(t)$ and, hence, $w(t+1) = \frac{(-1)^{t-1}}{t} w(2)$.

Therefore, $x_1(t) - x_2(t) = x_1(2) - x_2(2) + \sum_{s=2}^{t-1} w(s)$ does converge for $t \to \infty$. Thus swarm formation takes place although the assumption in part (ii)(b) of Corollary 8.6.3 that $\min^+ A(t) \geq \frac{\alpha}{t}$ for some $\alpha > 1$ is not satisfied. This example also shows that the convergence of velocities need not be monotone.

For the swarm model (8.6.5) the assumptions made in Theorem 8.6.2 and Corollary 8.6.3 do not stipulate any cause for the change of intensity or structure in the interaction. In particular there is no assumption on how the latter are connected to the positions of the birds. Such a connection is not unplausible and it is assumed often in the literature on swarms that the intensities $a_{ij}(t)$ depend on the distance $\|x^i(t) - x^j(t)\|$ between the birds. An example are the famous **Cucker–Smale model of bird flocking** and variations of it. The original articles of F. Cucker and S. Smale are [23] and [24]. Related contributions are [14, 19, 21, 38, 82, 94]. From Theorem 8.6.2 and Corollary 8.6.3, respectively, we obtain the following result which admits a rather general dependence of intensities on distances among birds and which, moreover, admits a structure of interaction much weaker than the one considered in the literature mentioned above. (All this, however, for discrete time only. See also the remarks at the end of this section.)

Theorem 8.6.5. *For the swarm model (8.6.5) let $(A(t))$ be a sequence of stochastic matrices such that*

$$a_{ij}(t) = f(\|x^i(t) - x^j(t)\|) \quad \textit{in case of } a_{ij}(t) > 0 \textit{ and } i \neq j, \tag{8.6.8}$$

where f is an antitone selfmapping of $\mathbb{R}_+$ with $f(0) < \frac{1}{n-1}$.

(i) *Suppose there is a sequence $r_0 = 0 \leq r_1 < r_2 < \cdots$ with $r = \sup_k (r_{k+1} - r_k) < \infty$ and such that $A(r_k)$ is coherent for $k = 1, 2, \dots$ and let*

$$p_k = |I_k| \quad \textit{for} \quad I_k = [r_{(k-1)(n-1)+1}, r_{k(n-1)}] (in \mathbb{N}).$$

  (a) *If $zf^{p_k}(z) \geq c > 0$ for all $z \geq \underline{z} \in \mathbb{R}_+$, all $k \geq \underline{k} \in \mathbb{N}$, then $\lim_{t\to\infty} v^i(t) = v^*(v(0))$ for all $i \in N$. (Thereby, f^p the p-fold product $f \dots f$.)*

  (b) *If $zf^{p_k}(z) \geq c > (n-1)r\Delta v(0)$ for $z \geq \underline{z}, k \geq \underline{k}$, then $\lim_{t\to\infty} (x^i(t) - x^j(t)) = x^*(i, j, x(0), v(0))$ for $i, j \in N$.*

(ii) *Suppose all matrices $A(t)$ are scrambling.*

  (a) *If $zf(z) \geq c > 0$ for all $z \geq \underline{z} \in \mathbb{R}_+$ then $\lim_{t\to\infty} v^i(t) = v^*(v(0))$ for all $i \in N$.*

  (b) *If $zf(z) \geq c > (n-1)\Delta v(0)$ for all $z \geq \underline{z} \in \mathbb{R}_+$ then $\lim_{t\to\infty} (x^i(t) - x^j(t)) = x^*(i, j, x(0), v(0))$ for $i, j \in N$.*

Proof. (i) From part (ii) (a) of Theorem 8.6.2 we have $\|x^i(t) - x^j(t)\| \leq c_1 + c_2 t$ for $i, j \in N$ where $c_1 = \Delta x(0)$, $c_2 = \Delta v(0)$. Fix $i \neq j \in N$ with $a_{ij}(t) > 0$. Since f is antitone, we obtain $a_{ij}(t) = f(\|x^i(t) - x^j(t)\|) \geq f(c_1 + c_2 t)$ and, using the assumption made in (a) and (b), respectively,

$$a_{ij}(t)^{p_k} \geq f(c_1 + c_2 t)^{p_k} \geq \frac{c}{c_1 + c_2 t} \quad \text{for } t \geq \underline{t}, k \geq \underline{k}. \tag{$*$}$$

$$\text{Choose } \alpha \text{ such that } 0 < \alpha < \frac{c}{(n-1)rc_2}. \tag{$**$}$$

In case of (b) we have that $\frac{c}{(n-1)rc_2} > 1$ and, hence, in that case $\alpha > 1$ can be chosen. Because of $(**)$ we can choose $\tilde{k} \geq \underline{k}$ such that $c_1 \leq k(\frac{c}{\alpha} - (n-1)rc_2)$ for $k \geq \tilde{k}$. For $t \in I_k$ we have that $t \leq r_{k(n-1)} \leq k(n-1)r$ and, hence, $t \leq k(n-1)r \leq \frac{\frac{c}{\alpha}k - c_1}{c_2}$ for $k \geq \tilde{k}$. Thus, $c_1 + c_2 t \leq \frac{c}{\alpha} k$ for $t \in I_k, k \geq \tilde{k}$. Choosing $\tilde{k}$ big enough we can assume that $t \geq \underline{t}$ for $t \in I_k, k \geq \tilde{k}$ and obtain from $(*)$

$$a_{ij}(t)^{p_k} \geq \frac{c}{c_1 + c_2 t} \geq \frac{\alpha}{k} \quad \text{for} \quad t \in I_k, \ k \geq \tilde{k}.$$

Since for $j \neq i$ we have $a_{ij}(t) \leq f(0)$ we obtain

$$a_{ii}(t) = 1 - \sum_{j \neq i} a_{ij}(t) \geq c_3 \quad \text{with} \quad c_3 = 1 - (n-1)f(0) > 0.$$

Furthermore, $p_k = r_{k(n-1)} - r_{(k-1)(n-1)+1} + 1 \leq (n-2)r + 1$ and, hence,

$$a_{ii}(t)^{p_k} \geq a_{ii}(t)^{(n-2)r+1} \geq c_3^{(n-2)r+1} \geq \frac{\alpha}{k} \quad \text{where} \quad k \geq k' \in \mathbb{N}.$$

Thus, we arrive at $m_k^{p_k} \geq \frac{\alpha}{k}$ for all $k \geq \tilde{k}, k'$. Therefore, statements (a) and (b) follow from parts (i) and (ii) (b) of Theorem 8.6.2.

(ii) This part follows from Corollary 8.6.3 in the same manner as in part (i) above. Now $I_t = \{t\}, p_t = 1$ and $r = 1$. As in $(**)$ choose α such that $0 < \alpha < \frac{c}{(n-1)c_2}$. In case of (b) we can choose $\alpha > 1$ since by assumption $\frac{c}{(n-1)c_2} > 1$. □

Theorem 8.6.5 we illustrate by discussing in some detail the **Cucker–Smale model of bird flocking in discrete time.** In this model $x(t), v(t) \in (\mathbb{R}^3)^n$ and

$$\begin{aligned} x(t+1) - x(t) &= v(t) \\ v^i(t+1) - v^i(t) &= \sum_{i \neq j=1}^{n} f_{ij}(x(t))(v^j(t) - v^i(t)) \end{aligned} \tag{8.6.9}$$

$$\text{with} \quad f_{ij}(x) = \frac{H}{(1+\|x^i - x^j\|^2)^\beta}, H > 0, \beta \geq 0 \text{ and } i \neq j \in N.$$

This model is of the form of our swarm model (8.6.5) but not exactly, since the intensities f_{ij}, though nonnegative, do not give a stochastic matrix. Since there are no f_{ii} in this model we will define those to obtain a stochastic matrix – for this, however, we shall assume that $H \leq \frac{1}{n}$. Let for $i \neq j, t \in \mathbb{N}$ $a_{ij}(t) = f_{ij}(x(t))$ or $a_{ij}(t) = 0$ and $a_{ii}(t) = 1 - \sum_{i \neq j=1}^{n} a_{ij}(t)$. Since $f_{ij}(x(t)) \leq H \leq \frac{1}{n}$ we have that $\sum_{i \neq j=1}^{n} a_{ij}(t) \leq \sum_{i \neq j=1}^{n} f_{ij}(x(t)) \leq (n-1)H < 1$. Thus, the matrix $A(t)$ of the $a_{ij}(t)$ is stochastic and has a positive diagonal. In case of $a_{ij}(t) > 0$ for $i \neq j$ we have $a_{ij}(t) = f_{ij}(x(t)) = f(\|x^i(t) - x^j(t)\|)$ where $f(z) = \frac{H}{(1+z^2)^\beta}$ is an antitone selfmapping of $\mathbb{R}_+$ with $f(0) = H < \frac{1}{n-1}$. Thus, Theorem 8.6.5 is applicable to discrete Cucker–Smale flocking for $H \leq \frac{1}{n}$. Part (i) of Theorem 8.6.2 generalizes the latter in that $A(t)$ is required to be coherent only for certain points in time. In the model (8.6.9) intensities are required to be strictly positive at each point in time and $A(t)$ is strictly positive for each $t \in \mathbb{N}$. To this quite strong structure of interaction part (ii) of Theorem 8.6.5 applies.The assumption in (ii) (a) means that $\frac{Hz}{(1+z^2)^\beta} \geq c > 0$ for $z \geq \underline{z} \in \mathbb{R}_+$, or equivalently, $\beta \leq \frac{1}{2}$. Thus, for $\beta \leq \frac{1}{2}$ (and $H \leq \frac{1}{n}$) in the model (8.6.9) velocities converge to $v^*(v(0))$ for any given $v(0)$. The assumption in (ii) (b) means that $\frac{Hz}{(1+z^2)^\beta} \geq c > (n-1)\Delta v(0)$ for $z \geq \underline{z}$, or, equivalently, $\beta < \frac{1}{2}$ or $\beta = \frac{1}{2}$ and $H > (n-1)\Delta v(0)$. Therefore, for model (8.6.9) we have in case of $\beta < \frac{1}{2}$ convergence of velocities to a common value as well as convergence of the relative positions $x^i(t) - x^j(t)$ and this holds for any initial conditions. In case of $\beta = \frac{1}{2}$ convergence of relative positions does, according to part (ii) (b) of Theorem 8.6.4, hold if relative initial velocities are not too big. In [23, 24] also for the case of $\beta > \frac{1}{2}$ conditions on the initial conditions are specified which guarantee convergence of velocities to a common value and convergence of relative positions.

In concluding this section we like to link it to the biological literature and to interprete the main result Theorem 8.6.2 and the consequences drawn from it. Thereby we concentrate on the "how" of organized flights and leave the "why" aside. A crucial assumption made in Theorem 8.6.2 is the one that the matrices $A(t)$ are coherent at certain points in time. According to Definition 8.1.6 a stochastic matrix A is coherent if any two saturated subsets of the set of agents N have a non-empty intersection. Equivalently, the intersection of all saturated subsets is non-empty. Calling the latter set the **core C for A** one verifies easily that $C = \bigcap_{i\in N} c(i)$ where $c(i)$ is the smallest saturated set containing i. Using the map $s(\cdot)$ defined for a subset $\emptyset \neq M \subseteq N$ by $s(M) = \{j \in N \mid a_{ij} > 0 \text{ for some } i \in M\}$ (see equation (8.1.6)) one finds that $c(i) = \bigcup_{k\in\mathbb{N}} s^k(i)$ (where $s(i) = s(\{i\})$ and $s^k(\cdot)$ is the k-th iterate of $s(\cdot)$). Thus, we obtain the following description of the core

$$C = \bigcap_{i\in N} \bigcup_{k\in\mathbb{N}} s^k(i). \tag{8.6.10}$$

Therefore, $j \in C$ if and only if for each $i \in N$ there exists $k \in \mathbb{N}$ such that $j \in s^k(i)$. Let G be the **directed graph** defined by A where N is the set of nodes and (i, j) is a (directed) edge for $i, j \in N$ precisely for $a_{ij} > 0$ or, equivalently, $j \in s(i)$. In the language of graph theory $j \in s^k(i)$ means there is a (directed) path of length k from i to j, especially j is reachable from i. In this language, $j \in N$ is in the core C if and only if it is reachable from each $i \in N$. The following picture depicts a typical case of a core for a coherent matrix A in the most simple case where C is a singleton.

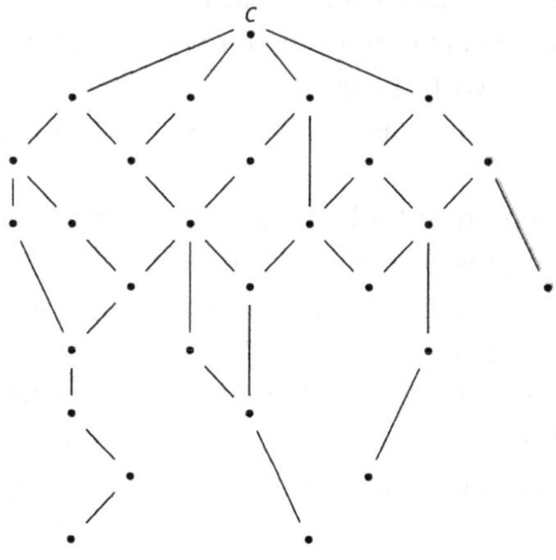

Fig. 8.1. Swarm formation.

In the picture the dots are the elements of N and an edge is represented by a dash, directed from bottom to top. The dot at the top of the figure is the core since it is reachable from any other dot and no other dot does have this property. Of course, Figure 8.1 is just a pictorial presentation of graph G for a coherent matrix A with C being a singleton. Thinking of swarms, however, one is tempted to look at this figure as a formation of birds called an **echelon** in biology [68]. The famous **V-formation** of birds and the **J-formation** as its asymmetric variant are cases of such an echelon. To substantiate this impression remember that a dash from i to j means an edge (i, j) that is $a_{ij} > 0$. With respect to the swarm model (8.6.5) this means that bird i matches its velocity with bird j. This requires a communication between the birds, say that "bird i sees bird j". Vision is considered to be an important communication channel among birds, sometimes as the appropriate one compatible with experimental data [4] Therefore, we interprete the neighborhood $N_i = \{j \in N \mid a_{ij} > 0\}$ of bird i as the birds seen by i. Since "seeing" refers to the positions of birds, Figure 8.1 can be interpreted as showing the positions of birds and a dash (i, j) means bird i sees bird j. The "wavy lines" in Figure 8.1 then are due to the conical field of vision of birds.

In case the core is a singleton as in Figure 8.1 one might think of this single bird as a **leader**. This applies in particular to the special cases of V- or J-formations. For a general line formation with many birds, however, such a leadership is rather indirect and mediated possibly by a lot of birds between leader and followers. Even less pronounced is **leadership** if the core consists of several birds. Of course, any two birds in the core do interact, either directly or indirectly. Therefore, two or more independent leaders are not possible. It has been observed, however, that big swarms of migratory birds are sometimes lead by a spherical sub-formation of birds. The opposite extreme to a core as in Figure 8.1 would be a core consisting of the whole swarm. (This is the case for the Cucker–Smale model (8.6.9).) In its most simple form this amounts to a **cycle**, where each bird is followed by just another one. For migrating birds such a cyclic formation would not make much sense, possibly with the exception of processes of starting and landing. Below we will argue that big cores and cycles in particular play a role for cluster formations.

Up to now a single matrix A and the formation of birds induced by interaction have been considered. In Theorem 8.6.2 a swarm is described by a sequence $(A(t))$ of matrices with $A(r_k)$ coherent at certain points in time r_k. Thus at each r_k there is a flight regime as discussed above. These regimes can differ and exhibit different cores. Even if the core is a singleton at each $r_k, k \in \mathbb{N}$, leaders (and followers) can change from time to time. Moreover, during flight different types of leadership according to cores of different size can occur. The **change in leader and leadership** enables the swarm to change the direction of flight, make even a turn. It is an important feature of Theorem 8.6.2 that the points in time r_k need not follow a specific rule and that, moreover, during the rest of time no particular flight formation is required. That means the swarm can change flight freely with irregular patterns of re-organization in between.

A major distinction concerning swarms is made in biology between **line formations** and **cluster formation** [68]. The interpretation put forward above was directed to line formations which are observed usually for migrating birds as various kinds of geese which fly in groups not too large. The above analysis of the core applies, however, also to huge groups of small birds as the European starling. The spectacular dynamics of the latter over a roost (as in Rome) has been empirically investigated recently employing modern stereo photography ([4], see also [50]). An important conclusion drawn in [4] is the one that interaction among the starlings does depend on the **"topological distance"** and not on the **metric distance**. This means that "each bird interacts on average with a fixed number of neighbors (six to seven), rather than with all neighbors within a fixed metric distance." [4, p. 1232]. Interaction as modeld in Theorem 8.6.2 does not presume any particular kind of concrete neighborhood as topological or metric distance or any other. (In the empirical study [77] on large flocks of surf scoters metric distance has found to be an appropriate tool.) In Theorem 8.6.5, as well as in the Cucker–Smale model, the intensity of interaction depends on metric distance without, however, specifying fixed distance neighborhoods. (A kind of neighborhood beyond topological or metric distance is discussed in Examples 8.5.22 (4) under the heading of "valuation".) In Theorem 8.6.2 as well as in Corollary 8.6.3 coherence of matrices may rest on topological distance (actually, maybe combinatorial distance would be a better name). Indeed, a stochastic matrix $A(t)$ is coherent if for each $i \in N$ the number of $j \in N$ such that $a_{ij}(t) > 0$ exceeds a fixed number m which depends on n.

Concerning the "turn and wheeling together" which is characteristic for huge swarms of the European starling over a roost, this is possibly due to a large core **within** the centre of a swarm. Actually, in 3 dimensions the interior of the polytope spanned by the swarm admits movements not only from the outside to the inside, but also in reverse directions, allowing for various kinds of spiraling and swirling. In contrast to a line formation the birds in a cluster formation are smaller and more swiftly, larger in number and more densely packed. Whereas in a line formation which is essentially 2-dimensional, cylces do not make sense, in 3 dimensions cycles of different orientation are to be expected. Furthermore, at times different from the r_k irregular patterns of re-organization, present also in line formations, become more dramatic in cluster formations due to small and swift birds densely packed in three dimensions. The cohesion of the swarm rests on the interaction according to coherent matrices which iteratively take hold at times r_k. Possible perturbations or re-organizations in between can be looked at as changes in "initial" conditions which after a while lead qualitatively to the same dynamics as before. In this view the point of Theorem 8.6.2 is that the tendency to common velocities and definite relative positions provides the law which governs the "reelin' and rockin" of the swarm.

Finally we add some more remarks concerning the literature.

Remarks 8.6.6. (1) For scalar opinion dynamics under bounded confidence an upper bound for convergence of $O(n^3)$) is given in [80] and in [7]. Lower bounds are obtained in [7, 66, 101]. A general framework to bound the time of convergence for a class of bidirectional multi-agreement systems, including opinion dynamics as well as bird flocking, has been developed in [19–21]. Surprisingly, the bound to reach a steady state is in general very high, a "tower-of-twos of hight linear in the number of birds". The framework developed is directed to a general study of "Natural Algorithms" [19].

(2) For the swarm model (8.6.5) and the role of coherent matrices in swarm dynamics see [64]. There [64, Theorem 3] a special version of Thorem 8.6.2 is proven.

(3) The Cucker–Smale model of bird flocking, for discrete time as well as for continuous time, has been developed in [23, 24]. This model and variants of it were further investigated in [14, 20, 38, 64, 82, 94]. Hierarchical leadership in the dicrete as well as in the continuous Cucker–Smale model is investigated in [94], where convergence rates for flocking are established. Results on flocking in the continuous Cucker–Smale model are obtained in [38] via a system of dissipative differential inequalities and in [14] by relating it to a Boltzmann-type equation. In [82] the continuous Cucker–Smale model is generalized by admitting non-symmetric interaction matrices based on relative distances. The proof makes use of an ineresting "energy functional" introduced in [38]. (For such a functional in the discrete case see Exercise 16.)

(4) The interpretation of Theorem 8.6.2 draws on the review of organized flight in birds [68] and the empirical studies [4, 77]. A model addressing the empirical data in [4] can be found in [50]. An informative general review of pattern formation in swarms as well as in other group-living species as ants, fishes, and humans is [83].

Exercises

1. Consider for a stochastic matrix $A \in \mathbb{R}_+^{n\times n}$ the property
   $Ax = x, x \in \mathbb{R}^n$ implies $x = (r, \ldots, r)'$ for some $r \in \mathbb{R}$.
   (a) Show that the above property holds if A is indecomposable or has a scrambling power.
   (b) Find an example A for which the above property holds though the condition in (a) is not satisfied.
2. Let $A \in \mathbb{R}_+^{n\times n}$ a stochastic matrix and $\bar{\lambda}$ a **second eigenvalue** of absolute value, that is $\bar{\lambda}$ is an eigenvalue of A for which $|\bar{\lambda}|$ is the maximal absolute value of all eigenvalues different from 1.
   (a) Obtain from Theorem 8.1.2 A that

$$|\bar{\lambda}| = \lim_{k\to\infty} c(A^k)^{\frac{1}{k}}$$

   if 1 is a simple eigenvalue.

(b) Show that for 1 not simple $c(A^k) = 1$ for all k. Find an example A for which the reverse implication does not hold.
(c) Derive from (a) and (b) that A has a scrambling power if and only if 1 is a simple eigenvalue and the only eigenvalue with absolute value 1.

3. Prove for a stochastic matrix $A \in \mathbb{R}_+^{n\times n}$ the following properties.
(a) A is coherent if and only if for any two $i, j \in \{1, \ldots, n\}$ exist $p_i, p_j \geq 0$ such that $s(i)^{p_i} \cap s(j)^{p_j} \neq \emptyset$.
(b) A has a power which is a Markov matrix if there exists $j \in s^{p_1}(1) \cap \ldots \cap s^{p_n}(n), p_i \geq 0$, with $a_{jj} > 0$.

4. (a) Prove that a (weighted) arithmetic mean $f(x_1, \ldots, x_n) = \sum_{i=1}^n a_i x_i$ is a strict abstract mean on int $\mathbb{R}_+^n$ if and only if $a_i > 0$ for all i.
(b) Let $Tx = Ax$ for a stochastic matrix $A \in \mathbb{R}_+^{n\times n}$ and $x \in$ int $\mathbb{R}_+^n$. Show by way of examples that for $\lim_{t\to\infty} T^t x = \bar{c}(x)$ to hold on int $\mathbb{R}_+^n$ it is neither necessary nor sufficient T_i is a strict abstract mean for some i.

5. Consider a ring of agents where each agent spends a fixed percentage of his money to his next neighbor and retains the rest, that is for $x_i(t) \in \mathbb{R}_+$ the amount of money of agent i at time $t \in \mathbb{N}$ one has $x_i(t+1) = a_{ii}x_i(t) + (1 - a_{i-1,i-1})x_{i-1}(t)$ where $a_{ii} \in [0, 1], 1 \leq i \leq n, a_{00} = a_{nn}, x_0(t) = x_n(t)$.
(a) Show that $\lim_{t\to\infty} x(t)$ exists for each $x(0) \in \mathbb{R}_+^n$ if $0 < a_{ii} < 1$ for all i.
(b) What can be said in (a) if one or more agents do not share their money with others, that is $a_{ii} = 1$?
(c) Under what conditions will in (a) the money be equally distributed in the limit among the agents?
(For the special case of the exercise where $a_{ii} = \frac{1}{2}$ for all i see [17, Theorem 7.1].)

6. Prove that the product of finitely many Sarymsakov $n\times n$-matrices is a Sarymsakov matrix again.

7. Let $\mathfrak{T}$ be the set of all mean maps on S^n for a non-empty convex subset S of $\mathbb{R}^d$. Prove the following properties of $\mathfrak{T}$.
(a) $\mathfrak{T}$ is a convex subset of the set $\mathfrak{F}$ of all selfmappings of S^n (for pointwise convex combination).
(b) $\mathfrak{T}$ is closed for the composition of maps.
(c) For $d = 1$, $\mathfrak{T}$ is closed for componentwise minima and maxima.
(d) If S is compact in $\mathbb{R}^d$ then $\mathfrak{T}$ is compact in $\mathfrak{F}$ with respect to the product topology.

8. Consider the selfmapping $Tx = A(x)x$ on S^n, $S \subseteq \mathbb{R}^d$ convex, with $A(x) = (a_{ij}(x)) \in \mathbb{R}_+^{n\times n}$ given by

$$a_{ij}(x) = \begin{cases} c_{ij} f(\|x^i - x^j\|) & \text{if } i \neq j \\ 1 - \sum\limits_{k\neq i} c_{ik} f(\|x^i - x^k\|) & \text{if } i = j \end{cases}$$

where $f\colon \mathbb{R}_+ \to [0,1]$ and $C = (c_{ij})$ is a stochastic $n \times n$-matrix ($\|\cdot\|$ any norm on $\mathbb{R}^d$).

(a) Find conditions on f and C for which

$$\lim_{t\to\infty} T^t x = \bar{c}(x) \text{ for all } x \in S^n.$$

(b) Investigate the case of $f(r) = e^{-\gamma r}, \gamma > 0$. (Cf. the ENDA Model in [78, pp. 125–128].)

9. Consider the **weighted Gini mean** on int $\mathbb{R}^n_+$ defined by

$$f(x) = \left[\frac{\sum_{k=1}^{n} a_k x_k^r}{\sum_{k=1}^{n} a_k x_k^s}\right]^{\frac{1}{r-s}}$$

for $r, s \in \mathbb{R}, r \neq s$ and weights $0 \le a_k, \sum_{k=1}^n a_k = 1$.

(a) Prove $\min_k x_k \le f(x) \le \max_k x_k$ on int $\mathbb{R}^n_+$.

(b) Prove that $a_i > 0$ for some i and $f(x) = \min_k x_k (f(x) = \max_k x_k)$ implies $x_i = \min_k x_k (x_i = \max_k x_k)$.

(c) Argue that part (i) of Theorem 8.3.12 holds true in case the soup has components given by weighted Gini means.

10. For the model from population biology considered in Examples 8.3.15 (3) let

$$M = \begin{bmatrix} a_1 & b_1 & c_1 & d_1 \\ a_2 & b_2 & c_2 & d_2 \\ a_3 & b_3 & c_3 & d_3 \\ a_4 & b_4 & c_4 & d_4 \end{bmatrix} \quad \begin{array}{l} \text{with } a_j, b_j, d_j \ge 0 \\ \text{and } a_j + \frac{1}{2}(b_j + c_j + d_j) = 1 \text{ for } 1 \le j \le 4. \end{array}$$

(a) Find further cases of M, besides the one given in Examples 8.3.15 (3) for which the mapping $T\colon \mathbb{R}^4_+ \to \mathbb{R}^4_+$ defined by M satisfies

$$\lim_{t\to\infty} T^t x = \bar{c}(x) \quad \text{on} \quad \text{int } \mathbb{R}^4_+.$$

(b) Find cases of M for which the above conclusion does not hold.

11. Consider the following **generalization of Borchardt's example** (cf. Remarks 8.3.16 (6)) where T is a selfmapping of $\mathbb{R}^n_+$ given by

$$T_i x = \sum_{j=1}^{n} a_{ij} \prod_{k \in I(i,j)} x_k^{\alpha_k(i,j)}, \quad 1 \le i \le n$$

with $A = (a_{ij})$ a stochastic matrix, $0 \le \alpha_k(i,j)$ and $\sum_{k\in I(i,j)} \alpha_k(i,j) = 1, \emptyset \subsetneq I(i,j) \subseteq \{1,\dots,n\}$ for all $1 \le i,j \le n$.

(a) Show that T is a mean map on $\mathbb{R}^n_+$.
(b) Prove $\lim_{t\to\infty} T^t x = \bar{c}(x)$ on int $\mathbb{R}^n_+$, provided $0 < \alpha_k(i,j)$ for $k \in I(i,j)$, $1 \le i,j \le n$ and the sets $I(i) = \bigcup_{a_{ij}>0} I(i,j)$ satisfy $I(i) \cap I(i') \neq \emptyset$ for any $i \neq i'$.
(c) Do you have any idea for a closed formula of the value $c(x)$?

12. Let a Gauss soup T on int $\mathbb{R}^n_+$ given by

$$T_i x = \prod_{j=1}^{n} x_j^{a_{ij}}, \quad 1 \le i \le n$$

for a stochastic matrix $A = (a_{ij})$.
(a) Show that $H(x) = \prod_{j=1}^n x_j^{v_j}$ on int $\mathbb{R}^n_+$ is an invariant for T, where $v \in \mathbb{R}^n_+$ is an eigenvector of the transpose A' for the eigenvalue 1.
(b) Show that for A scrambling $\lim_{t\to\infty} T^t x = \bar{c}(x)$ on int $\mathbb{R}^n_+$ and compute $c(x)$ using the invariant H (cf. Remarks 8.3.16 (7)).

13. [35, 63] Consider the variation of the arithmetic-geometric mean given for $(x_1, x_2) \in$ int $\mathbb{R}^2_+$ by

$$T_1(x_1,x_2) = \sqrt{x_1 \frac{x_1 + x_2}{2}}, T_2(x_1,x_2) = \sqrt{x_2 \frac{x_1 + x_2}{2}}.$$

(a) Verify that $H(x_1,x_2) = \frac{x_2^2 - x_1^2}{\log x_2 - \log x_1}, x_1 \neq x_2$ is an invariant for T.
(b) Prove $\lim_{t\to\infty} T^t x = \bar{c}(x)$ on int $\mathbb{R}^2_+$ and compute $c(x)$ using the invariant H.

14. Let T be the selfmapping of int $\mathbb{R}^3_+$ given by

$$Tx = \left(\frac{x_2 + x_3}{2}, \sqrt{x_1 x_3}, \sqrt{\frac{x_1^2 + x_2^2}{2}} \right).$$

(a) Show that the iterates of T converge for each $x \in$ int $\mathbb{R}^3_+$ to a value $\bar{c}(x)$.
(b) What can you say about the value $c(x)$?
(c) Check if T is strict, that is

$$\min_j x_j < T_i x < \max_j x_j$$

for $i = 1, 2, 3$ and not all components of $x \in$ int $\mathbb{R}^3_+$ being equal.

15. Let K be the convex cone $\mathbb{R}^{n\times n}_+$ and $\lambda(\cdot,\cdot)$ be the order function of K (see Definition 3.1.1).
(a) Verify that the part $[A]$ in K generated by $A \in K$ (see Definition 3.2.1) is given by $[A] = \{B \in K \mid B_{ij} = 0$ equivalent to $A_{ij} = 0$ for all $1 \le i,j \le n\}$.
(b) Let $\bar{A} \in [A]$ be a representative with $\bar{A}_{ij} = 1$ for $A_{ij} > 0$. Show for a stochastic scrambling matrix A that

$$1 - n\lambda(\bar{A}, A) \le c(A) \le 1 - \lambda(\bar{A}, A).$$

(c) Prove that condition (8.4.2) in Theorem 8.4.5 is equivalent to

$$\sum_{I \in \mathfrak{I}_+} \lambda\,(\bar{B}(I), B(I)) = \infty,$$

where $\mathfrak{I}_+$ is the set of $I \in \mathfrak{I}$ with $B(I)$ scrambling.

16. [54] Let $(x(t))$ be the mean process of opinion dynamics under bounded confidence in one dimension, that is

$$x_i(t+1) = |I(i, x(t))|^{-1} \sum_{j \in I(i,x(t))} x_j(t)$$

for $1 \le i \le n, t \in \mathbb{N}, x(0) \in \mathbb{R}^n, I(i,x) = \{1 \le j \le n \mid |x_i - x_j| \le \epsilon\}$ for $x = (x_1, \dots, x_n) \in \mathbb{R}^n, \epsilon > 0$.

(a) Prove for $2 \le n \le 4$ that convergence to consensus holds if and only if $x(0)$ is an ϵ-profile, that is for some numbering $(i_1, \dots, i_n)$ of $(1, \dots, n)$

$$x_{i_1}(0) \le \cdots \le x_{i_n}(0) \quad \text{and} \quad x_{i_{k+1}}(0) - x_{i_k}(0) \le \epsilon \quad \text{for} \quad 1 \le k \le n-1.$$

(b) Find an example for $n \ge 5$ that the condition on $x(0)$ in (a) is necessary but not sufficient for convergence to consensus.

(c) Verify that for $n = 6, \epsilon = 1$ and $x(0) = (0, 1, 2, 3, 4, 5)$ the dynamics reaches for $t = 6$ a stable configuration which is not a consensus. (Cf. [41] for the general case of equally spaced agents.)

17. Apply Theorem 8.5.3 to the following nonlinear and non-autonomous system

$$x(t+1) = T_t x(t), T_t x = A(t,x)x, A(t,x) \text{ stochastic } n \times n\text{-matrix}$$

for $t \in \mathbb{N}, x \in S^n, S \subseteq \mathbb{R}^d$ convex.

Let for $m \ge 1$

$$T_{t+m-1} \circ \cdots \circ T_t x = B_m(t,x)x$$

and

$$\delta_m(t,x) = \min_{1 \le i,j \le n} \sum_{k=1}^{n} \min\{B_m(t,x)_{ik}, B_m(t,x)_{jk}\}.$$

(a) Show that $\lim_{t \to \infty} x(t) = \bar{c}(x(0))$ under the condition $\sum_{t=0}^{\infty} \delta_m(t, x(t)) = \infty$ for some $m \ge 1$ and $x(0) \in S^n$.

(b) Verify the condition in (a) in case of $\delta_m(t,x) \ge \delta_m'(t).\ \delta_m''(x)$ (for some $m \ge 1$) with

- $\sum_{t=0}^{\infty} \delta_m'(t) = \infty$ and $\delta_m''(x) \ge \delta'' > 0$ on S^n

or

- $\delta_m'(t) \ge \delta' > 0$ on $\mathbb{N}$ and $\delta_m''(x) > 0$ continuous on S^n.

(Cf. [54, Theorem 1], [65, Theorem 9.5.4].)

18. Find an example of multiagent coordination of two agents for each of the following cases.
    (a) The assumptions for parts (i) and (ii) of Corollary 8.5.9 are satisfied but not the ones of part (iii).
    (b) The assumptions of Theorem 8.5.7 (and of Theorem 8.5.3) are satisfied but not the assumptions of any of the three parts of Corollary 8.5.9.
    (c) The assumptions of Theorem 8.5.7 can be fulfilled for sequences (t_k) and (β_k) such that $t_{k+1} - t_k$ tends to infinity and β_k tends to zero.
19. For the swarm model (8.6.5) given by

$$\begin{aligned} x(t+1) - x(t) &= v(t) \\ v(t+1) &= A(t)v(t) \qquad \text{with } x(0), v(0) \in S^n \end{aligned}$$

$$\text{let } d(t) = \max_{1 \le i,j \le n} \|x^i(t) - x^j(t)\|, p(t) = \max_{1 \le i,j \le n} \|v^i(t) - v^j(t)\|$$

where $\|\cdot\|$ is any norm on $\mathbb{R}^d$.
Prove for the **discrete energy functional**

$$E(t) = p(t+1) + \sum_{s=0}^{t} (1 - c(A(s)))(d(s+1) - d(s)), \quad t \in \mathbb{N},$$

that it decreases along the trajectory $(d(t), p(t))$.
(For a swarm model with an energy functional in continuous time see [38, 82].)

Bibliography

[1] D. Angeli and P. A. Bliman. Extension of a result by Moreau on stability of leaderless multi-agent systems. In *Proceedings of the 44th IEEE Conference on Decision and Control*, Seville, 2005.

[2] J. Arazy, T. Claesson, S. Janson, and J. Peetre. Means and their iterations. *Proceedings of the Nineteenth Nordic Congress of Mathematics*, Iceland Mathematical Society, Reykjavik, 191–212, 1985.

[3] R. Axelrod. The dissemination of culture: A model with local convergence and global polarization. *J. Conflict Resolution*, 41: 203–226, 1997.

[4] M. Ballerini, N. Cabibbo, R. Candelier, A. Cavagna, E. Cisbani, I. Giardina, V. Lecomte, A. Orlandi, G. Parisi, A. Procaccini, and M. Viala, and V. Zdravkovic. Interaction ruling animal collective behaviour depends on topological rather than metric distance: Evidence from a field study. *Proc. Nat. Acad. Sci. USA*, 105:1232–1237, 2008.

[5] E. Behrends. *Introduction to Markov Chains with Special Emphasis on Rapid Mixing*. Vieweg, Braunschweig/Wiesbaden, 2000.

[6] D. P Bertsekas and J. N. Tsitsiklis. Comments on "Coordination of groups of mobile autonomous agents using nearest neighbor rules". *IEEE Transactions on Automatic Control*, 52:968–969, 2007.

[7] A. Bhattacharayya, M. Braverman, B. Chazelle, and H. L. Nguyen. On the convergence of the Hegselmann–Krause system. *ITCS' 13, January 9–12, 2013, Berkeley, CA, USA.*

[8] V. D. Blondel, J. M. Hendrickx, A. Olshevsky, and J. N. Tsitsiklis. Convergence in multiagent coordination, consensus, and flocking. In *Proc. 44th IEEE Conference on Decision and Control*, Seville, Spain 2005, pp. 2996–3000.

[9] V. D. Blondel, J. M. Hendrickx, and J. N. Tsitsiklis. On the 2R conjecture for multi-agent systems. In *Proc. European Control Conf.*, Kos, Greece, 2007 pp. 874–881.

[10] V. D. Blondel, J. M. Hendrickx, and J. N. Tsitsiklis. On Krause's multi-agent consensus model with state-dependent connectivity. *IEEE Transactions on Automatic Control*, 54:2586–2597, 2009.

[11] C. W. Borchardt. Theorie des arithmetisch-geometrischen Mittels aus vier Elementen. *Monatshefte Berliner Akademie der Wissenschaften*, 611–621, 1876.

[12] J. M. Borwein and P. B. Borwein. *Pi and the AGM. A Study in Analytic Number Theory and Computational Complexity.* J. Wiley and Sons, New York, 1987.

[13] M. Cao, A. S. Morse, and B. D. O. Anderson. Reaching a consensus in a dynamically changing environment-convergence rates, measurement delays and asynchronous events. *SIAM J. Control. Optim.* 47:601–623, 2008.

[14] J. A. Carrillo, M. Fornasier, J. Rosado, and G. Toscani. Asymptotic flocking dynamics for the kinetic Cucker–Smale model. *SIAM J. Math. Anal.*, 42:218–236, 2010.

[15] B. C. Carlson. Algorithms involving arithmetic and geometric mean. *Amer. Math. Monthly*, 78:496–509, 1971.

[16] C. Castellano, S. Fortunato, and V. Loreto. Statistical physics of social dynamics. *Reviews of Modern Physics*, 81:591–646, 2009.

[17] G. Chang and T. W. Sederberg. *Over and Over Again.* The Mathematical Association of America, Washington, 1997.

[18] S. Chatterjee and E. Seneta. Towards consensus: some convergence theorems on repeated averaging. *J. Appl. Prob.*, 14:89–97, 1977.

[19] B. Chazelle. Natural algorithms. In *Proceedings of the 20th Annual ACM-SIAM Symposium Discrete Algorithms*, SIAM, Philadelphia, ACM, New York, 2009, pp. 422–431.

[20] B. Chazelle. The convergence of bird flocking. *Preprint* 2009. Online at http://arxiv.org/abs/0905.424v1.

[21] B. Chazelle. The total s-energy of a multiagent system. *SIAM J. Control Optim.* 49:1680–1706, 2011.

[22] J. E. Cohen, J. Hajnal, and C. M. Newman. Approaching consensus can be delicate when positions harden. *Stochastic Proc. Appl.*, 22:315–322, 1986.

[23] F. Cucker and S. Smale. Emergent behavior in flocks. IEEE *Trans. Autom. Control*, 52:852–862, 2007.

[24] F. Cucker and S. Smale. On the mathematics of emergence. *Japan J. Math.*, 2:197–227, 2007.

[25] N.C. Dalkey. The Delphi method: An experimental study of group opinion. *The Rand Corporation*, RM-5888, 1969.

[26] M. H. De Groot. Reaching a consensus. *J. Amer. Statist. Assoc.*, 69:118–121, 1974.

[27] G. Deffuant, D. Neau, F. Amblard, and G. Weisbuch. Mixing beliefs among ineracting agents. *Adv. Complex Systems* 3, 87–98, 2000.

[28] J. C. Dittmer. Consensus formation under bounded confidence. *Nonlin. Anal.*, 47:4615–4621, 2001.

[29] C.J. Everett and N. Metropolis. A generalization of the Gauss limit for iterated means. *Advances in Math.*, 7:297–300, 1971.

[30] N. E. Friedkin and E. C. Johnsen. Social influence and opinions. *J. Math. Soc.*, 15:193–206, 1990.

[31] S. Fortunato. The Krause–Hegselmann consensus model with discrete opinions. *Int. J. Modern Physics* C15, 1021-1029, 2004.

[32] S. Fortunato. On the consensus threshold for the dynamics of Krause–Hegselmann. *Int. J. Modern Physics* C16, 259–270, 2005.

[33] S. Fortunato, V. Latora, A. Pluchino, and A. Rapisarda. Vector Opinion dynamics in a bounded confidence consensus model. *Int. J. Modern Physics* C16, 1535–1551, 2005.

[34] F. R. Gantmacher. *The Theory of Matrices (Vol. II)*. Chelsea, New York, 1959.

[35] A. Gasull and V. Mañosa. A Darboux-type theory of integrability for discrete dynamical systems. *J. Difference Equ. Appl.*, 8:1171–1191, 2002.

[36] C. Genest and J. V. Zidek. Combining probability distributions: A critic and an annotated bibliography. *Statistical Science 1*, 114–148, 1986.

[37] W. Gustin. Gaussian means. *Amer. Math. Monthly*, 54:332–335, 1947.

[38] S. Y. Ha and J. G. Liu. A simple proof of the Cucker–Smale flocking dynamics and mean-field limit. *Commun. Math. Sci.*, 7:297–235, 2009.

[39] J. Hajnal. Weak ergodicity in non-homogeneous Markov chains. *Proc. Cambridge Philos. Soc.*, 54:233–246, 1958.

[40] D. S. Hartfiel. *Nonhomogeneous Matrix Products*. World Scientific, Singapore, 2002.

[41] P. Hegarty and E. Wedin. The Hegselmann–Krause dynamics for equally spaced agents. *Preprint* 2014. Online at http://arxiv.org/abs/1406.0819.

[42] R. Hegselmann. Opinion dynamics: Insights by radically simplifying models. In D. Gillies, editor *Laws and Models in Science*, London, 2004, pp. 1–29.

[43] R. Hegselmann and U. Krause. Opinion dynamics and bounded confidence: Models, analysis, and simulation. *J. Artificial Societies and Social Simulation* 5 (3), 2002. Online http://jasss.soc.surrey.ac.uk/5/3/2.html.

[44] R. Hegselmann and U. Krause. Opinion dynamics driven by various ways of averaging. *Computational Economics* 25, 381–405, 2005.

[45] R. Hegselmann and U. Krause. Truth and cognitive division of labor: First steps towards a computer aided social epistemology. *J. Artificial Societies and Social Simulation* 9 (3), 2006. Online http://jasss.soc.surrey.ac.uk/9/3/10.html.

[46] J. M. Hendrickx and V. B. Blondel. Convergence of different linear and non-linear Vicsek models. In *Proc. 17th Int. Symp. Math. Theory of Networks and Systems*, Kyoto, Japan 2006, pp. 1229–1240.

[47] J. M. Hendrickx. *Graphs and Networks for the Analysis of Autonomous Agent Systems*. Ph.D. thesis, Université Catholique de Louvain 2008.
Online http://www.inma.ucl.ac.be/~hendrickx/available publications/
Thesis_Julien_Hendrickx.pdf.

[48] J. M. Hendrickx. Order preservation in a generalized version of Krause's opinion dynamic model. *Physica A*, 387:5255–5262, 2008.

[49] J. M. Hendrickx and J.N. Tsitsiklis. A new condition for convergence in continuous-time consensus seeking systems. *50th IEEE Conference on Decision and Control and European Control Conference*, Orlando, USA 2011, pp. 5070–5075.

[50] H. Hildenbrandt, C. Carere, and C.K. Hemelrijk. Self-organized aerial displays of thousend starlings: a model. *Behavioral Ecology Advance Access*, 11:1–11, 2010

[51] M. O. Jackson. *Social and Economic Networks*. Princeton University Press, Princeton, 2008.

[52] A. Jadbabaie, J. Liu, and A. S. Morse. Coordination of groups of mobile autonomous agents using nearest neighbor rules. *IEEE Transactions on Automatic Control*, 48:988–1001, 2003.

[53] K. Knopp. *Infinite Sequences and Series*. Dover Publ. Inc., New York, 1956.

[54] U. Krause. A discrete nonlinear and non-autonomous model of consensus formation. In S. Elaydi et al., editor, *Communications in Difference Equations*, Gordon and Breach Publ., Amsterdam 2000, pp. 227–236.

[55] U. Krause. Positive particle interaction. In L. Benvenuti et al., editor, *Positive Systems, Lecture Notes Control Inform. Sciences* 294, 2003, pp. 199–206.

[56] U. Krause. Time-variant consensus formation in higher dimensions. In S. Elaydi et al., editor, *Proc. Eighth Int. Conf. Diff. Equ. Appl.*, Brno 2003, Chapman & Hall, Boca Raton 2005, pp. 185–191.

[57] U. Krause. Convergence of the multidimensional agreement algorithm when communication fades away. In C. Commault et al., editor, *Positive Systems, Lect. Notes Control Inform. Sciences* 341, 2006, pp. 217–222.

[58] U. Krause. Arithmetic-geometric discrete systems. *J. Difference Equ. Appl.*, 12:229–231, 2006.

[59] U. Krause. Compromise, consensus, and the iteration of means. *Elem. Math.*, 64:1–8, 2009.

[60] U. Krause. Markov chains, Gauss soups, and compromise dynamics. *Stochastic methods in modern mathematics and its applications* (Festschrift in honour of K. Krickeberg's 80th birthday, edited by H. Zessin), 44:59–66, 2009.

[61] U. Krause. Collective dynamics of Faustian agents. In J. Vint et al., editor, *Economic Theory and Economic Thought* (Essays in honour of I. Steedman), Routledge, London 2010, pp. 374–385.

[62] U. Krause. The production of customs by means of customs. In N. Salvadori and C. Gehrke, editor, *Keynes, Sraffa and the Criticism of Neoclassical Theory* (Essays in honour of H. Kurz), Routledge, London 2011, pp. 170–178.

[63] U. Krause. Opinion dynamics – local and global. In E. Liz and V. Mañosa, editor, *Proceedings of the Workshop "Future Directions in Difference Equations"*, Servicio de Publications da Universidade de Vigo, 2011, pp. 113–119.

[64] U. Krause. Swarm dynamics and positive dynamical systems. In S. Pinelas, M. Chipot,Z. Dosla, Editors, *Differential and Difference Equations with Applications*. Springer, New York, 2013, pp. 69–81.

[65] N. Kruse. Semizyklen und Kontraktivität nichtlinearer positiver Differenzengleichungen mit Anwendungen in der Populationsdynamik. Ph.D. thesis, Universität Bremen, dissertation.de, Berlin, 1999.

[66] S. Kurz. How long does it take to consensus in the Hegselmann–Krause model? *Preprint* 2014. Online at http://arxiv.org/abs/1405.5757.

[67] S. Kurz and J. Rambau. On the Hegselmann–Krause conjecture in opinion dynamics. *J. Difference Equ. Appl.*, 17:859–876, 2011.

[68] I. Lebar Bajec and F.H. Heppner. Organized flight in birds. *Animal Behaviour*, 78:777–789, 2009.

[69] K. Lehrer and C. G. Wagner. *Rational Consensus in Science and Society. A Philosophical and Mathematical Study.* D. Reidel Publ. Co., Dordrecht, 1981.

[70] B. Lemmens and R. Nussbaum. *Nonlinear Perron–Frobenius Theory*. Cambridge University Press, 2012.

[71] J. Lorenz. A stabilization theorem for dynamics of continuous opinions. *Physica* A, 335:217–223, 2005.

[72] J. Lorenz. *Repeated Averaging and Bounded Confidence. Modeling, Analysis and Simulation of Continuous Opinion Dynamics.* Ph.D. thesis, Universität Bremen, 2007. http://nbn-resolving.de/urn:nbn:de:gbv:46-diss000106688.

[73] J. Lorenz. Continuous opinion dynamics under bounded confidence: A survey. *Int. J. Modern Physics* C18:1819–1838, 2007.

[74] J. Lorenz. Heterogeneous bounds of confidence: Meet, discuss and find consensus! *Complexity*, 15:43–52, 2009.

[75] D. A. Lorenz and J. Lorenz. Convergence to consensus by general averaging. In R. Bru and S. Romero-Vivo, editor, *Positive Systems, Lect. Notes Control Inform. Sciences* 389, 2009, pp. 91–99.

[76] D. G. Luenberger. *Introduction to Dynamic Systems. Theory, Models, and Applications.* John Wiley & Sons, New York, 1979.

[77] R. Lukeman, Y. X. Li, and L. Edelstein–Keshet. Inferring individual rules from collective behavior. *Proc. Nat. Acad. Sci. Early Edition*, 1–5, 2010 www.pnas.org/cgi/doi/10.1073/pnas.1001763107

[78] M. Meiler. *Analytic Advances in Difference Equations of Diffusion processes.* Ph.D. thesis, Technische Universität München, Sierke Verlag, Göttingen, 2009.

[79] A. Mirtabatabaei and F. Bullo. Opinion dynamics in heterogeneous networks: Convergence conjectures and theorems. SIAM J. Control Optim., 50:2763–2785, 2012.

[80] S. Mohajer and B. Touri. On convergence rate of scalar Hegselmann–Krause dynamics. arXiv:1211.4189v1[math.DS]18Nov2012.

[81] L. Moreau. Stability of multiagent systems with time-dependent communication links. *IEEE Trans. Autom. Control*, 50:169–182, 2005.

[82] S. Motsch and E. Tadmor. A new model for self-organized dynamics and its flocking behavior. *J. Stat. Phys.*, published online 19 August 2011, doi10.1007/s10955-011-0285-9.

[83] M. Moussaid, S. Garnier, G. Theraulaz, D. Helbing. Collective information processing and pattern formation in swarms, flocks, and crowds. *Topics in Cognitive Science*, 1:469–497, 2009.

[84] R. D. Nussbaum Hilbert's projective metric and iterated nonlinear maps. *Mem. Amer. Math. Soc.*, 391:1–137, 1988.

[85] R. D. Nussbaum. Iterated nonlinear maps and Hilbert's projective metric, II. *Mem. Amer. Math. Soc.*, 401:1–118, 1989.

[86] R. Olfati-Saber, J. A. Fax, and R. M. Murray. Consensus and cooperation in networked multi-agent systems. *Proceedings of the IEEE*, 95:215–233, 2007.

[87] V. Pluchino, V. Latora, and A. Rapisarda. Compromise and synchronisation in opinion dynamics. *Europ. Phys. J.*B, 50:169–176 (2006).

[88] W. Ren and R. W. Beard. Consensus seeking in multiagent systems under dynamically changing interaction topologies. *IEEE Transactions on Automatic Control*, 50:655–661, 2005.

[89] C. W. Reynolds. Flocks, herds, and schools: a distributed behavioral model. *Computer Graphics*, 21:25–34, 1987.

[90] L. Scardovi, A. Sarlette, R. Sepulchre. Synchronization and balancing on the N-torus. *Systems and Control Letters*, 56:335–341, 2007.

[91] E. Seneta. On the historical development of the theory of finite inhomogeneous Markov chains. *Proc. Camb. Phil. Soc.*, 74:507–513, 1973.

[92] E. Seneta. Coefficients of ergodicity: Structure and applications. *Adv. Appl. Prob.*, 11:576–590, 1979.

[93] E. Seneta. *Non-negative Matrices and Markov Chains.* Springer, Berlin, Revised Printing, 2006.

[94] J. Shen. Cucker–Smale flocking under hierarchical leaderhip. *SIAM J. Appl. Math.*, 68:694–719, 2008.

[95] M. Sieveking. Condensing. In C. Commault et. al., editors, *Positive Systems, Lect. Notes Control Inform. Sciences*, 341, 2006, pp. 247–254.

[96] F. Slanina. Dynamical phase transition in Hegselmann–Krause model of opinion dynamics and consensus. *Europ. Phys. J.* B, 79:99–106, 2011.

[97] M. Stone. The opinion pool. *Ann. Math. Stat.*, 32:1339–1342, 1961.

[98] J. N. Tsitsiklis. *Problems in Decentralized Decision Making and Computation*. Ph.D. thesis, Massachusetts Institute of Technology, 1984. Online http://web.mit.edu/jnt/www/PhD-84-jnt.pdf

[99] J. N. Tsitsiklis, D. P. Bertsekas, and M. Athans. Distributed asynchronous deterministic and stochastic gradient optimization algorithms. *IEEE Trans. Autom. Control*, 31:803–812, 1986.

[100] T. Vicsek, A. Czirók, E. Ben-Jacob, I. Cohen, and O. Shochet. Novel type of phase transition in a system of self-driven particles. *Physical Review Letters*, 75:1226–1229, 1995.

[101] E. Wedin and P. Hegarty. A quadratic lower bound for the convergence rate in the one-dimensional Hegselmann–Krause bounded confidence dynamics. *Preprint* 2014. Online at http://arxiv.org/abs/1406.0769.

[102] G. Weisbuch. Bounded confidence and social networks. *Europ. Phys. J.* B, 38:339–343, 2004.

[103] J. Wolfowitz. Products of indecompsable, aperiodic, stochastic matrices. *Proc. Amer. Math. Soc.*, 14:733–737, 1963.

[104] M. Zahri. Mathematical modeling of condensing on metric spaces. *Eur. J. Pure Appl. Math.*, 6:172–188, 2013.

Index

De Gruyter Studies in Mathematics

Volume 61
Francesco Altomare, Mirella Cappelletti Montano, Vita Leonessa, Ioan Rasa
Markov Operators, Positive Semigroups and Approximation Processes, 2015
ISBN 978-3-11-037274-8, e-ISBN 978-3-11-036697-6, Set-ISBN 978-3-11-036698-3

Volume 60
Vladimir A. Mikhailets, Alexandr A. Murach, Peter V. Malyshev
Hörmander Spaces, Interpolation, and Elliptic Problems, 2014
ISBN 978-3-11-029685-3, e-ISBN 978-3-11-029689-1, Set-ISBN 978-3-11-029690-7

Volume 59
Jan de Vries
Topological Dynamical Systems, 2014
ISBN 978-3-11-034073-0, e-ISBN 978-3-11-034240-6, Set-ISBN 978-3-11-034241-3

Volume 58
Lubomir Banas, Zdzislaw Brzezniak, Mikhail Neklyudov, Andreas Prohl
Stochastic Ferromagnetism: Analysis and Numerics, 2014
ISBN 978-3-11-030699-6, e-ISBN 978-3-11-030710-8, Set-ISBN 978-3-11-030711-5

Volume 57
Dmitrii S. Silvestrov
American-Type Options: Stochastic Approximation Methods, Volume 2, 2014
ISBN 978-3-11-032968-1, e-ISBN 978-3-11-032984-1, Set-ISBN 978-3-11-032985-8

Volume 56
Dmitrii S. Silvestrov
American-Type Options: Stochastic Approximation Methods, Volume 1, 2013
ISBN 978-3-11-032967-4, e-ISBN 978-3-11-032982-7, Set-ISBN 978-3-11-032983-4

Volume 55
Lucio Boccardo, Gisella Croce
Elliptic Partial Differential Equations, 2013
ISBN 978-3-11-031540-0, e-ISBN 978-3-11-031542-4, Set-ISBN 978-3-11-031543-1

Bei Fragen zur Produktsicherheit wenden Sie sich bitte an:
If you have any questions regarding product safety,
please contact:

Walter de Gruyter GmbH
Genthiner Straße 13
10785 Berlin
productsafety@degruyterbrill.com